I0056623

Fertilizers: Soil Improvement and Plant Growth

Fertilizers: Soil Improvement and Plant Growth

Editor: Kye Young

R CALLISTO REFERENCE

www.callistoreference.com

Callisto Reference,
118-35 Queens Blvd., Suite 400,
Forest Hills, NY 11375, USA

Visit us on the World Wide Web at:
www.callistoreference.com

© Callisto Reference, 2017

This book contains information obtained from authentic and highly regarded sources. Copyright for all individual chapters remain with the respective authors as indicated. All chapters are published with permission under the Creative Commons Attribution License or equivalent. A wide variety of references are listed. Permission and sources are indicated; for detailed attributions, please refer to the permissions page and list of contributors. Reasonable efforts have been made to publish reliable data and information, but the authors, editors and publisher cannot assume any responsibility for the validity of all materials or the consequences of their use.

ISBN: 978-1-63239-792-8 (Hardback)

The publisher's policy is to use permanent paper from mills that operate a sustainable forestry policy. Furthermore, the publisher ensures that the text paper and cover boards used have met acceptable environmental accreditation standards.

Trademark Notice: Registered trademark of products or corporate names are used only for explanation and identification without intent to infringe.

Printed in the United States of America.

Cataloging-in-publication Data

Fertilizers : soil improvement and plant growth / edited by Kye Young.
    p. cm.
Includes bibliographical references and index.
ISBN 978-1-63239-792-8
1. Soil fertility. 2. Fertilizers. 3. Crops--Nutrition. 4. Growth (Plants). 5. Plants--Effect of agricultural chemicals on.
I. Young, Kye.
S596.7 .F47 2017
631.422--dc23

Table of Contents

Permissions

List of Contributors

Index

Preface

This book outlines the benefits and applications of fertilizers in soil and plant growth in detail. Fertilizers are natural or synthetic substances used for enhancing the soil with essential nutrients. They are necessary for plant growth and crop yield quality. This book elucidates the concepts and innovations and respective developments with respect to this field. The topics included in this book are of utmost significance and are bound to provide incredible insights to readers about fertilizers. It strives to provide a fair idea about this discipline and to help develop a better understanding of the latest advances within this field. This text is an essential guide for both academicians and those who wish to pursue this discipline further.

The main aim of this book is to educate learners and enhance their research focus by presenting diverse topics covering this vast field. This is an advanced book which compiles significant studies by distinguished experts. This book addresses successive solutions to the challenges arising in the area of application, along with it; the book provides scope for future developments.

It was a great honour to edit this book, though there were challenges, as it involved a lot of communication and networking between me and the editorial team. However, the end result was this all-inclusive book covering diverse themes in the field.

Finally, it is important to acknowledge the efforts of the contributors for their excellent chapters, through which a wide variety of issues have been addressed. I would also like to thank my colleagues for their valuable feedback during the making of this book.

Editor

Carbon and Nitrogen Isotopic Survey of Northern Peruvian Plants: Baselines for Paleodietary and Paleoecological Studies

Paul Szpak[1]*, **Christine D. White**[1], **Fred J. Longstaffe**[2], **Jean-François Millaire**[1], **Víctor F. Vásquez Sánchez**[3]

1 Department of Anthropology, The University of Western Ontario, London, Ontario, Canada, **2** Department of Earth Sciences, The University of Western Ontario, London, Ontario, Canada, **3** Centro de Investigaciones Arqueobiológicas y Paleoecológicas Andinas (ARQUEOBIOS), Trujillo, Peru

Abstract

The development of isotopic baselines for comparison with paleodietary data is crucial, but often overlooked. We review the factors affecting the carbon (δ^{13}C) and nitrogen (δ^{15}N) isotopic compositions of plants, with a special focus on the carbon and nitrogen isotopic compositions of twelve different species of cultivated plants ($n=91$) and 139 wild plant species collected in northern Peru. The cultivated plants were collected from nineteen local markets. The mean δ^{13}C value for maize (grain) was -11.8 ± 0.4 ‰ ($n=27$). Leguminous cultigens (beans, Andean lupin) were characterized by significantly lower δ^{15}N values and significantly higher %N than non-leguminous cultigens. Wild plants from thirteen sites were collected in the Moche River Valley area between sea level and ~4,000 meters above sea level (masl). These sites were associated with mean annual precipitation ranging from 0 to 710 mm. Plants growing at low altitude sites receiving low amounts of precipitation were characterized by higher δ^{15}N values than plants growing at higher altitudes and receiving higher amounts of precipitation, although this trend dissipated when altitude was >2,000 masl and MAP was >400 mm. For C_3 plants, foliar δ^{13}C was positively correlated with altitude and precipitation. This suggests that the influence of altitude may overshadow the influence of water availability on foliar δ^{13}C values at this scale.

Editor: John P. Hart, New York State Museum, United States of America

Funding: This research was supported by the Wenner-Gren Foundation (PS, Dissertation Fieldwork Grant #8333), the Social Sciences and Humanities Research Council of Canada, the Natural Sciences and Engineering Research Council of Canada, the Canada Research Chairs Program, the Canada Foundation for Innovation and the Ontario Research Infrastructure Program. The funders had no role in study design, data collection and analysis, decision to publish, or preparation of the manuscript.

Competing Interests: The authors have declared that no competing interests exist.

* E-mail: pszpak@uwo.ca

Introduction

Stable isotope analysis is an important tool for reconstructing the diet, local environmental conditions, migration, and health of prehistoric human and animal populations. This method is useful because the carbon and nitrogen isotopic compositions of consumer tissues are directly related to the carbon and nitrogen isotopic compositions of the foods consumed [1,2], after accounting for the trophic level enrichments of ^{13}C and ^{15}N for any particular tissue [3,4].

In all cases, interpretations of isotopic data depend on a thorough understanding of the range and variation in isotopic compositions of source materials [5]. For instance, studies of animal migrations using oxygen and hydrogen isotopic analyses require a thorough understanding of the spatial variation in surface water and precipitation isotopic compositions [6], and in that avenue of research, there has generally been an emphasis on establishing good baselines. With respect to diet and local environmental conditions, the interpretation of isotopic data (typically the carbon and nitrogen isotopic composition of bone or tooth collagen) depends upon a thorough knowledge of the range and variation in isotopic compositions of foods that may have been consumed. Although several authors have attempted to

develop such isotopic baselines for dietary reconstruction [7–10], these studies have typically focused on vertebrate fauna.

Despite the fact that plants are known to be characterized by extremely variable carbon and nitrogen isotopic compositions [11,12], few studies have attempted to systematically document this variability in floral resources at a regional scale using an intensive sampling program, although there are exceptions [13–15]. This is problematic, particularly in light of the development and refinement of new techniques (e.g. isotopic analysis of individual amino acids), which will increase the resolution with which isotopic data can be interpreted. If isotopic baselines continue to be given marginal status, the power of new methodological advancements will never be fully realized.

With respect to the Andean region of South America, the isotopic composition of plants is very poorly studied, both from ecological and paleodietary perspectives. The most comprehensive study of the latter type was conducted by Tieszen and Chapman [14] who analyzed the carbon and nitrogen isotopic compositions of plants collected along an altitudinal transect (~0 to 4,400 masl) following the Lluta River in northern Chile. Ehleringer et al. [16] presented δ^{13}C values for plants along a more limited altitudinal transect in Chile (Atacama Desert). A number of other studies

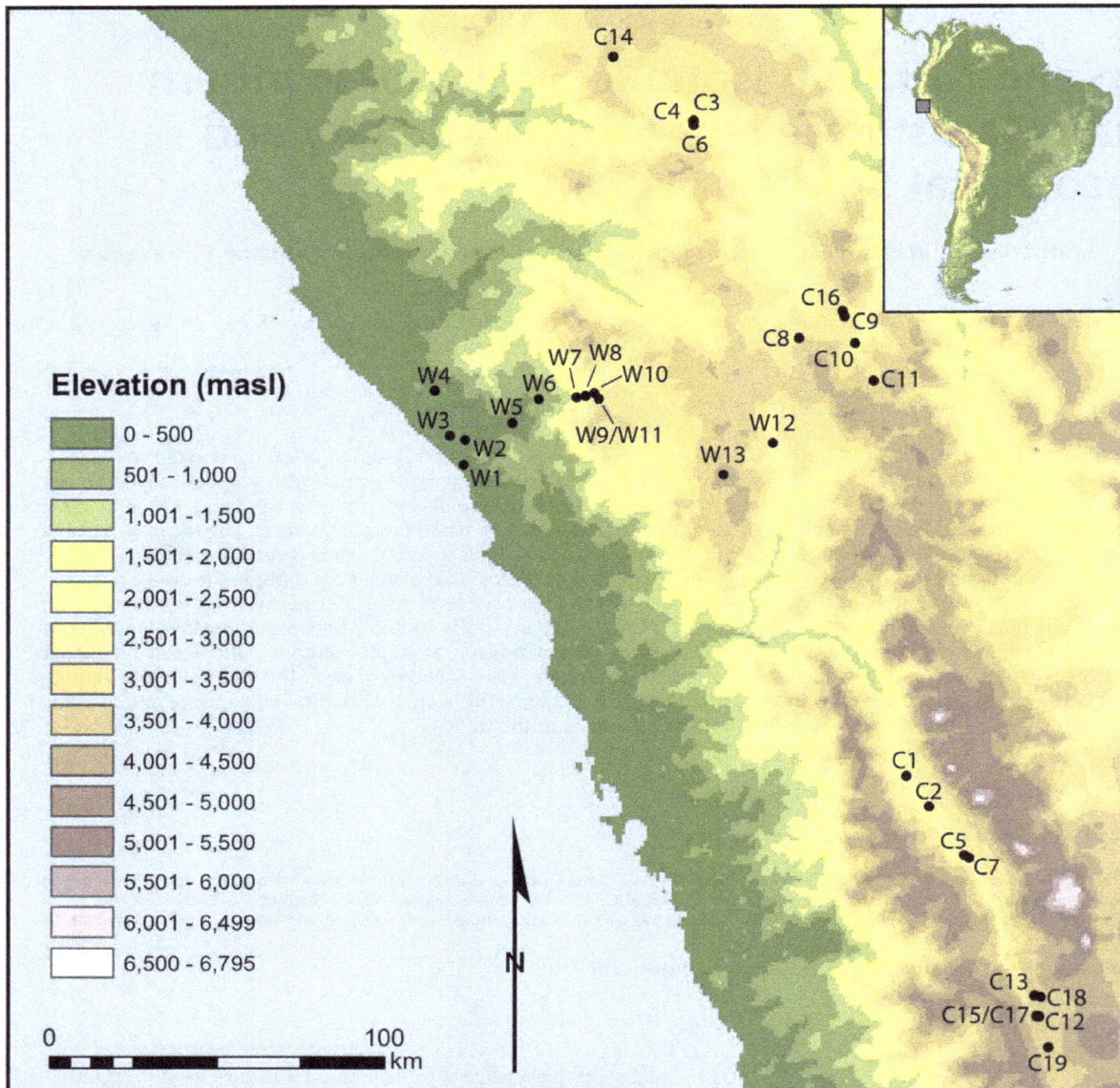

Figure 1. Digital elevation model of the study region derived from the Global 30 Arc-Second Elevation (GTOPO30) data set.

have provided isotopic data on a much more limited scale from various sites in Argentina [17–21], Chile [22–24], Bolivia [25,26], Ecuador [26], Colombia [26], and Peru [26–30].

The number of carbon and nitrogen isotopic studies in the Andean region has increased dramatically in the last ten years, facilitated by outstanding organic preservation in many areas. The majority of these studies have been conducted in Peru [31–42] and Argentina [18–21,43–47]. With respect to northern Peru in particular, a comparatively small number of isotopic data have been published [40,48,49], although this will certainly rise in coming years as biological materials from several understudied polities (e.g. Virú, Moche, Chimú) in the region are subjected to isotopic analysis.

The purpose of this study is to systematically examine the carbon and nitrogen isotopic compositions of plants from the Moche River Valley in northern Peru collected at various altitudes from the coast to the highlands. These data provide a robust

baseline for paleodietary, paleoecological, and related investigations in northern Peru that will utilize the carbon and nitrogen isotopic compositions of consumer tissues.

Study Area

The Andes are an area of marked environmental complexity and diversity. This diversity is driven largely by variation in altitude (Figure 1). As one proceeds from the Pacific coast to the upper limits of the Andes, mean daily temperature declines, typically by ~5°C per 1,000 m [50], and mean annual precipitation increases (Figure 2). The eastern slope of the Andes, which connects to the Amazon basin, is environmentally very different from the western slope. Because this study deals exclusively with the western slope, the eastern slope is not discussed further. Many authors have addressed the environment of the central Andes [51–58], hence only a brief review is necessary here.

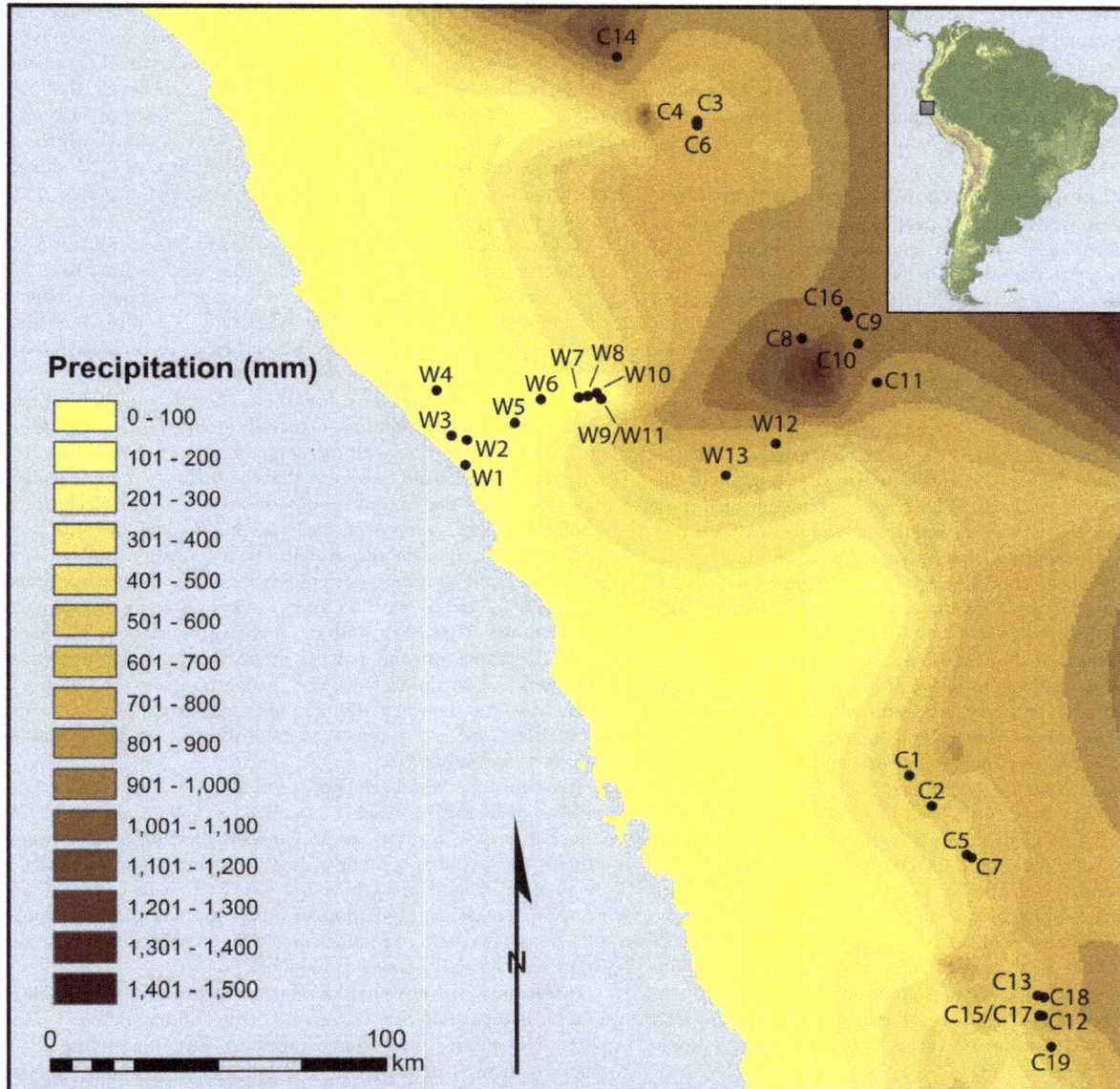

Figure 2. Extrapolated mean annual precipitation for study area. Mean annual precipitation data from 493 monitoring stations in Peru [218] were extrapolated using the natural neighbor method in ArcMap (ArcGIS 10.0, ESRI).

The coastal region of Peru is dominated by the hyper-arid Peruvian desert. Cool sea-surface temperatures created by the northward flowing Peruvian Current, combined with a subtropical anticyclone, create remarkably stable and relatively mild temperatures along the roughly 2,000 km north-south extent of the Peruvian desert [55]. The phytogeography of the coastal region of Peru is fairly homogenous, although the composition of the vegetation varies in accordance with local topography [59]. Except in El Niño years, precipitation is extremely low or non-existent along much of the Peruvian coast, but in areas where topography is steep close to the coast, a fog zone forms (typically between 600 and 900 masl), which allows for the development of ephemeral plant communities (*lomas*) [60–62]. Aside from these *lomas*, riparian vegetation grows in the relatively lush river valleys that cut into the Andes, although the vast majority of this land is cultivated. Thickets of the leguminous algarroba tree regularly occur at low

altitudes, and it is generally believed that much more extensive forests of these trees existed in the past [63,64]. The coastal zone usually ends where the oceanic influence becomes minimal, typically about 1,000 masl [58].

Immediately above the area of oceanic influence and up to an altitude of ~1,800 m, the environment is cooler, although generally similar, in comparison to the coastal zone. Although mean annual precipitation increases, this zone can still be characterized as dry, with most locations receiving less than 400 mm of annual precipitation. In some circumstances, *lomas* may form within this zone [52], although this is not common. In the Moche River Valley of northern Peru, the vegetation is dominated by xerophytic scrub vegetation from 500 to 1,800 masl, and transitions to thorny steppe vegetation between 1,800 and 2,800 masl. Again, the area is still characterized by relatively low annual precipitation, although water availability is greater close to major watercourses and other ground water sources. Ascending further, mean annual precipitation increases, and average daily temperature decreases. Night frost begins to occur. Vegetation is largely dominated by low-growing shrubs, herbs, and grasses, as well as open stands of some tree species (*Acacia, Polylepis*) [56]. Pastures dominated by dense bunchgrasses occur in moister areas.

Natural Variation in Plant Carbon Isotopic Composition

Photosynthetic pathway and taxonomy. The most salient mechanism influencing the carbon isotopic composition ($\delta^{13}C$) of terrestrial plants is the photosynthetic pathway utilized. Plants that fix carbon using the C_3 pathway (Calvin cycle) are characterized by lower $\delta^{13}C$ values (ca. -26 ‰) than plants utilizing the C_4 (Hatch-Slack) pathway (ca. -12 ‰) [65,66]. This is because carbon isotope discrimination ($\Delta^{13}C$) is smaller in C_4 plants than in C_3 plants. In other words, C_3 plants discriminate more strongly against the heavier isotope (^{13}C) than C_4 plants. The vast majority of C_4 plants are tropical grasses, the most significant of which in New World archaeological contexts is maize (*Zea mays*), but also amaranth (*Amaranthus caudatus*). With respect to human diet, most wild C_4 plants are not significant, and thus a large body of research has focused on assessing and quantifying the contribution of C_4 cultigens (mostly maize, but also millet) to the diet [67]. Some desert plants and succulents exhibit carbon isotopic compositions that are intermediate between C_3 and C_4 plants. Referred to as CAM (Crassulacean acid metabolism) plants, these species fix carbon in a manner analogous to C_4 plants overnight, but utilize the C_3 photosynthetic pathway during the afternoon [68].

Additional plant groups that are not readily assigned into the aforementioned categories include mosses and lichens. Mosses, which are non-vascular plants, utilize the C_3 photosynthetic pathway [69,70], but are distinct from vascular plants in that they lack stomata and CO_2 availability is influenced primarily by the thickness of the water film accumulated on the leaves. Lichens are composite organisms, consisting of two parts: a mycobiont (fungi) and photobiont or phycobiont (algae). The carbon isotopic composition of lichens is determined largely by the type of photobiont involved. Lichens with green algae as the photobiont exhibit a wide range of carbon isotopic compositions (-35 to -17 ‰), while lichens with cyanobacteria as the photobiont tend to have higher, and a more restricted range of carbon isotopic compositions (-23 to -14 ‰) [71–73].

Environmental factors affecting plant $\delta^{13}C$. Aside from the differences in carbon isotopic composition resulting from variable carbon fixation, a number of environmental factors have also been demonstrated to influence the carbon isotopic composition of plant tissues. For example, low-growing plants under dense forest cover tend to exhibit lower $\delta^{13}C$ values relative to canopy plants and plants growing in more open environments. Often referred to as the 'canopy effect', this is attributed to relatively ^{13}C-depleted CO_2 in the understory due to the utilization of recycled CO_2 [74–78], and/or lower irradiance and higher [CO_2] relative to the canopy [79,80]. The magnitude of differences in plant carbon isotopic composition observed due to the canopy effect typically range between 2 and 5 ‰ [81]. It has been posited that the canopy effect significantly impacts the carbon isotopic composition of consumer tissues and thus reflects the use of closed and open habitats [82–84]. None of the sites sampled in this study were characterized by sufficiently dense forest for a canopy effect to have been significant.

Water availability has been observed to be negatively correlated with the carbon isotopic composition of plants [85–91]. In most instances, these effects are limited to C_3 plants, with most studies finding little or no correlation between rainfall and/or water

availability and plant $\delta^{13}C$ for C_4 plants [86,92]. Murphy and Bowman [87] found a positive correlation between rainfall and C_4 plant $\delta^{13}C$ over a continental (Australia) rainfall gradient, although this relationship is atypical. It is believed that the relationship between aridity and plant $\delta^{13}C$ is caused by increased stomatal closure when water availability is low, which is accompanied by decreased discrimination against ^{13}C during photosynthesis and, in turn, comparatively less negative $\delta^{13}C$ values [93,94].

Soil salinity has also been demonstrated to influence plant $\delta^{13}C$ values. In a manner somewhat analogous to drought stress, salt stress induces increased stomatal closure, and therefore reduces discrimination against ^{13}C by the plant [95]. A number of studies have observed this relationship, which occurs in both halophytic (salt-tolerant) [96,97] and non-halophytic species [98,99].

A number of studies have found elevational gradients in plant carbon isotopic composition. Generally, foliar $\delta^{13}C$ values have been found to increase with increasing altitude [88,100,101]. It is important to point out, however, that the majority of these studies have examined the isotopic composition of a single species or a small number of species over an elevational gradient of $\sim 1,000$ m. The exact mechanism responsible for the relationship between plant $\delta^{13}C$ and altitude is not entirely clear. Some have suggested exceptionally high carboxylation rates relative to stomatal conductance [102,103] and/or high carboxylation efficiency [104] for plants growing at high altitudes, resulting in decreased discrimination against ^{13}C. A very strong positive correlation has been observed between altitude and leaf mass per unit area [100,101], which is thought to be instrumental in increasing carboxylation capacity.

Irradiance has also been shown to influence foliar $\delta^{13}C$ values, with higher irradiance being associated with less negative $\delta^{13}C$ values in leaves. Such variation can occur within a single plant (usually trees), and even along a single branch, with leaves growing in interior, shaded areas having lower $\delta^{13}C$ values than leaves growing in exterior, exposed areas [105,106]. These differences in $\delta^{13}C$ associated with irradiance have been attributed to differences in intercellular CO_2 concentration [94].

Intraplant and temporal variation in plant $\delta^{13}C$. Carbon isotopic composition is not necessarily equal among different plant parts. Numerous studies have observed variation in the $\delta^{13}C$ values of leaves, stems, roots, and other tissues [107–109]. The vast majority of studies examining the carbon isotopic compositions of multiple plant tissues have found that leaves are slightly depleted of ^{13}C relative to non-photosynthetic tissues, typically by 2 to 4 ‰ [108,110,111]. These differences are only consistent among C_3 plants, with C_4 plants often showing little variation between leaves and non-photosynthetic tissues, or leaves with relatively high $\delta^{13}C$ values in some cases [107,108]. There are several potential variables contributing to intraplant variation in tissue $\delta^{13}C$. First, different tissues may contain variable proportions of molecules that are relatively enriched or depleted of ^{13}C compared to total organic matter. Most notably, lipids [112] and lignin [113] are known to be characterized by relatively low $\delta^{13}C$ values, while the opposite is true for cellulose, sugars, and starches [114]. Because some studies have found significant differences in the $\delta^{13}C$ of specific compounds (e.g. cellulose, sucrose) between different plant parts [110,111], it is thought that additional mechanisms are responsible for the observed patterns in intraplant $\delta^{13}C$ variation. Damesin and Lelarge [110] suggest that some discrimination occurs during the translocation of sugars, particularly when certain plasma membrane proteins are involved in phloem transport. Potential mechanisms causing intraplant variation in $\delta^{13}C$ are treated at length by Cernusak et al. [109].

In addition to variation among plant parts, a number of studies have found variation in $\delta^{13}C$ within plant parts, over time. Specifically, emerging leaves, which are not yet photosynthetic and therefore more closely resemble other non-photosynthetic or heterotrophic plants parts, tend to have less negative $\delta^{13}C$ values (by about 1 to 3 ‰) relative to fully emerged, photosynthetic leaves [91,110,111]. Products assimilated via photosynthesis will tend to have lower $\delta^{13}C$ values than those acquired heterotrophically, and this is likely partly responsible for the decrease in leaf $\delta^{13}C$ over time [115].

Marine plants. For the purpose of this paper, 'marine plants' refers specifically to macroalgae, or plants that are typically classified as kelps, seaweeds, and seagrasses. One of the most commonly reported distinctions in carbon isotopic composition is that marine animals tend to have higher $\delta^{13}C$ values than terrestrial animals, except in cases where C_4 plants dominate the diet of the latter. While this distinction holds in the vast majority of circumstances [8,116,117], the same relationship is not necessarily true for marine and terrestrial plants.

Marine plants are characterized by a high degree of variability in carbon isotopic composition. Figure 3 presents the carbon isotopic compositions for the four major classes of marine macroalgae. In general, marine plants are characterized by carbon isotopic compositions that are intermediate in comparison to terrestrial C_3 and C_4 plants, with two notable exceptions. Seagrasses (*Zostera* sp.), have extremely high $\delta^{13}C$ values, typically higher than most terrestrial C_4 plants (Figure 3d). There is evidence to suggest C_4 photosynthetic activity in a few species of marine algae [118], but the comparatively high $\delta^{13}C$ values observed in many species, including seagrasses, cannot typically be explained in this way [119]. The variable use of dissolved $CO_{2(aq)}$ and $HCO_3^-{}_{(aq)}$ is a significant factor, as $\delta^{13}C$ of $HCO_3^-{}_{(aq)}$ is ~9 ‰ less negative than that of $CO_{2(aq)}$ [120]. Moreover, for intertidal plants, which are exposed to the atmosphere for a portion of the day, the utilization of atmospheric CO_2 further complicates matters [119]. The thickness of the diffusive boundary layer is also a potentially important factor with respect to $\Delta^{13}C$ as it may differ due to variable water velocity [121,122]. Other environmental factors have also been demonstrated to influence aquatic plant $\delta^{13}C$ values, such as: salinity [123], extracellular CO_2 concentration [124,125], light intensity [123], algal growth rate [126], water velocity [122], and water temperature [127].

Some red algae (Floridiophyceae) are characterized by consistently very low $\delta^{13}C$ values (<-30 ‰). In general, the brown algae (kelps) have been noted to contribute significantly to nearshore ecosystems in terms of secondary production, with numerous studies examining the relative contributions of offshore phytoplankton and nearshore macroalgae [128].

Natural Variation in Plant Nitrogen Isotopic Composition

Nitrogen Source. Unlike carbon, which is obtained by plants as atmospheric CO_2, nitrogen is actively taken up from the soil in the vast majority of cases. The two most important nitrogenous species utilized by plants are nitrate (NO_3^-) and ammonium (NH_4^+). In general, nitrate is the most abundant form of mineralized nitrogen available to plants, but in some instances, such as waterlogged or acidic soils, ammonium may predominate [129,130]. Additionally, some plants rely, at least to some extent, on atmospheric nitrogen (N_2), which is obtained by symbiotic bacteria residing in root nodules (rhizobia) [131]. Plants may also take up organic nitrogen (e.g. free amino acids) from the soil [132], although the relative importance of such processes is not well understood and relatively poorly documented [133,134]. The extent to which plants rely on these N sources is significant because they may have distinct nitrogen isotopic compositions due to fractionations associated with different steps in the nitrogen cycle (e.g. ammonification, nitrification, denitrification), as well as the uptake and eventual incorporation of mineralized N into organic N [135–137].

There are two important aspects of variation in N source pertinent to the present study. The first relates to N_2-fixation by plants (mostly members of Fabaceae), which are common in both wild and domestic contexts in many parts of the central Andes. Plants that utilize significant amounts of atmospheric N_2 are characterized by comparatively low $\delta^{15}N$ values, typically ~0 ‰ [27,138–140]. These plants acquire such compositions because the $\delta^{15}N$ of atmospheric N_2 is ~0 ‰ [141] and the assimilation of N from N_2-fixation is not associated with significant fractionation of ^{15}N [138–140]. By comparison, soil NO_3^- and NH_4^+ tend to have $\delta^{15}N$ values >0 ‰ [142], and non N_2-fixing plants have $\delta^{15}N$ values that tend to be >0 ‰, although these are highly variable for a number of reasons as discussed in more detail below.

The second potentially significant source-related cause of plant $\delta^{15}N$ variation is the uptake of fertilizer-derived N by plants. Animal fertilizers are characterized by extremely variable $\delta^{15}N$ values depending on the relative proportions of N-bearing species in the fertilizer (e.g. urea, uric acid, ammonium, organic matter) [143]. Manures consisting primarily of solid waste derived from terrestrial herbivores tend to have $\delta^{15}N$ values between 2 and 8 ‰ [144], while those that contain a mix of solid and liquid waste (slurry fertilizers) tend to have higher $\delta^{15}N$ values, often between 6 and 15 ‰ [145,146]. The highest $\delta^{15}N$ values for animal fertilizers (>25 ‰) have been recorded for seabird guano [143,147], which consists primarily of uric acid and is subject to significant NH_4^+ volatilization. The addition of animal fertilizer N to the soil therefore adds an N source with an isotopic composition that is usually enriched in ^{15}N relative to endogenous soil N. This results in higher $\delta^{15}N$ values for plants growing in soils fertilized with animal waste than those plants growing in unfertilized soil or soils fertilized with chemical fertilizers [143,145–147].

Animal-derived N may be delivered to plants by means other than purposeful application of manures. Several studies have documented that the addition of N from animal carcasses (salmon in particular) provide substantial quantities of N taken up by plants. These plants tend to be characterized by relatively high $\delta^{15}N$ values [148,149]. Increased grazing intensity has also been suggested to influence plant $\delta^{15}N$ values due to the concentrated addition of animal waste, but studies have produced conflicting results, with some finding grazing to: increase plant $\delta^{15}N$ values [150,151], decrease plant $\delta^{15}N$ values [152,153], have little or no impact on plant $\delta^{15}N$ values [154,155], or increase $\delta^{15}N$ in plant roots, but decrease $\delta^{15}N$ in shoots [156].

Taxonomic variation. Strong distinctions in plant $\delta^{15}N$ have been related to mycorrhizal (fungal) associations [12,157,158]. In some ecosystems, particularly those at high latitudes characterized by soils with low N content, this facilitates the distinction between plant functional types – trees, shrubs, and grasses [159–161]. In a global survey of foliar $\delta^{15}N$ values, Craine et al. [12] found significant differences in plant $\delta^{15}N$ on the basis of mycorrhizal associations, with the following patterns (numbers in parentheses are differences relative to non-mycorrhizal plants): ericoid (-2 ‰), ectomycorrhizal (-3.2 ‰), arbuscular (-5.9 ‰). The comparatively low $\delta^{15}N$ values of plants with mycorrhizal associations has been attributed to a fractionation of 8 to 10 ‰ against ^{15}N during the transfer of N from fungi to plants [162,163], with the lowest values indicating higher retention of N in the fungi compared to the plant [164].

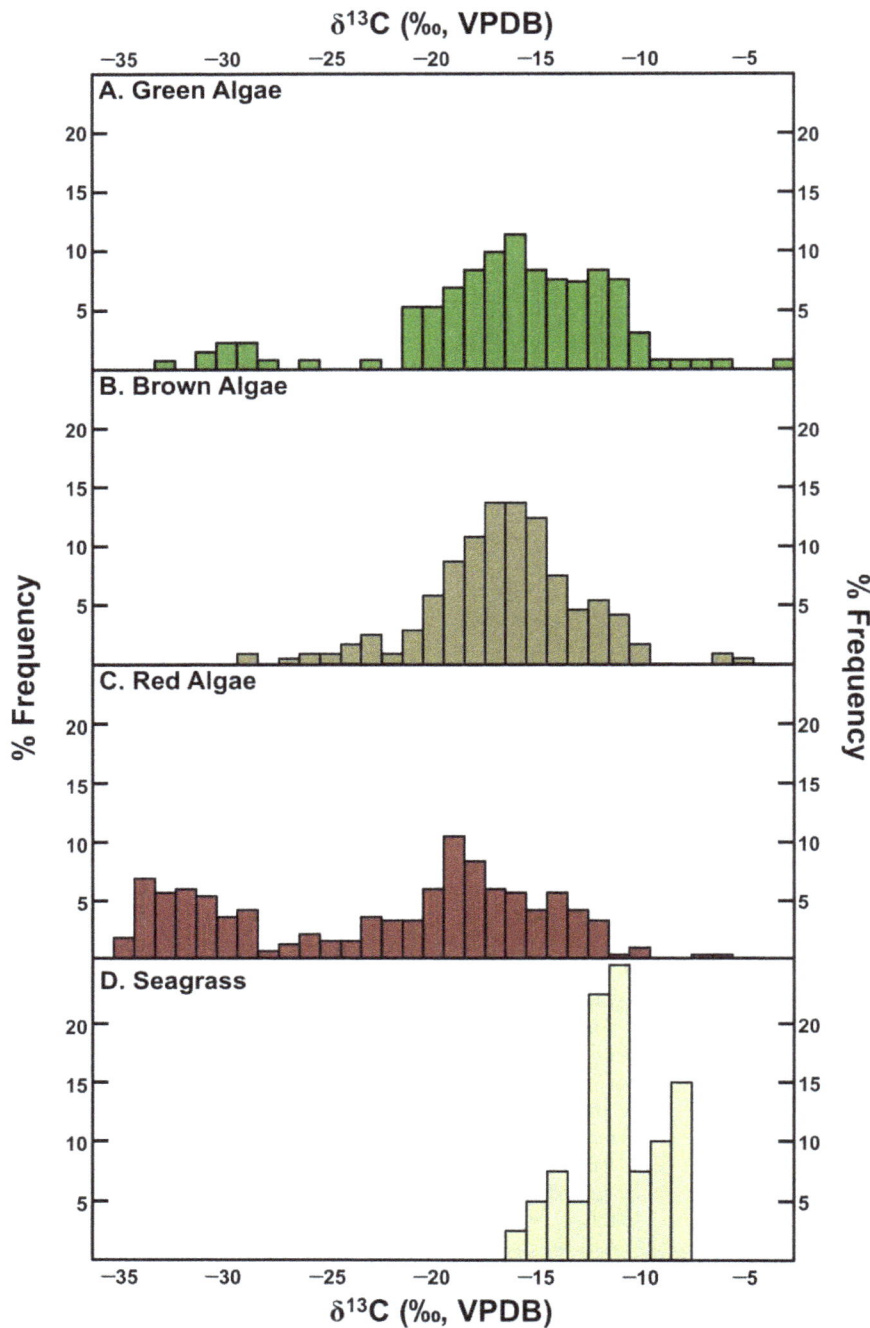

Figure 3. Frequency distributions of carbon isotopic compositions of marine macroalgae. Data are taken from published literature [119,219–235].

Intraplant and temporal variation in plant $\delta^{15}N$. There are three main reasons that plants exhibit intraplant and temporal variation in their tissue $\delta^{15}N$ values: (1) fractionations associated with NO_3^- assimilation in the root vs. shoot, (2) movement of nitrogenous compounds between nitrogen sources and sinks, (3) reliance on isotopically variable N sources as tissue forms over time.

Both NO_3^- and NH_4^+ are taken up by plant roots. NO_3^- can be immediately assimilated into organic N in the root, or it may be routed to the shoot and assimilated there. The assimilation of NO_3^- into organic N is associated with a fractionation of ^{15}N of up to -20 ‰ [137,165]. Therefore, the NO_3^- that is moved to

Table 1. Ecological zones used for sampling in this study [54].

Zone	Altitude
Coastal desert	0 − 500 masl
Premontane desert scrub	500 − 1,800 masl
Premontane thorny steppe	1,800 − 2,800 masl
Montane moist pasture	2,800 − 3,700 masl
Montane wet pasture	3,700 − 4,200 masl

Figure 4. Images of eight of the wild plant sampling locations. Corresponding geographical data for these sites can be found in Table 6.

the shoot has already been exposed to some fractionation associated with assimilation and is enriched in ^{15}N compared to the NO_3^- that was assimilated in the root. On this basis, it is expected that shoots will have higher $\delta^{15}N$ values than roots in plants fed with NO_3^- [166]. Because NH_4^+ is assimilated only in the root, plants with NH_4^+ as their primary N source are not expected to have significant root/shoot variation in $\delta^{15}N$ [136].

As plants grow they accumulate N in certain tissues (sources) and, over time, move this N to other tissues (sinks). In many species, annuals in particular, large portions of the plant's resources are allocated to grain production or flowering. In these cases, significant portions of leaf and/or stem N is mobilized and allocated to the fruits, grains, or flowers [167]. When stored proteins are hydrolyzed, moved, and synthesized, isotopic fractionations occur [168,169]. Theoretically, nitrogen sources (leaves, stems) should be comparatively enriched in ^{15}N in relation to sinks (grains, flowers), which has been observed in several studies [143,145,147].

In agricultural settings, the variation within a plant over time may become particularly complex due to the application of nitrogenous fertilizers. The availability of different N-bearing species from the fertilizer (NH_4^+, NO_3^-) and the nitrogen isotopic composition of fertilizer-derived N changes over time as various soil processes (e.g. ammonification, nitrification) occur. The nature of this variation is complex and will depend on the type of fertilizer applied [147].

Environmental factors affecting plant $\delta^{15}N$. Plant nitrogen isotopic compositions are strongly influenced by a series of environmental factors. The environmental variation in plant $\delta^{15}N$ can be passed on to consumers and cause significant spatial variation in animal isotopic compositions at regional and continental scales [170–175].

Plant $\delta^{15}N$ values have been observed to be positively correlated with mean annual temperature (MAT) [176,177], although this relationship appears to be absent in areas where MAT $\leq -0.5°C$ [12]. A large number of studies have found a negative correlation between plant $\delta^{15}N$ values and local precipitation and/or water availability. These effects have been demonstrated at regional or

Table 2. Environmental data for market plant sampling sites.

Site ID	Site Name	Latitude	Longitude	Altitude (masl)
C1	Caraz	−9.0554	−77.8101	2233
C2	Yungay	−9.1394	−77.7481	2468
C3	Jesus	−7.2448	−78.3797	2530
C4	Jesus II	−7.2474	−78.3821	2573
C5	Ampu	−9.2757	−77.6558	2613
C6	Shuto	−7.2568	−78.3807	2629
C7	Carhuaz	−9.2844	−77.6422	2685
C8	Yamobamba	−7.8432	−78.0956	3176
C9	Huamachuco	−7.7846	−77.9748	3196
C10	Curgos	−7.8599	−77.9475	3220
C11	Poc Poc	−7.9651	−77.8964	3355
C12	Recuay	−9.7225	−77.4531	3400
C13	Olleros	−9.6667	−77.4657	3437
C14	Hierba Buena	−7.0683	−78.5959	3453
C15	Mirador II	−9.7220	−77.4601	3466
C16	Yanac	−7.7704	−77.9799	3471
C17	Mirador I	−9.7224	−77.4601	3477
C18	Conray Chico	−9.6705	−77.4484	3530
C19	Catac	−9.8083	−77.4282	3588

continental [15,85–87,172,178], and global [12,176,179] scales. Several authors have hypothesized that relatively high $\delta^{15}N$ values in herbivore tissues may be the product of physiological processes within the animal related to drought stress [171,173,174], although controlled experiments have failed to provide any evidence supporting this hypothesis [180]. More recent research has demonstrated a clear link between herbivore tissue $\delta^{15}N$ values

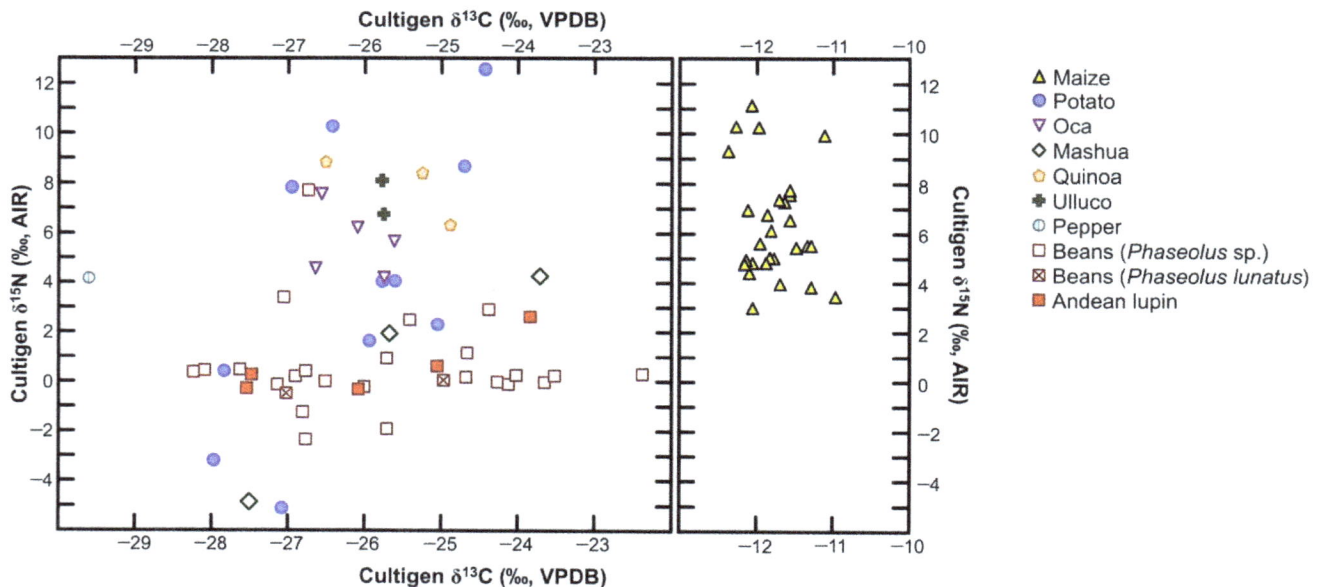

Figure 5. Carbon and nitrogen isotopic compositions of cultigens. Note that the x-axis is not continuous.

Table 3. Mean carbon and nitrogen isotopic compositions for cultigens ($\pm 1\sigma$).

Common Name	Taxonomic Name	n	δ^{13}C (‰, VPDB)	δ^{15}N (‰, AIR)	%C	%N
Beans	*Phaseolus* sp.	24	-25.7 ± 1.6	0.7 ± 2.0	39.8 ± 0.7	3.7 ± 0.6
Beans (Lima)	*Phaseolus lunatus*	2	-26.0 ± 1.4	-0.2 ± 0.4	39.0 ± 0.3	2.7 ± 0.5
Chocho (Andean lupin)	*Lupinus mutabilis*	5	-26.0 ± 1.6	0.6 ± 1.2	48.3 ± 2.8	6.8 ± 1.3
Coca	*Erythroxylum coca*	4	-29.8 ± 0.9	–	45.4 ± 1.5	–
Maize (Grain)	*Zea mays*	27	-11.8 ± 0.4	6.4 ± 2.2	40.4 ± 0.5	1.2 ± 0.2
Maize (Leaf)	*Zea mays*	2	-12.9 ± 0.4	4.5 ± 1.6	41.9 ± 4.6	1.3 ± 1.3
Mashua	*Tropaeolum tuberosum*	3	-25.6 ± 1.9	0.5 ± 4.7	41.5 ± 2.8	3.0 ± 0.7
Oca	*Oxalis tuberosa*	6	-26.4 ± 0.7	5.7 ± 1.3	43.1 ± 3.2	1.6 ± 0.6
Pepper	*Capsicum annuum*	1	-29.6	4.2	48.3	2.1
Potato	*Solanum tuberosum*	12	-26.3 ± 1.3	4.0 ± 5.5	40.5 ± 1.5	1.4 ± 0.4
Quinoa	*Chenopodium quinoa*	3	-25.6 ± 0.9	7.9 ± 1.3	39.9 ± 2.1	2.6 ± 0.3
Ulluco	*Ullucus tuberosus*	2	-25.8 ± 0.0	7.5 ± 1.0	40.6 ± 0.4	3.4 ± 1.0

and plant δ^{15}N values, while providing no support for the 'physiological stress hypothesis' [172,181].

The nature of the relationship between rainfall and plant δ^{15}N values appears to be extremely complex, with numerous variables contributing to the pattern. Several authors, including Handley et al. [179], have attributed this pattern to the relative 'openness' of the nitrogen cycle. In comparison to hot and dry systems, which are prone to losses of excess N, colder and wetter systems more efficiently conserve and recycle mineral N [176] and are thus considered less open. With respect to ecosystem δ^{15}N, ^{15}N enrichment will be favored for any process that increases the flux of organic matter to mineral N, or decreases the flux of mineral N into organic matter [178]. For instance, low microbial activity, or high NH_3 volatilization would cause an overall enrichment in ^{15}N of the soil-plant system.

Marine plants. In comparison to terrestrial plants, the factors affecting the nitrogen isotopic composition of marine plants have not been investigated intensively other than the influence of anthropogenic nitrogen. As is the case with terrestrial plants, marine plant δ^{15}N values are strongly influenced by the forms and isotopic composition of available N [182,183]. Specifically, the relative reliance on upwelled NO_3^- relative to recycled NH_4^+ will strongly influence the δ^{15}N of marine producers, including macroalgae. Systems that are nutrient poor (oligotrophic) tend to be more dependent on recycled NH_4^+, and

systems that are nutrient rich (eutrophic) tend to be more dependent on upwelled NO_3^-. This results in nutrient-rich, upwelling systems being enriched in ^{15}N relative to oligotrophic systems [184].

Materials and Methods

Sample Collection

Wild plants were collected between 2011/07/18 and 2011/08/03. We used regional ecological classifications defined by Tosi [54], which are summarized in Table 1. In each of these five zones, two sites were selected that typified the composition of local vegetation. Sampling locations were chosen to minimize the possibility of significant anthropogenic inputs; in particular, areas close to agricultural fields and disturbed areas were avoided. Sampling locations were fairly open and did not have significant canopy cover. At each sampling location, all plant taxa within a 10 m radius were sampled. Wherever possible, three individuals of each species were sampled and were later homogenized into a single sample for isotopic analysis. Images for eight of the wild plant sampling locations are presented in Figure 4.

Cultigens (edible portions) were collected from local markets between 2008/10/08 and 2008/11/09 (Table 2). Plants introduced to the Americas were not collected (e.g. peas, barley), even though these species were common. Entire large cultigens (e.g.

Table 4. Results of ANOVA post-hoc tests (Dunnett's T3) for cultigen δ^{15}N.

Cultigen %N	Bean (*P. lunatus*)	Andean lupin	Maize	Mashua	Oca	Potato	Quinoa	Ulluco
Bean (*Phaseolus* sp.)	0.860	1.000	**<0.001**	1.000	**0.003**	0.798	**0.028**	0.880
Bean (P. lunatus)	–	0.971	**<0.001**	1.000	**0.005**	0.479	**0.037**	0.121
Andean lupin	–	–	**<0.001**	1.000	**0.006**	0.802	**0.020**	0.060
Maize	–	–	–	0.696	1.000	0.983	0.855	0.917
Mashua	–	–	–	–	0.788	1.000	0.626	0.723
Oca	–	–	–	–	–	1.000	0.626	0.723
Potato	–	–	–	–	–	–	0.688	0.780
Quinoa	–	–	–	–	–	–	–	1.000

Values in boldface are statistically significant ($p<0.05$).

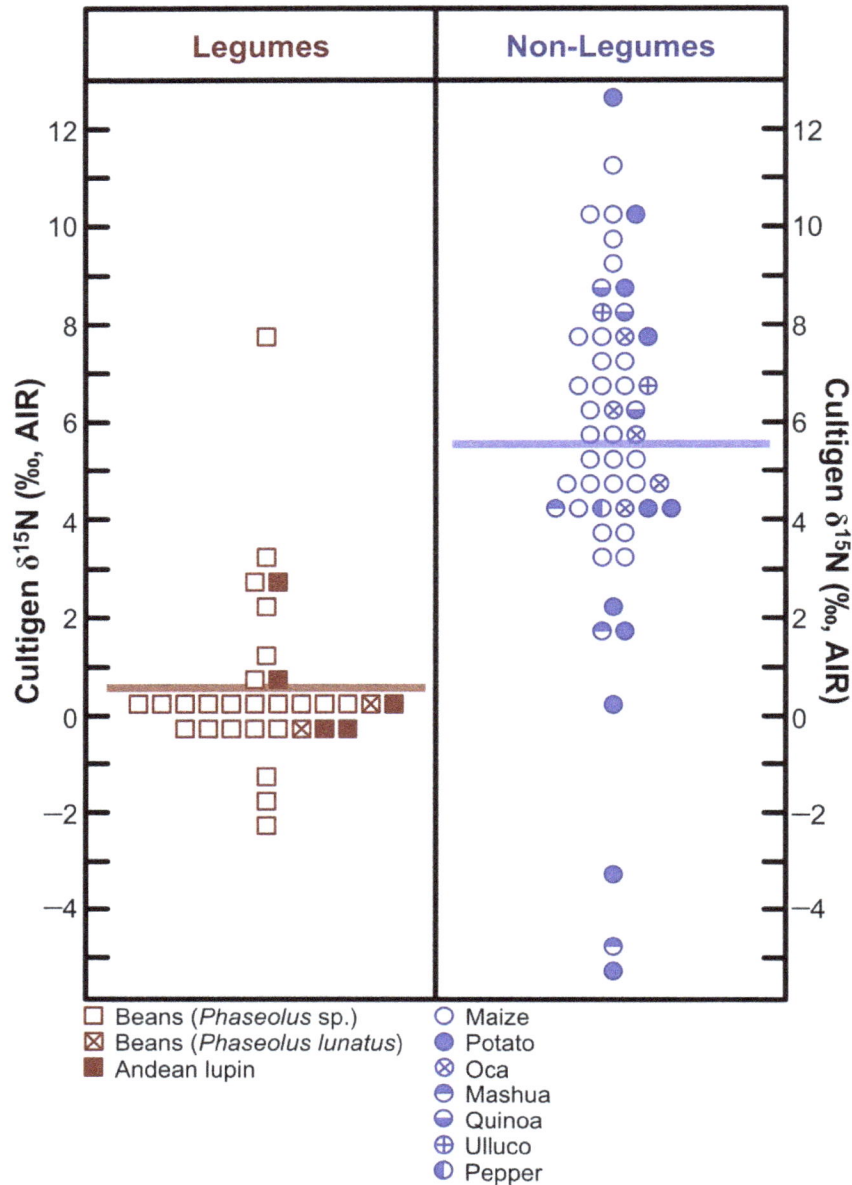

Figure 6. Dot-matrix plot of nitrogen isotopic compositions of legumes and non-legumes. Horizontal bars represent means. Increment = 0.5 ‰.

tubers) were selected and subsequently, a thin (ca. 0.5 cm) slice was sampled. For smaller cultigens (e.g. maize, beans, quinoa) one handful of material was sampled.

For both wild plants and cultigens, geospatial data were recorded using a Garmin® Oregon® 450 portable GPS unit (Garmin®, Olathe, KS, USA). After collection, plants were air-dried on site. Prior to shipping, plants were dried with a Salton® DH–1171 food dehydrator (Salton Canada, Dollard-des-Ormeaux, QC, Canada). Plants were separated according to tissue (leaf, stem, seed, flower). For grasses, all aboveground tissues were considered to be leaf except where significant stem development was present, in which case, leaf and stem were differentiated. All geospatial data associated with these sampling sites are available as a Google Earth.kmz file in the Supporting Information (Dataset S1).

Plants were not sampled from privately-held land or from protected areas. Endangered or protected species were not sampled. Plant materials were imported under permit #2011–03853 from the Canadian Food Inspection Agency. No additional specific permissions were required for these activities.

Sample Preparation

Samples were prepared according to Szpak et al. [143] with minor modifications. As described above, plant material was dried prior to arrival in the laboratory. Whole plant samples were first homogenized using a Magic Bullet® compact blender (Homeland Housewares, Los Angeles, CA, USA). Ground material was then sieved, with the <180 µm material retained for analysis in glass vials. If insufficient material was produced after sieving, the remaining material was further ground using a Wig-L-Bug

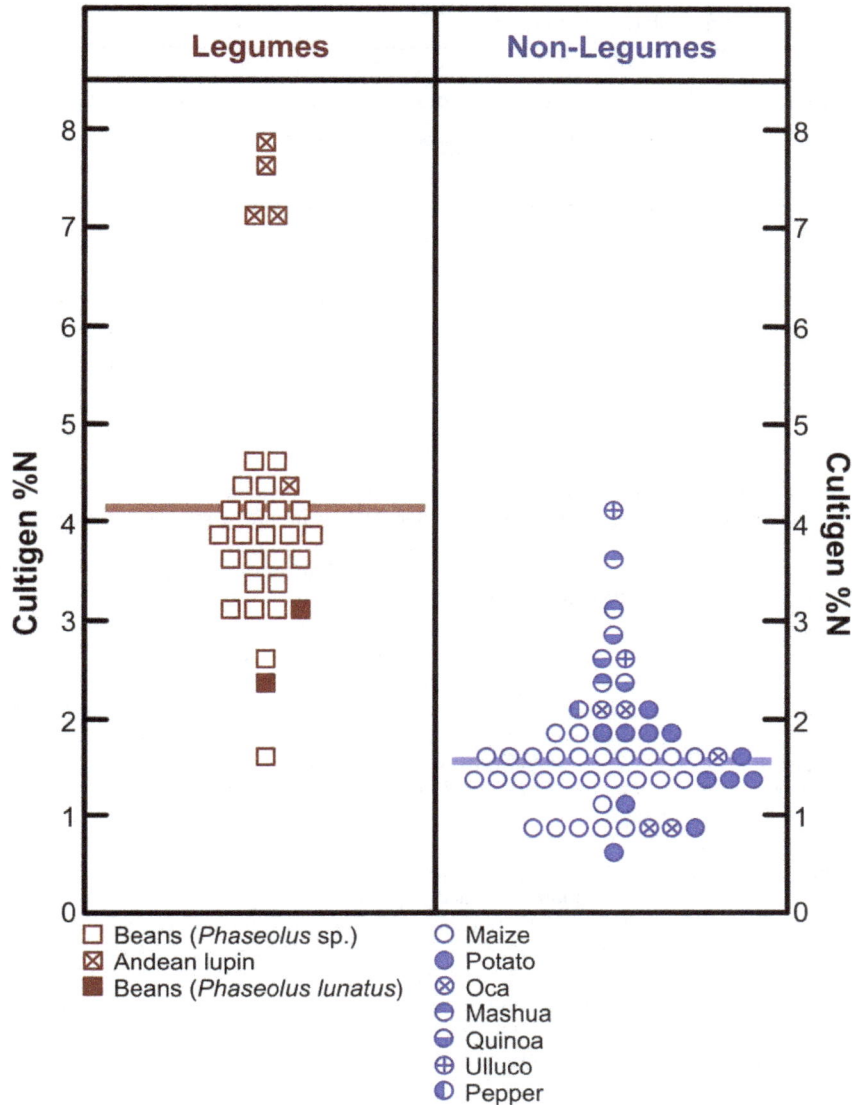

Figure 7. Dot-matrix plot of nitrogen content of legumes and non-legumes. Horizontal bars represent means. Increment = 0.25%.

mechanical shaker (Crescent, Lyons, IL, USA) and retained for analysis in glass vials. Glass vials containing the ground material were dried at 90°C for at least 48 h under normal atmosphere.

Stable Isotope Analysis

Isotopic (δ^{13}C and δ^{15}N) and elemental compositions (%C and %N) were determined using a Delta V isotope ratio mass spectrometer (Thermo Scientific, Bremen, Germany) coupled to an elemental analyzer (Costech Analytical Technologies, Valencia, CA, USA), located in the Laboratory for Stable Isotope Science (LSIS) at the University of Western Ontario (London, ON, Canada). For samples with <2% N, nitrogen isotopic compositions were determined separately, with excess CO_2 being removed with a Carbo-Sorb trap (Elemental Microanalysis, Okehampton, Devon, UK) prior to isotopic analysis.

Sample δ^{13}C and δ^{15}N values were calibrated to VPDB and AIR, respectively, with USGS40 (accepted values: δ^{13}C = −26.39 ‰, δ^{15}N = −4.52 ‰) and USGS41 (accepted values: δ^{13}C = 37.63 ‰,

δ^{15}N = 47.6 ‰). In addition to USGS40 and USGS41, internal (keratin) and international (IAEA-CH-6, IAEA-N-2) standard reference materials were analyzed to monitor analytical precision and accuracy. A δ^{13}C value of −24.03±0.14 ‰ was obtained for 81 analyses of the internal keratin standard, which compared well with its average value of −24.04 ‰. A δ^{13}C value of −10.46±0.09 ‰ was obtained for 46 analyses of IAEA-CH-6, which compared well with its accepted value of −10.45 ‰. Sample reproducibility was ±0.10 ‰ for δ^{13}C and ±0.50% for %C (50 replicates). A δ^{15}N value of 6.37±0.13 ‰ was obtained for 172 analyses of an internal keratin standard, which compared well with its average value of 6.36 ‰. A δ^{15}N value of 20.3±0.4 ‰ was obtained for 76 analyses of IAEA-N-2, which compared well with its accepted value of 20.3 ‰. Sample reproducibility was ±0.14 ‰ for δ^{15}N and ±0.10% for %N (84 replicates).

Data Treatment and Statistical Analyses

Plants were grouped into the following major functional categories for analysis: herb/shrub, tree, grass/sedge, vine. Plants

Table 5. Results of ANOVA post-hoc tests (Dunnett's T3) for cultigen N content.

Cultigen %N	Bean (*P. lunatus*)	Andean lupin	Maize	Mashua	Oca	Potato	Quinoa	Ulluco
Bean (*Phaseolus* sp.)	0.637	0.072	**<0.001**	0.869	**0.009**	**<0.001**	0.123	1.000
Bean (P. lunatus)	–	**0.037**	0.462	1.000	0.619	0.505	1.000	0.995
Andean lupin	–	–	**0.009**	**0.034**	**0.005**	**0.008**	**0.021**	0.295
Maize	–	–	–	0.232	0.981	0.992	0.101	0.566
Mashua	–	–	–	–	**0.019**	1.000	0.216	**0.009**
Oca	–	–	–	–	–	1.000	**0.033**	**0.001**
Potato	–	–	–	–	–	–	**0.033**	**0.001**
Quinoa	–	–	–	–	–	–	–	0.885

Values in boldface are statistically significant ($p<0.05$).

that are invasive and/or introduced species were included in the calculation of means for particular sites since their isotopic compositions should still be impacted by the same environmental factors as other plants. For all statistical analyses of carbon isotopic composition, grass/sedge and herb/shrub were further separated into C_3 and C_4 categories. For comparisons among plant functional types, and sampling sites, foliar tissue was used since other tissues were not as extensively sampled.

Correlations between foliar isotopic compositions and environmental parameters (altitude, mean annual precipitation) were assessed using Spearman's rank correlation coefficient (ρ). One-way analysis of variance (ANOVA) followed by either a Tukey's HSD test (if variance was homoscedastic) or a Dunnett's T3 test (if variance was not homoscedastic) was used to compare means. All statistical analyses and regressions were performed in SPSS 16 for Windows.

Results

Cultigens

The carbon and nitrogen isotopic compositions were analyzed for a total of 85 cultigen samples from eleven species. Carbon and

nitrogen isotopic compositions for cultigens are presented in Figure 5. Mean $\delta^{13}C$ and $\delta^{15}N$ values for cultigens are presented in Table 3. Isotopic and elemental data, as well as corresponding geospatial data for individual cultigens are presented in Table S1. All isotopic and elemental compositions for cultigens are for consumable portions of the plant, with one exception (maize leaves), which is excluded from Table 3 and Figure 5. Mean $\delta^{13}C$ values for C_3 cultigens ranged from -29.8 ± 0.9 ‰ (coca) to -25.6 ± 1.9 ‰ (mashua). The mean $\delta^{13}C$ value for maize, which was the only C_4 plant examined, was -11.8 ± 0.4 ‰. Mean $\delta^{15}N$ values for cultigens were typically more variable than $\delta^{13}C$ values, ranging from -0.2 ± 0.4 ‰ (*Phaseolus lunatus*) to 7.9 ± 1.3 ‰ (quinoa).

When maize is excluded, there were no significant differences in $\delta^{13}C$ among cultigens ($F_{[7,49]}=0.3$, $p=0.93$), but there were for $\delta^{15}N$ (maize included) ($F_{[8,73]}=9.7$, $p<0.001$). Results of post-hoc Dunnett's T3 test for $\delta^{15}N$ differences among individual cultigen species are presented in Table 4. The three leguminous species were generally characterized by significantly lower $\delta^{15}N$ values than non-leguminous species (Table 4); collectively, legumes were characterized by significantly lower $\delta^{15}N$ values than non-legumes (Figure 6; $F_{[1,80]}=51.8$, $p<0.001$).

Table 6. Environmental data for wild plant sampling sites and summary of number of C_3 and C_4 plant species sampled.

Site ID	Site Name	Latitude	Longitude	Altitude (masl)	MAP (mm)[1]	C_3 Plant Taxa Sampled	C_4 Plant Taxa Sampled
W1	Las Delicias	−8.1956	−78.9996	10	7	7	2
W2	Río Moche	−8.1267	−78.9963	33	5	9	1
W3	Ciudad Universitaria	−8.1137	−79.0373	38	6	2	0
W4	Cerro Campana	−7.9900	−79.0768	164	11	4	1
W5	La Carbonera	−8.0791	−78.8681	192	56	5	3
W6	Poroto	−8.0137	−78.7972	447	113	17	6
W7	Salpo 5	−8.0089	−78.6962	1181	143	0	2
W8	Salpo 4	−8.0047	−78.6726	1557	140	9	0
W9	Salpo 3	−8.0132	−78.6355	2150	141	16	0
W10	Salpo 2	−7.9973	−78.6481	2421	142	8	0
W11	Salpo 1	−8.0132	−78.6355	2947	171	9	1
W12	Stgo de Chuco	−8.1361	−78.1685	3041	702	21	1
W13	Cahuide	−8.2235	−78.3013	4070	591	15	0

[1]Mean annual precipitation (MAP) estimated as described in the text.

Table 7. Carbon and nitrogen isotopic compositions for all wild plant taxa sampled.

Taxonomic Name	Site ID	Altitude	Type	Leaf δ¹³C (‰)	Leaf δ¹⁵N (‰)	Stem δ¹³C (‰)	Stem δ¹⁵N (‰)	Root δ¹³C (‰)	Root δ¹⁵N (‰)	Flowers δ¹³C (‰)	Flowers δ¹⁵N (‰)	Seeds δ¹³C (‰)	Seeds δ¹⁵N (‰)
Eriochloa mutica	W1	10	Grass	−11.6	−1.5	−11.7	1.6	–	–	–	–	−12.1	−0.2
Distichia spicata	W1	10	Grass	−14.9	−3.2	–	–	–	–	–	–	–	–
Baccharis glutinosa	W1	10	Shrub	−27.4	3.3	−26.6	5.1	–	–	−27.0	4.1	−26.9	4.2
Rauvolfia sp.	W1	10	Shrub	−28.0	9.8	−27.9	11.0	–	–	–	–	–	–
Plantago major[1]	W1	10	Herb	−28.6	7.5	−27.5	7.9	−26.7	8.6	–	–	–	–
Typha angustifolia	W1	10	Herb	−29.3	1.3	–	–	−28.7	2.6	–	–	–	–
Blumea crispata[1]	W1	10	Herb	−29.8	13.7	−30.4	13.7	−30.5	11.6	–	–	–	–
Rosippa nastrutium aquaticum[1]	W1	10	Herb	−30.1	12.5	–	–	−30.0	11.4	–	–	–	–
Oxalis corniculata	W1	10	Herb	−30.6	7.1	−31.0	6.0	−31.2	4.7	–	–	–	–
Paspalum racemosum	W2	33	Grass	−12.7	0.8	−12.8	11.7	–	–	–	–	–	–
Salix humboldtiana	W2	33	Tree	−26.4	5.2	−26.5	4.4	–	–	–	–	–	–
Phyla nodiflora	W2	33	Herb	−27.7	6.5	−26.8	5.1	−27.1	2.8	−26.7	7.9	−27.3	8.9
Melochia lupulina	W2	33	Shrub	−28.3	6.9	−27.6	6.4	–	–	−28.6	6.8	–	–
Ipomoea alba	W2	33	Herb	−28.7	9.3	−28.1	8.1	–	–	–	–	–	–
Persea americana	W2	33	Tree	−28.8	1.9	−26.8	7.0	–	–	–	–	–	–
Ambrosia peruviana	W2	33	Herb	−29.6	2.2	−30.0	1.7	–	–	–	–	–	–
Arundo donax[1]	W2	33	Grass	−30.3	8.5	−30.1	10.2	–	–	–	–	–	–
Acacia huarango[2]	W2	33	Shrub	−31.0	3.5	−30.0	2.3	–	–	−29.8	3.4	–	–
Psittacanthus obovatus	W2	33	Shrub (Parasitic)	−31.9	5.1	−30.6	6.2	–	–	–	–	–	–
Prosopis pallida[2]	W3	38	Tree	−27.9	4.0	−28.9	1.5	–	–	−29.1	5.8	–	–
Acacia macracantha[2]	W3	38	Tree	−30.7	8.3	−30.1	6.8	–	–	−30.5	8.6	−28.9	5.1
Tillandsia usneoides	W4	164	Epiphyte	−13.6	3.7	−14.2	1.9	−13.9	14.5	−13.6	0.0	–	–
Cryptocarpus pyriformis	W4	164	Shrub	−22.5	10.3	−22.2	10.4	–	–	–	–	−22.4	12.1
Trixis cacalioides	W4	164	Shrub	−26.6	9.2	−26.0	7.6	–	–	–	–	−25.7	9.4
Scutia spicata	W4	164	Shrub	−27.1	4.9	−25.7	4.4	–	–	–	–	–	–
Capparis angulata	W4	164	Shrub	−27.3	10.0	−27.7	10.7	–	–	–	–	−26.0	11.6
Paspalidium paladivagum	W5	192	Grass	−12.5	10.5	−12.8	11.0	−12.0	11.7	–	–	−12.3	13.4
Amaranthus celosiodes	W5	192	Herb	−13.1	9.1	−12.5	11.0	–	–	−13.5	10.9	−12.2	8.6
Tribulus terrestris	W5	192	Herb	−15.6	11.8	−16.2	14.4	–	–	–	–	−14.0	13.6
Hydrocotyle bonariensis	W5	192	Herb	−26.5	9.0	–	–	–	–	–	–	–	–
Cestrum auriculatum	W5	192	Shrub	−26.9	10.6	−26.8	8.5	–	–	–	–	−27.0	12.1
Cucumis dipsaceus	W5	192	Herb	−27.4	5.6	−26.8	4.2	–	–	−27.0	5.5	−28.1	6.5
Argemone subfusiformis	W5	192	Herb	−28.8	6.9	−28.1	6.3	−28.8	6.2	−28.9	5.7	–	–
Picrosia longifolia	W5	192	Herb	−30.6	5.3	−30.5	1.1	–	–	−30.0	9.3	–	–

Table 7. Cont.

Taxonomic Name	Site ID	Altitude	Type	Leaf δ¹³C (‰)	Leaf δ¹⁵N (‰)	Stem δ¹³C (‰)	Stem δ¹⁵N (‰)	Root δ¹³C (‰)	Root δ¹⁵N (‰)	Flowers δ¹³C (‰)	Flowers δ¹⁵N (‰)	Seeds δ¹³C (‰)	Seeds δ¹⁵N (‰)
Cyperus corymbosus	W6	447	Sedge	−13.1	8.3	−11.2	8.8	−11.7	7.8	−14.2	9.1	–	–
Echinochloa crusgalli[1]	W6	447	Grass	−13.4	2.8	−13.8	2.8	–	–	–	–	−13.7	3.9
Cynodon dactylon[1]	W6	447	Grass	−13.9	0.8	–	–	–	–	−14.1	1.2	–	–
Sorghum halepense[1]	W6	447	Grass	−14.0	2.5	−15.0	4.7	–	–	–	–	−13.1	3.7
Trianthema portulacastrum	W6	447	Herb	−14.2	17.3	−13.7	12.3	–	–	–	–	–	–
Amaranthus spinosus	W6	447	Herb	−14.4	13.3	−13.8	16.1	–	–	−14.0	15.3	−14.4	15.3
Gynerium sagittatum	W6	447	Grass	−25.8	2.7	−25.1	2.3	–	–	−25.6	5.3	–	–
Alternanthera halimifolia	W6	447	Herb	−26.0	8.4	−26.1	9.2	–	–	−26.1	8.2	–	–
Cissus sicyoides	W6	447	Vine	−26.6	10.9	−26.1	12.4	–	–	–	–	−25.1	11.9
Dalea onobrychis[2]	W6	447	Herb	−27.2	8.7	−27.4	7.8	–	–	–	–	−26.8	7.4
Cleome spinosa	W6	447	Herb	−27.3	9.0	−27.1	9.8	–	–	–	–	−26.9	9.8
Crotalaria incae[2]	W6	447	Shrub	−27.3	0.2	−26.6	−2.4	–	–	−25.4	1.0	−26.0	−0.8
Ludwigia octovalvis[2]	W6	447	Herb	−27.5	0.6	−26.9	1.3	–	–	–	–	–	–
Passiflora foetida	W6	447	Vine	−27.5	9.5	−27.2	1.7	–	–	−27.5	7.8	–	–
Wedelia latifolia	W6	447	Shrub	−28.0	6.4	−27.3	4.8	–	–	−26.4	8.1	–	–
Baccharis salicifolia	W6	447	Shrub	−28.3	6.5	−27.2	8.4	–	–	–	–	–	–
Waltheria ovata	W6	447	Shrub	−28.4	6.1	−28.2	5.9	–	–	−27.7	6.0	−27.1	8.0
Verbena littoralis	W6	447	Herb	−28.8	7.9	−28.4	5.8	–	–	–	–	−27.7	7.3
Cyperus odoratus	W6	447	Sedge	−28.8	9.2	−27.6	10.1	–	–	−28.0	10.2	–	–
Mimosa pigra	W6	447	Shrub	−29.3	1.7	−28.5	0.3	–	–	–	–	−29.1	1.3
Cajanus cajan[1,2]	W6	447	Tree	−29.6	−1.4	−28.4	–	–	–	−28.3	0.3	−27.6	−0.7
Polygonum hydropiperoides	W6	447	Herb	−30.2	6.8	−30.6	6.7	–	–	–	–	−27.2	8.1
Mimosa albida[2]	W6	447	Shrub	−30.5	−0.8	−30.1	−1.5	–	–	–	–	−28.8	1.2
Melinis repens[1]	W7	1181	Grass	−13.3	5.6	−13.4	7.3	–	–	–	–	−14.5	3.1
Cenchrus myosuroides	W7	1181	Grass	−13.3	5.7	–	–	–	–	–	–	–	–
Dicliptera peruviana	W8	1557	Herb	−24.7	3.9	−26.3	3.2	–	–	–	–	−25.0	3.2
Tournefortia microcalyx	W8	1557	Shrub	−26.0	6.4	−26.2	6.0	–	–	−25.6	7.2	–	–
Ophryosporus peruvianus	W8	1557	Shrub	−26.7	2.9	−23.9	1.8	–	–	–	–	−24.0	2.6
Alternanthera porrigens	W8	1557	Herb	−27.8	2.8	−27.2	2.2	–	–	–	–	−25.7	6.7
Asclepias curassavica	W8	1557	Shrub	−28.9	2.6	−28.7	4.2	–	–	−28.8	0.4	−28.0	0.2
Boerhavia erecta	W8	1557	Herb	−29.3	9.1	−28.3	9.2	–	–	–	–	–	–
Centaurea melitensis	W8	1557	Herb	−29.5	0.5	−29.8	0.2	–	–	–	–	–	–
Mentzelia aspera	W8	1557	Herb	−30.0	1.0	−27.6	6.8	–	–	−29.5	1.3	−28.9	1.7
Sida spinosa	W8	1557	Herb	−30.1	3.1	−30.1	4.7	–	–	−31.3	1.7	–	–

Table 7. Cont.

Taxonomic Name	Site ID	Altitude	Type	Leaf δ¹³C (‰)	Leaf δ¹⁵N (‰)	Stem δ¹³C (‰)	Stem δ¹⁵N (‰)	Root δ¹³C (‰)	Root δ¹⁵N (‰)	Flowers δ¹³C (‰)	Flowers δ¹⁵N (‰)	Seeds δ¹³C (‰)	Seeds δ¹⁵N (‰)
Rubus robustus	W9	2150	Shrub	−25.0	3.0	−24.4	2.7	–	–	–	–	–	–
Puya sp.	W9	2150	Succulent	−25.4	−0.7	–	–	–	–	–	–	–	–
Barnadesia dombeyana	W9	2150	Shrub	−26.0	−2.0	−25.4	−0.3	–	–	−25.7	−2.7	–	–
Iochroma edule	W9	2150	Shrub	−26.1	8.5	−25.8	7.6	–	–	–	–	−25.4	7.3
Eupatorium sp.	W9	2150	Herb	−26.8	2.5	–	–	–	–	–	–	–	–
Capparis scabrida	W9	2150	Shrub	−26.8	1.3	−26.3	0.9	–	–	−26.5	2.2	–	–
Vasquezia oppositifolia	W9	2150	Herb	−27.0	−1.6	–	–	–	–	–	–	−26.9	−1.3
Stipa ichu	W9	2150	Grass	−27.0	0.3	–	–	−27.4	0.2	−27.3	0.6	–	–
Lupinus sp.[2]	W9	2150	Herb	−27.1	1.4	−27.1	3.4	–	–	−26.6	3.2	−26.7	0.8
Alonsoa meridionalis	W9	2150	Herb	−27.5	1.3	−26.4	−1.9	–	–	–	–	−25.9	0.1
Bromus catharticus	W9	2150	Grass	−27.8	1.1	−29.3	−1.3	–	–	–	–	−27.5	−0.7
Baccharis sp.	W9	2150	Shrub	−28.9	−1.1	−28.8	0.1	–	–	−29.5	0.5	–	–
Minthostachys mollis	W9	2150	Herb	−29.0	0.5	−28.1	−1.6	–	–	−27.2	0.1	–	–
Satureja sp.	W9	2150	Herb	−30.2	−3.2	–	–	–	–	−29.8	−2.3	–	–
Achyrocline alata	W9	2150	Shrub	−30.3	0.3	−27.8	1.2	–	–	−27.5	2.0	–	–
Polypogon sp.	W9	2150	Grass	−31.1	−5.3	−27.9	−4.4	−31.0	2.4	−27.8	−3.7	–	–
Browallia americana	W10	2421	Herb	−25.4	−1.6	−26.8	−2.5	–	–	−25.7	−0.8	–	–
Coniza sp.	W10	2421	Herb	−26.7	6.1	−26.1	4.0	–	–	–	–	–	–
Heliotropium sp.	W10	2421	Herb	−26.9	3.7	−28.4	2.2	–	–	–	–	−28.2	3.2
Caesalpina spinosa[2]	W10	2421	Tree	−27.4	2.7	−27.7	−0.4	–	–	–	–	−25.1	0.0
Oenothera rosea	W10	2421	Herb	−27.4	4.9	−27.9	4.6	–	–	–	–	−27.4	2.9
Avena sterilis[1]	W10	2421	Grass	−27.5	2.3	−27.2	2.1	−27.0	0.0	–	–	−22.5	2.2
Berberis sp.	W10	2421	Shrub	−27.7	1.1	−24.6	1.9	–	–	–	–	−26.7	1.9
Alternanthera sp.	W10	2421	Herb	−28.3	−2.9	−27.5	−3.0	–	–	–	–	−27.2	−0.8
Pennisetum purpurem[1]	W11	2947	Grass	−12.5	7.2	−12.8	6.6	–	–	–	–	−15.5	7.1
Ruellia floribunda	W11	2947	Herb	−23.7	4.5	−24.0	1.9	–	–	−23.5	4.9	–	–
Schinus molle	W11	2947	Tree	−24.6	2.3	−23.4	0.3	–	–	−21.3	0.8	–	–
Spartium junceum[1,2]	W11	2947	Shrub	−26.5	1.1	−27.1	−1.1	–	–	−23.7	−1.3	−25.4	0.8
Acacia aroma[2]	W11	2947	Tree	−26.8	9.6	−26.6	9.6	–	–	−26.6	10.1	–	–
Croton ovalifolius	W11	2947	Shrub	−27.0	7.4	−27.6	5.8	–	–	–	–	–	–
Leonotis nepetifolia[1]	W11	2947	Shrub	−28.0	2.2	–	–	–	–	−27.2	3.0	−26.1	2.0
Lycianthes lycioides	W11	2947	Shrub	−28.0	−0.3	–	–	–	–	–	–	−24.3	2.0
Phenax hirtus	W11	2947	Shrub	−28.3	2.5	−29.1	7.1	–	–	–	–	−28.5	6.9
Inga feulleu[2]	W11	2947	Tree	−28.9	0.3	−27.6	−0.8	–	–	–	–	−27.1	1.1

Table 7. Cont.

Taxonomic Name	Site ID	Altitude	Type	Leaf $\delta^{13}C$ (‰)	Leaf $\delta^{15}N$ (‰)	Stem $\delta^{13}C$ (‰)	Stem $\delta^{15}N$ (‰)	Root $\delta^{13}C$ (‰)	Root $\delta^{15}N$ (‰)	Flowers $\delta^{13}C$ (‰)	Flowers $\delta^{15}N$ (‰)	Seeds $\delta^{13}C$ (‰)	Seeds $\delta^{15}N$ (‰)
Andropogon sp.	W12	3041	Grass	−13.5	−1.6	–	–	−13.2	−1.0	–	–	–	–
Sebastiania obtusifolia	W12	3041	Shrub	−23.7	0.8	−24.7	0.0	–	–	–	–	−24.0	2.5
Lupinus aridulus[2]	W12	3041	Herb	−24.3	2.0	−23.7	2.2	–	–	−22.7	4.0	−22.0	5.4
Silybum marianum[1]	W12	3041	Herb	−25.9	2.2	−25.8	1.6	–	–	–	–	−25.1	2.2
Phrygilanthus sp.	W12	3041	Shrub (Parasitic)	−25.9	−0.5	−24.7	7.3	–	–	–	–	–	–
Solanum amotapense	W12	3041	Shrub	−25.9	7.9	−25.2	5.1	–	–	–	–	−24.7	8.3
Acacia sp.[2]	W12	3041	Tree	−26.2	−1.0	−25.0	−2.5	–	–	–	–	–	–
Baccharis serpifolia	W12	3041	Shrub	−26.4	2.2	−27.1	1.5	–	–	−26.9	1.0	–	–
Aristida adsensionis	W12	3041	Grass	−26.5	−2.6	−26.2	−2.0	–	–	−26.7	−1.0	–	–
Baccharis emarginata	W12	3041	Shrub	−26.5	−0.2	−25.4	0.1	–	–	–	–	–	–
Brassica campestris	W12	3041	Herb	−27.1	2.3	–	–	–	–	–	–	−25.4	4.3
Mauria sp.	W12	3041	Tree	−27.2	6.1	−25.8	3.1	–	–	–	–	–	–
Solanum agrimoniaefolium	W12	3041	Shrub	−28.0	6.0	−28.2	3.8	–	–	–	–	−26.6	4.1
Salvia punctata	W12	3041	Herb	−28.1	−3.5	–	–	–	–	−27.7	−2.1	−27.3	−1.7
Duranta sp.	W12	3041	Shrub	−28.4	1.3	−27.6	0.8	–	–	–	–	–	–
Flourensia cajabambensis	W12	3041	Shrub	−28.6	2.9	−27.5	2.7	–	–	–	–	−29.4	2.5
Marrubium vulgare	W12	3041	Herb	−28.8	3.8	−26.6	1.4	−27.9	1.0	−26.9	4.0	–	–
Scutellaria sp.	W12	3041	Herb	−28.8	2.2	−28.6	0.6	–	–	–	–	–	–
Viguiera peruviana	W12	3041	Shrub	−28.9	5.3	−27.3	4.6	–	–	–	–	−26.6	5.4
Jungia rugosa	W12	3041	Shrub	−28.9	1.4	−26.8	1.1	–	–	−27.1	2.7	–	–
Saccellium sp.	W12	3041	Shrub	−29.0	2.0	−27.3	1.0	–	–	–	–	–	–
Baccharis libertadensis	W12	3041	Shrub	−29.6	3.8	−28.4	1.9	–	–	–	–	–	–
Usnea andina	W13	4070	Lichen	−20.5	−6.5	–	–	–	–	–	–	–	–
Astragalus garbancillo[2]	W13	4070	Shrub	−24.6	4.2	−25.3	3.0	−25.1	3.8	−23.8	3.9	−22.5	5.4
Luzula sp.	W13	4070	Sedge	−25.1	0.9	–	–	−25.0	3.9	−25.1	3.2	–	–
Distichia muscoides	W13	4070	Grass	−25.3	4.4	–	–	−25.2	2.9	–	–	–	–
Muehlenbeckia sp.	W13	4070	Herb	−25.3	6.3	–	–	−25.7	4.9	–	–	–	–
Urtica sp.	W13	4070	Shrub	−25.5	11.9	−25.1	9.0	−26.0	9.4	–	–	−26.6	11.9
Agrostis breviculmis	W13	4070	Grass	−25.9	2.1	–	–	−26.0	4.1	−25.5	2.4	–	–
Chuquiraga spinosa	W13	4070	Shrub	−26.0	−0.5	−24.9	−1.1	−24.4	−1.4	−24.3	−0.7	−24.4	−0.2
Werneria nubigena	W13	4070	Herb	−26.2	1.3	–	–	−25.8	1.8	–	–	–	–
Festuca dolichopylla	W13	4070	Grass	−26.3	−1.8	–	–	−25.4	−0.3	–	–	−26.5	3.6
Hypochaeris sp.	W13	4070	Herb	−26.6	7.3	–	–	−26.9	8.2	–	–	–	–
Plantago tubulosa	W13	4070	Herb	−26.9	−5.2	–	–	−26.0	−3.0	–	–	–	–

Table 7. Cont.

Taxonomic Name	Site ID	Altitude	Type	Leaf δ¹³C (‰)	Leaf δ¹⁵N (‰)	Stem δ¹³C (‰)	Stem δ¹⁵N (‰)	Root δ¹³C (‰)	Root δ¹⁵N (‰)	Flowers δ¹³C (‰)	Flowers δ¹⁵N (‰)	Seeds δ¹³C (‰)	Seeds δ¹⁵N (‰)
Stipa mucronata	W13	4070	Grass	−27.7	−1.4	–	–	−26.2	1.5	–	–	−26.5	1.5
Stenandrium dulce	W13	4070	Herb	−28.4	0.6	–	–	–	–	–	–	–	–
Senecio nutans	W13	4070	Shrub	−29.4	6.2	−28.6	5.0	−27.6	5.3	–	–	–	–

1. Species is invasive or introduced.
2. Member of the family Fabaceae (legume).

Cultigen N content is presented in Table 3 and Figure 7. Mean %N for cultigens ranged from 1.2±0.2% (maize) to 6.8±1.3% (Andean lupin). Results of post-hoc Dunnett's T3 test for differences between individual cultigen species in N content are presented in Table 5. The three leguminous species were characterized by significantly higher N contents than non-leguminous species (Table 5); collectively, legumes were characterized by significantly higher %N values than non-legumes (Figure 7; $F_{[1,80]} = 116.0$, $p<0.001$).

Wild Plants

A total of 139 species were sampled primarily from ten sites distributed along an altitudinal transect from 10 to 4,070 masl. The number of taxa sampled and environmental variables for each of the sampling locations are presented in Table 6. The number of C_4 plant taxa was generally higher at lower altitude sites receiving low amounts of rainfall. This fits with what is known about the global distribution of C_4 plants [185].

The carbon and nitrogen isotopic compositions were measured for all 139 species. Foliar tissue was analyzed from all species, and additional tissues analyzed included: 112 stems, 28 roots, 51 flowers, and 62 seeds. Carbon and nitrogen isotopic compositions for wild plants are presented in Table 7 according to plant part. Foliar $\delta^{13}C$ values for C_3 plants ranged from −31.9 to −22.5 ‰, with a mean value of −27.6±1.9 ‰ ($n = 122$). Foliar $\delta^{13}C$ values for C_4 plants ranged from −15.6 to −11.6 ‰, with a mean value of −13.5±1.0 ‰ ($n = 17$). Foliar $\delta^{15}N$ values for C_3 plants ranged from −4.1 to 13.0‰, with a mean value of 3.7±4.0 ‰. Foliar $\delta^{15}N$ values for C_4 plants ranged from −3.2 to 15.0 ‰, with a mean value of 5.5±5.7 ‰. The single lichen analyzed (*Usnea andina*) was characterized by a $\delta^{13}C$ value intermediate between C_3 and C_4 plants (−20.5 ‰) and a very low $\delta^{15}N$ value (−6.5 ‰), consistent with previously reported results for lichens [71–73].

There were no significant differences in foliar $\delta^{15}N$ among plant functional groups ($F_{[3,132]} = 1.8$, $p = 0.15$). Foliar $\delta^{13}C$ differed significantly among plant functional groups ($F_{[5,130]} = 195.0$, $p<0.001$), although this was driven by differences between C_3 and C_4 groups; there were no significant differences in foliar $\delta^{13}C$ between plant functional groups within C_3 and C_4 groups (Table 8).

There was no clear pattern of intraplant variation in $\delta^{15}N$ (Figure 8) with differences in $\delta^{15}N$ between tissues ($\Delta^{15}N$) being highly variable: $\Delta^{15}N_{stem-leaf} = −0.3±2.3$ ‰, $\Delta^{15}N_{root-leaf} = 0.4±3.1$ ‰, $\Delta^{15}N_{flower-leaf} = 0.5±1.4$ ‰, $\Delta^{15}N_{seed-leaf} = 0.5±1.7$ ‰. Conversely, foliar tissue was typically characterized by lower $\delta^{13}C$ values than all other tissues analyzed (Figure 9), and intraplant variation was generally smaller: $\Delta^{13}C_{stem-leaf} = 0.5±0.9$ ‰, $\Delta^{13}C_{root-leaf} = 0.4±0.8$ ‰, $\Delta^{13}C_{flower-leaf} = 0.6±1.0$ ‰, $\Delta^{13}C_{seed-leaf} = 0.5±1.7$ ‰. For C_4 plants ($n = 17$), there was no clear pattern of intraplant variation in $\delta^{13}C$: $\Delta^{13}C_{stem-leaf} = 0.0±0.8$ ‰, $\Delta^{13}C_{root-leaf} = 0.5±0.7$ ‰, $\Delta^{13}C_{flower-leaf} = −0.3±0.6$ ‰, $\Delta^{13}C_{seed-leaf} = −0.2±1.3$ ‰.

Foliar nitrogen isotopic compositions for wild legumes (Fabaceae) were highly variable, ranging from −1.4 to 9.6 ‰. Among *Acacia* trees and shrubs alone, foliar $\delta^{15}N$ values ranged from −1.0 to 9.6 ‰, suggesting that some species are not engaged in active N_2-fixation. While wild legumes were characterized by lower foliar $\delta^{15}N$ values relative to non-legumes (4.1±4.4 ‰, $n = 119$ for non-legumes; 2.7±3.4 ‰, $n = 17$ for legumes), this difference was not statistically significant ($F_{[1,134]} = 1.8$, $p = 0.18$).

Mean wild C_3 plant foliar $\delta^{13}C$ and $\delta^{15}N$ values for sampling locations with ≥5 species sampled are presented in Table 9. Mean foliar carbon and nitrogen isotopic compositions for these sites are plotted against altitude in Figure 10 and estimated mean annual precipitation in Figure 11. Mean foliar $\delta^{15}N$ values at low altitude

Table 8. Results of ANOVA post-hoc tests (Dunnett's T3) for foliar $\delta^{13}C$ between plant functional groups.

Foliar $\delta^{13}C$	C$_3$ Grass/Sedge	C$_4$ Herb/Shrub	C$_3$ Herb/Shrub	Tree	Vine
C$_4$ Grass/Sedge	**<0.001**	0.999	**<0.001**	**<0.001**	**<0.001**
C$_3$ Grass/Sedge	–	**<0.001**	0.993	1.000	1.000
C$_4$ Herb/Shrub	–	–	0.999	**<0.001**	**<0.001**
C$_3$ Herb/Shrub	–	–	–	0.997	0.994
Tree	–	–	–	–	1.000

Values in boldface are statistically significant ($p<0.05$).

sites were 2 to 8 ‰ higher than mean foliar $\delta^{15}N$ values at high altitude sites. Foliar $\delta^{15}N$ was negatively correlated with mean annual precipitation (Spearman's $\rho = -0.770$, $p = 0.009$) and altitude (Spearman's $\rho = -0.782$, $p = 0.008$). Foliar $\delta^{13}C$ was positively correlated with mean annual precipitation (Spearman's $\rho = 0.879$, $p = 0.001$) and altitude (Spearman's $\rho = 0.903$, $p<0.001$). For comparative purposes, mean plant $\delta^{13}C$ values for sites sampled along an altitudinal transect in northern Chile are presented in Figure 12 [14].

Marine Plants

The carbon and nitrogen isotopic compositions were determined for a total of 25 marine plant samples from five species. Mean $\delta^{13}C$ and $\delta^{15}N$ values for marine plants are presented in Table 10. Mean $\delta^{13}C$ values for marine plants ranged from -18.7 ± 0.7 ‰ (*Gymnogongrus furcellatus*) to -14.2 ± 1.2 ‰ (*Grateloupia doryphora*). Mean $\delta^{15}N$ values for marine plants ranged from 2.5 ± 0.9 ‰

(*Gymnogongrus furcellatus*) to 7.8 ± 0.1 ‰ (*Cryptopleura cryptoneuron*). Overall, marine plants were characterized by $\delta^{13}C$ values that were intermediate between C$_3$ and C$_4$ plant isotopic compositions, although more similar to the latter. In comparison to wild plants growing at the three sites located closest to the coast, marine plants were not characterized by significantly higher $\delta^{15}N$ values when the plants from the three terrestrial sites are treated separately ($F_{[3,39]} = 0.5$, $p = 0.71$) or grouped together ($F_{[1,41]}<0.1$, $p = 0.91$).

Discussion

Cultigens

The carbon isotopic composition of maize was ~2 ‰ more enriched in ^{13}C than wild C$_4$ plants (all tissues), similar to previously determined values for other parts of the world [186,187]. This suggests that a $\delta^{13}C$ value of -10.3 ‰ (adjusted by $+1.5$ ‰ for the Suess Effect [188,189]) would be appropriate for

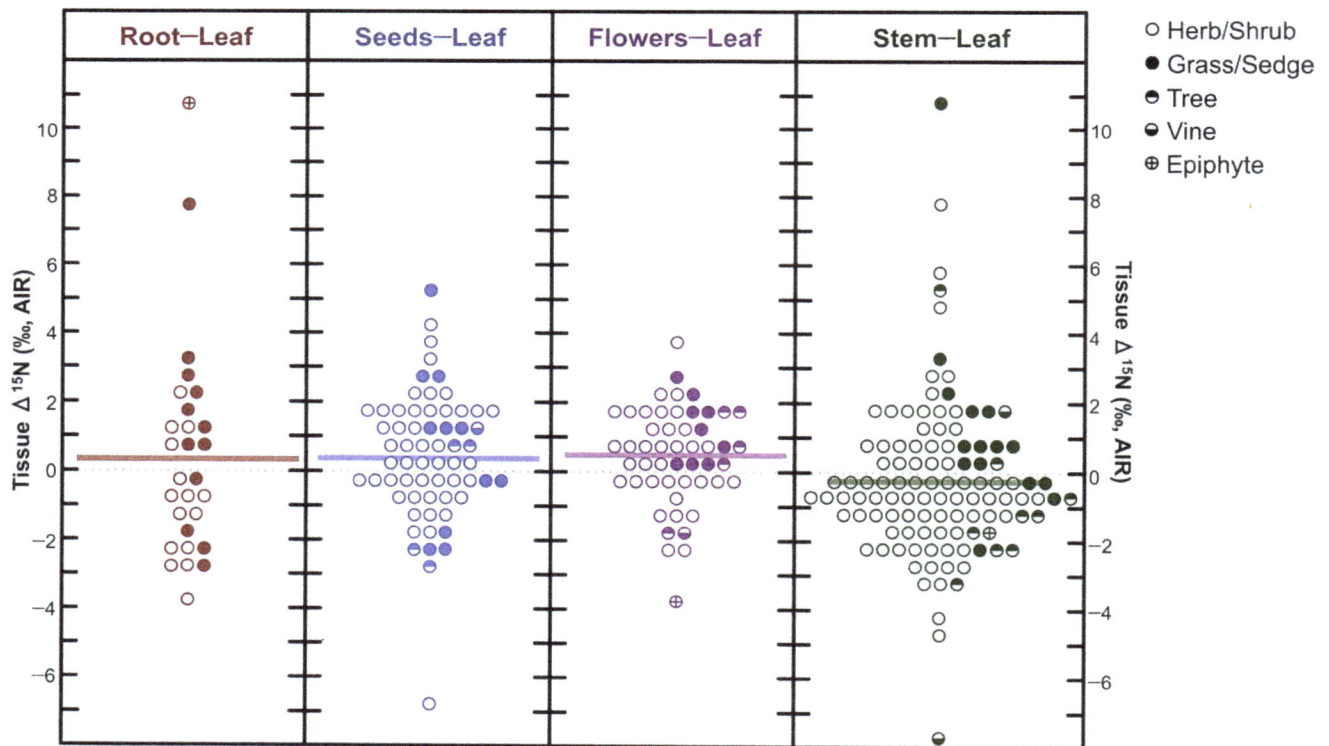

Figure 8. Dot-matrix plot of differences in nitrogen isotopic composition between foliar and other tissues ($\Delta^{15}N$). Horizontal bars represent means. Increment = 0.5 ‰.

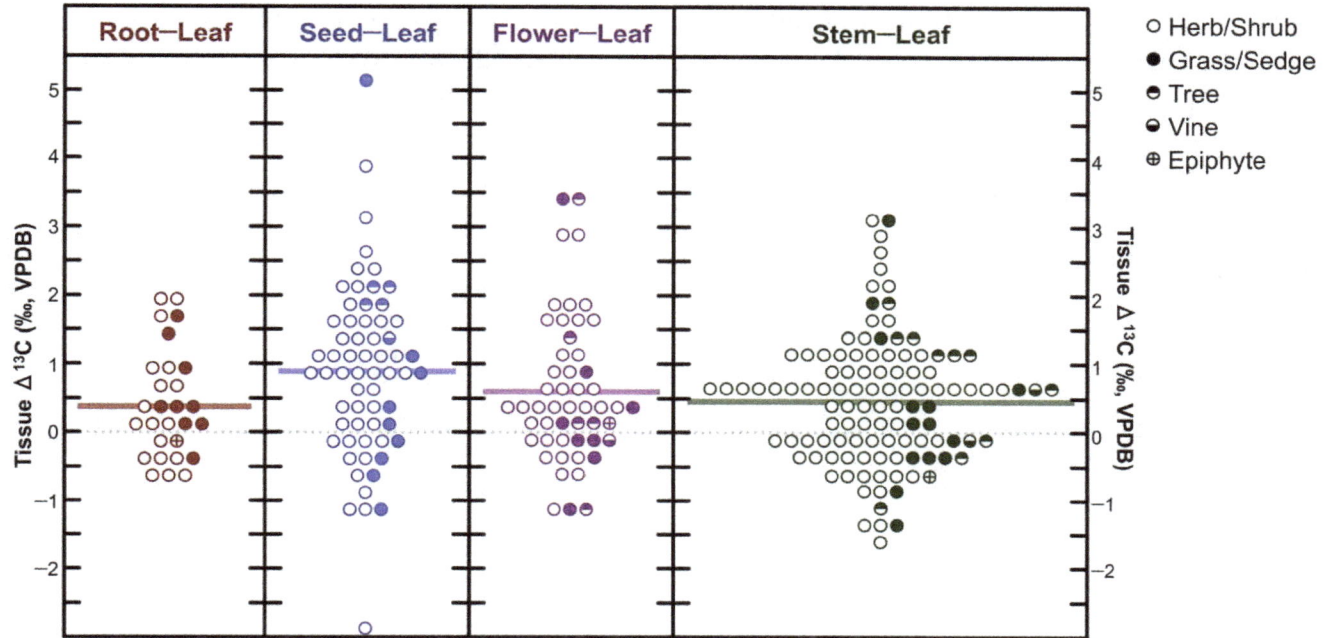

Figure 9. Dot-matrix plot of differences in carbon isotopic composition between foliar and other tissues (Δ^{13}C). Horizontal bars represent means. Increment = 0.5 ‰.

paleodietary models in the central Andes. There may, however, be some small-scale environmental effects on maize δ^{13}C values along an altitudinal gradient as discussed in more detail below.

For the most part, the δ^{15}N values of the modern cultigens presented in this study should be interpreted cautiously with respect to paleodietary studies. The primary factor influencing the nitrogen isotopic composition of plant tissues is the N source, and it cannot be assumed that modern N sources are directly analogous to those used in antiquity. The nitrogen isotopic composition of locally grown produce sold in Andean markets today may be influenced by chemical fertilizers (which cause plants to have relatively low nitrogen isotopic compositions) or by animal manures (e.g. sheep, cow, pig) that would not have been available

in the region prior to the arrival of the Spanish. The same is true for nitrogen isotopic data obtained from modern agricultural plants globally, and as a general rule, the limitations of these data must be recognized. Nevertheless, some patterns are likely to be broadly applicable.

In contrast to the vast majority of published literature [27,138–140,190–200], Warinner et al. [187] showed very little distinction between the nitrogen isotopic composition of Mesoamerican legumes and non-legumes, suggesting that the assumption of lower δ^{15}N values in legumes in that region is tenuous. Where the potential effects of nitrogenous fertilizers on legume δ^{15}N values are unknown (as is the case for the data presented by Warriner et al. [187]), the interpretation of δ^{15}N values in legumes and

Table 9. Mean ($\pm 1\sigma$) isotopic and elemental compositions for sampling locations with >3 plant species sampled (data for C$_3$ plants only).

Site ID	Latitude	Longitude	Altitude (masl)	MAP (mm)[1]	n^2	δ^{13}C (‰, VPDB)	δ^{15}N (‰, AIR)
W1	−8.1956	−78.9996	10	7	7	−29.1±1.2	7.9±4.5
W2	−8.1267	−78.9963	33	5	9	−29.2±1.7	5.5±2.6
W5	−8.0791	−78.8681	192	56	5	−28.1±1.7	7.5±2.2
W6	−8.0137	−78.7972	447	113	17	−28.1±1.4	5.4±4.0
W8	−8.0047	−78.6726	1557	140	9	−28.1±1.9	3.6±2.7
W9	−8.0132	−78.6355	2150	141	16	−27.6±1.8	0.4±3.0
W10	−7.9973	−78.6481	2421	142	8	−27.2±0.8	2.0±3.1
W11	−8.0132	−78.6355	2947	171	9	−26.9±1.8	3.3±3.3
W12	−8.1361	−78.1685	3041	702	21	−27.3±1.6	2.1±2.8
W13	−8.2235	−78.3013	4070	591	15	−26.0±2.0	2.0±4.9

1. Mean annual precipitation (MAP) estimated as described in the text.
2. Number of C$_3$ plant species sampled.

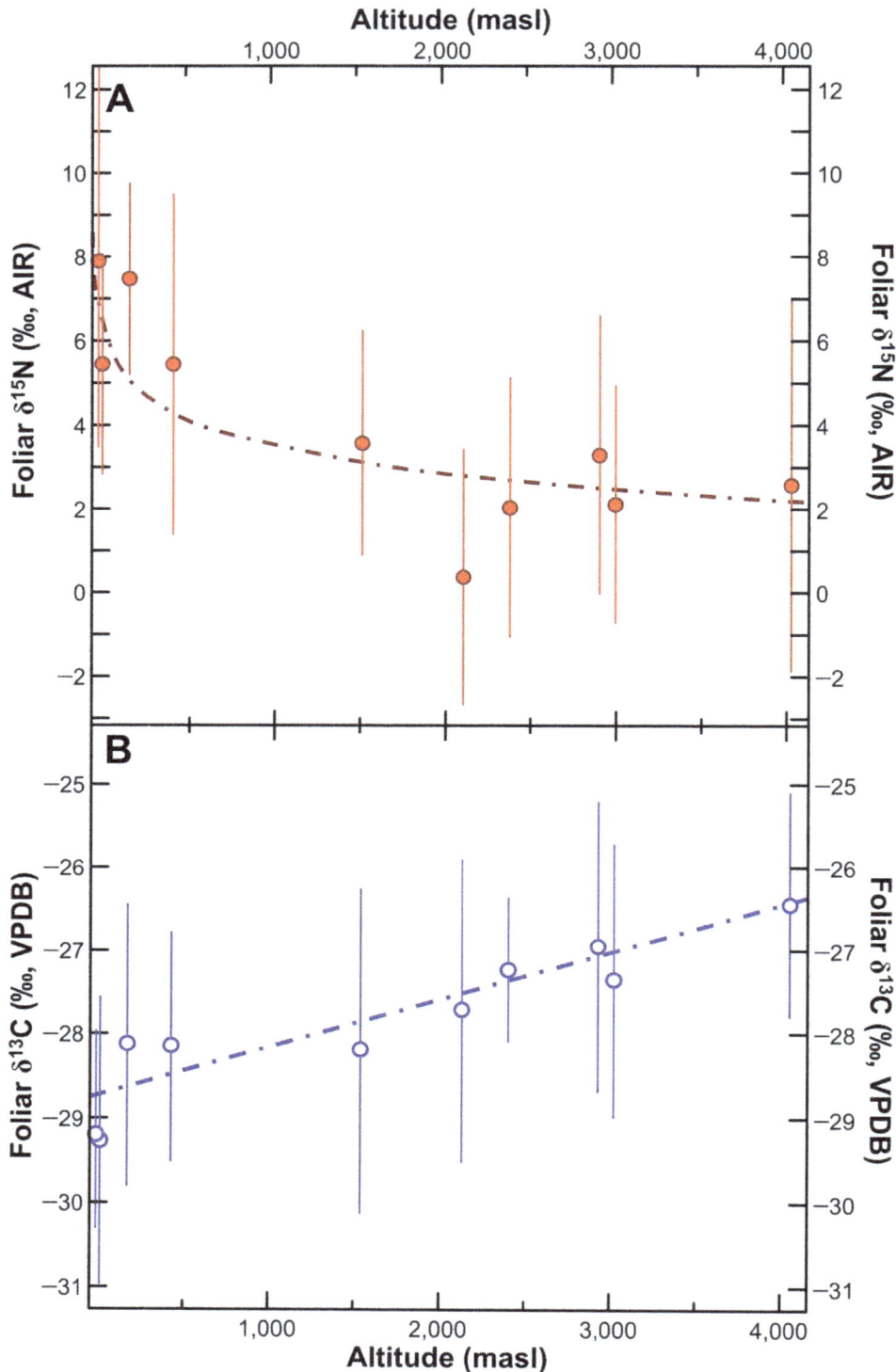

Figure 10. Bivariate plots of foliar $\delta^{15}N$ and altitude (A) and foliar $\delta^{13}C$ (B) for C$_3$ plants only. Points represent means $\pm 1\sigma$ for sites with ≥ 5 C$_3$ plant species sampled. Equation for $\delta^{15}N$ and altitude: $y = 10.3 - \log x$, $r^2 = 0.71$; $p = 0.002$. Equation for $\delta^{13}C$ and altitude: $y = x/1,733 - 28.8$, $r^2 = 0.85$; $p < 0.001$.

non-legumes is not straightforward. While there was some overlap in $\delta^{15}N$ values between legumes and non-legumes in this study, leguminous cultigens had significantly higher N contents (Figure 7; Table 5) and significantly lower $\delta^{15}N$ values (Figure 6; Table 4) than non-legumes.

Aside from the differences in $\delta^{15}N$ between legumes and non-legumes, it is very difficult to generalize the $\delta^{15}N$ values for

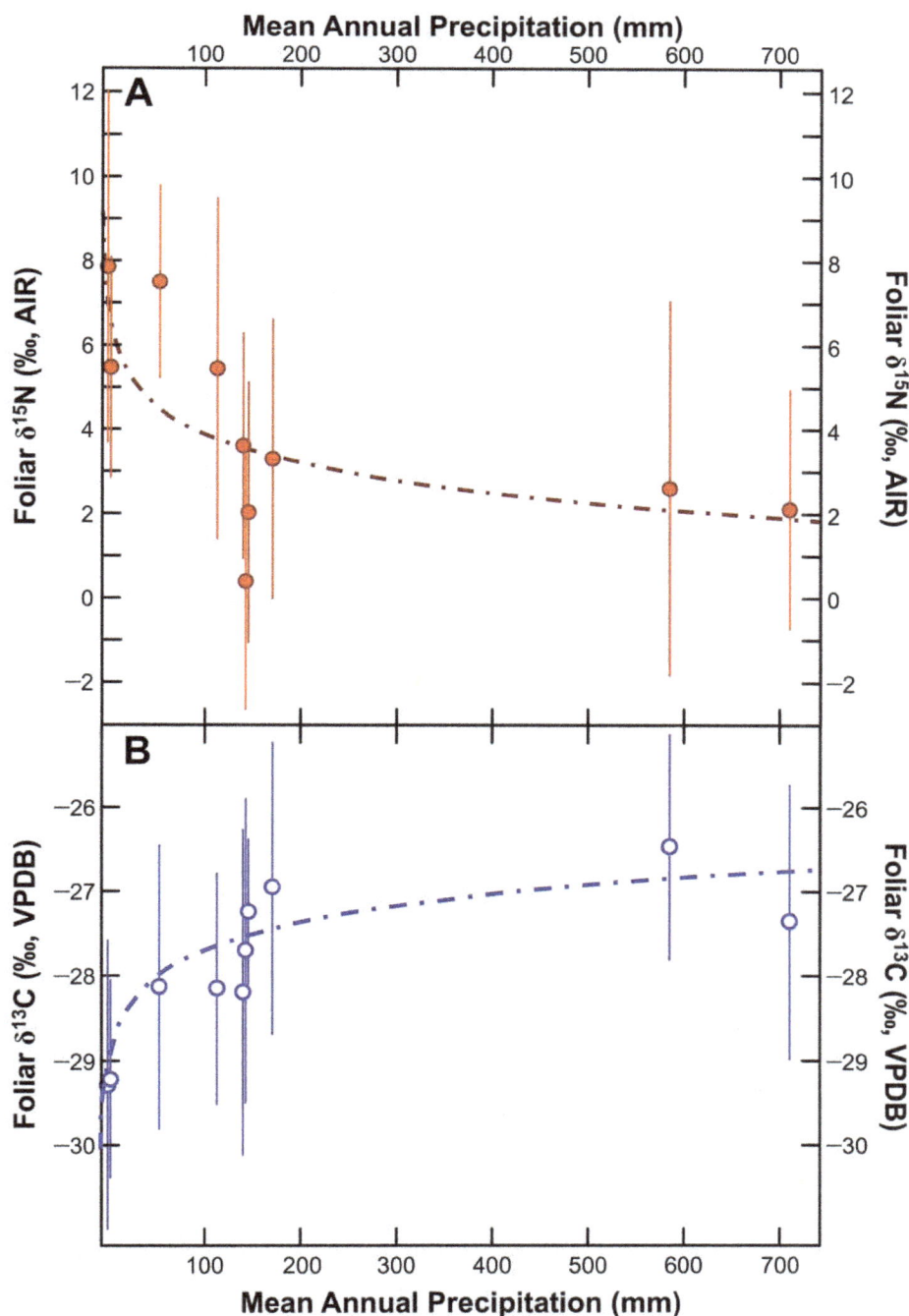

Figure 11. Bivariate plots of foliar δ¹⁵N and mean annual precipitation (A) and foliar δ¹³C (B) for C₃ plants only. Points represent means ±1σ for sites with ≥5 C₃ plant species sampled. Equation for δ¹⁵N and MAP: $y = 8.8 - 1.1 \log x$, $r^2 = 0.49$; $p = 0.03$. Equation for δ¹³C and MAP: $y = -30.1 + 0.5 \log x$, $r^2 = 0.81$; $p < 0.001$.

cultigens in this study. Nitrogen isotopic compositions were highly variable, particularly for potato, which most likely reflected variable local growing conditions (soil fertility, type of manure used) rather than any biochemical or physiological process specific to any particular plant species. Ultimately, the best source of baseline isotopic data for paleodietary studies may be from archaeobotanical remains [27,201–203], provided that preservation of original carbon and nitrogen isotopic compositions can be demonstrated. Considerable work has been done in this regard for the isotopic composition of bone collagen [204–208] and to a lesser extent hair keratin [209], but a solid set of parameters for detecting preservation versus alteration of original plant carbon and nitrogen isotopic compositions have not yet been determined. The excellent organic preservation at many archaeological sites on the coasts of Peru and Chile provides the potential for such analyses to be conducted on botanical remains.

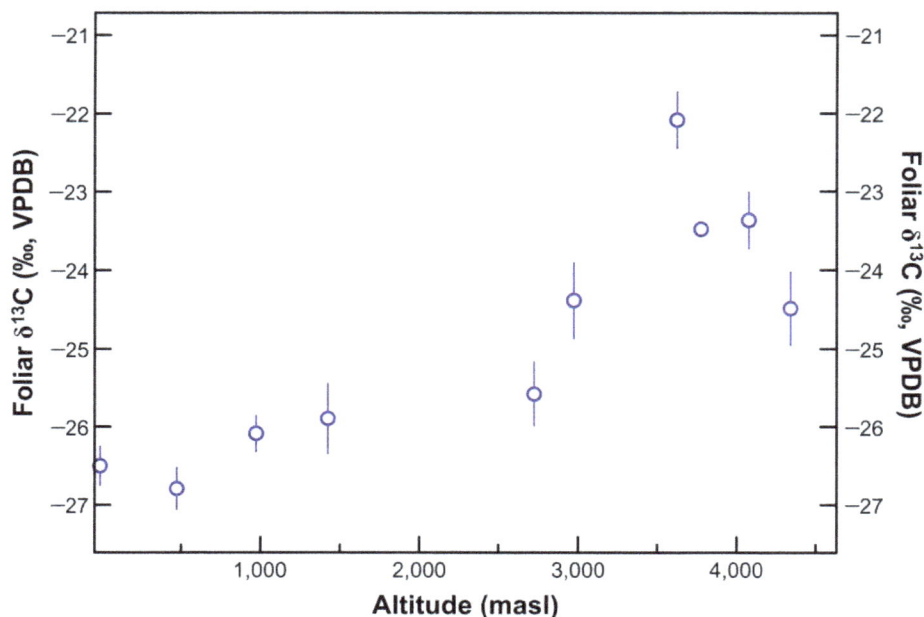

Figure 12. Bivariate plot of altitude and foliar δ¹³C for plants collected in northern Chile [14].

Wild Plants

Plant Functional Group. There were no clear distinctions between different plant functional groups (grass, herb, shrub, tree, vine) with respect to either carbon or nitrogen isotopic compositions. While some systematic variation may be expected due to variable nitrogen acquisition strategies (e.g. rooting depth) or differential distribution of biomolecules with distinct isotopic compositions, the diverse range of environmental conditions from which plants were sampled likely served to blur any isotopic distinctions between functional groups. Moreover, the sample sizes for different plant functional groups within any one site were too small for meaningful comparisons to be made.

There was no consistent pattern in plant $\delta^{15}N$ with respect to leguminous trees and shrubs, with some species having foliar $\delta^{15}N$ values close to 0 ‰, and others having relatively high $\delta^{15}N$ values. Previous studies have similarly found conflicting patterns of relatively high and low $\delta^{15}N$ values in leguminous trees. Codron et al. [13] found no clear distinction between leguminous and non-leguminous trees at a regional scale in South Africa. Aranibar et al. [178] did not observe significant amounts of N_2-fixation among leguminous trees in an arid region of southern Africa, with trees growing at the most arid sites showing no evidence of N_2-fixation. Fruit-bearing trees of the genus *Prosopis* (often called huarango or algarrobo) are suggested to have been an important food source for various groups in the Andean region [64,210]. Catenazzi and Donnelly [28] found $\delta^{15}N$ values typical of N_2-fixing trees (ca. 0 ‰) in *Prosopis pallida* from the Sechura Desert of northern Peru. Conversely, on the basis of the isotopic data recorded in this study for leguminous trees in the Moche River Valley, the assumption that *Prosopis* would be characterized by significantly lower $\delta^{15}N$ values relative to other plants is tenuous. Given the potential importance of these foods in the diet, a more extensive study of the nitrogen isotopic composition of central Andean leguminous trees would be beneficial.

Intraplant Variation in Carbon and Nitrogen Isotopic Compositions. Plant nitrogen isotopic composition did not systematically vary between different tissues sampled. On the basis of hydroponic studies, significant intraplant variation (between roots and shoots) is only expected when plants are fed with NO_3^- as the N source [166]. Additionally, plant $\delta^{15}N$ may vary considerably among tissues due to biochemical processes associated with growth and senescence over time [143,211–213]. The lack of any clear pattern of intraplant variation in $\delta^{15}N$ likely relates to a number of factors, including: variable reliance on different N sources (nitrate, ammonium, organic N) by different plant taxa and between sampling locations, differences in plant life

Table 10. Mean ($\pm 1\sigma$) isotopic and elemental compositions for marine algae.

Taxonomic Name	Type	n	δ¹³C (‰, VPDB)	δ¹⁵N (‰, AIR)	%C	%N
Ulva lactuca	Chlorophyta	5	−14.3±0.4	6.4±0.1	29.2±0.4	3.6±0.2
Gymnogongrus furcellatus	Rhodophyta	5	−18.7±0.7	2.5±0.9	23.3±2.4	2.1±0.2
Grateloupia doryphora	Rhodophyta	5	−14.2±1.2	6.8±0.3	29.9±1.0	3.2±0.1
Gigartina chamissoi	Rhodophyta	5	−16.7±1.0	5.4±0.5	25.7±0.4	2.7±0.1
Cryptopleura cryptoneuron	Rhodophyta	5	−18.4±0.4	7.8±0.1	21.2±1.6	2.8±0.4

cycles between different taxa, and spatial variation in the influence of environmental factors on the isotopic composition of source N.

Foliar tissues tended to be more depleted of ^{13}C than other tissues (Figure 9). The magnitude of this difference was typically ≤1 ‰, but was absent for C_4 plants. This fits with previously described data for other plants. The small difference in $\delta^{13}C$ among plant tissues is not likely to be significant with respect to the interpretation of isotopic data in the context of paleodietary studies.

Geographic Variation in Carbon and Nitrogen Isotopic Compositions. There were strong relationships between sampling site and foliar carbon and nitrogen isotopic compositions. Foliar $\delta^{15}N$ was negatively correlated with altitude (Figure 10a) and mean annual precipitation (Figure 11a), although based on the large number of studies finding a strong relationship between rainfall amount and soil, plant, and animal $\delta^{15}N$ [12,15,85–87,172,176–179], this relationship is likely driven by rainfall. This suggests that arid sites are characterized by a fairly open nitrogen cycle, as described in previous studies [179]. It is unclear to what extent these processes would act on agricultural plants growing in relatively arid versus wet sites. Even on the hyper-arid coast where rainfall is negligible, agriculture is made possible by substantial irrigation networks. Hence, water availability in agricultural contexts is markedly higher than in non-irrigated areas. Agricultural products grown in coastal regions of the central Andes may therefore not be characterized by higher $\delta^{15}N$ values relative to those growing at wetter, higher altitude sites. For instance, maize grown as part of a controlled experiment (no fertilization) located ~6 km from the coast, had grain $\delta^{15}N$ values of 6.3±0.3 ‰ [147], comparable to results for maize growing at higher altitudes in this study (6.4±2.2 ‰). Aside from issues of irrigation, agricultural plants analyzed in this study were sampled along a relatively limited altitudinal transect (2233 to 3588 masl) where effects on tissue $\delta^{15}N$ values would be expected to be more limited (Figure 10a).

The positive relationship found between rainfall and foliar $\delta^{13}C$ in C_3 plants contrasts with most other studies, which have typically found a negative relationship between rainfall and foliar $\delta^{13}C$. The majority of these studies, however, sampled plants along a large rainfall gradient (>1,000 mm), but with little difference in elevation between sites. Conversely, we sampled along a more restricted rainfall gradient (~700 mm), but a very large altitudinal gradient (~4,000 m). Increased altitude and increased rainfall have opposing effects on foliar $\delta^{13}C$ values, and the results of this study suggest the predominance of altitudinal effects on foliar carbon isotopic compositions in northern Peru. A similar pattern was observed along a comparable altitudinal gradient in northern Chile (Figure 12). This pattern is most likely related to high carboxylation rates relative to stomatal conductance at high altitudes resulting in lower ^{13}C discrimination. Such effects should be equally apparent in cultivated plants, although they were not observed in this study because of the limited altitudinal range from which cultigens were sampled (Table 2).

Variation in plant isotopic compositions along environmental gradients is particularly important with respect to the reconstruction of the diet of humans and animals using isotopic data. While the majority of wild plants analyzed in this study would not have been consumed by humans, the results are very relevant to the reconstruction of animal management practices. There is considerable debate in the Andean region with respect to the herding practices of South American camelids (llama and alpaca), and whether or not animals recovered from coastal sites were raised locally, or imported from elsewhere [214]. The results of this study suggest that animals feeding on wild plants at drier, low altitude sites

would be characterized by higher tissue $\delta^{15}N$ values than animals feeding on wild plants at wetter, high altitude sites. The magnitude of this difference could easily be 4 to 6 ‰, although the consumption of agricultural plants dependent on irrigation at lower altitudes could serve to obscure this difference (as discussed above).

The potential consequences of altitudinal variation in plant $\delta^{13}C$ values are more difficult to evaluate. While the positive linear relationship between altitude and foliar $\delta^{13}C$ is strong, the relative distribution of C_3 and C_4 plants would serve to counter these effects. Because there will be proportionally more C_4 plants at dry, low altitude sites relative to moister, high altitude sites, the average $\delta^{13}C$ value of available forage would still be higher at low altitude sites. Thus, markedly higher $\delta^{13}C$ and $\delta^{15}N$ values observed in some camelids from low altitude sites [38,215] can be satisfactorily explained by the consumption of local terrestrial vegetation.

Marine Plants

Marine algae are known to have been an important dietary resource for many groups of people in the coastal regions of Peru and Chile [216], but the lack of preservation of marine algae in all but the most exceptional archaeological contexts makes evaluating the potential importance of marine algae in the diet extremely difficult. Marine plants were characterized by $\delta^{13}C$ values intermediate between C_3 and C_4 plants, with $\delta^{15}N$ values comparable to terrestrial plants growing on the coast. DeNiro [215] has suggested that consumption of marine algae may have been responsible for relatively high $\delta^{13}C$ and $\delta^{15}N$ values in coastal Peruvian camelids. While the number of macroalgal species sampled in this study is not extensive, the isotopic data presented here are not consistent with this explanation. With the exception of instances in which marine plants grow in areas of exceptionally high influence of marine bird and/or mammalian excreta [217], there is no reason to expect marine algal $\delta^{15}N$ values to be higher than the $\delta^{15}N$ values of plants growing along the arid coast of Peru.

Conclusions

Maize from the study area has a mean $\delta^{13}C$ value of −11.8±0.4‰, which suggests that a $\delta^{13}C$ value (adjusted for the Suess Effect) of ca. −10.3 ‰ would be appropriate for paleodietary models in the region. Leguminous cultigens were characterized by significantly lower $\delta^{15}N$ values and higher N contents than non-leguminous cultigens; this distinction was not as clear for wild legumes. Marine plants were characterized by $\delta^{13}C$ values intermediate between wild terrestrial C_3 and C_4 vegetation and $\delta^{15}N$ values that were very similar to terrestrial plants growing at low altitudes. C_4 plants were generally more abundant at lower altitude sites. Carbon and nitrogen isotopic compositions of wild plants were strongly influenced by local environmental factors. Foliar $\delta^{13}C$ was positively correlated with altitude and negatively correlated with mean annual precipitation. Foliar $\delta^{15}N$ was negatively correlated with altitude and mean annual precipitation.

While the last twenty years have seen a proliferation of studies utilizing the isotopic analysis of archaeological materials for the purpose of reconstructing diet, the development of isotopic baselines for interpreting such data has lagged behind these investigations. This hampers our ability to realize the full potential of isotopic data. This study begins to fill part of that gap by providing an initial understanding of the baseline isotopic variation in plants from northern Peru. Further studies of this nature are required to better understand baseline isotopic variation in other regions.

Supporting Information

Dataset S1 Sampling site locations for wild and market plants. This.kmz file can be executed in Google Earth (http://www.earth.google.com)

Acknowledgments

The authors wish to thank Kim Law and Li Huang of UWO's Laboratory for Stable Isotope Science (LSIS) for technical assistance, Sharon Buck and Tessa Plint for assistance with sample preparation, and Estuardo La Torre for assisting with sample collection. This is LSIS contribution # 294.

Author Contributions

Conceived and designed the experiments: PS CDW FJL JFM VFVS. Performed the experiments: PS. Analyzed the data: PS CDW FJL. Wrote the paper: PS.

References

1. DeNiro MJ, Epstein S (1978) Influence of diet on the distribution of carbon isotopes in animals. Geochimica et Cosmochimica Acta 42: 495–506.
2. DeNiro MJ, Epstein S (1981) Influence of diet on the distribution of nitrogen isotopes in animals. Geochimica et Cosmochimica Acta 45: 341–351.
3. Szpak P, Orchard TJ, McKechnie I, Gröcke DR (2012) Historical ecology of late Holocene sea otters (*Enhydra lutris*) from northern British Columbia: isotopic and zooarchaeological perspectives. Journal of Archaeological Science 39: 1553–1571.
4. Caut S, Angulo E, Courchamp F (2009) Variation in discrimination factors (Δ^{15}N and Δ^{13}C): the effect of diet isotopic values and applications for diet reconstruction. Journal of Applied Ecology 46: 443–453.
5. Casey MM, Post DM (2011) The problem of isotopic baseline: Reconstructing the diet and trophic position of fossil animals. Earth-Science Reviews 106: 131–148.
6. Hobson KA (1999) Tracing origins and migration of wildlife using stable isotopes: a review. Oecologia 120: 314–326.
7. Katzenberg MA, Weber A (1999) Stable Isotope Ecology and Palaeodiet in the Lake Baikal Region of Siberia. Journal of Archaeological Science 26: 651–659.
8. Szpak P, Orchard TJ, Gröcke DR (2009) A Late Holocene vertebrate food web from southern Haida Gwaii (Queen Charlotte Islands, British Columbia). Journal of Archaeological Science 36: 2734–2741.
9. Bösl C, Grupe G, Peters J (2006) A Late Neolithic vertebrate food web based on stable isotope analyses. International Journal of Osteoarchaeology 16: 296–315.
10. Grupe G, Heinrich D, Peters J (2009) A brackish water aquatic foodweb: trophic levels and salinity gradients in the Schlei fjord, Northern Germany, in Viking and medieval times. Journal of Archaeological Science 36: 2125–2144.
11. Kohn MJ (2010) Carbon isotope compositions of terrestrial C3 plants as indicators of (paleo)ecology and (paleo)climate. Proceedings of the National Academy of Sciences 107: 19691–19695.
12. Craine JM, Elmore AJ, Aidar MPM, Bustamante M, Dawson TE, et al. (2009) Global patterns of foliar nitrogen isotopes and their relationships with climate, mycorrhizal fungi, foliar nutrient concentrations, and nitrogen availability. New Phytologist 183: 980–992.
13. Codron J, Codron D, Lee-Thorp JA, Sponheimer M, Bond WJ, et al. (2005) Taxonomic, anatomical, and spatio-temporal variations in the stable carbon and nitrogen isotopic compositions of plants from an African savanna. Journal of Archaeological Science 32: 1757–1772.
14. Tieszen LL, Chapman M (1992) Carbon and nitrogen isotopic status of the major marine and terrestrial resources in the Atacama Desert of northern Chile. Proceedings of the First World Congress on Mummy Studies. Santa Cruz de Tenerife: Museo Arqueológico y Etnográfico de Tenerife. 409–425.
15. Hartman G, Danin A (2010) Isotopic values of plants in relation to water availability in the Eastern Mediterranean region. Oecologia 162: 837–852.
16. Ehleringer JR, Rundel PW, Palma B, Mooney HA (1998) Carbon isotope ratios of Atacama Desert plants reflect hyperaridity of region in northern Chile. Revista Chilena de Historia Natural 71: 79–86.
17. Panarello HO, Fernández J (2002) Stable carbon isotope measurements on hair from wild animals from altiplanic environments of Jujuy, Argentina. Radiocarbon 44: 709–716.
18. Gil AF, Neme GA, Tykot RH, Novellino P, Cortegoso V, et al. (2009) Stable isotopes and maize consumption in central western Argentina. International Journal of Osteoarchaeology 19: 215–236.
19. Izeta AD, Laguens AG, Marconetto MB, Scattolin MC (2009) Camelid handling in the meridional Andes during the first millennium AD: a preliminary approach using stable isotopes. International Journal of Osteoarchaeology 19: 204–214.
20. Gil AF, Tykot RH, Neme G, Shelnut NR (2006) Maize on the Frontier: Isotopic and Macrobotanical Data from Central-Western Argentina. In: Staller JE, Tykot RH, Benz BF, editors. Histories of Maize. Amsterdam: Elsevier Academic Press. pp. 199–214.
21. Martínez G, Zangrando AF, Prates L (2009) Isotopic ecology and human palaeodiets in the lower basin of the Colorado River, Buenos Aires province, Argentina. International Journal of Osteoarchaeology 19: 281–296.
22. Falabella F, Planella MT, Aspillaga E (2007) Dieta en sociedades alfareras de Chile Central: Aporte de análisis de isótopos estables. Chungará (Arica) 39: 5–27.
23. Rundel PW, Gibson AC, Midgley GS, Wand SJE, Palma B, et al. (2002) Ecological and ecophysiological patterns in a pre-altiplano shrubland of the Andean Cordillera in northern Chile. Plant Ecology 169: 179–193.
24. Latorre C, González AL, Quade J, Fariña JM, Pinto R, et al. (2011) Establishment and formation of fog-dependent *Tillandsia landbeckii* dunes in the Atacama Desert: Evidence from radiocarbon and stable isotopes. Journal of Geophysical Research 116: G03033.
25. Miller MJ, Capriles JM, Hastorf CA (2010) The fish of Lake Titicaca: implications for archaeology and changing ecology through stable isotope analysis. Journal of Archaeological Science 37: 317–327.
26. Ehleringer JR, Cooper DA, Lott MJ, Cook CS (1999) Geo-location of heroin and cocaine by stable isotope ratios. Forensic Science International 106: 27–35.
27. DeNiro MJ, Hastorf CA (1985) Alteration of ^{15}N/^{14}N and ^{13}C/^{12}C ratios of plant matter during the initial stages of diagenesis: studies utilizing archaeological specimens from Peru. Geochimica et Cosmochimica Acta 49: 97–115.
28. Catenazzi A, Donnelly MA (2007) Distribution of geckos in northern Peru: Long-term effect of strong ENSO events? Journal of Arid Environments 71: 327–332.
29. Townsend-Small A, McClain ME, Brandes JA (2005) Contributions of Carbon and Nitrogen from the Andes Mountains to the Amazon River: Evidence from an Elevational Gradient of Soils, Plants, and River Material. Limnology and Oceanography 50: 672–685.
30. Turner BL, Kingston JD, Armelagos GJ (2010) Variation in dietary histories among the immigrants of Machu Picchu: Carbon and nitrogen isotope evidence. Revista Chungara. Revista de Antropologia Chilena 42: 515–524.
31. Finucane B, Agurto PM, Isbell WH (2006) Human and animal diet at Conchopata, Peru: stable isotope evidence for maize agriculture and animal management practices during the Middle Horizon. Journal of Archaeological Science 33: 1766–1776.
32. Finucane BC (2007) Mummies, maize, and manure: multi-tissue stable isotope analysis of late prehistoric human remains from the Ayacucho Valley, Peru. Journal of Archaeological Science 34: 2115–2124.
33. Finucane BC (2008) Trophy heads from Nawinpukio, Peru: Physical and chemical analysis of Huarpa-era modified human remains. American Journal of Physical Anthropology 135: 75–84.
34. Finucane BC (2009) Maize and Sociopolitical Complexity in the Ayacucho Valley, Peru. Current Anthropology 50: 535–545.
35. Kellner CM, Schoeninger MJ (2008) Wari's imperial influence on local Nasca diet: The stable isotope evidence. Journal of Anthropological Archaeology 27: 226–243.
36. Knudson KJ, Aufderheide AE, Buikstra JE (2007) Seasonality and paleodiet in the Chiribaya polity of southern Peru. Journal of Archaeological Science 34: 451–462.
37. Slovak NM, Paytan A (2011) Fisherfolk and farmers: Carbon and nitrogen isotope evidence from Middle Horizon Ancón, Peru. International Journal of Osteoarchaeology 21: 253–267.
38. Thornton EK, Defrance SD, Krigbaum J, Williams PR (2011) Isotopic evidence for Middle Horizon to 16th century camelid herding in the Osmore Valley, Peru. International Journal of Osteoarchaeology 21: 544–567.
39. Tomczak PD (2003) Prehistoric diet and socioeconomic relationships within the Osmore Valley of southern Peru. Journal of Anthropological Archaeology 22: 262–278.
40. White CD, Nelson AJ, Longstaffe FJ, Grupe G, Jung A (2009) Landscape bioarchaeology at Pacatnamu, Peru: inferring mobility from δ^{13}C and δ^{15}N values of hair. Journal of Archaeological Science 36: 1527–1537.
41. Williams JS, Katzenberg MA (2012) Seasonal fluctuations in diet and death during the late horizon: a stable isotopic analysis of hair and nail from the central coast of Peru. Journal of Archaeological Science 39: 41–57.
42. Tykot RH, Burger RL, van der Merwe NJ (2006) The Importance of Maize in Initial Period and Early Horizon Peru. In: Staller JE, Tykot RH, Benz BF, editors. Histories of Maize. Amsterdam: Elsevier Academic Press. pp. 187–197.
43. Berón MA, Luna LH, Barberena R (2009) Isotopic archaeology in the western Pampas (Argentina): preliminary results and perspectives. International Journal of Osteoarchaeology 19: 250–265.

44. Politis GG, Scabuzzo C, Tykot RH (2009) An approach to pre-Hispanic diets in the Pampas during the Early/Middle Holocene. International Journal of Osteoarchaeology 19: 266–280.

45. Tessone A, Zangrando AF, Barrientos G, Goñi R, Panarello H, et al. (2009) Stable isotope studies in the Salitroso Lake Basin (southern Patagonia, Argentina): assessing diet of Late Holocene hunter-gatherers. International Journal of Osteoarchaeology 19: 297–308.

46. Yacobaccio HD, Morales MR, Samec CT (2009) Towards an isotopic ecology of herbivory in the Puna ecosystem: new results and patterns on *Lama glama*. International Journal of Osteoarchaeology 19: 144–155.

47. Gil AF, Neme GA, Tykot RH (2011) Stable isotopes and human diet in central western Argentina. Journal of Archaeological Science 38: 1395–1404.

48. Ericson JE, West M, Sullivan CH, Krueger HW (1989) The Development of Maize Agriculture in the Viru Valley, Peru. In: Price TD, editor. The Chemistry of Prehistoric Human Bone. Cambridge: Cambridge University Press. pp. 68–104.

49. Verano JW, DeNiro MJ (1993) Locals or foreigners? Morphological, biometric and isotopic approaches to the question of group affinity in human skeletal remains recovered from unusual archaeological context. In: Sandford MK, editor. Investigations of Ancient Human Tissue: Chemical Analysis in Anthropology. Langhorne: Gordon and Breach. pp. 361–386.

50. Bush MB, Hansen BCS, Rodbell DT, Seltzer GO, Young KR, et al. (2005) A 17 000-year history of Andean climate and vegetation change from Laguna de Chochos, Peru. Journal of Quaternary Science 20: 703–714.

51. Brush SB (1982) The Natural and Human Environment of the Central Andes. Mountain Research and Development 2: 19–38.

52. Sandweiss DH, Richardson JB, III (2008) Central Andean environments. In: Silverman H, Isbell WH, editors. Handbook of South American Archaeology. New York: Springer. pp. 93–104.

53. Troll C (1968) The cordilleras of the tropical Americas. In: Troll C, editor. Geoecology of the Mountainous Regions of the Tropical Americas. Proceedings of the UNESCO Mexico Symposium Colloqium Geographicum Volume 9. Bonn: Geographisches Institut der Universtat. pp. 15–56.

54. Tosi JA, Jr. (1960) Zonas de vida natural en el Peru. Lima, Peru: Instituto de Ciencias Agrícolas de la OEA, Zona Andina.

55. Rundel PW, Dillon MO, Palma B, Mooney HA, Gulmon SL (1991) The phytogeography and ecology of the coastal Atacama and Peruvian deserts. Aliso 13: 1–49.

56. Winterhalder BP, Thomas RB (1978) Geoecology of southern highland Peru. Boulder: Occasional Paper of the University of Colorado, Institute of Arctic and Alpine Research No.27.

57. Koepcke H-W (1961) Synökologische Studien an der Westseite der Peruanischen Anden. Bonn: Donner Geographische Abhandlingen Heft 29.

58. Koepcke M (1954) Corte ecológica transversal en los Andes del Perú Central con especial consideración de los aves. Lima, Peru: Universidad Nacional Mayor de San Marcos, Museo de Historia Natural "Javier Pradom" Memorias no.3.

59. de Mera AG, Orellana JAV, Garcia JAL (1997) Phytogeographical Sectoring of the Peruvian Coast. Global Ecology and Biogeography Letters 6: 349–367.

60. Ono M, editor (1986) Taxonomic and Ecological Studies on the Lomas Vegetation in the Pacific Coast of Peru. Tokyo: Makino Herbarium, Tokyo Metropolitan University. 88 p.

61. Péfaur JE (1982) Dynamics of plant communities in the Lomas of southern Peru. Plant Ecology 49: 163–171.

62. Oka S, Ogawa H (1984) The distribution of lomas vegetation and its climatic environments along the Pacific coast of Peru. In: Matsuda I, Harayama M, Suzuki K, editors. Geographical Reports of the Tokyo Metropolitan University, Number 19. Tokyo: Department of Geography, Tokyo Metropolitan University. pp. 113–125.

63. West M (1971) Prehistoric Human Ecology in the Virú Valley. California Anthropologist 1: 47–56.

64. Beresford-Jones DG, T SA, Whaley OQ, Chepstow-Lusty AJ (2009) The role of *Prosopis* in ecological and landscape change in the Samaca Basin, Lower Ica Valley, South Coast Peru from the Early Horizon to the Late Intermediate Period. Latin American Antiquity 20: 303–332.

65. Smith BN, Epstein S (1971) Two Categories of $^{13}C/^{12}C$ Ratios for Higher Plants. Plant Physiology 47: 380–384.

66. O'Leary MH (1981) Carbon isotope fractionation in plants. Phytochemistry 20: 553–567.

67. Schwarcz HP (2006) Stable Carbon Isotope Analysis and Human Diet. In: Staller JE, Tykot RH, Benz BF, editors. Histories of Maize. Amsterdam: Elsevier Academic Press. pp. 315–321.

68. O'Leary MH (1988) Carbon Isotopes in Photosynthesis. Bioscience 38: 328–336.

69. Rice SK (2000) Variation in carbon isotope discrimination within and among *Sphagnum* species in a temperate wetland. Oecologia 123: 1–8.

70. Rundel PW, Stichler W, Zander RH, Ziegler H (1979) Carbon and hydrogen isotope ratios of bryophytes from arid and humid regions. Oecologia 44: 91–94.

71. Lange O, Green T, Ziegler H (1988) Water status related photosynthesis and carbon isotope discrimination in species of the lichen genus *Pseudocyphellaria* with green or blue-green photobionts and in photosymbiodemes. Oecologia 75: 494–501.

72. Lee Y, Lim H, Yoon H (2009) Carbon and nitrogen isotope composition of vegetation on King George Island, maritime Antarctic. Polar Biology 32: 1607–1615.

73. Huiskes A, Boschker H, Lud D, Moerdijk-Poortvliet T (2006) Stable Isotope Ratios as a Tool for Assessing Changes in Carbon and Nutrient Sources in Antarctic Terrestrial Ecosystems. Plant Ecology 182: 79–86.

74. van der Merwe NJ, Medina E (1989) Photosynthesis and $^{13}C/^{12}C$ ratios in Amazonian rain forests. Geochimica et Cosmochimica Acta 53: 1091–1094.

75. van der Merwe NJ, Medina E (1991) The canopy effect, carbon isotope ratios and foodwebs in Amazonia. Journal of Archaeological Science 18: 249–259.

76. Sonesson M, Gehrke C, Tjus M (1992) CO_2 environment, microclimate and photosynthetic characteristics of the moss *Hylocomium splendens* in a subarctic habitat. Oecologia 92: 23–29.

77. Medina E, Sternberg L, Cuevas E (1991) Vertical stratification of $\delta^{13}C$ values in closed natural and plantation forests in the Luquillo mountains, Puerto Rico. Oecologia 87: 369–372.

78. Vogel JC (1978) Recycling of CO_2 in a forest environment. Oecologia Plantarum 13: 89–94.

79. Buchmann N, Guehl JM, Barigah TS, Ehleringer JR (1997) Interseasonal comparison of CO_2 concentrations, isotopic composition, and carbon dynamics in an Amazonian rainforest (French Guiana). Oecologia 110: 120–131.

80. Broadmeadow MSJ, Griffiths H (1993) Carbon isotope discrimination and the coupling of CO_2 fluxes within forest canopies. In: Ehleringer JR, Hall AE, Farquhar GD, editors. Stable Isotopes and Plant Carbon-Water Relations. San Diego: Academic Press. pp. 109–129.

81. Heaton THE (1999) Spatial, species, and temporal variations in the $^{13}C/^{12}C$ ratios of C_3 plants: implications for palaeodiet studies. Journal of Archaeological Science 26: 637–649.

82. Drucker DG, Bridault A, Hobson KA, Szuma E, Bocherens H (2008) Can carbon-13 in large herbivores reflect the canopy effect in temperate and boreal ecosystems? Evidence from modern and ancient ungulates. Palaeogeography, Palaeoclimatology, Palaeoecology 266: 69–82.

83. Voigt CC (2010) Insights into Strata Use of Forest Animals Using the 'Canopy Effect'. Biotropica 42: 634–637.

84. Schoeninger MJ, Iwaniec UT, Glander KE (1997) Stable isotope ratios indicate diet and habitat use in New World monkeys. American Journal of Physical Anthropology 103: 69–83.

85. Austin AT, Vitousek PM (1998) Nutrient dynamics on a precipitation gradient in Hawai'i. Oecologia 113: 519–529.

86. Swap RJ, Aranibar JN, Dowty PR, Gilhooly WP, III, Macko SA (2004) Natural abundance of ^{13}C and ^{15}N in C_3 and C_4 vegetation of southern Africa: patterns and implications. Global Change Biology 10: 350–358.

87. Murphy BP, Bowman DMJS (2009) The carbon and nitrogen isotope composition of Australian grasses in relation to climate. Functional Ecology 23: 1040–1049.

88. Lajtha K, Getz J (1993) Photosynthesis and water-use efficiency in pinyon-juniper communities along an elevation gradient in northern New Mexico. Oecologia 94: 95–101.

89. Scartazza A, Lauteri M, Guido MC, Brugnoli E (1998) Carbon isotope discrimination in leaf and stem sugars, water-use efficiency and mesophyll conductance during different developmental stages in rice subjected to drought. Australian Journal of Plant Physiology 25: 489–498.

90. Syvertsen JP, Smith ML, Lloyd J, Farquhar GD (1997) Net Carbon Dioxide Assimilation, Carbon Isotope Discrimination, Growth, and Water-use Efficiency of Citrus Trees in Response to Nitrogen Status. Journal of the American Society for Horticultural Science 122: 226–232.

91. Damesin C, Rambal S, Joffre R (1997) Between-tree variations in leaf $\delta^{13}C$ of *Quercus pubescens* and *Quercus ilex* among Mediterranean habitats with different water availability. Oecologia 111: 26–35.

92. Schulze ED, Ellis R, Schulze W, Trimborn P, Ziegler H (1996) Diversity, metabolic types and $\delta^{13}C$ carbon isotope ratios in the grass flora of Namibia in relation to growth form, precipitation and habitat conditions. Oecologia 106: 352–369.

93. Farquhar G, Richards R (1984) Isotopic Composition of Plant Carbon Correlates With Water-Use Efficiency of Wheat Genotypes. Australian Journal of Plant Physiology 11: 539–552.

94. Farquhar G, O'Leary M, Berry J (1982) On the Relationship Between Carbon Isotope Discrimination and the Intercellular Carbon Dioxide Concentration in Leaves. Australian Journal of Plant Physiology 9: 121–137.

95. Farquhar GD, Ehleringer JR, Hubick KT (1989) Carbon Isotope Discrimination and Photosynthesis. Annual Review of Plant Physiology and Plant Molecular Biology 40: 503–537.

96. Guy RD, Reid DM, Krouse HR (1980) Shifts in carbon isotope ratios of two C_3 halophytes under natural and artificial conditions. Oecologia 44: 241–247.

97. Farquhar GD, Ball MC, Caemmerer S, Roksandic Z (1982) Effect of salinity and humidity on $\delta^{13}C$ value of halophytes–Evidence for diffusional isotope fractionation determined by the ratio of intercellular/atmospheric partial pressure of CO_2 under different environmental conditions. Oecologia 52: 121–124.

98. van Groenigen J-W, van Kessel C (2002) Salinity-induced Patterns Of Natural Abundance Carbon-13 And Nitrogen-15 In Plant And Soil. Soil Science Society of America Journal 66: 489–498.

99. Isla R, Aragüés R, Royo A (1998) Validity of various physiological traits as screening criteria for salt tolerance in barley. Field Crops Research 58: 97–107.

100. Vitousek PM, Field CB, Matson PA (1990) Variation in foliar $\delta^{13}C$ in Hawaiian *Metrosideros polymorpha*: a case of internal resistance? Oecologia 84: 362–370.

101. Hultine KR, Marshall JD (2000) Altitude trends in conifer leaf morphology and stable carbon isotope composition. Oecologia 123: 32–40.

102. Körner C, Diemer M (1987) In situ Photosynthetic Responses to Light, Temperature and Carbon Dioxide in Herbaceous Plants from Low and High Altitude. Functional Ecology 1: 179–194.

103. Friend AD, Woodward FI, Switsur VR (1989) Field Measurements of Photosynthesis, Stomatal Conductance, Leaf Nitrogen and $\delta^{13}C$ Along Altitudinal Gradients in Scotland. Functional Ecology 3: 117–122.

104. Körner C, Farquhar GD, Wong SC (1991) Carbon isotope discrimination by plants follows latitudinal and altitudinal trends. Oecologia 88: 30–40.

105. Ehleringer JR, Field CB, Lin Z-f, Kuo C-y (1986) Leaf carbon isotope and mineral composition in subtropical plants along an irradiance cline. Oecologia 70: 520–526.

106. Zimmerman J, Ehleringer J (1990) Carbon isotope ratios are correlated with irradiance levels in the Panamanian orchid *Catasetum viridiflavum*. Oecologia 83: 247–249.

107. Badeck F-W, Tcherkez G, Nogués S, Piel C, Ghashghaie J (2005) Post-photosynthetic fractionation of stable carbon isotopes between plant organs–a widespread phenomenon. Rapid Communications in Mass Spectrometry 19: 1381–1391.

108. Hobbie EA, Werner RA (2004) Intramolecular, compound-specific, and bulk carbon isotope patterns in C_3 and C_4 plants: a review and synthesis. New Phytologist 161: 371–385.

109. Cernusak LA, Tcherkez G, Keitel C, Cornwell WK, Santiago LS, et al. (2009) Why are non-photosynthetic tissues generally ^{13}C enriched compared with leaves in C_3 plants? Review and synthesis of current hypotheses. Functional Plant Biology 36: 199–213.

110. Damesin C, Lelarge C (2003) Carbon isotope composition of current-year shoots from *Fagus sylvatica* in relation to growth, respiration and use of reserves. Plant, Cell & Environment 26: 207–219.

111. Leavitt SW, Long A (1982) Evidence for $^{13}C/^{12}C$ fractionation between tree leaves and wood. Nature 298: 742–744.

112. DeNiro MJ, Epstein S (1977) Mechanism of Carbon Isotope Fractionation Associated with Lipid Synthesis. Science 197: 261–263.

113. Benner R, Fogel ML, Sprague EK, Hodson RE (1987) Depletion of ^{13}C in lignin and its implications for stable carbon isotope studies. Nature 329: 708–710.

114. Gleixner G, Danier HJ, Werner RA, Schmidt HL (1993) Correlations between the ^{13}C Content of Primary and Secondary Plant Products in Different Cell Compartments and That in Decomposing Basidiomycetes. Plant Physiology 102: 1287–1290.

115. Terwilliger VJ, Huang J (1996) Heterotrophic whole plant tissues show more 13C enrichment than their carbon sources. Phytochemistry 43: 1183–1188.

116. Ambrose SH, Butler BM, Hanson DB, Hunter-Anderson RL, Krueger HW (1997) Stable isotopic analysis of human diet in the Marianas Archipelago, Western Pacific. American Journal of Physical Anthropology 104: 343–361.

117. Schoeninger MJ, DeNiro MJ (1984) Nitrogen and carbon isotopic composition of bone collagen from marine and terrestrial animals. Geochimica et Cosmochimica Acta 48: 625–639.

118. Xu J, Fan X, Zhang X, Xu D, Mou S, et al. (2012) Evidence of Coexistence of C_3 and C_4 Photosynthetic Pathways in a Green-Tide-Forming Alga, *Ulva prolifera*. PLoS One 7: e37438.

119. Raven JA, Johnston AM, Kübler JE, Korb R, McInroy SG, et al. (2002) Mechanistic interpretation of carbon isotope discrimination by marine macroalgae and seagrasses. Functional Plant Biology 29: 355–378.

120. Kroopnick PM (1985) The distribution of ^{13}C of ΣCO_2 in the world oceans. Deep Sea Research Part A. Oceanographic Research Papers 32: 57–84.

121. France RL (1995) Carbon-13 enrichment in benthic compared to planktonic algae: foodweb implications. Marine Ecology Progress Series 124: 307–312.

122. Osmond CB, Valaane N, Haslam SM, Uotila P, Roksandic Z (1981) Comparisons of $\delta^{13}C$ values in leaves of aquatic macrophytes from different habitats in Britain and Finland; some implications for photosynthetic processes in aquatic plants. Oecologia 50: 117–124.

123. Cornelisen CD, Wing SR, Clark KL, Bowman MH, Frew RD, et al. (2007) Patterns in the $\delta^{13}C$ and $\delta^{15}N$ signature of *Ulva pertusa*: Interaction between physical gradients and nutrient source pools. Limnology and Oceanography 52: 820–832.

124. Burkhardt S, Riebesell U, Zondervan I (1999) Effects of growth rate, CO_2 concentration, and cell size on the stable carbon isotope fractionation in marine phytoplankton. Geochimica et Cosmochimica Acta 63: 3729–3741.

125. Kopczyńska EE, Goeyens L, Semeneh M, Dehairs F (1995) Phytoplankton composition and cell carbon distribution in Prydz Bay, Antarctica: relation to organic particulate matter and its $\delta^{13}C$ values. Journal of Plankton Research 17: 685–707.

126. Laws EA, Popp BN, Bidigare RR, Kennicutt MC, Macko SA (1995) Dependence of phytoplankton carbon isotopic composition on growth rate and $[CO_2]_{aq}$: Theoretical considerations and experimental results. Geochimica et Cosmochimica Acta 59: 1131–1138.

127. Wiencke C, Fischer G (1990) Growth and stable carbon isotope composition of cold-water macroalgae in relation to light and temperature. Marine Ecology Progress Series 65: 283–292.

128. Miller R, Page H (2012) Kelp as a trophic resource for marine suspension feeders: a review of isotope-based evidence. Marine Biology 159: 1391–1402.

129. Pilbeam DJ (2010) The Utilization of Nitrogen by Plants: A Whole Plant Perspective. Annual Plant Reviews 42: 305–351.

130. Yoneyama T, Ito O, Engelaar WMHG (2003) Uptake, metabolism and distribution of nitrogen in crop plants traced by enriched and natural ^{15}N: Progress over the last 30 years. Phytochemistry Reviews 2: 121–132.

131. Vitousek PM, Cassman K, Cleveland C, Crews T, Field CB, et al. (2002) Towards an ecological understanding of biological nitrogen fixation. Biogeochemistry 57–58: 1–45.

132. Persson J, Näsholm T (2001) Amino acid uptake: a widespread ability among boreal forest plants. Ecology Letters 4: 434–438.

133. Jones DL, Healey JR, Willett VB, Farrar JF, Hodge A (2005) Dissolved organic nitrogen uptake by plants–an important N uptake pathway? Soil Biology and Biochemistry 37: 413–423.

134. Näsholm T, Kielland K, Ganeteg U (2009) Uptake of organic nitrogen by plants. New Phytologist 182: 31–48.

135. Högberg P (1997) Tansley Review No. 95 ^{15}N natural abundance in soil-plant systems. New Phytologist 137: 179–203.

136. Evans RD (2001) Physiological mechanisms influencing plant nitrogen isotope composition. Trends in Plant Science 6: 121–126.

137. Robinson D (2001) $\delta^{15}N$ as an integrator of the nitrogen cycle. Trends in Ecology & Evolution 16: 153–162.

138. Delwiche CC, Steyn PL (1970) Nitrogen isotope fractionation in soils and microbial reactions. Environmental Science & Technology 4: 929–935.

139. Shearer G, Kohl DH (1986) N_2-Fixation in Field Settings: Estimations Based on Natural ^{15}N Abundance. Australian Journal of Plant Physiology 13: 699–756.

140. Mariotti A, Mariotti F, Amargar N, Pizelle G, Ngambi JM, et al. (1980) Fractionnements isotopiques de l'azote lors des processus d'absorption des nitrates et de fixation de l'azote atmosphérique par les plantes. Physiologie Végétale 18: 163–181.

141. Mariotti A (1983) Atmospheric nitrogen is a reliable standard for natural ^{15}N abundance measurements. Nature 303: 685–687.

142. Shearer G, Kohl DH, Chien S-H (1978) The Nitrogen-15 Abundance In A Wide Variety Of Soils. Soil Science Society of America Journal 42: 899–902.

143. Szpak P, Longstaffe FJ, Millaire J-F, White CD (2012) Stable Isotope Biogeochemistry of Seabird Guano Fertilization: Results from Growth Chamber Studies with Maize (*Zea mays*). PLoS One 7: e33741.

144. Bateman AS, Kelly SD (2007) Fertilizer nitrogen isotope signatures. Isotopes in Environmental and Health Studies 43: 237–247.

145. Choi W-J, Lee S-M, Ro H-M, Kim K-C, Yoo S-H (2002) Natural ^{15}N abundances of maize and soil amended with urea and composted pig manure. Plant and Soil 245: 223–232.

146. Yun S-I, Ro H-M, Choi W-J, Chang SX (2006) Interactive effects of N fertilizer source and timing of fertilization leave specific N isotopic signatures in Chinese cabbage and soil. Soil Biology and Biochemistry 38: 1682–1689.

147. Szpak P, Millaire J-F, White CD, Longstaffe FJ (2012) Influence of seabird guano and camelid dung fertilization on the nitrogen isotopic composition of field-grown maize (*Zea mays*). Journal of Archaeological Science 39: 3721–3740.

148. Ben-David M, Hanley TA, Schell DM (1998) Fertilization of Terrestrial Vegetation by Spawning Pacific Salmon: The Role of Flooding and Predator Activity. Oikos 83: 47–55.

149. Hilderbrand GV, Hanley TA, Robbins CT, Schwartz CC (1999) Role of brown bears (*Ursus arctos*) in the flow of marine nitrogen into a terrestrial ecosystem. Oecologia 121: 546–550.

150. Li C, Hao X, Willms WD, Zhao M, Han G (2010) Effect of long-term cattle grazing on seasonal nitrogen and phosphorus concentrations in range forage species in the fescue grassland of southwestern Alberta. Journal of Plant Nutrition and Soil Science 173: 946–951.

151. Coetsee C, Stock WD, Craine JM (2011) Do grazers alter nitrogen dynamics on grazing lawns in a South African savannah? African Journal of Ecology 49: 62–69.

152. Golluscio R, Austin A, García Martínez G, Gonzalez-Polo M, Sala O, et al. (2009) Sheep Grazing Decreases Organic Carbon and Nitrogen Pools in the Patagonian Steppe: Combination of Direct and Indirect Effects. Ecosystems 12: 686–697.

153. Frank DA, Evans RD (1997) Effects of native grazers on grassland in cycling in Yellowstone National Park. Ecology 78: 2238–2248.

154. Wittmer M, Auerswald K, Schönbach P, Bai Y, Schnyder H (2011) ^{15}N fractionation between vegetation, soil, faeces and wool is not influenced by stocking rate. Plant and Soil 340: 25–33.

155. Xu Y, He J, Cheng W, Xing X, Li L (2010) Natural ^{15}N abundance in soils and plants in relation to N cycling in a rangeland in Inner Mongolia. Journal of Plant Ecology 3: 201–207.

156. Frank D, Evans RD, Tracy B (2004) The role of ammonia volatilization in controlling the natural ^{15}N abundance of a grazed grassland. Biogeochemistry 68: 169–178.

157. Högberg P (1990) 15N natural abundance as a possible marker of the ectomycorrhizal habit of trees in mixed African woodlands. New Phytologist 115: 483–486.

158. Michelsen A, Quarmby C, Sleep D, Jonasson S (1998) Vascular plant ^{15}N natural abundance in heath and forest tundra ecosystems is closely correlated with presence and type of mycorrhizal fungi in roots. Oecologia 115: 406–418.

159. Högberg P, Högbom L, Schinkel H, Högberg M, Johannisson C, et al. (1996) ^{15}N abundance of surface soils, roots and mycorrhizas in profiles of European forest soils. Oecologia 108: 207–214.

160. Michelsen A, Schmidt IK, Jonasson S, Quarmby C, Sleep D (1996) Leaf ^{15}N abundance of subarctic plants provides field evidence that ericoid, ectomycorrhizal and non-and arbuscular mycorrhizal species access different sources of soil nitrogen. Oecologia 105: 53–63.

161. Schulze ED, Chapin FS, Gebauer G (1994) Nitrogen nutrition and isotope differences among life forms at the northern treeline of Alaska. Oecologia 100: 406–412.

162. Hobbie E, Jumpponen A, Trappe J (2005) Foliar and fungal [15]N:[14]N ratios reflect development of mycorrhizae and nitrogen supply during primary succession: testing analytical models. Oecologia 146: 258–268.

163. Hobbie EA, Macko SA, Shugart HH (1999) Insights into nitrogen and carbon dynamics of ectomycorrhizal and saprotrophic fungi from isotopic evidence. Oecologia 118: 353–360.

164. Hobbie EA, Colpaert JV (2003) Nitrogen availability and colonization by mycorrhizal fungi correlate with nitrogen isotope patterns in plants. New Phytologist 157: 115–126.

165. Ledgard SF, Woo KC, Bergersen FJ (1985) Isotopic fractionation during reduction of nitrate and nitrite by extracts of spinach leaves. Australian Journal of Plant Physiology 12: 631–640.

166. Evans RD, Bloom AJ, Sukrapanna SS, Ehleringer JR (1996) Nitrogen isotope composition of tomato (Lycopersicon esculentum Mill. cv. T-5) grown under ammonium or nitrate nutrition. Plant, Cell & Environment 19: 1317–1323.

167. Crawford TW, Rendig VV, Broadbent FE (1982) Sources, Fluxes, and Sinks of Nitrogen during Early Reproductive Growth of Maize (Zea mays L.). Plant Physiology 70: 1654–1660.

168. Bada JL, Schoeninger MJ, Schimmelmann A (1989) Isotopic fractionation during peptide bond hydrolysis. Geochimica et Cosmochimica Acta 53: 3337–3341.

169. Silfer JA, Engel MH, Macko SA (1992) Kinetic fractionation of stable carbon and nitrogen isotopes during peptide bond hydrolysis: Experimental evidence and geochemical implications. Chemical Geology 101: 211–221.

170. Szpak P, Gröcke DR, Debruyne R, MacPhee RDE, Guthrie RD, et al. (2010) Regional differences in bone collagen δ^{13}C and δ^{15}N of Pleistocene mammoths: Implications for paleoecology of the mammoth steppe. Palaeogeography, Palaeoclimatology, Palaeoecology 286: 88–96.

171. Gröcke DR, Bocherens H, Mariotti A (1997) Annual rainfall and nitrogen-isotope correlation in macropod collagen: application as a palaeoprecipitation indicator. Earth and Planetary Science Letters 153: 279–285.

172. Murphy BP, Bowman DMJS (2006) Kangaroo metabolism does not cause the relationship between bone collagen δ^{15}N and water availability. Functional Ecology 20: 1062–1069.

173. Sealy JC, van der Merwe NJ, Lee-Thorp JA, Lanham JL (1987) Nitrogen isotopic ecology in southern Africa: implications for environmental and dietary tracing. Geochimica et Cosmochimica Acta 51: 2707–2717.

174. Ambrose SH, DeNiro MJ (1986) The isotopic ecology of east African mammals. Oecologia 69: 395–406.

175. Schwarcz HP, Dupras TL, Fairgrieve SI (1999) ^{15}N enrichment in the Sahara: in search of a global relationship. Journal of Archaeological Science 26: 629–636.

176. Amundson R, Austin AT, Schuur EAG, Yoo K, Matzek V, et al. (2003) Global patterns of the isotopic composition of soil and plant nitrogen. Global Biogeochemical Cycles 17: 1031.

177. Martinelli LA, Piccolo MC, Townsend AR, Vitousek PM, Cuevas E, et al. (1999) Nitrogen stable isotopic composition of leaves and soil: Tropical versus temperate forests. Biogeochemistry 46: 45–65.

178. Aranibar JN, Otter L, Macko SA, Feral CJW, Epstein HE, et al. (2004) Nitrogen cycling in the soil-plant system along a precipitation gradient in the Kalahari sands. Global Change Biology 10: 359–373.

179. Handley LL, Austin AT, Stewart GR, Robinson D, Scrimgeour CM, et al. (1999) The ^{15}N natural abundance (δ^{15}N) of ecosystem samples reflects measures of water availability. Australian Journal of Plant Physiology 26: 185–199.

180. Ambrose SH (2000) Controlled diet and climate experiments on nitrogen isotope ratios of rats. In: Ambrose SH, Katzenberg MA, editors. Biogeochemical Approaches to Paleodietary Analysis. New York: Kluwer Academic. 243–259.

181. Hartman G (2011) Are elevated δ^{15}N values in herbivores in hot and arid environments caused by diet or animal physiology? Functional Ecology 25: 122–131.

182. Ostrom NE, Macko SA, Deibel D, Thompson RJ (1997) Seasonal variation in the stable carbon and nitrogen isotope biogeochemistry of a coastal cold ocean environment. Geochimica et Cosmochimica Acta 61: 2929–2942.

183. Waser NAD, Harrison PJ, Nielsen B, Calvert SE, Turpin DH (1998) Nitrogen Isotope Fractionation During the Uptake and Assimilation of Nitrate, Nitrite, Ammonium, and Urea by a Marine Diatom. Limnology and Oceanography 43: 215–224.

184. Wu J, Calvert SE, Wong CS (1997) Nitrogen isotope variations in the subarctic northeast Pacific: relationships to nitrate utilization and trophic structure. Deep Sea Research Part I: Oceanographic Research Papers 44: 287–314.

185. Sage R, Pearcy R (2004) The Physiological Ecology of C_4 Photosynthesis. In: Leegood RC, Sharkey TD, von Caemmerer S, editors. Photosynthesis: Physiology and Metabolism. New York: Kluwer Academic. 497–532.

186. Tieszen LL, Fagre T (1993) Carbon Isotopic Variability in Modern and Archaeological Maize. Journal of Archaeological Science 20: 25–40.

187. Warinner C, Garcia NR, Tuross N (2012) Maize, beans and the floral isotopic diversity of highland Oaxaca, Mexico. Journal of Archaeological Science doi: 10.1016/j.jas.2012.07.003.

188. Keeling CD (1979) The Suess effect: [13]Carbon-[14]Carbon interrelations. Environment International 2: 229–300.

189. Yakir D (2011) The paper trail of the ^{13}C of atmospheric CO_2 since the industrial revolution period. Environmental Research Letters 6: 034007.

190. Belane A, Dakora F (2010) Symbiotic N_2 fixation in 30 field-grown cowpea (Vigna unguiculata L. Walp.) genotypes in the Upper West Region of Ghana measured using ^{15}N natural abundance. Biology and Fertility of Soils 46: 191–198.

191. Sprent JI, Geoghegan IE, Whitty PW, James EK (1996) Natural abundance of ^{15}N and ^{13}C in nodulated legumes and other plants in the cerrado and neighbouring regions of Brazil Oecologia 105: 440–446.

192. Yoneyama T, Fujita K, Yoshida T, Matsumoto T, Kambayashi I, et al. (1986) Variation in Natural Abundance of ^{15}N among Plant Parts and in ^{15}N/[14]N Fractionation during N_2 Fixation in the Legume-Rhizobia Symbiotic System. Plant and Cell Physiology 27: 791–799.

193. Spriggs AC, Stock WD, Dakora FD (2003) Influence of mycorrhizal associations on foliar δ^{15}N values of legume and non-legume shrubs and trees in the fynbos of South Africa: Implications for estimating N_2 fixation using the ^{15}N natural abundance method Plant and Soil 255: 495–502.

194. Yoneyama T, Muraoka T, Murakami T, Boonkerd N (1993) Natural abundance of ^{15}N in tropical plants with emphasis on tree legumes Plant and Soil 153: 295–304.

195. Shearer G, Kohl DH, Virginia RA, Bryan BA, Skeeters JL, et al. (1983) Estimates of N_2-fixation from variation in the natural abundance of ^{15}N in Sonoran desert ecosystems. Oecologia 56: 365–373.

196. Delwiche CC, Zinke PJ, Johnson CM, Virginia RA (1979) Nitrogen isotope distribution as a presumptive indicator of nitrogen fixation. Botanical Gazette 140: S65–S69.

197. Virginia RA, Delwiche CC (1982) Natural ^{15}N abundance of presumed N_2-fixing and non-N_2-fixing plants from selected ecosystems. Oecologia 54: 317–325.

198. Steele KW, Bonish PM, Daniel RM, O'Hara GW (1983) Effect of Rhizobial Strain and Host Plant on Nitrogen Isotopic Fractionation in Legumes. Plant Physiology 72: 1001–1004.

199. Kohl DH, Shearer G (1980) Isotopic Fractionation Associated With Symbiotic N_2 Fixation and Uptake of NO_3^- by Plants. Plant Physiology 66: 51–56.

200. Gathumbi SM, Cadisch G, Giller KE (2002) ^{15}N natural abundance as a tool for assessing N_2-fixation of herbaceous, shrub and tree legumes in improved fallows. Soil Biology and Biochemistry 34: 1059–1071.

201. Lightfoot E, Stevens RE (2012) Stable isotope investigations of charred barley (Hordeum vulgare) and wheat (Triticum spelta) grains from Danebury Hillfort: implications for palaeodietary reconstructions. Journal of Archaeological Science doi: 10.1016/j.jas.2011.10.026.

202. Aguilera M, Araus JL, Voltas J, Rodríguez-Ariza MO, Molina F, et al. (2008) Stable carbon and nitrogen isotopes and quality traits of fossil cereal grains provide clues on sustainability at the beginnings of Mediterranean agriculture. Rapid Communications in Mass Spectrometry 22: 1653–1663.

203. Fiorentino G, Caracuta V, Casiello G, Longobardi F, Sacco A (2012) Studying ancient crop provenance: implications from δ^{13}C and δ^{15}N values of charred barley in a Middle Bronze Age silo at Ebla (NW Syria). Rapid Communications in Mass Spectrometry 26: 327–335.

204. DeNiro MJ (1985) Postmortem preservation and alteration of in vivo bone collagen isotope ratios in relation to palaeodietary reconstruction. Nature 317: 806–809.

205. Ambrose SH (1990) Preparation and characterization of bone and tooth collagen for isotopic analysis. Journal of Archaeological Science 17: 431–451.

206. van Klinken GJ (1999) Bone Collagen Quality Indicators for Palaeodietary and Radiocarbon Measurements. Journal of Archaeological Science 26: 687–695.

207. Nehlich O, Richards M (2009) Establishing collagen quality criteria for sulphur isotope analysis of archaeological bone collagen. Archaeological and Anthropological Sciences 1: 59–75.

208. Szpak P (2011) Fish bone chemistry and ultrastructure: implications for taphonomy and stable isotope analysis. Journal of Archaeological Science 38: 3358–3372.

209. O'Connell TC, Hedges REM, Healey MA, Simpson AHRW (2001) Isotopic Comparison of Hair, Nail and Bone: Modern Analyses. Journal of Archaeological Science 28: 1247–1255.

210. Towle MA (1961) The Ethnobotany of Pre-Columbian Peru. Chicago: Aldine.

211. Choi W-J, Chang SX, Ro H-M (2005) Seasonal Changes of Shoot Nitrogen Concentrations and ^{15}N/[14]N Ratios in Common Reed in a Constructed Wetland. Communications in Soil Science and Plant Analysis 36: 2719–2731.

212. Kolb KJ, Evans RD (2002) Implications of leaf nitrogen recycling on the nitrogen isotope composition of deciduous plant tissues. New Phytologist 156: 57–64.

213. Näsholm T (1994) Removal of nitrogen during needle senescence in Scots pine (Pinus sylvestris L.). Oecologia 99: 290–296.

214. Shimada M, Shimada I (1985) Prehistoric llama breeding and herding on the north coast of Peru. American Antiquity 50: 3–26.

215. DeNiro MJ (1988) Marine food sources for prehistoric coastal Peruvian camelids: isotopic evidence and implications. British Archaeological Reports, International Series 427. In: Wing ES, Wheeler JC, editors. Economic Prehistory of the Central Andes. Oxford: Archaeopress. pp. 119–128.

216. Masuda S (1985) Algae Collectors and *Lomas*. In: Masuda S, Shimada I, Morris C, editors. Andean Ecology and Civilization: An Interdisciplinary Perspective on Andean Ecological Complementarity. Tokyo: University of Tokyo Press. pp. 233–250.

217. Wainright SC, Haney JC, Kerr C, Golovkin AN, Flint MV (1998) Utilization of nitrogen derived from seabird guano by terrestrial and marine plants at St. Paul, Pribilof Islands, Bering Sea, Alaska. Marine Biology 131: 63–71.

218. Peterson TC, Vose RS (1997) An Overview of the Global Historical Climatology Network Temperature Database. Bulletin of the American Meteorological Society 78: 2837–2849.

219. Bode A, Alvarez-Ossorio MT, Varela M (2006) Phytoplankton and macrophyte contributions to littoral food webs in the Galician upwelling estimated from stable isotopes. Marine Ecology Progress Series 318: 89–102.

220. Corbisier TN, Petti MV, Skowronski RSP, Brito TS (2004) Trophic relationships in the nearshore zone of Martel Inlet (King George Island, Antarctica): $\delta^{13}C$ stable-isotope analysis. Polar Biology 27: 75–82.

221. Filgueira R, Castro BG (2011) Study of the trophic web of San Simón Bay (Ría de Vigo) by using stable isotopes. Continental Shelf Research 31: 476–487.

222. Fredriksen S (2003) Food web studies in a Norwegian kelp forest based on stable isotope ($\delta^{13}C$ and $\delta^{15}N$) analysis. Marine Ecology Progress Series 260: 71–81.

223. Gillies C, Stark J, Smith S (2012) Small-scale spatial variation of $\delta^{13}C$ and $\delta^{15}N$ isotopes in Antarctic carbon sources and consumers. Polar Biology 35: 813–827.

224. Golléty C, Riera P, Davoult D (2010) Complexity of the food web structure of the *Ascophyllum nodosum* zone evidenced by a $\delta^{13}C$ and $\delta^{15}N$ study. Journal of Sea Research 64: 304–312.

225. Grall J, Le Loc'h F, Guyonnet B, Riera P (2006) Community structure and food web based on stable isotopes ($\delta^{15}N$ and $\delta^{13}C$) analysis of a North Eastern Atlantic maerl bed. Journal of Experimental Marine Biology and Ecology 338: 1–15.

226. Kang C-K, Choy E, Son Y, Lee J-Y, Kim J, et al. (2008) Food web structure of a restored macroalgal bed in the eastern Korean peninsula determined by C and N stable isotope analyses. Marine Biology 153: 1181–1198.

227. Mayr CC, Försterra G, Häussermann V, Wunderlich A, Grau J, et al. (2011) Stable isotope variability in a Chilean fjord food web: implications for N- and C-cycles. Marine Ecology Progress Series 428: 89–104.

228. Nadon MO, Himmelman JH (2010) The structure of subtidal food webs in the northern Gulf of St. Lawrence, Canada, as revealed by the analysis of stable isotopes. Aquatic Living Resources 23: 167–176.

229. Olsen YS, Fox SE, Teichberg M, Otter M, Valiela I (2011) $\delta^{15}N$ and $\delta^{13}C$ reveal differences in carbon flow through estuarine benthic food webs in response to the relative availability of macroalgae and eelgrass. Marine Ecology Progress Series 421: 83–96.

230. Riera P, Escaravage C, Leroux C (2009) Trophic ecology of the rocky shore community associated with the *Ascophyllum nodosum* zone (Roscoff, France): A $\delta^{13}C$ vs $\delta^{15}N$ investigation. Estuarine, Coastal and Shelf Science 81: 143–148.

231. Schaal G, Riera P, Leroux C (2009) Trophic significance of the kelp *Laminaria digitata* (Lamour.) for the associated food web: a between-sites comparison. Estuarine, Coastal and Shelf Science 85: 565–572.

232. Schaal G, Riera P, Leroux C (2010) Trophic ecology in a Northern Brittany (Batz Island, France) kelp (*Laminaria digitata*) forest, as investigated through stable isotopes and chemical assays. Journal of Sea Research 63: 24–35.

233. Schaal G, Riera P, Leroux C (2012) Food web structure within kelp holdfasts (*Laminaria*): a stable isotope study. Marine Ecology 33: 370–376.

234. Vizzini S, Mazzola A (2003) Seasonal variations in the stable carbon and nitrogen isotope ratios ($^{13}C/^{12}C$ and $^{15}N/^{14}N$) of primary producers and consumers in a western Mediterranean coastal lagoon. Marine Biology 142: 1009–1018.

235. Wang WL, Yeh HW (2003) $\delta^{13}C$ values of marine macroalgae from Taiwan. Botanical Bulletin of Academia Sinica 44: 107–112.

2

Arbuscular-Mycorrhizal Networks Inhibit *Eucalyptus tetrodonta* Seedlings in Rain Forest Soil Microcosms

David P. Janos[1]*, **John Scott**[2], **Catalina Aristizábal**[1], **David M. J. S. Bowman**[3]

1 Department of Biology, University of Miami, Coral Gables, Florida, United States of America, **2** Research Institute for the Environment and Livelihoods, Charles Darwin University, Darwin, Northern Territory, Australia, **3** School of Plant Science, The University of Tasmania, Hobart, Tasmania, Australia

Abstract

Eucalyptus tetrodonta, a co-dominant tree species of tropical, northern Australian savannas, does not invade adjacent monsoon rain forest unless the forest is burnt intensely. Such facilitation by fire of seedling establishment is known as the "ashbed effect." Because the ashbed effect might involve disruption of common mycorrhizal networks, we hypothesized that in the absence of fire, intact rain forest arbuscular mycorrhizal (AM) networks inhibit *E. tetrodonta* seedlings. Although arbuscular mycorrhizas predominate in the rain forest, common tree species of the northern Australian savannas (including adult *E. tetrodonta*) host ectomycorrhizas. To test our hypothesis, we grew *E. tetrodonta* and *Ceiba pentandra* (an AM-responsive species used to confirm treatments) separately in microcosms of ambient or methyl-bromide fumigated rain forest soil with or without severing potential mycorrhizal fungus connections to an AM nurse plant, *Litsea glutinosa*. As expected, *C. pentandra* formed mycorrhizas in all treatments but had the most root colonization and grew fastest in ambient soil. *E. tetrodonta* seedlings also formed AM in all treatments, but severing hyphae in fumigated soil produced the least colonization and the best growth. Three of ten *E. tetrodonta* seedlings in ambient soil with intact network hyphae died. Because foliar chlorosis was symptomatic of iron deficiency, after 130 days we began to fertilize half the *E. tetrodonta* seedlings in ambient soil with an iron solution. Iron fertilization completely remedied chlorosis and stimulated leaf growth. Our microcosm results suggest that in intact rain forest, common AM networks mediate belowground competition and AM fungi may exacerbate iron deficiency, thereby enhancing resistance to *E. tetrodonta* invasion. Common AM networks–previously unrecognized as contributors to the ashbed effect–probably help to maintain the rain forest–savanna boundary.

Editor: Minna-Maarit Kytöviita, Jyväskylä University, Finland

Funding: The Cooperative Research Centre (CRC) for Tropical Savannas, and the Parks and Wildlife Commission of the Northern Territory funded this research. The University of Tasmania supported DPJ as a visiting scholar to prepare the manuscript. The funders had no role in study design, data collection and analysis, decision to publish, or preparation of the manuscript.

Competing Interests: The authors have declared that no competing interests exist.

* E-mail: davidjanos@miami.edu

Introduction

Eucalypts predominate across much of monsoon tropical, northern Australia. *Eucalyptus tetrodonta* F.Muell. and *E. miniata* Cunn. ex Schauer are canopy co-dominants of coastal savannas [1] throughout which are scattered patches of rain forest [2,3]. The savannas are highly susceptible to fire, but only after very dry conditions can fires cross the abrupt ecotone between savanna and rain forest [4]. If high-intensity fires penetrate the rain forest, then *E. tetrodonta* seedlings can invade, but otherwise they cannot, even after canopy destruction by tropical cyclones [5]. Reciprocally, in the absence of fire, rain forest plants can colonize savanna, albeit very slowly [6]. Nevertheless, the mechanisms by which fire facilitates rain forest invasion by *E. tetrodonta* are uncertain.

The apparent necessity of fire to facilitate establishment of some species' seedlings, especially those of eucalypts and pines, is known as the "ashbed effect" (a tenet which underpins much Australian forestry practice [7]). Although the ashbed effect has been investigated for more than half a century, its mechanisms remain ambiguous because it likely involves multiple phenomena associated with fire and soil desiccation. Those phenomena may include at different times and places, direct fertilization by ash [7], soil physical and chemical changes that diminish P adsorption [7,8],

release of mineral nutrients from heat-killed soil microorganisms [8] (but see [9]), partial soil sterilization that eliminates pathogenic microbes [10] (but see [11]), or other alterations of the soil microflora, especially ectomycorrhizal fungi [12,13]. Notwithstanding uncertainty about the mechanisms behind the ashbed effect, empirical evidence from across Australia shows that without fire, rain forest resists invasion by savanna plant species, just as fire contributes to savannas' resistance to replacement by rain forest [4].

Bowman and Fensham [5] demonstrated that soil fumigation alone can mimic the ashbed effect in facilitating *E. tetrodonta* seedling growth in rain forest soil. Although the success of fumigation suggests a primary role of soil microflora in the ashbed effect, whether that role mainly is attributable to mineral nutrient release from killed saprotrophs, elimination of pathogens and parasites, or changes in mycorrhizal fungi is uncertain. *E. tetrodonta*, *E. miniata*, and other common tree species of the savanna are ectomycorrhizal (ECM) as adults [14], but Australian rain forest tree species almost exclusively comprise arbuscular mycorrhizal (AM) hosts [15]. So, we hypothesized that rain forest AM fungi are detrimental to *E. tetrodonta* seedlings, and that, much like intense burns, soil fumigation or partial sterilization (as in a pilot experiment that we conducted; Figure 1) relieves the detriment.

Figure 1. *Eucalyptus tetrodonta* **seedlings in ambient savanna, ambient rain forest, and microwaved rain forest soil.** The seedlings were transplanted 58 d previously to this pilot experiment (data not shown). True leaves of the seedlings in their native savanna soil are dark green, but those of seedlings growing in ambient rain forest soil are chlorotic and extremely stunted. Seedlings in microwave-heated (μ-waved) rain forest soil are not stunted versus those in savanna soil, but their leaves are chlorotic, exhibiting the dark green veins and yellow inter-vein areas characteristic of iron deficiency. All root systems were sparse, and no mycorrhizas of any type were apparent.

Although fire more often negatively affects ECM than AM fungi [16,17], intense fires burning into rain forest kill AM host plants and thereby might disrupt otherwise persistent networks of mycorrhizal interconnections among those plants.

Accumulating evidence supports the inference that plants can be interconnected by mycorrhizal fungus hyphal networks, often called "common mycorrhizal networks" [18–20]. These constellations of hyphal bridges among root systems have been suggested to be pathways for inter-plant movement of fixed-carbon by both ECM [21] and AM [22] fungi, mineral nutrients such as nitrogen [23,24] and phosphorus [25,26] by both types of mycorrhizal fungi, and water, the latter especially by ECM networks [27].

Ectomycorrhizal networks in the savanna might facilitate seedling establishment by redistributing fixed carbon (possibly in the form of nitrogenous compounds [28]), mineral nutrients [20], and water [29]. Additionally, ECM networks may favor seedlings simply by maintaining a high density of fungi that lead to rapid mycorrhiza formation, or by influencing mycorrhizal fungus species composition [20,30]. Arbuscular mycorrhizal networks, however, are not likely to supply fixed carbon to host plants because transported carbon probably remains as storage lipids within the fungi in receiver plant roots [31]. Moreover, because AM fungi generally do not produce rhizomorphs or mycelial strands, they likely do not redistribute water to the same extent as ECM networks. In the savanna, unless coexisting ECM and AM networks include plant species such as eucalypts that can form both types of mycorrhizas [32] and thereby might link their networks, ECM and AM networks probably represent distinct niches [30].

Arbuscular mycorrhizal fungus networks may be as likely to intensify belowground competition as they are to enhance mycorrhiza formation and overall mineral nutrient acquisition. Although Chiariello et al. [25] found in a short-term, pulse-chase, field experiment with P^{32} that it was distributed to plants surrounding a decapitated donor without relationship to distance to recipient, recipient size, or recipient species, several greenhouse experiments have suggested that plants interconnected by AM networks can compete strongly belowground [33–35]. Our hypothesis of AM fungus detriment to *E. tetrodonta* seedlings anticipates that within a rain forest competitive milieu, *E. tetrodonta* seedlings will be disadvantaged by inclusion within a common AM network. Thereby, instead of AM and eventual ECM hosts

coexisting relatively benignly as often tacitly assumed, we propose that in some situations they may interact antagonistically.

We designed a microcosm experiment to examine our hypothesis that failure of *E. tetrodonta* seedlings to grow in rain forest soil is a consequence of inhibition by common AM networks. Because *Eucalyptus* species are susceptible to iron deficiency [36] and we observed symptoms suggestive of iron deficiency in rain forest soil in a pilot experiment with *E. tetrodonta* seedlings (Figure 1), as an additional treatment 130 days after transplant (DAT), we applied soluble iron to one-half of the *E. tetrodonta* seedlings in ambient soil to test an additional hypothesis that iron deficiency limited their growth.

Materials and Methods

Experiment design

We grew 40 *E. tetrodonta* seedlings (10 in each of four treatments) in the end compartments of rectangular microcosms divided into three equal compartments by nylon mesh root barriers through which AM fungus hyphae could pass (Figure 2). In the central compartments, "nurse" plants in ambient (= not fumigated) rain forest soil sustained AM fungi. Hyphae could extend from the nurse plant into the two end compartments potentially to establish common mycorrhizal networks with seedlings at both ends either in ambient soil or in soil fumigated initially to eliminate AM fungi. Hyphae extending into one end compartment of each microcosm could be severed repeatedly to disrupt connections to the nurse plant. Thus, there were four treatments in a split-plot, factorial design: ambient or fumigated soil in both end compartments of individual microcosm "plots" crossed with intact (hereafter called "networked") or severed common mycorrhizal networks.

In order to determine if soil fumigation and repeated hypha severing succeeded in diminishing AM formation as intended, we grew 40 seedlings of an AM-responsive [37] tropical tree species, *Ceiba pentandra* (L.) Gaertn. (Malvaceae) in microcosms to which we applied the same four treatments as for *E. tetrodonta*. We did not grow *C. pentandra* together with *E. tetrodonta* in the same microcosms because we did not wish them to compete with one another (although both could compete with their respective nurse plants).

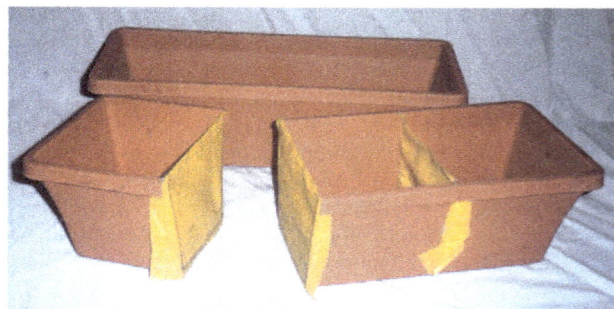

Figure 2. A disassembled microcosm. The three-compartmented planter box is divided by double layers of nylon silk-screen cloth mesh root barriers (45 μm pore size). A double layer was cemented into a slot on the right side, and single layers were cemented across each cut end of the completely separated left side. When assembled, the compartmented box nested tightly within the fully-intact outer box, but the separated end periodically could be lifted vertically, sliding the layers of cloth mesh against one another and severing hyphae that crossed them.

Establishment of microcosms

We constructed the microcosms from plastic planter boxes (15 cm×45 cm×15 cm deep; Figure 2). Each was divided into thirds, at one side by a slit sawn to, but not through, the rim, and at the opposite side by entirely separating the end one-third. In the slit, we cemented with silicone sealer a double layer of nylon silk-screen cloth mesh (45 μm pore size) that was a barrier to roots but not to AM fungus hyphae or other microorganisms. Similarly, across the fully-cut ends we cemented single pieces of mesh. Thereby, both ends were separated from the central compartment by two layers of mesh. Each modified box was nested within an intact box in which it fit tightly, but which allowed the cut end to be lifted momentarily in order to sever fungus hyphae that crossed the double mesh root barrier. Soil pressed the layers of mesh together tightly so there was no air gap between them.

We filled the microcosm central compartments with rain forest soil (Table 1) collected to no more than 20 cm depth at Gunn Point (12°09.258′ S, 131°02.186′ E) near Darwin, NT, Australia. We filled the two end compartments with either the same ambient soil, or with that soil fumigated with methyl bromide gas at 1 kg m^{-3}. As supplemental inoculum of indigenous AM fungi, we collected fine roots from the Gunn Point rain forest one-day prior to use. The roots were hand cut into 1–2 cm pieces and were layered in central compartments at half their depth. Above this layer of fresh roots, we transplanted a 15–20 cm tall, field-collected root sprout of the Australian rain forest tree species *Litsea glutinosa* (Lour.) C. B. Rob. (Lauraceae). Before transplant, each sprout's source root was treated by dipping it into powdered rooting hormone and micronutrients ("Clonex"). Then, the compartments were filled with ambient soil topped with 2–3 cm of fumigated soil to reduce the chance of spores splashing into end compartments. We filled both end compartments of 20 microcosms with ambient soil, and filled those of an additional 20 microcosms with fumigated soil. The end compartments were neither actively inoculated with AM fungi, nor were they treated with a microbial filtrate [38,39] because the central compartments likely would serve as a source of bacteria and saprotrophic fungi as well as AM fungi.

After transplanting *L. glutinosa* sprouts on 20 June, 1997, we randomized all 40 microcosms with no more than 2 cm separation in two rows on a wire mesh screenhouse bench at the CSIRO Berrimah campus (12°24.800′ S, 130°55.280′ E) in Darwin. The screenhouse provided light shade. The nurse plants grew in the microcosms for six months before we transplanted *E. tetrodonta* and *C. pentandra* seedlings to begin the experiment. This pre-treatment allowed AM fungus hyphae to extend into stationary end compartments, and allowed hyphae from propagules in stationary, ambient-soil end compartments potentially to reach nurse plant roots. In order to disrupt hypha spread, three days after transplanting the *L. glutinosa* sprouts, we began lifting the detached end compartments of each microcosm by 8–10 cm every two or three days. We continued this periodic severing of hyphae until all plants were harvested.

Seedling growth, measurement, and harvest

Seeds of *E. tetrodonta* obtained from Top End Seeds (Nightcliff, NT) and of *C. pentandra* obtained from the George Brown Botanic Gardens (Darwin) were sown in germination flats of fumigated rain forest soil in the screenhouse. On 20 December, 1997, we transplanted two 1–2 week-old *E. tetrodonta* or three just-emerged *C. pentandra* seedlings into both end compartments of separate microcosms. One week later on 28 December, we replaced *C. pentandra* of which all had died in 22 microcosm ends. A week before the first growth measurement, we clipped the weakest

Table 1. Attributes of Australian Northern Territory monsoon rain forest and *Eucalyptus* savanna soils.

Attribute (units)	Rain forest	Savanna	t, P
pH	**6.3** (0.1)	5.0 (0.3)	5.01, 0.038
EC (mS cm^{-1})	**0.26** (0.03)	0.10 (0.02)	4.44, 0.047
Ca (mg kg^{-1})	**3050** (50)	330 (130)	19.53, 0.003
Mg (mg kg^{-1})	**305** (5)	130 (30)	5.75, 0.029
K (mg kg^{-1})	**145** (5)	27 (3)	20.24, 0.002
Na (mg kg^{-1})	<25 (0)	<25 (0)	NA
P (Colwell; mg kg^{-1})	**23** (1)	8 (1)	13.86, 0.005
S (mg kg^{-1})	39 (5)	37 (5)	0.28, 0.804
TKN (%)	**0.31** (0.04)	0.11 (0.02)	4.80, 0.041
Cu (mg kg^{-1})	0.7 (0.0)	0.6 (0.1)	3.01, 0.204
Fe (DTPA; mg kg^{-1})	17 (1)	**43** (2)	−12.85, 0.006
Mn (mg kg^{-1})	101 (12)	156 (60)	−0.91, 0.460
Zn (mg kg^{-1})	**7.8** (0.5)	1.0 (0.7)	8.60, 0.013
OC (%)	**7.20** (0.46)	1.69 (0.07)	11.96, 0.007
Ex Al (me%)	0.01 (0.00)	0.31 (0.29)	−1.03, 0.489

Values are means ±1 standard error in parentheses with significantly highest values in bold. Soils (n = 2 for each) were compared by t-test (approximate t-test for unequal variances for Cu and exchangeable Al); t statistics and probabilities are shown. EC = electrical conductivity; TKN = total Kjeldahl N; OC = organic carbon; Ex Al = exchangeable Al; NA = not tested (below detection limit).

individuals of *C. pentandra* and *E. tetrodonta*, leaving only one seedling in each end compartment. Thus, there were 40 individuals (10 per treatment) in 20 microcosms for each species (40 microcosms in total).

During the rainy season (October–April) all microcosms received natural rainfall and were watered only if necessary, but during the dry season they were watered three times daily by an overhead sprinkler system. Mean monthly rainfall from December, 1997 through April, 1998 ranged from 75 mm to 789 mm, and mean monthly maximum and minimum air temperatures ranged from 31.6 to 33.9 and 24.9 to 25.5 °C, respectively. While we continued to grow *E. tetrodonta* seedlings in ambient soil with or without iron fertilization from May through July, 1998, mean monthly rainfall was less than 0.4 mm, and the maximum and minimum mean monthly temperatures were 33.4 and 21.4 °C, respectively (all weather information for Darwin airport from www.bom.gov.au).

Beginning on 19 January, 1998, one month after transplant, and continuing every two weeks until harvest, we measured height from the soil surface to the shoot apex, length of the longest leaflet or leaf, and counted the number of leaves of every seedling. Additionally, we categorically recorded foliar chlorosis of *E. tetrodonta*. We measured *L. glutinosa* nurse plants only twice, on 22 December, 1997, at the start of the experiment, and on 10 May, 1998. We measured *L. glutinosa* stem diameter, height from the soil surface to the tallest shoot apex, longest leaf length, and counted the number of leaves.

On 10 May, 1998, 141 DAT, we harvested all *C. pentandra* and only the *E. tetrodonta* grown in fumigated soil. We harvested at this time because of declining health and mortality of *E. tetrodonta*, especially in networked, ambient soil. Root systems of both species were extracted from the soil, rinsed gently over a 1 mm sieve, and preserved in 50% ethanol for assessment of mycorrhizal colonization. For *E. tetrodonta* and their accompanying *L. glutinosa* nurse plants, we separated leaf blades from petioles and stems, dried all to constant weight in an oven at 60 °C, and then weighed them.

Iron fertilization

We did not harvest the *E. tetrodonta* seedlings grown in ambient soil when we harvested the seedlings from fumigated soil. Instead, we continued to grow them for an additional 69 days with or without iron fertilization. We randomly assigned half of each group of hyphae-severed and networked seedlings to receive 81 mg of iron (100 mL of 3.24 g L^{-1} of "Librel Fe–Lo", Allied Colloids, England; water soluble Fe = 13.2% and Fe chelated by EDTA = 12.5%) eight times at irregular intervals (i.e., on 29 April, 13, 21, 28 May, 4, 11, 24 June, and 11 July, 1998). This addition was intended transiently to elevate the available iron concentration of rain forest soil approximately to that of savanna soil (Table 1). We monitored these seedlings' growth, and processed them and their nurse plants at harvest (210 DAT) as described previously.

Mycorrhizal colonization and leaf tissue analyses

Preserved root systems of all *E. tetrodonta* seedlings were blotted dry and weighed before haphazardly removing a fine-root sample for assessment of mycorrhizal colonization. Afterwards, the remaining roots were weighed again before being dried to constant weight. We used dry-weight to fresh-weight ratios to calculate the dry weights of entire root systems.

For both *C. pentandra* and *E. tetrodonta*, we cleared fine-root samples in 10% KOH at room temperature for 48 h before acidifying them in 5% HCl for 15 min, and then stained them in 0.05% trypan blue in lactoglycerol at room temperature for 6 hr. We mounted ten to twenty, 1–2 cm long root segments on microscope slides and scored them for percentage length colonized by AM fungi (typical hyphae and vesicles in the root cortex) with a compound microscope according to the magnified gridline intersection method [40]. We usually examined 160 to more than 500 intersections per plant, but examined fewer than 100 intersections for seven *E. tetrodonta* with very small root systems. We excluded colonization data of two small individuals from the fumigated soil, networked treatment because they were outliers (below the lower 99.9% confidence limit for the other plants of that treatment).

E. tetrodonta leaf tissues (petioles excluded) were analyzed for element concentrations by the Department of Primary Industries & Fisheries, Berrimah Agricultural Research Centre, Chemistry Section (Berrimah). Leaf tissues of as many individuals as possible were analyzed separately, but because some *E. tetrodonta* had tiny leaves, we composited tissue of 2–4 plants as necessary. There was one composite of 2 fumigated soil, hyphae-severed plants; three composites (two of 2 and one of 3 plants) of fumigated soil, networked plants; one composite of 4 ambient soil, hyphae-severed plants; and two composites of 2 ambient soil, networked plants. The Chemistry Section also analyzed ambient and fumigated composited samples of the rain forest soil, and for comparison, ambient and fumigated samples of savanna soil collected 0.57 km distant from the rain forest. Soil pH and electrical conductivity were determined in a 1:5 soil:water extract; K, Ca, Mg, and Na in ammonium chloride; Colwell P in sodium bicarbonate; S in calcium dihydrogen phosphate; Zn, Cu, Mn, and Fe in DTPA; organic carbon by modified Walkley-Black digestion; and exchangeable Al in calcium chloride.

Statistical analyses

We used *t*-tests (approximate *t*-tests for Cu and exchangeable Al) to compare rain forest and savanna soil (n = 2 for each soil), and we report as significantly different (Table 1) those attributes for which $P{\leq}0.05$. All statistical tests except those noted otherwise were performed with Statistix v. 9.0 (Analytical Software, Tallahassee, FL).

We analyzed *C. pentandra* percentage root length colonized by AM fungi by split-plot, two-factor ANOVA with soil fumigation as the whole-plot factor, but because of different harvest dates, we analyzed *E. tetrodonta* AM colonization by separate two-sample *t*-tests for each harvest. Percentage colonized root length was arcsine-square root transformed before analysis. *E. tetrodonta* dry weight variables were compared by paired *t*-tests (pairing plants in the same microcosm) for plants in fumigated soil, and by *t*-tests (because of unequal numbers of surviving plants) for plants in ambient soil after Bonferroni correction for the number of analyzed response variables ($P{\leq}0.0125 = 0.05/4$ dry weight response variables).

We analyzed morphological responses of *C. pentandra* and *E. tetrodonta* by split-plot, two-factor, repeated-measures ANOVAs after Levene's test of heteroscedasticity. Applying a Bonferroni correction for three morphometric response variables, we report as significant, effects for which $P{\leq}0.017$. Mortality of *E. tetrodonta* in the ambient, networked treatment unbalanced the numbers of seedlings per treatment, so we used JMP Pro v. 10.0.0 (SAS Institute, Inc., Cary, NC) with restricted maximum-likelihood calculations to perform the *E. tetrodonta* repeated-measures analyses.

We used a one-way MANOVA performed with JMP Pro v. 10.0.0 followed by univariate analyses of the four *L. glutinosa* response variables to compare the effects on *L. glutinosa* of being in a microcosm with *C. pentandra* versus being with *E. tetrodonta*. Because *L. glutinosa* individuals were 15–20 cm tall at transplant, we relativized final morphological measurements by using their differences from initial measurements. We examined hypothesized negative associations of *C. pentandra* and *E. tetrodonta* with *L. glutinosa* for effects of severing common mycorrhizal networks with Pearson's correlations. Because we did not determine the dry weights of *C. pentandra*, we analyzed relativized final morphological measurements for *C. pentandra* (longest leaflet length change) versus *L. glutinosa* number of leaves. For *E. tetrodonta* versus *L. glutinosa*, however, we analyzed shoot dry weights. One ambient, networked *E. tetrodonta* seedling with a dry weight (1.67 g) almost twice the upper 99.9% confidence limit for its treatment was excluded.

We analyzed morphological responses of *E. tetrodonta* to iron fertilization over 69 d following the initial fertilization by repeated-measures ANOVAs performed with JMP Pro v. 10.0.0, followed by Bonferroni correction ($P{\leq}0.017$) for having examined three response variables. We used Fisher exact tests to examine the effects of iron fertilization on recovery from chlorosis.

We analyzed element concentrations in *E. tetrodonta* foliage across both harvests by one-way ANOVA or non-parametric Kruskal-Wallis tests if variances were heteroscedastic. We used a Bonferroni-corrected P≤0.0045 (= 0.05/11 elements) to assess the significance of these ANOVAs and of Pearson's correlation coefficients between *E. tetrodonta* leaf dry weights and element concentrations.

Results

Soil attributes

Among the 15 soil attributes shown in Table 1, ten differed significantly, but only one, iron, was greater in savanna soil than in rain forest soil. Those mineral nutrient concentrations that differed significantly were 2.3 (Mg) to 9.2 (Ca) times higher in rain forest than in savanna soil, except for available iron which was 2.6 times more abundant in savanna than in rain forest soil.

C. pentandra mycorrhizas and growth

Fumigation significantly reduced *C. pentandra* AM-colonized root length ($F_{1,9} = 7.13$, $P = 0.026$) but colonization was not affected significantly by severing hyphae ($F_{1,18} = 0.00$, $P = 0.957$) or by the interaction of fumigation and severing hyphae ($F_{1,18} = 1.66$, $P = 0.214$; Figure 3A). At harvest, 141 DAT, *C. pentandra* seedlings averaged 50% AM-colonized root length in ambient soil, but only 39% in fumigated soil.

Fumigation significantly diminished height increase (fumigation×time $F_{9,162} = 4.96$, $P < 0.001$; Figure 3B) and number of leaves increase (fumigation×time $F_{9,162} = 3.02$, $P = 0.002$). At 141 DAT, mean seedling height differed by 3.7 cm and mean number of leaves by 0.7 leaves in fumigated versus ambient soil. Increase of longest leaflet length, which appeared to have reached an asymptote at 8.0 cm, was not significantly affected by soil fumigation (fumigation×time $F_{9,162} = 1.85$, $P = 0.063$). We found no significant effects of severing hyphae on *C. pentandra* growth rates (height: severing×time $F_{9,162} = 1.05$, $P = 0.399$; leaflet length: severing×time $F_{9,162} = 2.30$, $P = 0.018$; number of leaves: severing×time $F_{9,162} = 1.21$, $P = 0.293$), nor did fumigation, severing, and time interact for any morphological response variable (all $P > 0.544$).

E. tetrodonta mycorrhizas and seedling performance

Severing common mycorrhizal network hyphae significantly reduced *E. tetrodonta* AM-colonized root length in fumigated soil 141 DAT (df = 16, $t = 2.15$, $P = 0.047$), but in ambient soil 210 DAT, it did not (df = 15, $t = 0.39$, $P = 0.701$; Figure 4A). In ambient soil, 45% of *E. tetrodonta* root length was colonized by AM fungi, which was similar to the 47% colonized root length of seedlings in fumigated soil with intact networks. Hyphae-severed seedlings in fumigated soil had only 29% colonization. We did not find any ectomycorrhizas.

Three of ten *E. tetrodonta* seedlings in ambient soil with intact common mycorrhizal networks failed to survive until 141 DAT. Among surviving seedlings, severing hyphae significantly elevated height increase (severing×time $F_{9,147.7} = 5.96$, $P < 0.001$), longest leaf length increase (severing×time $F_{9,150.4} = 4.53$, $P < 0.001$; Figure 4B), and number of leaves increase (severing×time $F_{9,146.4} = 3.59$, $P < 0.001$). Hyphae-severed seedlings in fumigated soil exceeded the mean height of seedlings in all other treatments by 3.4 cm, the mean longest leaf length by 2.2 cm, and the mean number of leaves by 3.4. We found no significant effects of fumigation on any morphological variable (height: fumigation×

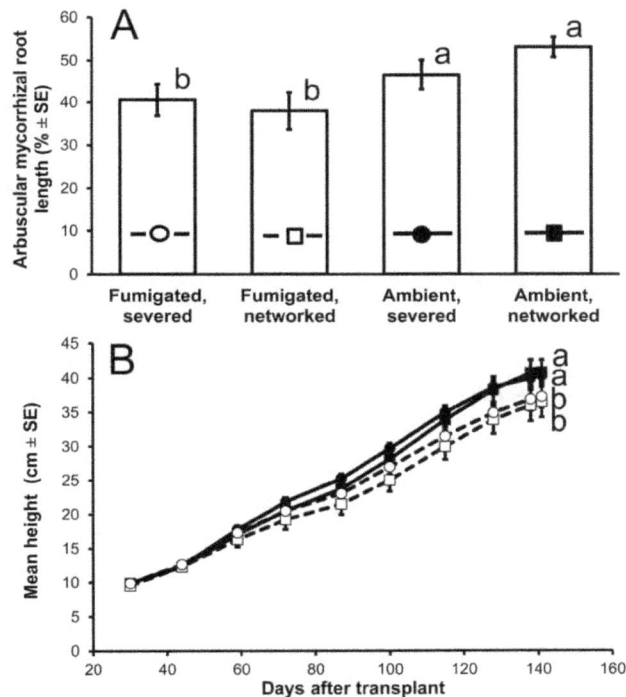

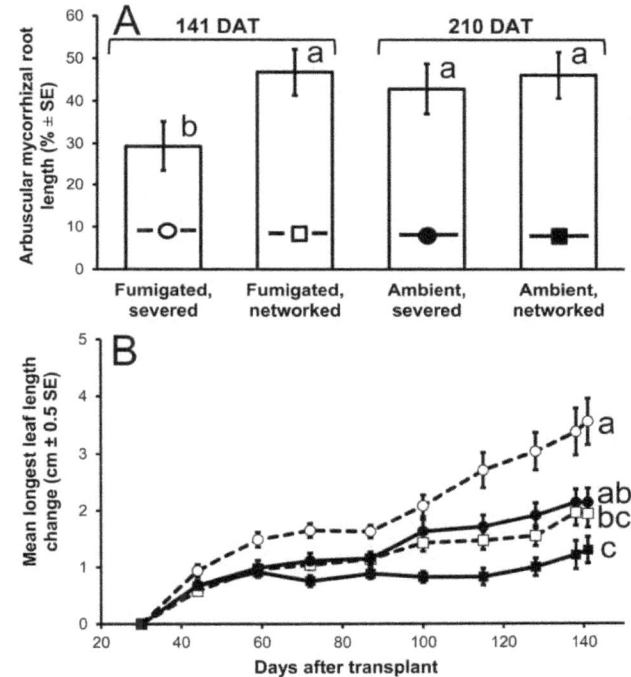

Figure 3. *Ceiba pentandra* **arbuscular mycorrhizas and height growth responses.** (A) arbuscular mycorrhizal root length (%±1 standard error) at 141 days after transplant, and (B) mean height (cm±1 standard error) versus days after transplant to microcosms with *Litsea glutinosa* AM nurse plants containing ambient (filled symbols; solid lines) or methyl-bromide fumigated (open symbols; dashed lines) rain forest soil in which hyphae repeatedly were severed (circles) or not (squares). Symbols on bars in Figure 3A correspond to those in Figure 3B. Bars topped by the same lowercase letter in 3A do not differ significantly; in 3B, letters similarly denote interactions with time. Fumigation reduced colonized root length and diminished mean height increase. (Neither hypha severing nor its interaction with fumigation significantly affected either root colonization or growth.)

Figure 4. *Eucalyptus tetrodonta* **arbuscular mycorrhizas and longest leaf length growth responses.** (A) arbuscular mycorrhizal root length (%±1 standard error) at 141 days after transplant (DAT) to methyl-bromide fumigated rain forest soil and at 210 DAT to ambient soil, and (B) mean longest leaf length change from the initial measurement (cm±0.5 standard error) versus days after transplant to microcosms with *Litsea glutinosa* AM nurse plants containing ambient (filled symbols; solid lines) or methyl-bromide fumigated (open symbols; dashed lines) soil in which hyphae repeatedly were severed (circles) or not (squares). Symbols on bars in Figure 4A correspond to those in Figure 4B. Bars topped by the same lowercase letter in 4A do not differ significantly within a harvest; in 4B, letters similarly denote interactions with time. Severing network hyphae reduced colonized root length in fumigated soil, but not in ambient soil. Severing network hyphae increased mean longest leaf length change, but neither fumigation nor its interaction with severing significantly affected longest leaf length change.

time $F_{9,163.8} = 0.26$, $P = 0.985$; leaf length: fumigation×time $F_{9,158.3} = 1.82$, $P = 0.069$; number of leaves: fumigation×time $F_{9,161.7} = 0.47$, $P = 0.895$), nor did fumigation, hyphae severing, and time interact significantly for any variable (all $P > 0.456$).

Leaf and stem dry weights of *E. tetrodonta* seedlings grown in fumigated soil were significantly increased by severing hyphae, but total root dry weights and fine root-to-leaf dry weight ratio did not differ (Table 2). Hyphae-severed seedlings grown an additional 69 days in ambient soil without iron fertilization consistently exceeded the dry weights of networked seedlings, although not significantly (Table S1).

Common mycorrhizal network effects on plant interactions

C. pentandra seedlings were conspicuously larger than *E. tetrodonta* seedlings at harvest. A one-way MANOVA with all four *L. glutinosa* response variables showed that the *L. glutinosa* nurse plants grew significantly less ($F_{4,35} = 2.95$, $P = 0.033$) when accompanied by *C. pentandra* than when accompanied by *E. tetrodonta*. Univariate analyses of *L. glutinosa* nurse plant height change ($F_{1,38} = 3.19$, $P = 0.082$), largest leaf length change ($F_{1,38} = 4.32$, $P = 0.044$), number of leaves change ($F_{1,38} = 9.96$, $P < 0.001$), and stem diameter change ($F_{1,38} = 6.57$, $P = 0.014$) revealed significant effects on the latter three response variables.

When we examined associations between *C. pentandra* longest leaflet length change (a proxy for growth rate that was not affected significantly by soil fumigation) and *L. glutinosa* number of leaves on 10 May, 1998 (a proxy for plant size), although there was no significant association for hyphae-severed plants (n = 20, $r = -0.05$, $P = 0.832$; Figure 5A), we found a significant negative association (n = 20, $r = -0.66$, $P = 0.002$; Figure 5B) among networked plants. Similarly, for shoot dry weights of *E. tetrodonta* versus *L. glutinosa* harvested with them, although there was no significant association for hyphae-severed plants (n = 20, $r = -0.10$, $P = 0.688$; Figure 6A), there was a significant negative association among networked plants (n = 16, $r = -0.50$, $P = 0.047$; Figure 6B).

Iron fertilization of *E. tetrodonta*

Iron fertilization of *E. tetrodonta* seedlings in ambient soil stimulated growth and remedied chlorosis. Iron fertilization significantly elevated longest leaf length increase (fertilization× time $F_{6,90} = 4.02$, $P = 0.001$; Figure 7), although neither height increase (fertilization×time $F_{6,90} = 1.87$, $P = 0.095$) nor number of leaves increase (fertilization×time $F_{6,90} = 1.52$, $P = 0.179$) was affected significantly. At harvest, mean longest leaf length of iron-fertilized plants exceeded that of not-fertilized plants by

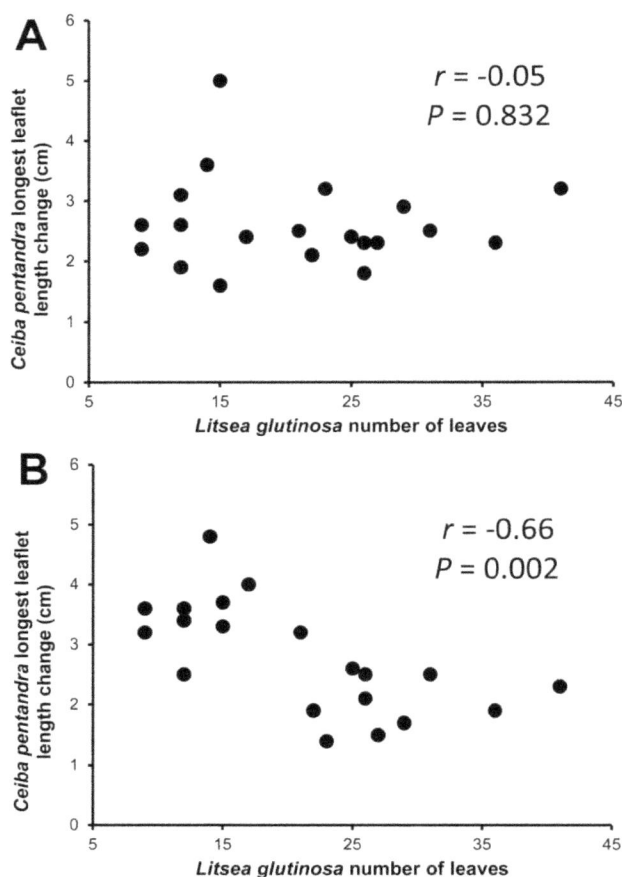

Figure 5. *Ceiba pentandra* **growth versus** *Litsea glutinosa* **nurse plant size.** (A) with potential hyphal network interconnections repeatedly severed, and (B) without hypha severing. Longest leaflet length change from initial measurement (cm) for *C. pentandra* is shown versus the number of leaves on *Litsea glutinosa* nurse plants 141 days after transplant (DAT). With hypha severing, there is no significant association, but in the likely presence of common AM networks a significant negative association suggests belowground competitive interactions.

2.0 cm, and iron fertilization had remedied chlorosis in all 5 chlorotic plants among the 8 plants randomly allocated to be fertilized. In contrast, all 4 chlorotic among 9 not-fertilized plants remained chlorotic. Although the two random groups did not

Table 2. Dry weight responses of *Eucalyptus tetrodonta* seedlings to common mycorrhizal network hypha severing in fumigated rain forest soil and to soluble iron fertilization in non-fumigated, ambient soil.

Response (units)	Fumigated soil (141 DAT)			Ambient soil (210 DAT)		
	Hyphae severed	Networked	DF, *t, P*	No fertilization	Iron fertilization	DF, *t, P*
Leaf weight (g)	**0.480** (0.338)	0.202 (0.053)	9, 3.14, 0.012	0.380 (0.118)	0.504 (0.168)	15, 0.61, 0.549
Stem weight (g)	**0.187** (0.033)	0.085 (0.022)	9, 4.05, 0.003	0.175 (0.044)	0.204 (0.058)	15, 0.40, 0.695
Root weight (g)	0.090 (0.032)	0.037 (0.013)	9, 2.02, 0.074	0.208 (0.088)	0.206 (0.082)	15, 0.02, 0.988
Fine roots:leaf	0.084 (0.025)	0.065 (0.012)	9, 0.91, 0.384	0.319 (0.090)	0.181 (0.033)	10.1, 1.44, 0.181

Values are means ± 1 standard error in parentheses with significantly highest values in bold. Degrees of freedom (DF), *t* statistics, and associated two-tail probabilities (*P*) are shown. Satterthwaite's *t*-test (with fractional degrees of freedom) was used when variances were not homogenous. In methyl-bromide fumigated soil, n = 10 for both groups, but in non-fumigated, ambient soil, 9 plants were not fertilized versus 8 plants fertilized. DAT = days after transplant.

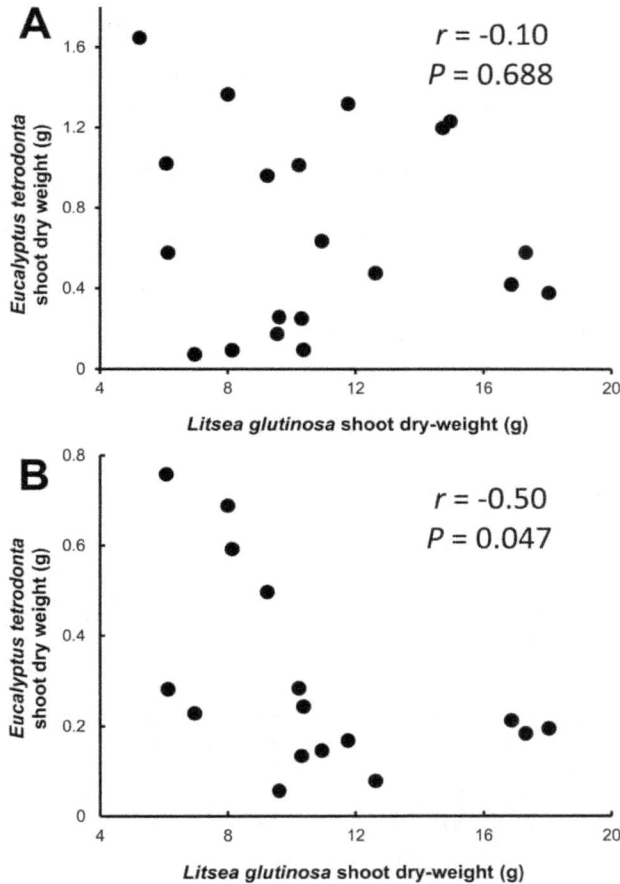

Figure 6. *Eucalyptus tetrodonta* versus *Litsea glutinosa* **nurse plant dry weights.** (A) with potential hyphal network interconnections repeatedly severed, and (B) without hypha severing. Shoot dry weights (g) of both *E. tetrodonta* and *L. glutinosa* at harvest are shown. With hypha severing, there is no significant association, but in the likely presence of common AM networks a significant negative association suggests belowground competitive interactions.

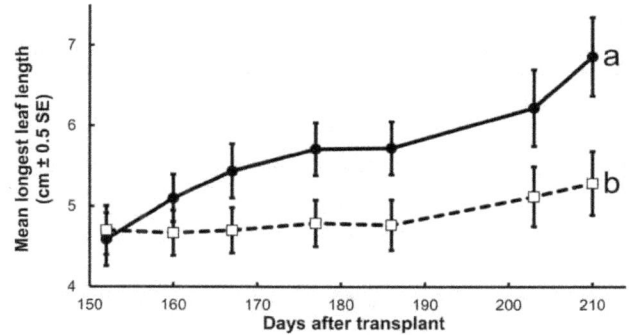

Figure 7. *Eucalyptus tetrodonta* **growth response to iron fertilization.** Mean longest leaf length (cm±0.5 standard error) is shown from 152 to 210 days after transplant (DAT) to microcosms containing ambient rain forest soil either fertilized with soluble iron (filled circles; solid lines) beginning 130 DAT or not fertilized (open squares; dashed lines). Different lower case letters denote significantly different interactions with time. Soluble iron fertilization improved *E. tetrodonta* longest leaf length increase.

significant effects either with or without iron fertilization (Table S2). Among all treatments and eleven elements, the only element that was correlated significantly (negatively) with leaf dry weight was iron (n = 27, $r = -0.62$, $P < 0.001$). There were no significant differences among treatments in the total foliar content of any element.

Discussion

Our results strongly support our hypotheses that AM fungi are detrimental to *E. tetrodonta* seedlings, and that the detriment involves iron limitation. The detriment likely is a consequence of two complementary phenomena: common mycorrhizal network interconnections mediate plant competition [33,41,42] to the disadvantage of *E. tetrodonta*, and active AM fungus mycelia may lead to iron sequestration whether or not those mycelia interconnect plants. Although our study did not trace hyphal interconnections between plants, negative associations between *C. pentandra* or *E. tetrodonta* and *L. glutinosa* only when hyphae were not severed are consistent with common mycorrhizal networks being present, as is the enhancement of *E. tetrodonta* mean root colonization in fumigated soil when networked [41].

The negative effects of potentially networked hyphae on the rates of increase of all *E. tetrodonta* morphological parameters are somewhat surprising because of the extent to which *E. tetrodonta* seedlings formed AM (seedlings in all treatments except fumigated, hyphae-severed, had 45% mean root length colonized, similar to the 45% mean colonization of *C. pentandra* across all treatments). AM improved *C. pentandra* growth in ambient versus fumigated soil, but in distinct contrast, the growth of *E. tetrodonta* seedlings with the lowest mean root colonization (29%) in hyphae-severed, fumigated soil exceeded growth in all other treatments. Consequently, little-colonized seedlings might be the most likely to establish successfully in the stressful milieu of rain forest gaps. Indeed, under our experiment's conditions that simulated non-burnt rain forest gaps, three networked *E. tetrodonta* seedlings in ambient soil died. Although AM previously have been reported to cause growth depressions of plants grown singly in pots, those negative effects usually have reflected AM failing to repay their cost to the host under conditions of low light, extremely low available phosphorus, or high phosphorus fertilization initiated after abundant mycorrhizas already had formed [43]. None of

differ in the proportion of chlorotic plants before fertilization began (Fisher exact two-tail $P = 0.637$), they differed significantly in the proportion of plants that recovered from chlorosis (Fisher exact two-tail $P = 0.008$). Iron fertilization, however, did not significantly elevate *E. tetrodonta* seedling dry weights (Table 2), nor did hyphae-severed plants differ from networked plants after fertilization (Table S1).

E. tetrodonta foliar mineral nutrient concentrations

After Bonferroni correction, only leaf tissue manganese concentration was affected significantly among the four treatments shown in Table 3. Both hyphae-severed and networked *E. tetrodonta* seedlings grown in fumigated soil (at a mean of 204 mg kg^{-1}) had 2.6 times the mean manganese concentration of those in ambient soil. Without Bonferroni correction, boron differed only for fumigated, hyphae-severed versus networked plants (plants in ambient soil are intermediate and cannot be distinguished statistically from either of those groups). Although not significantly different among these four groups ($P = 0.061$), iron had a conspicuously high mean concentration in fumigated soil for networked plants which also had the smallest mean leaf dry weight of all groups (Table 2). In ambient soil, severing hyphae had no

Table 3. Leaf element concentrations of *Eucalyptus tetrodonta* seedlings in response to common mycorrhizal network hypha severing in fumigated rain forest soil and to soluble iron fertilization in non-fumigated, ambient soil.

Element (units)	Fumigated soil (141 DAT)		Ambient soil (210 DAT)		$F_{3,23}$ (KW), P
	Hyphae-severed	Networked	No fertilization	Iron fertilization	
P (%)	0.14 (0.02)	0.26 (0.12)	0.14 (0.03)	0.16 (0.01)	(1.46), 0.692
S (%)	0.14 (0.01)	0.24 (0.08)	0.15 (0.02)	0.14 (0.01)	1.48, 0.246
K (%)	0.54 (0.04)	1.35 (0.62)	0.56 (0.08)	0.55 (0.05)	1.93, 0.153
Ca (%)	1.17 (0.15)	1.26 (0.35)	1.18 (0.14)	0.97 (0.04)	(1.62), 0.655
Mg (%)	0.45 (0.04)	0.48 (0.13)	0.50 (0.05)	0.40 (0.03)	0.34, 0.800
Na (mg kg^{-1})	0.2 (0.02)	0.3 (0.08)	0.4 (0.03)	0.3 (0.02)	(7.28), 0.064
Zn (mg kg^{-1})	57 (7.1)	60 (21.4)	52 (8.9)	43 (2.4)	0.40, 0.755
Cu (mg kg^{-1})	113 (8.6)	116 (32.8)	125 (24.7)	71 (11.1)	1.35, 0.283
Mn (mg kg^{-1})	**204** (17.4)	**204** (62.0)	88 (8.1)	69 (5.4)	5.81, 0.004
Fe (mg kg^{-1})	335 (68.9)	647 (162)	256 (65.0)	317 (94.6)	2.83, 0.061
B (mg kg^{-1})	29 (1.5)	45 (10.6)	39 (2.5)	31 (1.9)	(9.89), 0.020

Values are means ± 1 standard error in parentheses with significantly highest values after Bonferroni correction ($P \leq 0.0046$) in bold. All elements with homoscedastic variances were compared by one-way analyses of variance (F and P shown), but P, Ca, Na, and B, for which variances were heteroscedastic were compared by non-parametric Kruskal-Wallis analyses (KW = Kruskal-Wallis statistics shown in parentheses). In methyl-bromide fumigated soil, n = 9 for hyphae-severed and 6 for networked treatments; in non-fumigated, ambient soil, there were 6 samples in each group. DAT = days after transplant.

those effects is likely to explain the suppression of *E. tetrodonta* growth, because under the same conditions in which AM were detrimental to *E. tetrodonta*, AM benefitted *C. pentandra*. Even though mycorrhizas usually are considered to be archetypical mutualisms, our work underscores that mutualism is not constitutive but is context-dependent [44].

It is peculiar that *E. tetrodonta* seedlings sustain AM that might contribute to their deaths in rain forest soil. Natural selection may not favor rejection of AM because AM are beneficial in other contexts or if formed by a different suite of fungus species. For instance, many shrub-layer species in Northern Territory savannas form AM [14], so *E. tetrodonta* seedlings could be connected to common AM networks in savanna. Available iron is two-and-a-half times more abundant there than in rain forest soil, however, perhaps mitigating potential negative effects of AM. Alternatively, if *E. tetrodonta* seedlings quickly form ectomycorrhizas [45–47] amidst dominant ectomycorrhizal adults [14], they may avoid AM networks and diminish selection pressure to reject AM. Nevertheless, a price paid for lacking the capacity to reject AM might be inability to invade rain forest.

E. tetrodonta growth enhancement by hypha severing

Our treatments reduced but did not eliminate AM colonization of either host species. Nevertheless, fumigated, hyphae-severed compartments probably had the least AM inoculum of any treatment while, ambient, networked compartments probably had the most inoculum. Reduction of colonization of *C. pentandra* by fumigation even when hyphae were not severed implies that fumigation diminished viable AM fungus propagules to an extent not entirely compensated by hyphae from the nurse plant. Repeated severing did not significantly affect *C. pentandra* root colonization, however, probably because roots of these relatively large plants came sufficiently close to mesh root barriers to be colonized in the 2–3 d interval between hypha severing. Once established, colonization could spread within a root system to reach an asymptote [48]. In contrast, in fumigated soil, repeated hypha severing did retard root colonization of *E. tetrodonta* which was smaller than *C. pentandra*, thereby probably prolonging the

time needed for roots to closely approach mesh barriers. Not severing hyphae in fumigated soil compartments, however, elevated AM colonization of *E. tetrodonta* at 141 DAT to a level similar to that at 210 DAT, thereby supporting that seedlings and nurse plants indeed may have been connected by common mycorrhizal networks. No effect of hypha severing at 210 DAT, suggests that *E. tetrodonta* seedlings had attained asymptotic colonization.

In fumigated soil, repeated hypha severing doubled *E. tetrodonta* whole plant dry weight versus that of networked plants. Fumigation, however, neither significantly affected *E. tetrodonta* growth rates (assessed with morphological measurements), nor did fumigation interact significantly with hypha severing. Therefore, any fertilization effect of fumigation such as N and P release from killed microbes [8] or an increase in the ratio of ammonium to nitrate that facilitated iron reduction [49] little affected our results. Furthermore, we failed to detect inhibitory effects on *E. tetrodonta* of non-networked AM. That differs from the pot experiments reported by Stocker [50] and Bowman and Fensham [5] in which non-fumigated rain forest soil was inhibitory to singly-grown eucalypt seedlings. Both those studies' ambient soils, however, had not been maintained plant-free for six months prior to eucalypt planting as had ours. In our microcosms, repeated severing of potential network connections to nurse plants maximized *E. tetrodonta* growth rates only when combined with initial elimination of AM inoculum by fumigation, so it is possible that in ambient soil there may have been an inhibitory effect of non-networked AM that we could not detect statistically.

When hyphae were not severed we found significant negative associations between *C. pentandra* and *L. glutinosa*, and between *E. tetrodonta* and *L. glutinosa* which suggest that belowground competition was mediated across common mycorrhizal networks. The belowground competition was not sufficiently strong to produce a significant beneficial effect of hypha severing for *C. pentandra*, however, because *C. pentandra* were the largest plants and probably the strongest competitors overall. Aboveground, because of close spacing of the completely randomized microcosms, tall *C. pentandra* were as likely to shade adjacent *E. tetrodonta* seedlings as they were to shade their accompanying nurse plants, thereby probably distributing aboveground competition relatively evenly across our entire experiment. Belowground, however, hyphal interconnections likely influenced competition within individual microcosms. Even though root system overlap was prevented by mesh barriers, AM fungus hyphae could cross the mesh and might have redistributed mineral nutrients [20,30,51]. Similar to our results, other greenhouse experiments have found that plants interconnected by AM networks can compete strongly belowground [33–35,41,42].

Iron deficiency of *E. tetrodonta* in rain forest soil

The response of *E. tetrodonta* seedlings to iron fertilization of ambient soil unambiguously demonstrated that iron was a growth-limiting mineral nutrient. Fertilization not only stimulated leaf growth but also completely eliminated chlorosis. Iron limitation of *E. tetrodonta* in rain forest soil is consistent with only two-fifths as much iron being available in rain forest as in savanna soil. Moreover, iron fertilization has been reported to remedy chlorosis of several eucalypt species [36]. Although the highest mean foliar iron concentration was found for the smallest plants overall in our experiment, that does not contraindicate iron as a growth-limiting mineral nutrient because iron can remain in leaf veins and be physiologically ineffective [49]. Extreme chlorosis that greatly reduces leaf growth can result in exceptionally high iron concentrations [52–54], as we found.

E. tetrodonta leaf analyses do not suggest that any element other than iron was limiting. Although manganese concentration was highest for all *E. tetrodonta* grown in fumigated soil, iron-fertilized seedlings at 210 DAT attained the highest mean whole plant dry weight with only one-third the manganese concentration of plants in fumigated soil. Furthermore, no analyzed element's total foliar content differed significantly among treatments. Even though we did not analyze nitrogen, it is unlikely that nitrogen limited growth because iron fertilization stimulated growth without supplemental nitrogen.

AM have been reported to improve the iron nutrition of woody plants under some conditions [55–57], but improved iron uptake was unlikely in our experiment because the best *E. tetrodonta* growth was associated with the lowest mean AM colonization. Alternatively, under some conditions AM fungi might exacerbate iron deficiency by producing the glycoprotein glomalin [58] which can contain 8.8% iron by weight [59]. If glomalin sequesters iron, then for such an effect to have operated in our experiment, glomalin in fumigated, hyphae-severed compartments would have had to diminish during the six months before *E. tetrodonta* seedlings were planted. Such a rapid decline of immunoreactive, easily-extractable glomalin is supported by findings of Preger et al. [60].

Fire, mycorrhiza networks, and the rain forest–savanna boundary

If the hypothesized decline of glomalin contributed to *E. tetrodonta* seedling growth in fumigated, hyphae-severed rain forest soil, then intense fires must be capable of leading to similar diminution of glomalin in the field. Knorr et al. [61], however, found no effect of fire on glomalin in Ohio, U.S.A. oak forest soils. Nevertheless, their soils had been stored for 1–7 years at room temperature, and loss of immunoreactivity during storage might have obscured effects of fire. Alternatively, some studies report direct effects of fire on AM fungi, but others report none or only indirect effects through reduced numbers of live host plants [16].

If a severe fire burns into rain forest, microbe death and glomalin decline might elevate iron availability, and AM host death might help *E. tetrodonta* seedlings avoid inclusion in common

AM networks that exacerbate belowground competition with rapidly-growing rain forest species. Once ECM fungi invade to associate with establishing seedlings that have survived sufficiently long for the fungi to encounter them [62], ECM fungus siderophores [27,63] could maintain iron availability, a positive feedback.

ECM networks in savanna may enhance the resistance of savanna to invasion by AM rain forest plants. ECM networks probably have far greater potential for hydraulic redistribution [64] than common AM networks because ECM canopy trees have deep roots [65] and some ECM fungi produce rhizomorphs or mycelial strands. Indeed, the open savanna canopy might elevate ground-level temperature [66] and reduce humidity to levels with which rain forest plant species have difficulty. Nevertheless, in the absence of fire, rain forest species slowly may invade savanna, perhaps by sharing AM fungus associates with AM savanna species [14] and thereby coping with low mineral nutrient availability. If such associations extend common AM networks and elevate glomalin, then phosphorus availability might increase because of iron sequestration [67], further favoring invasion by rain forest species.

Fire–mycorrhiza–vegetation feedbacks likely provide resilience to rain forest and savanna alternative stable-state systems, although rain forest resilience is overcome by the ashbed effect. The mechanisms of that effect are unresolved, however, and in our experiments, most suggested ashbed mechanisms played no role. No ECM [12,13] formed, nor were pathogen effects [10] apparent (because iron fertilization resulted in seedlings in ambient soil recovering full health). We detected no fertilization effect of soil fumigation that killed microbes [8], but any release of iron–possibly because of glomalin degradation–likely would have been more important than elevation of nitrogen or phosphorus. The most important feature of our experiment for facilitating *E. tetrodonta* growth was disruption of common AM networks, not previously recognized as part of the ashbed effect, but a potential consequence of fire-caused mortality of AM hosts.

Our results sharply distinguish the possible roles of different mycorrhiza types in influencing plant community composition. ECM and AM should not be viewed simply as alternative plant adaptations that minimize niche overlap and foster coexistence of their hosts. We have shown that common AM networks can be actively antagonistic to an eventual ECM host.

Acknowledgments

DPJ and DMJSB are grateful to Harvard University for Bullard Fellowships in Forest Research that led to this collaboration. We thank Tania Wyss for helpful criticism of the manuscript.

Author Contributions

Conceived and designed the experiments: DPJ JS DMJSB. Performed the experiments: JS CA. Analyzed the data: DPJ CA. Wrote the paper: DPJ DMJSB.

References

1. Wilson BA, Brocklehurst PS, Clark MJ, Dickinson KJM (1990) Vegetation of the Northern Territory, Australia. Darwin: Conservation Commission of the Northern Territory, Technical Report No. 49.

2. Russell-Smith J (1991) Classification, species richness, and environmental relations of monsoon rain forest in northern Australia. Journal of Vegetation Science 2: 259–278.

3. Bowman DMJS (1992) Monsoon forests in North-western Australia. II. Forest-savanna transitions. Australina Journal of Botany 40: 89–102.

4. Bowman DMJS (2000) Australian Rainforests: Islands of Green in a Land of Fire. Cambridge: Cambridge University Press.

5. Bowman DMJS, Fensham RJ (1995) Growth of Eucalyptus tetrodonta seedlings on savanna and monsoon rainforest soils in the Australian monsoon tropics. Australian Forestry 58: 46–47.

6. Bowman DMJS, Murphy BP, Banfai DS (2010) Has global environmental change caused monsoon rainforests to expand in the Australian monsoon tropics? Landscape Ecology 25: 1247–1260.

7. Humphreys FR, Lambert MJ (1965) An examination of a forest site which has exhibited the ash-bed effect. Australian Journal of Soil Research 3: 81–94.

8. Chambers DP, Attiwill PM (1994) The ash-bed effect in Eucalyptus regnans forest: chemical, physical and microbiological changes in soil after heating or partial sterilisation. Australian Journal of Botany 42: 739–749.

9. Ashton DH, Kelliher KJ (1996) Effects of forest soil dessication on the growth of Eucalyptus regnans F. Muell. seedlings. Journal of Vegetation Science 7: 487–496.

10. Florence RG, Crocker RL (1962) Analysis of Blackbutt (Eucalyptus pilularis Sm.) seedling growth in a Blackbutt forest soil. Ecology 43: 670–679.

11. Iles TM, Ashton DH, Kelliher KJ, Keane PJ (2010) The effect of Cylindrocarpon destructans on the growth of Eucalyptus regnans seedlings in air-dried and undried forest soil. Australian Journal of Botany 58: 133–140.

12. Launonen TM, Ashton DH, Kelliher KJ, Keane PJ (2004) The growth and P acquisition of Eucalyptus regnans F. Muell. seedlings in air-dried and undried forest soil in relation to seedling age and ectomycorrhizal infection. Plant and Soil 267: 179–189.

13. Warcup JH (1991) The fungi forming mycorrhizas on eucalypt seedlings in regeneration coupes in Tasmania. Mycological Research 95: 329–332.

14. Reddell P, Milnes AR (1992) Mycorrhizas and other specialized nutrient-acquisition strategies: their occurrence in woodland plants from Kakadu and their role in rehabilitation of waste rock dumps at a local uranium mine. Australian Journal of Botany 40: 223–242.

15. Hopkins MS, Reddell P, Hewett RK, Graham AW (1996) Comparison of root and mycorrhizal characteristics in primary and secondary rainforest on a metamorphic soil in North Queensland, Australia. Journal of Tropical Ecology 12: 871–885.

16. McMullan-Fisher SJM, May TW, Robinson RM, Bell TL, Lebel T, et al. (2011) Fungi and fire in Australian ecosystems: a review of current knowledge, management implications and future directions. Australian Journal of Botany 59: 70–90.

17. Brundrett MC, Ashwath N, Jasper DA (1996) Mycorrhizas in the Kakadu region of tropical Australia. II. Propagules of mycorrhizal fungi in disturbed habitats. Plant and Soil 184: 173–184.

18. Leake JR, Johnson D, Donnelly DP, Muckle GE, Boddy L, et al. (2004) Networks of power and influence: the role of mycorrhizal mycelium in controlling plant communities and agroecosystem functioning. Canadian Journal of Botany 82: 1016–1045.

19. Newman EI (1988) Mycorrhizal links between plants: their functioning and ecological significance. Advances in Ecological Research 18: 243–270.

20. Simard SW, Durall DM (2004) Mycorrhizal networks: a review of their extent, function, and importance. Botany 82: 1140–1165.

21. Simard SW, Perry DA, Jones MD, Myrold DD, Durall DM, et al. (1997) Net transfer of carbon between ectomycorrhizal tree species in the field. Nature 388: 579–582.

22. Francis R, Read DJ (1984) Direct transfer of carbon between plants connected by vesicular arbuscular mycorrhizal mycelium. Nature 307: 53–56.

23. Arnebrant K, Ek H, Finlay RD, Soderstrom B (1993) Nitrogen translocation between Alnus glutinosa (L.) Gaertn. seedlings inoculated with Frankia sp. and Pinus contorta Dougl. ex Loud. seedlings connected by a common ectomycorrhizal mycelium. New Phytologist 124: 231–242.

24. van Kessel C, Singleton PW, Hoben HJ (1985) Enhanced N-transfer from soybean to maize by vesicular arbuscular mycorrhizal (VAM) fungi. Plant Physiology 79: 562–563.

25. Chiariello N, Hickman JC, Mooney HA (1982) Endomycorrhizal role for interspecific transfer of phosphorus in a community of annual plants. Science 217: 941–943.

26. Woods FW, Brock K (1964) Interspecific transfer of Ca-45 and P-32 by root systems. Ecology 45: 886–889.

27. Courty P-E, Buée M, Diedhiou AG, Frey-Klett P, Le Tacon F, et al. (2010) The role of ectomycorrhizal communities in forest ecosystem processes: new perspectives and emerging concepts. Soil Biology and Biochemistry 42: 679–698.

28. Teste FP, Simard SW, Durall DM, Guy RD, Berch SM (2010) Net carbon transfer between Pseudotsuga menziesii var. glauca seedlings in the field is influenced by soil disturbance. Journal of Ecology 98: 429–439.

29. Booth MG, Hoeksema JD (2010) Mycorrhizal networks counteract competitive effects of canopy trees on seedling survival. Ecology 91: 2294–2302.

30. Bever JD, Dickie IA, Facelli E, Facelli JM, Klironomos J, et al. (2010) Rooting theories of plant community ecology in microbial interactions. TRENDS in Ecology and Evolution 25: 468–478.

31. Lekberg Y, Hammer EC, Olsson PA (2010) Plants as resource islands and storage units - adopting the mycocentric view of arbuscular mycorrhizal networks. FEMS Microbial Ecology 74: 336–345.

32. Adams F, Reddell P, Webb MJ, Shipton WA (2006) Arbuscular mycorrhizas and ectomycorrhizas on Eucalyptus grandis (Myrtaceae) trees and seedlings in native forests of tropical north-eastern Australia. Australian Journal of Botany 54: 271–281.

33. Kytöviita MM, Vestberg M, Tuomi J (2003) A test of mutual aid in common mycorrhizal networks: established vegetation negates benefit in seedlings. Ecology 84: 898–906.

34. Wilson GWT, Hartnett DC, Rice CW (2006) Mycorrhizal-mediated phosphorus transfer between tallgrass prairie plants Sorghastrum nutans and Artemisia ludoviciana. Functional Ecology 20: 427–435.

35. Moora M, Zobel M (1998) Can arbuscular mycorrhiza change the effect of root competition between conspecific plants of different ages? Canadian Journal of Botany 76: 613–619.

36. Parsons RF, Uren NC (2007) The relationship between lime chlorosis, trace elements and Mundulla Yellows. Australasian Plant Pathology 36: 415–418.

37. Allen EB, Allen MF, Egerton-Warburton L, Corkidi L, Gómez-Pompa A (2003) Impacts of early- and late-seral mycorrhizae during restoration in seasonal tropical forest, Mexico. Ecological Applications 13: 1701–1717.

38. Allen EB, Cannon JP, Allen MF (1993) Controls for rhizosphere microorganisms to study effects of VA mycorrhizae on Artemisia tridentata. Mycorrhiza 2: 147–152.

39. Koide RT, Li M (1989) Appropriate controls for vesicular-arbuscular mycorrhiza research. New Phytologist 111: 35–44.

40. McGonigle TP, Miller MH, Evans DG, Fairchild GL, Swan JA (1990) A new method which gives an objective measure of colonization of roots by vesicular-arbuscular mycorrhizal fungi. New Phytologist 115: 495–501.

41. Eissenstat DM, Newman EI (1990) Seedling establishment near large plants: Effects of vesicular-arbuscular mycorrhizas on the intensity of plant competition. Functional Ecology 4: 95–99.

42. Janouskova M, Rydlova J, Pueschel D, Szakova J, Vosatka M (2011) Extraradical mycelium of arbuscular mycorrhizal fungi radiating from large plants depresses the growth of nearby seedlings in a nutrient deficient substrate. Mycorrhiza 21: 641–650.

43. Janos DP (2007) Plant responsiveness to mycorrhizas differs from dependence upon mycorrhizas. Mycorrhiza 17: 75–91.

44. Hoeksema JD, Chaudhary VB, Gehring CA, Johnson NC, Karst J, et al. (2010) A meta-analysis of context-dependency in plant response to inoculation with mycorrhizal fungi. Ecology Letters 13: 394–407.

45. McGuire KL (2007) Common ectomycorrhizal networks may maintain monodominance in a tropical rain forest. Ecology 88: 567–574.

46. Nuñez MA, Horton TR, Simberloff D (2009) Lack of belowground mutualisms hinders pinaceae invasions. Ecology 90: 2352–2359.

47. Collier FA, Bidartondo MI (2009) Waiting for fungi: the ectomycorrhizal invasion of lowland heathlands. Journal of Ecology 97: 950–963.

48. McGonigle TP (2001) On the use of non-linear regression with the logistic equation for changes with time of percentage root length colonized by arbuscular mycorrhizal fungi. Mycorrhiza 10: 249–254.

49. Mengel K (1994) Iron availability in plant tissues - iron chlorosis on calcareous soils. Plant and Soil 165: 275–283.

50. Stocker GC (1969) Fertility differences between the surface soils of monsoon and eucalypt forests in the Northern Territory. Australian Forest Research 4: 31–38.

51. Selosse M-A, Richard F, He X, Simard SW (2006) Mycorrhizal networks: des liaisons dangereuses? TRENDS in Ecology and Evolution 21: 621–628.

52. Bavaresco L, Giachino E, Colla R (1999) Iron chlorosis paradox in grapevine. Journal of Plant Nutrition 22: 1589–1597.

53. Morales F, Grasa R, Abadía A, Abadía J (1998) Iron chlorosis paradox in fruit trees. Journal of Plant Nutrition 21: 815–825.

54. Römheld V (2000) The chlorosis paradox: Fe inactivation as a secondary event in chlorotic leaves of grapevine. Journal of Plant Nutrition 23: 1629–1643.

55. Janos DP, Schroeder MS, Schaffer B, Crane JH (2001) Inoculation with arbuscular mycorrhizal fungi enhances growth of Litchi chinensis Sonn. trees after propagation by air-layering. Plant and Soil 233: 85–94.

56. Treeby M (1992) The role of mycorrhizal fungi and non-mycorrhizal micro-organisms in iron nutrition of citrus. Soil Biology & Biochemistry 24: 857–864.

57. Wang M, Christie P, Xiao Z, Qin C, Wang P, et al. (2008) Arbuscular mycorrhizal enhancement of iron concentration by *Poncirus trifoliata* L. Raf and *Citrus reticulata* Blanco grown on sand medium under different pH. Biology and Fertility of Soils 45: 65–72.

58. Janos DP, Garamszegi S, Beltran B (2008) Glomalin extraction and measurement. Soil Biology and Biochemistry 40: 728–739.

59. Wright SF, Upadhyaya A (1998) A survey of soils for aggregate stability and glomalin, a glycoprotein produced by hyphae of arbuscular mycorrhizal fungi. Plant and Soil 198: 97–107.

60. Preger AC, Rillig MC, Johns AR, Du Preez CC, Lobe I, et al. (2007) Losses of glomalin-related soil protein under prolonged arable cropping: a chronose-quence study in sandy soils of the South African Highveld. Soil Biology & Biochemistry 39: 445–453.

61. Knorr MA, Boerner REJ, Rillig MC (2003) Glomalin content of forest soils in relation to fire frequency and landscape position. Mycorrhiza 13: 205–210.

62. Janos DP (1980) Mycorrhizae influence tropical succession. Biotropica 12 (supplement): 56–64.

63. Szaniszlo PJ, Powell PE, Reid CPP, Cline GR (1981) Production of hydroxamate siderophore iron chelators by ectomycorrhizal fungi. Mycologia 73: 1158–1174.

64. Egerton-Warburton LM, Querejeta JI, Allen MF (2007) Common mycorrhizal networks provide a potential pathway for the transfer of hydraulically lifted water between plants. Journal of Experimental Botany 58: 1473–1483.

65. Janos DP, Scott J, Bowman DMJS (2008) Temporal and spatial variation of fine roots in a northern Australian *Eucalyptus tetrodonta* savanna. Journal of Tropical Ecology 24: 177–188.

66. Turton SM, Duff GA (1992) Light environments and floristic composition across an open forest-rainforest boundary in northeastern Queensland. Australian Journal of Ecology 17: 415–423.

67. Cardoso IM, Boddington CL, Janssen BH, Oenema O, Kuyper TW (2006) Differential access to phosphorus pools of an Oxisol by mycorrhizal and nonmycorrhizal maize. Communications in Soil Science and Plant Analysis 37: 1537–1551.

3

Effect of Optimal Daily Fertigation on Migration of Water and Salt in Soil, Root Growth and Fruit Yield of Cucumber (*Cucumis sativus* L.) in Solar-Greenhouse

Xinshu Liang, Yinan Gao, Xiaoying Zhang, Yongqiang Tian, Zhenxian Zhang, Lihong Gao*

Beijing Key Laboratory of Growth and Developmental Regulation for Protected Vegetable Crops, Department of Vegetable Science, China Agricultural University, Beijing, P.R. China

Abstract

Inappropriate and excessive irrigation and fertilization have led to the predominant decline of crop yields, and water and fertilizer use efficiency in intensive vegetable production systems in China. For many vegetables, fertigation can be applied daily according to the actual water and nutrient requirement of crops. A greenhouse study was therefore conducted to investigate the effect of daily fertigation on migration of water and salt in soil, and root growth and fruit yield of cucumber. The treatments included conventional interval fertigation, optimal interval fertigation and optimal daily fertigation. Generally, although soil under the treatment optimal interval fertigation received much lower fertilizers than soil under conventional interval fertigation, the treatment optimal interval fertigation did not statistically decrease the economic yield and fruit nutrition quality of cucumber when compare to conventional interval fertigation. In addition, the treatment optimal interval fertigation effectively avoided inorganic nitrogen accumulation in soil and significantly ($P<0.05$) increased the partial factor productivity of applied nitrogen by 88% and 209% in the early-spring and autumn-winter seasons, respectively, when compared to conventional interval fertigation. Although soils under the treatments optimal interval fertigation and optimal daily fertigation received the same amount of fertilizers, the treatment optimal daily fertigation maintained the relatively stable water, electrical conductivity and mineral nitrogen levels in surface soils, promoted fine root (<1.5 mm diameter) growth of cucumber, and eventually increased cucumber economic yield by 6.2% and 8.3% and partial factor productivity of applied nitrogen by 55% and 75% in the early-spring and autumn-winter seasons, respectively, when compared to the treatment optimal interval fertigation. These results suggested that optimal daily fertigation is a beneficial practice for improving crop yield and the water and fertilizers use efficiency in solar greenhouse.

Editor: Jin-Song Zhang, Institute of Genetics and Developmental Biology, Chinese Academy of Sciences, China

Funding: This work was supported by the earmarked fund for Modern Agro-industry Technology Research System in China (CARS-25) and the Special Fund for Research in the Public Interest provided by the Chinese Ministry of Water Resources (201001061). The funders had no role in study design, data collection and analysis, decision to publish, or preparation of the manuscript.

Competing Interests: The authors have declared that no competing interests exist.

* E-mail: gaolh@cau.edu.cn

Introduction

Protected vegetable production systems have been rapidly developed in recent decades in China. Solar-greenhouse, an unheated plastic greenhouse using solar light energy to make sure that vegetables grow normally, plays more and more important roles in China's vegetable production and supplication during the winter [1]. However, excessive irrigation and fertilization are commonly practiced in the solar-greenhouse production systems. For instance, some investigations have revealed that irrigation water rate is 1000 mm each year and fertilizer N apparent recovery efficiency can be less than 10% using conventional management practices [2,3]. Consequently, redundant water and fertilizers affect environmental protection by nutrient accumulation and soil salinization [4–6].

It has become the focus of agricultural field that the efficient water and fertilizer management methods are applied in vegetable production in greenhouse. For example, Cabello et al. [7] demonstrated that under moderate deficit irrigation (90% evapotranspiration) condition, reducing the inputs of nitrogen fertilizer did not reduce yield of melon, but increased water and nitrogen fertilizer use efficiencies. Similarly, Mahajan and Singh [8] found that fertigation through reducing rates of water and nitrogen fertilizer promoted fruit yield and quality, and water and fertilizer use efficiencies in greenhouse tomato cultivation system. In addition, both subsurface drip irrigation and alternate furrow irrigation can significantly improve the root growth, and thus enhance the yield and water and fertilizer use efficiencies [9–12].

However, conventional interval fertigation is still common in intensive vegetable production systems in China. In general, the period between successive irrigation and fertilization events is 7 to 10 days, or even longer. Consequently, the nutrient concentration in the root-zone soil may be in excess of plant requirement for growth on the first day after fertigation, and then decreases gradually to reach deficit levels before the day preceding the next fertigation event, and eventually inhibit crop growth and development [13]. Therefore, there is a need for ensuring water and nutrient in root-zone soil stable to reduce adverse effects caused by their fluctuation on crop growth under intensive vegetable management practices.

Drip fertigation, a technique used to manage soil water and nutrition supplies according to the actual water and nutrient requirement of the plants, can be applied to vegetable production to maintain stable water and nutrition contents in the root zone of crops. It applies frequent and small amounts of soluble fertilizers along with water by exerting the soil buffer characteristic and by reducing the time interval between successive irrigations [14,15]. This technique is still rarely implemented in soil cultivation systems in developing countries, despite of the fact that it has been applied for many years in developed countries.

The efficiency of a fertigation treatment can be evaluated with a combination of the availability and distribution of soil water and nutrients, the growth and development of crop root, and the formation of crop yield. These parameters are relatively simple, rapid, easy to obtain and cost effective. Among these parameters, particular attention should be paid to plant root since the morphology and spatial configuration of root can significantly affect soil water and nutrient transformation, mobilization and use efficiency by plant and crop yield [16–18]. However, due to the complexity of root growth environment and the limitations of root research methods, little information is available regarding the vegetable root growth under daily fertigation based on crop requirement under greenhouse soil cultivation conditions.

Cucumber is one of the major greenhouse vegetables in China. However, inappropriate irrigation and fertilizing practices have caused soil nutrient imbalance that negatively affects cucumber growth and the reduction of soil water and fertilizer use efficiencies. It is important for farmers to manage soil water and nutrition supplies according to the actual water and nutrient requirement of the plants during greenhouse cucumber production seasons. Thus, the objectives of this study were to investigate how the migration of water and salt in soil, and root growth and fruit yield of cucumber are affected by the daily fertigation based upon the actual water and nutrient requirement of the plants.

Materials and Methods

Ethics statement

No specific permissions were required for the described field study. The experiment was carried out in 2011 in a typical solar greenhouse of Fangshan District Agricultural Science Research Institute, Beijing, China (39.7°N; 116.1°E), which is an experimental station of China Agricultural University. The field study did not involve endangered or protected species.

Site description and experimental design

The greenhouse was covered with polyethylene film (ground area 56 m×7 m) without supplementary lighting and heating. Daily average soil temperatures and air temperature in solar-greenhouse were shown in Fig. 1. Both the two temperatures in April gradually increased until each reached the highest value in July and August, and then gradually decreased. Furthermore, Daily average soil temperatures at 10 cm depth from conventional interval fertigation (CK), optimal interval fertigation (OIF) and optimal daily fertigation (ODF) were not significantly different. Field experiments were conducted on a sandy loam soil (36% sand, 48% silt and 16% clay). The topsoil (0–30 cm layer) had a pH (1:2.5 soil/water) value of 7.37, an electrical conductivity (EC) (1:5 soil/water) value of 1.17 dS m^{-1}, a field capacity of 20.8% and a bulk density of 1.36 g cm^{-3}, and contained 15.8 g kg^{-1} organic matter, 0.96 g kg^{-1} total nitrogen (N), 195 mg kg^{-1} inorganic N, 193 mg kg^{-1} available phosphorus (P) and 275 mg kg^{-1} available potassium (K).

The experimental period comprised two growth cycles including the early-spring (ES) (February 1-July 29) and autumn-winter (AW) (August 1-December 25) seasons. The varity of cucumber was Zhongnong No. 26. This varity was a high-quality and largely fruit-bearing cucumber hybrid. The average fruit length and width of this varity were 30 and 3.3 cm, respectively. Due to its high resistance to several diseases (including powdery mildew, downy mildew and gray mold) and tolerance to both low temperature and weak light, this varity was very suitable for cultivation in solar-greenhouse. Cucumber seedlings with two leaves were transplanted by hand, before which, soils were incorporated with organic fertilizer (basal fertilizer) at designed rates (Table 1), and were ploughed and harrowed to a depth of 30 cm. Plant pruning was performed as follows: all lateral branches were removed by hand, however, the axial shoot of cucumber was remained and it climbed upward along a vertical rope. Planting density of cucumber was 5 plants m^{-2}.

The experiment consisted of three treatments:

(1) Conventional interval fertigation (CK): Conventional organic manure (basal fertilizer) and chemical fertilizer were applied at rates based on average fertilization level used by greenhouse cucumber growers in the suburb of Beijing (Table 1). A detailed description of chemical fertilizer topdressing/irrigation rates under the CK was given in Fig. 2.

(2) Optimal interval fertigation (OIF): The organic manure (basal fertilizer) application rate was half the rate conventionally applied per hectare (Table 1). Based on the N requirement for cucumber growth [19–21] and N fertilizer recommendation, the total N rates applied by topdressing were 427.5 and 151.5 kg N ha^{-1} in the ES and AW seasons (Table 1), respectively. These total mineral N (N$_{min}$) application rates were calculated using the method of soil N balance, where expected yield of solar greenhouse cucumber were 120 and 60 t ha^{-1} in the ES and AW seasons, respectively. The equation [22] was as follows:

$$N_{recommend} = N_{crop} + N_{safty} + N_{loss}$$
$$- N_{initial} - N_{manure} - N_{mineralization} \qquad (1)$$

Where N$_{recommend}$ is Recommended fertilizer N; N$_{crop}$ is crop N uptake; N$_{safty}$ is soil N$_{min}$ safety margin; N$_{loss}$ is N loss; N$_{initial}$ is soil N$_{min}$ in the root zone before transplanting; N$_{manure}$ is N$_{min}$ from the mineralization of organic fertilizer; N$_{mineralization}$ is N$_{min}$ from the mineralization of organic nitrogen in soil.

Chemical fertilizer topdressing/irrigation frequencies and irrigation rates were the same as the CK, except fertilizer topdressing rates. A detailed description of chemical fertilizer topdressing/irrigation rates under the treatment OIF was given in Fig. 2.

(3) Optimal daily fertigation (ODF): The total amounts of organic manure (basal fertilizer) and chemical fertilizer were the same as treatment OIF (Table 1), however, chemical fertilizer was applied automatically daily according to the actual water and nutrient requirement of the plants. A detailed description of chemical fertilizer topdressing/irrigation rates under the treatment ODF was given in Fig. 2.

The total irrigation amount was the same for three treatments in whole growth period of cucumber. Fertigation was used under transparent thin film for all treatment. Irrigations were applied 16

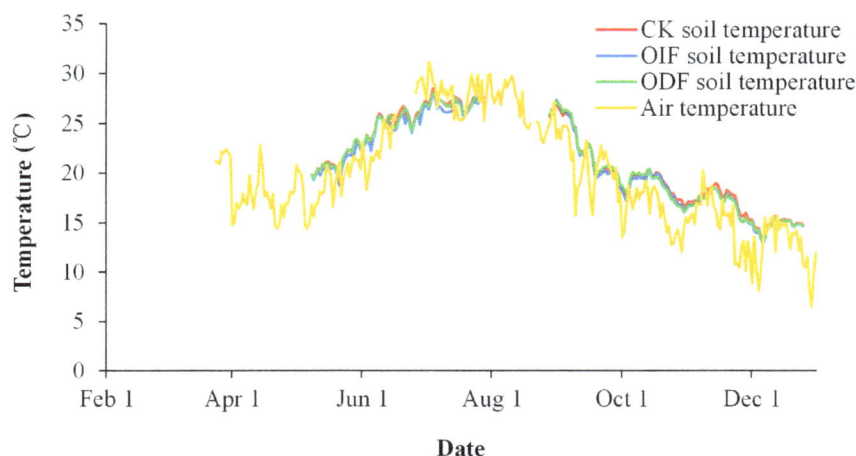

Figure 1. Daily average soil temperatures at 10 cm depth from conventional interval fertigation (CK), optimal interval fertigation (OIF) and optimal daily fertigation (ODF) and daily average air temperature at 150 cm height in cucumber greenhouse cropping system in 2011 at Fangshan, Beijing suburbs. These data were determined by RTH-1010 TPE rensin-shield sensor and RT-12 Thermo Recorder made in Japan. Time interval for data recording was set to 10 minutes.

and 7 times in the ES and AW seasons, respectively. The irrigation rate each time for CK and OIF teatments was 22.5–30 mm according to requirement rule of water for cucumber [23–25].

The chemical fertilizer used in this study was a special water-soluble fertilizer for cucumber called Shengdanshu (a local fertilizer). This fertilizer had three $N:P_2O_5:K_2O$ formulations, i.e., 20:20:20, 19:8:27 and 18:6:34, which were used at the early harvest stage (April 21 – May 12 during ES season, September 25 – October 11 during AW season), the middle harvest stage (May 13 – July 5 during ES season, October 12 – November 30 during AW season) and the late harvest stage (July 6 – July 29 during ES season, December 1 – December 25 during AW season), respectively. Fertilizers were applied 12 and 6 times in the ES and AW seasons, respectively. Except different fertigation methods related to different treatments, the same local management practices were applied in all treatments. The experiment was a randomized block design with four replications and the size of each replicate plot was 3.9 m×4.8 m. Each replicate plot had three cultivation furrows and was separated from the adjacent plots by plastic films buried at a depth of 50 cm.

Soil water content, EC and mineral N content

To evaluate the migration of soil water and salt under different fertigation treatments, soil samples from five cores per subplot were collected five times within a single fertigation cycle during the middle fruit harvest period when daily fruit production was very high in each cropping season. For the ES season, soils were sampled on May 27, May 28, May 31, June 2 and June 4. For the AW season, the corresponding sampling times were October 26, October 27, November 1, November 6 and November 12. Soil samples were taken at 0–15, 15–30 and 30–45 cm depth. Soil samples of each plot at each depth were mixed thoroughly and passed through a 2-mm sieve. Sub-samples of 20 g fresh soil were dried for 12 h at 105°C, and then soil water content was determined at the ratio of water and dry soil weight. Sub-samples of about 300 g fresh soil used to measure EC were air-dried passed through a 1-mm sieve. Soil EC was analyzed from a 1:5 (w/v) soil (air-dried) to water ratio using an EC meter and combination glass electrodes (FE30, METTLER TOLEDO, Shanghai, China). Sub-samples of 12 g fresh soil used to estimate mineral N were submerged into 100 ml 0.01 M $CaCl_2$ solution and shaken for 1 hour to extract inorganic N. The extracts were filtered and

Table 1. Amounts of fertilizers used in the treatments CK (conventional interval fertigation), OIF (optimal interval fertigation) and ODF (optimal daily fertigation) in the ES (early-spring) and AW (autumn-winter) seasons.

Cropping season	Treatment	Organic manure[a] (t ha^{-1})	Chemical fertilizer (kg ha^{-1})			Irrigation water (mm)
			N	P_2O_5	K_2O	
ES	CK	45.0	819.0	690.4	1006.2	349.2
	OIF	22.5	427.5	361.3	524.1	349.2
	ODF	22.5	427.5	361.3	524.1	349.2
AW	CK	60.0	465.0	312.0	593.4	147.2
	OIF	30.0	151.5	123.1	183.1	147.2
	ODF	30.0	151.5	123.1	183.1	147.2

[a]For the ES season, total N, total P and total K contents of the organic manure were 2.09% (N), 2.06% (P_2O_5) and 1.34% (K_2O), respectively. For the AW season, total N, total P and total K contents of the organic manure were 1.30% (N), 0.5% (P_2O_5) and 1.06% (K_2O), respectively.

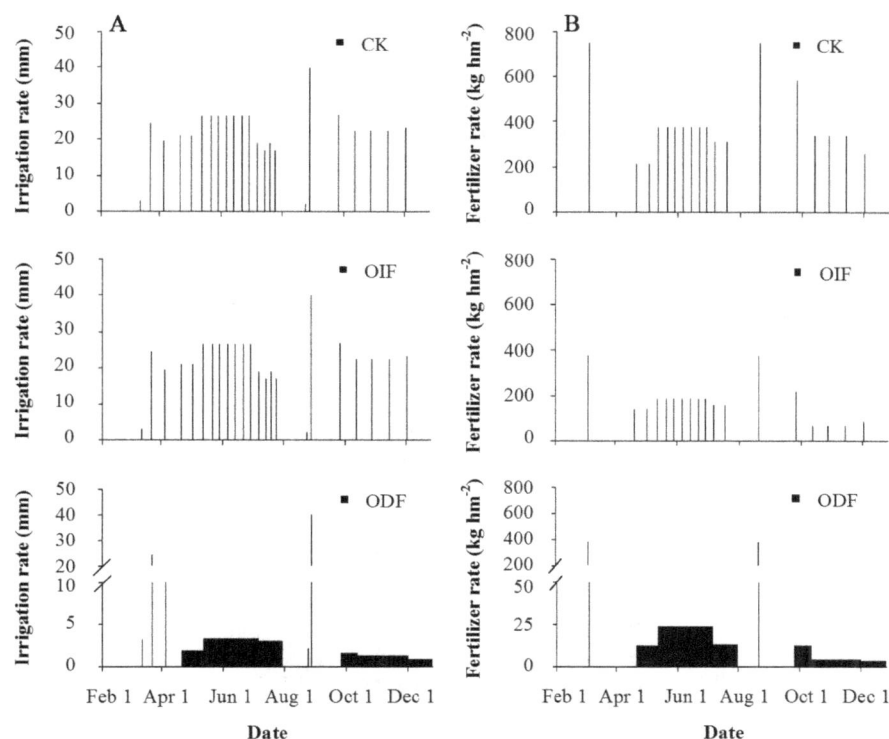

Figure 2. Irrigation scheduling (A) and fertilization scheduling (B) under three different fertigation ways in a solar greenhouse cucumber cultivation system. All fertilizer varity were compound fertilizer. The two $N:P_2O_5:K_2O$ formulations of basal fertilizer were respectively (18:46:0) and (15:15:15) during ES and AW seasons. The three $N:P_2O_5:K_2O$ formulations of topdressing fertilizer were respectively (20:20:20), (19:8:27) and (18:6:34) for early fruit stage, middle fruit stage, late fruit stage during ES and AW seasons.

analyzed using an continuous flowing analyzer (TRAACS2000, USA) to determine NO_3^--N and NH_4^+-N contents [26].

Cucumber root morphologic characters

Once the final harvest was completed, one typical cucumber root in each plot was collected with an Eijelkamp root auger (length = 0.15 m, diameter = 0.08 m) from the 0–15, 15–30 and 30–45 cm soil layers. For each layer, 9 holes were drilled around the cucumber main root in a shape of 3×3 cross square. The roots in the soil were carefully selected and washed to acquire the roots from different soil layers. Root morphology was analyzed by using fresh roots and a root scanner system (EPSON EXPRESSION 4990, Japan). Data were then analyzed with the WinRHIZO root analysis software (LC4800-II LA2400; Saint foy, Canada) to determine the root characteristics, including root length, root surface, root volume and average diameter. The root characteristics could be divided into four classification according to the root diameters aggregated into classes of 0.0–0.5, 0.5–1.0, 1.0–1.5 and ≥1.5 mm. The scanned roots were oven-dried at 65°C until weight constancy and weighed.

Cucumber fruit yield, irrigation water use efficiency and partial factor productivity of applied nitrogen

Economic yield was measured for whole cucumber growth cycles in each plot and translated into economic yield weight per hectare. The ratio of yield to water supply was referred to as irrigation water use efficiency (*IWUE*, kg mm^{-1}):

$$IWUE = Y/W \qquad (2)$$

where Y and W represent the economic yield (kg ha^{-1}) and the amount of water (mm) applied to the cucumber during the growing cycle [7], respectively. The ratio of yield to N supply is referred to as partial factor productivity of applied N (*PFP$_N$*, kg kg^{-1}):

$$PFP_N = Y/F \qquad (3)$$

where F is the amount of fertilizer N (kg) applied to the cucumber during the growing cycle [27].

Vitamin C, soluble sugar and nitrate contents in cucumber fruit

To estimate cucumber fruit quality under different fertigation treatments, three quality parameters which were concerned by local residents were made. Fresh fruit samples of cucumber in each plot were collected when daily fruit production was very high, and their appearance should be similar and marketable. Fresh fruits were washed, chopped and mixed in the lab to assess different quality indexes. Contents of Vitamin C and soluble sugar were determined by the methods of 2, 6-dichloro-indophenol titration [28] and anthrone ethyl acetate colorimetic [29], respectively. Content of nitrate was measured by the method of sulfuric acid-acid [30].

Statistical analysis

SPSS 17.0 was used to analyse data. Treatment means were separated using the least significant difference (LSD) test at $P<0.05$. Principal component analysis (PCA) was done to comprehensively determine the whole effect of different fertigation methods on several root morphologic characters (Root dry weight, length, surface area, average diameter and volume).

Results

Soil water content, EC and mineral N content

Soil water content under the treatments OIF and CK increased rapidly and then decreased gradually within an irrigation cycle (Fig. 3). However, it maintained a relatively stable level under the treatment ODF. No significant difference was found in soil water content between the treatments OIF and CK in all tested soil layers in both ES and AW seasons. The treatment ODF decreased the amplitude (i.e. the difference between maximum and minimum values) of soil water content in all tested soil layers during ES season and that in the 0–15 cm soil layer during AW season in an irrigation cycle, when compared to the treatments OIF and CK (Table 2). In addition, the relative soil water content under the treatment ODF was higher in the 0–15 cm soil layer than in the 15–30 and 30–45 cm soil layers, suggesting that the soil water was concentrated by the treatment ODF in the 0–15 cm soil layer.

The EC values in the 0–15 cm soil layer were higher and more variable than those in the 15–30 and 30–45 cm soil layers, suggesting the salts were concentrated in the 0–15 cm soil layer (Fig. 3). The CK showed the highest and most variable EC values

in the 0–15 cm soil layer, resulting in the highest amplitude of soil EC value (Table 2). Generally, in the ES season, the treatments OIF and ODF did not show significant differences in EC values in the 0–15 cm soil layer. However, in the AW season, EC values under the treatment OIF were significantly higher and more variable than those under the treatment ODF in most sampling times. The soil EC values were not significantly different ($P > 0.05$) between any two treatments in the 15–30 and 30–45 soil layers in both ES and AW seasons.

Similar to soil EC values, generally, soil mineral N (N_{min}) contents in the 0–15 cm soil layer were higher and more variable than those in the 15–30 and 30–45 cm soil layers (Fig. 3). In addition, the CK had the higher and more variable soil N_{min} contents in the 0–15 cm soil layer. However, the amplitude of soil N_{min} content was significantly higher under the CK than under the treatments OIF and ODF in the 15–30 and 30–45 soil layers in ES and AW seasons, respectively (Table 2).

Principal component analysis of root growth properties

Root dry weight, length, surface area, average diameter and volume were used to evaluate the plant root growth. Since there were statistical links among each other of these root properties (all $P < 0.05$), we used the principal component analysis (PCA) to transform these correlated root variables (Table 3) into two principal components (i.e. PC_1 and PC_2) to obtain two princioal expressions as follow:

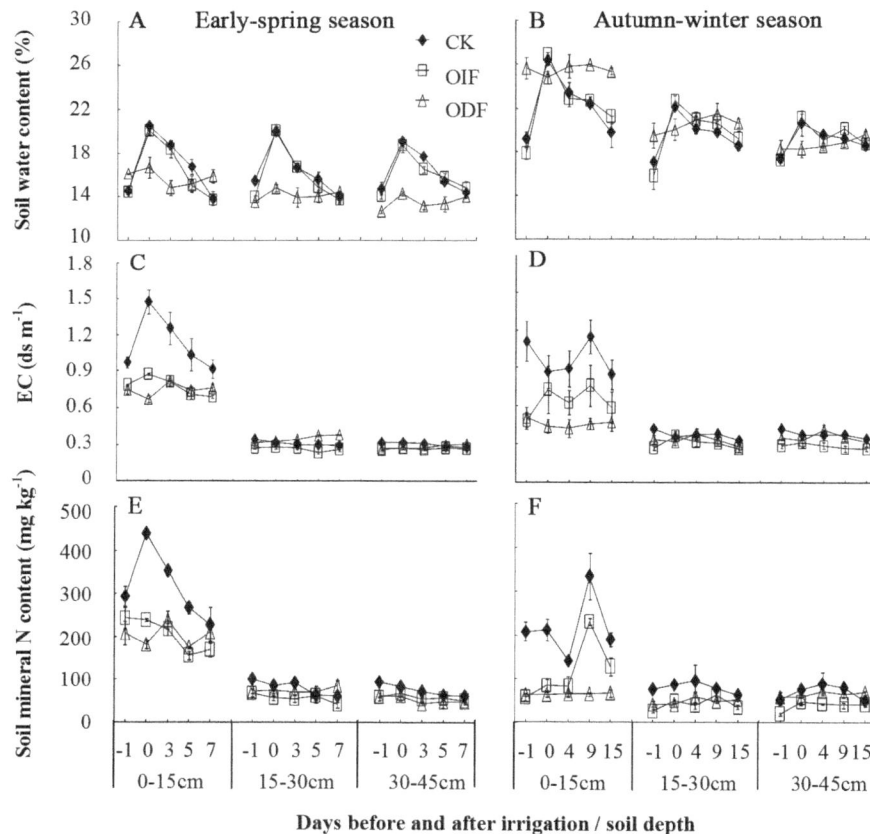

Figure 3. Effects of conventional interval fertigation (CK), optimal interval fertigation (OIF) and optimal daily fertigation (ODF) on changes of soil water content (A and B), EC value (C and D) and mineral N content (E and F) in the early-spring (ES) and autumn-winter (AW) seasons. The numbers on the abscissa represent the days before (negative value), during (zero) and after (positive value) irrigation. Bars represent standard errors.

Table 2. Effects of conventional interval fertigation (CK), optimal interval fertigation (OIF) and optimal daily fertigation (ODF) on the amplitudes (the difference between maximum and minimum values) of soil water content, EC value and mineral N content in the early-spring (ES) and autumn-winter (AW) seasons.

Cropping season	Treatment	Soil water (%)			Soil EC (mS cm^{-1})			Soil mineral N (mg kg^{-1})		
		0–15 cm[a]	15–30 cm	30–45 cm	0–15 cm	15–30 cm	30–45 cm	0–15 cm	15–30 cm	30–45 cm
ES	CK	7.0 a[b]	6.1 a	5.0 a	0.717 a	0.066 a	0.071 a	230.7 a	59 a	43.9 a
	OIF	6.5 a	6.6 a	5.0 a	0.204 b	0.062 a	0.033 a	114.5 b	39.9 b	35.6 a
	ODF	3.4 b	2.2 b	2.1 b	0.154 b	0.087 a	0.077 a	91.6 b	32.6 b	41.8 a
AW	CK	8.4 a	5.2 a	3.8 a	0.682 a	0.108 a	0.094 a	205.2 a	77.5 a	76.1 a
	OIF	9.0 a	7.0 a	4.0 a	0.495 a	0.130 a	0.076 a	177.6 a	38.7 a	36.4 b
	ODF	3.5 b	4.0 a	2.2 a	0.223 b	0.112 a	0.097 a	23.9 b	29.6 a	25.2 b

[a]Soil sampling layers
[b]The same letter in the same data column denotes no significant difference (P≤0.05) by LSD.

$$PC_1 = 0.452X_1 + 0.489X_2 + 0.544X_3 - 0.243X_4 + 0.449X_5 \quad (4)$$

$$PC_2 = 0.418X_1 - 0.357X_2 - 0.076X_3 + 0.705X_4 + 0.441X_5 \quad (5)$$

where X_1, X_2, X_3, X_4 and X_5 represent the root dry weight, length, surface area, average diameter and volume, respectively. The eigenvalue contribution rates of PC_1 and PC_2 were 65.95% and 31.60%, respectively.

Since the greater values of root dry weight, length, surface area and volume and the thinner root diameter, the root system is more powerful for absorption of water and nutrient, the comprehensive principal component (CPC) was received as follow:

$$CPC = 0.6774PC_1 - 0.3155PC_2 =$$
$$0.201X_1 + 0.444X_2 + 0.389X_3 - 0.349X_4 + 0.219X_5 \quad (6)$$

The treatment CK showed the lowest PC_1 and CPC values in the ES season, but the highest PC_2 values in the ES and AW seasons and PC_1 and CPC values in the AW season (Table 4). However, the treatment OIF showed the lowest PC_1 and CPC values in the AW season. The treatment ODF showed the lowest

Table 4. Effects of conventional interval fertigation (CK), optimal interval fertigation (OIF) and optimal daily fertigation (ODF) on the first (PC1) and second (PC2) principal components and comprehensive principal component (CPC) values of root characteristic parameters in the early-spring (ES) and autumn-winter (AW) seasons.

Principal component	ES season			AW season		
	CK	OIF	ODF	CK	OIF	ODF
PC$_1$	−0.09	1.05	1.61	0.50	−2.62	−0.46
PC$_2$	1.28	1.26	−0.81	−0.22	−0.58	−0.93
CPC	−0.46	0.30	1.32	0.40	−0.93	−0.01

PC_2 value in the ES and AW seasons, but the highest PC_1 and CPC values in the ES season.

Root length and distribution in different soil layers

In general, root length of four diameter classes (i.e. 0.0–0.5, 0.5–1.0, 1.0–1.5 and >1.5 mm) decreased with rooting depth under all fertigation treatments (Fig. 4). However, while the fine roots (<0.5 mm diameter) mainly concentrated at 0–15 cm depth, the thicker roots (>0.5 mm diameter) were more evenly distributed

Table 3. Effects of conventional interval fertigation (CK), optimal interval fertigation (OIF) and optimal daily fertigation (ODF) on the root weight, length, surface area, average diameter and volume in the early-spring (ES) and autumn-winter (AW) seasons.

Cropping season	Treatment	X$_1$	X$_2$	X$_3$	X$_4$	X$_5$
		Root weight (g)	Root length (cm)	Root surface area (cm^2)	Root average diameter (mm)	Root volume (cm^3)
ES	CK	1.68 a[a]	4913.69 b	653.73 a	0.416 a	8.58 a
	OIF	1.86 a	5641.86 ab	732.78 a	0.407 a	9.31 a
	ODF	1.67 a	7368.76 a	781.54 a	0.370 b	8.47 a
AW	CK	1.51 a	6104.37 a	715.41 a	0.388 a	8.35 a
	OIF	1.09 b	3870.42 b	481.92 b	0.399 a	5.92 b
	ODF	1.31 a	5855.82 ab	659.25 a	0.384 a	7.08 b

[a]The same letter in the same data column denotes no significant difference (P≤0.05) by LSD.

with soil depth. Root length of 0.0–0.5 mm diameter was statistically affected by fertigation treatment within the uppermost 15 cm of the soil depth, however, root length of 0.5–1.5 mm diameter was statistically affected by fertigation treatment within the uppermost 30 cm of the soil depth ($P<0.05$; Fig. 4A,B). Root length of >1.5 mm diameter was only statistically affected by fertigation treatment in the 15–30 cm soil layer in the AW season ($P<0.05$; Fig. 4B). The treatment OIF significantly increased the root length of 0.5–1.0 mm diameter in the 0–15 cm soil layer in the ES season, however, the treatment ODF significantly increased the root length of 0.0–1.5 mm diameter in the 0–15 cm soil layer in the ES season and the root length of 1.0–1.5 mm diameter in the 0–15 cm soil layer in the AW season, when compared to the CK. Generally, the root lengths were not significantly different ($P>0.05$) between the treatment OIF and CK in the 15–30 and 30–45 soil layers in both ES and AW seasons. However, the treatment ODF significantly decreased the root length in the 15–30 and 30–45 soil layers in both ES and AW seasons, when compared to the CK.

Cucumber economic yield, IWUE and PFP$_N$

In both ES and AW seasons, cucumber economic yield, IWUE, PFP$_N$ from the treatment ODF were significantly ($P<0.05$) higher than those from the CK (Table 5). No significant ($P>0.05$) difference was found in both cucumber economic yield and IWUE between the treatments OIF and CK, however, PFP$_N$ from the

treatment OIF was significantly ($P<0.05$) higher than that from the CK in both ES and AW seasons. The treatment ODF increased cucumber economic yield, IWUE and PFP$_N$ by 6.2%, 6.1% and 103% in the ES season, and 8.3%, 8.4% and 232% in the AW season when compared to the CK, respectively. The treatment OIF increased PFP$_N$ by 88% and 209% in the ES and AW seasons when compared to the CK, respectively.

Vitamin C, soluble sugar and nitrate in cucumber fruit

In general, there was no significant ($P>0.05$) difference in soluble sugar in cucumber fruit between the optimal fertigation treatments (i.e. OIF and ODF) and the CK (Table 6). However, the nitrate in cucumber fruit was significantly ($P<0.05$) decreased by the treatment OIF in the ES and AW seasons, and by the treatment ODF in the AW season, when compared to the CK. In addition, vitamin C in cucumber fruit was significantly ($P<0.05$) increased by the treatment ODF in both ES and AW seasons, when compared to the treatment OIF.

Discussion

Effects of different fertigation treatments on soil water and nutrients

The spatial and temporal distribution of water and nutrients is very heterogeneous in soil. Plant root growth not only can be induced by nutrient supply intensity on the whole, but also can be effected by spatiotemporal variation of water and nutrients [31,32]. Thus, it is important to maintain a relative stable nutrient supply in plant root-zone. Our results clearly showed that increasing the frequency of fertigation could decrease the amplitudes of water and nutrient contents in soils (Table 2). In general, optimal daily fertigation significantly decreased the amplitudes of water content, EC value and mineral N content in the 0–15 cm soil layer in the AW season, when compared to optimal interval fertigation (ODF vs OIF; Table 2). Thus, the water and nutrient limitation for plant root growth is probably not likely in soils under optimal daily fertigation. In addition, since the topsoil (0–15 cm) received more water and nutrient than the subsoil (15–45 cm) under fertigation conditions, optimal daily fertigation significantly promoted the root length of <1.5 mm diameter in the 0–15 cm soil layer when compared to optimal interval fertigation (ODF vs OIF; Fig. 4). These results are in accordance with previous researches [33–36] and further demonstrate that using the traditional interval fertigation may lead to larger fluctuation of water and nutrients in soil and inhibit plant root growth [37,13]. Furthermore, since soil relative water contents in the 0–15 cm soil layer generally exceeded 25.5% and were higher than soil field capacity (20.8%) under optimal daily fertigation in the AW season (Fig. 3), there is a possibility for our study to further reduce the irrigation rate.

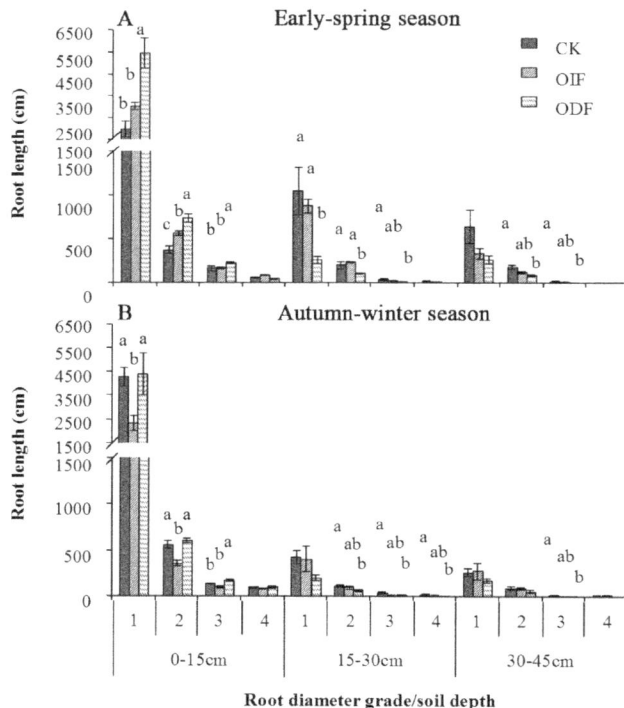

Figure 4. Effects of conventional interval fertigation (CK), optimal interval fertigation (OIF) and optimal daily fertigation (ODF) on the length and distribution of root with different diameter grades in different soil layers in the early-spring (A) and autumn-winter (B) seasons. The numbers 1, 2, 3 and 4 on the abscissa represent four root diameter scales, 0.0–0.5, 0.5–1.0, 1.0–1.5 and ≥1. 5 mm, respectively. Bars represent standard errors. The same letter in the same data column denotes no significant difference (P≤0.05) by LSD.

Effects of different fertigation treatments on plant root growth

Plant roots are able to make different responses to different water and nutrient contents of soils. Although excessive fertilizer application can increase crop yield, it can inhibit plant root growth due to the high nutrient concentration in root-zone. Furthermore, unused nutrients will accumulate in the soils and finally enhance the potential threat to the environment. In contrast, root growth can be depressed when soil nutrient is in deficit [38,2,39,40]. In general, optimal fertilizer application can promote plant root growth and improve fertilizer use efficiency. In this study, however, optimal fertilizer application had no promoting effect on both cucumber root growth (Fig. 4B) and economic yield

Table 5. Effects of conventional interval fertigation (CK), optimal interval fertigation (OIF) and optimal daily fertigation (ODF) on cucumber economic yield, irrigation water use efficiency (IWUE) and partial factor productivity of applied nitrogen (PFP$_N$) in the early-spring (ES) and autumn-winter (AW) seasons.

Cropping season	Treatment	Economic yield (t ha^{-1})	IWUE (kg mm^{-1})	PFP$_N$ (kg kg N^{-1})
ES	CK	132.6b[a]	379. 8b	161. 9c
	OIF	130.2b	372. 8b	304. 5b
	ODF	140.8a	403. 1a	329. 2a
AW	CK	41.2b	279. 8b	88. 6c
	OIF	41.5b	281. 7b	273. 7b
	ODF	44.6a	303. 2a	294. 5a

[a]The same letter in the same data column denotes no significant difference (P≤0.05) by LSD.

(Table 5) in the AW season under same fertilizer application ways (OIF *vs* CK). The explanation is that although the amount of inorganic fertilizer application was sufficient for plant root growth under the treatment OIF during middle fruit harvest period, it was insufficient for plant root growth during the late fruit harvest period, resulting in significant inhibition of root growth and no significant effect on the cucumber economic yield. For instance, contents of mineral nitrogen in soil laye of 0–30 cm of the treatments CK, OIF and ODF during the late fruit harvest period were respectively 93, 35 and 61 mg kg^{-1}. The contents of mineral nitrogen of the treatment OIF was obviousely less than 50 mg kg^{-1}, which is the lower limit of soil N$_{min}$ content for depth 0–30 cm for normal growth and development of cucumber [41], leading to significantly inhibit root growth in the AW seasnon. The results presented here suggested that fertilizer should be applied according to the actual nutrient requirement of the crops at different growing stages.

Effects of different fertigation treatments on cucumber fruit yield

Root growth is closely coordinated with shoot growth [42]. Generally, spatial distribution of roots in soils, together with distribution of water and nutrients in root-zone and water and nutrients requirements for root and shoot growth, determine the water and nutrient use efficiency of crops and the economic yield formation [43]. The data on economic yield in this study showed that although the amount of fertilizer applied to soil was visibly lower under optimal interval fertigation than under conventional interval fertigation (Table 1), optimal interval fertigation still met the water and nutrient requirements of cucumber (Fig. 3) and

maintained root growth (Table 3 and Fig. 4), resulting in the higher efficiency of roots as absorbing organs and no significant effect on cucumber economic yield (Table 5). Thus, excessive fertilizer applied to soil under conventional interval fertigation did not make much contribution to increasing crop yield. In contrast, it can inhibit plant root growth due to the high nutrient concentration in root-zone.

Although soils under the treatments OIF and ODF received the same amount of fertilizer (Table 1), the treatment ODF significantly increased economic yield when compared to the treatment OIF (Table 5). This can be explained by the fact that ODF maintained sufficient and relatively stable water and nutrients in soil (Table 2) and promoted proliferation of fine roots which increased root surface area (Table 3). In addition, ODF reduced the spatiotemporal variation of water in root-zone when compare to OIF (Fig. 3). It was also supported by two recent studies: Gao et al. [14] found that the technology of solution daily application based on the requirement of crops had a promoting effect on tomato yield in a plotted system; Yoshida [44] reported that this technology could increase both the yield of tomato fruit and the water and fertilizers use efficiency by plant.

Root function is related to the morphological and physiological characteristics of root. However, this study did not consider the root physiological function, hence, further work is required test the root physiological function to elucidate the mechanism of communication of root and shoot growth.

Table 6. Effects of conventional interval fertigation (CK), optimal interval fertigation (OIF) and optimal daily fertigation (ODF) on the contents of vitamin C, soluble sugar and nitrate in cucumber fruit in the early-spring (ES) and autumn-winter (AW) seasons.

Cropping season	Treatment	Vitamin C (mg 100 g^{-1})	Soluble sugar (mg g^{-1})	Nitrate (ug g^{-1})
ES	CK	6. 94ab[a]	20. 78a	92. 23a
	OIF	6. 71b	19. 81a	76. 97b
	ODF	7. 31a	21. 12a	84. 94ab
AW	CK	5. 51ab	32. 56a	137. 68a
	OIF	5. 33b	32. 31a	78. 65b
	ODF	5. 76a	31. 02a	63. 89b

[a]The same letter in the same data column denotes no significant difference (P≤0.05) by LSD.

Effects of different fertigation treatments on nitrate content in cucumber fruit

In general, vegetables can provide 80% of nitrate absorbed into the human body [45]. However, part of these nitrate can be reduced to nitrite by bacteria, which may induce lower oxygen-carrying capacity of blood and methemoglobinemia. In addition, nitrite can produce a reaction with secondary amines, such as amides and amino acids, possibly causing the formation of nitrosamines, which may induce cancer in the digestive system of human body [46]. Therefore, it is important to control and maintain low nitrate content in edible part of vegetables. In this study, both OIF and ODF treatments significantly decreased nitrite content, compared to CK treatment. The relative high nitrate content in fruit under CK can be explained by the previous research that surplus nitrate can be partly absorbed and stored by the plant than its need, in order to maintain normal growth requirement when nitrate supply is in deficit [47]. The results presented here suggested that reducing the fertilizer amounts, according to the actual nutrient requirement of the crops at different growing stages, is beneficial to fruit quality and human healthy without having a significant effect on cucumber fruit yield.

Conclusions

On the premise of optimal management of irrigation rate for cucumber, optimal interval fertilization based on nutrient target values, including soil mineral nutrient safety margin and crop uptake, and soil nutrients in root-zone had no significant effect on economic yield, but increased partial factor productivity of applied N. Optimal daily fertigation based on the actual requirement of crops maintained the relatively stable water and nutrients in soils, reduced the spatiotemporal variation of water and nutrients in root-zone and promoted cucumber fine root (<1.5 mm diameter) growth. In addition, optimal daily fertigation maintained significantly higher cucumber economic yield, irrigation water use efficiency and partial factor productivity of applied N than optimal interval fertigation.

Acknowledgments

The authors thank you very much to all the staff of Fangshan District Agricultural Science Research Institute.

Author Contributions

Conceived and designed the experiments: XL ZZ LG. Performed the experiments: XL XZ. Analyzed the data: XL YG. Wrote the paper: XL YT ZZ LG.

References

1. Gao LH, Qu M, Ren HZ, Sui XL, Chen QY, et al. (2010) Structure, function, application and ecological benefit of single-slope, energy-efficient solar greenhouse in China. Horttechnology 20: 626–631.
2. He F, Chen Q, Jiang R, Chen X, Zhang F (2007) Yield and nitrogen balance of greenhouse tomato (Lycopersicum esculentum Mill) with conventional and site-specific nitrogen management in Northern China. Nutrient Cycling in Agroecosystems 77: 1–14.
3. Hvistendahl M (2010) China's Push to Add by Subtracting Fertilizer. Sci. 327: 801–801.
4. Shi WM, Yao J, Yan F (2009) Vegetable cultivation under greenhouse conditions leads to rapid accumulation of nutrients, acidification and salinity of soils and groundwater contamination in South-Eastern China. Nutrient Cycling in Agroecosystems 83: 73–84.
5. Thompson R, Martinez-Gaitan C, Gallardo M, Gimenez C, Fernandez M (2007) Identification of irrigation and N management practices that contribute to nitrate leaching loss from an intensive vegetable production system by use of a comprehensive survey. Agricultural Water Management 89: 261–274.
6. Tian Y, Liu J, Zhang X, Gao L (2010) Effects of summer catch crop, residue management, soil temperature and water on the succeeding cucumber rhizosphere nitrogen mineralization in intensive production systems. Nutrient Cycling in Agroecosystems 88: 429–446.
7. Cabello MJ, Castellanos MT, Romojaro F, Martinez-Madrid C, Ribas F (2009) Yield and quality of melon grown under different irrigation and nitrogen rates. Agricultural Water Management 96: 866–874.
8. Mahajan G, Singh KG (2006) Response of Greenhouse tomato to irrigation and fertigation. Agricultural Water Management 84: 202–206.
9. Jokinen K, Sarkka LE, Nakkila J, Tahvonen R (2011) Split root fertigation enhances cucumber yield in both an open and a semi-closed greenhouse. Sci Hortic (Amsterdam) 130: 808–814.
10. Wang Y, Liu F, Andersen MN, Jensen CR (2010) Improved plant nitrogen nutrition contributes to higher water use efficiency in tomatoes under alternate partial root-zone irrigation. Funct Plant Biol 37: 175–182.
11. Zhang L, Gao L, Zhang L, Wang S, Sui X, et al.(2012) Alternate furrow irrigation and nitrogen level effects on migration of water and nitrate-nitrogen in soil and root growth of cucumber in solar-greenhouse. Sci Hortic (Amsterdam) 138: 43–49.
12. Zotarelli L, Scholberg JM, Dukes MD, Munoz-Carpena R, Icerman J (2009) Tomato yield, biomass accumulation, root distribution and irrigation water use efficiency on a sandy soil, as affected by nitrogen rate and irrigation scheduling. Agricultural Water Management 96: 23–34.
13. Matsuo N, Mochizuki T (2009) Assessment of three water-saving cultivations and different growth responses among six rice cultivars. Plant Prod Sci 12: 514–525.
14. Gao YM, Li JS, Cao YE (2006) Study on the fertigation of tomato soil culture by drip irrigation in greenhouse. Acta Agriculturae Boreali-Occidentalis Sinica 15: 121–126. (in Chinese with English abstract)
15. Zhao SM, Li BM (2001) Introduction to present status,development and new techniques of nutri-culture systems in Japan. Transactions of the Chinese Society of Agricultural Engineering 17: 171-173. (in Chinese with English abstract)
16. Lynch JP (1995) Root architecture and plant productivity. Plant Physiol 109: 7–13.
17. Lynch JP (2007) Roots of the second green revolution. Australian Journal of Botany 55: 493–512.
18. Lynch JP, Brown KM (2012) New roots for agriculture: exploiting the root phenome. Philos Trans R Soc Lond B Biol Sci 367: 1598–1604.
19. Pei XB, Zhang FM, Wang L (2002) Effect of light and temperature on uptake and distribution of nitrogen,phosphorus and potassium of solar greenhouse cucumber. Zhongguo Nong Ye Ke Xue 35: 1510–1513. (in Chinese with English abstract)
20. Wang JF, Xing SZ, Chen SY, Yu QY, Li XL (2005) Dry matter accumulation in and NPK uptake by non-polluted cucumber in protected cultivation. Chinese Journal of Soil Science 36: 708–711. (in Chinese with English abstract)
21. Yu SF, Gao XB, Lu LP (2000) The study on nutrient absorption of cucumber in chinese lean-to solar greenhouse. China Vegetables: 10–11. (in Chinese with English abstract)
22. Chen Q, Zhang FS, Li XL (2005) Application of N management strategies in vegatable production. China Vegetables: 57–63. (in Chinese with English abstract)
23. An SW, Wang YQ, Li HL, Wang SJ, Gao LH (2010) Effects of different irrigation quantities on growth, yield and fruit qual ity of tomato in solar greenhouse. Acta Agriculturae Boreali-Occidentalis Sinica: 188–192. (in Chinese with English abstract)
24. Sun LP, Wen YG, Wang SZ, Zhao JW, Gao LH (2010) Effects of irrigation on water distribution of cucumber in solar greenhouse. Acta Agriculturae Boreali-Occidentalis Sinica: 173–178. (in Chinese with English abstract)
25. Wei Y, Sun L, Wang S, Wang Y, Zhang Z, et al. (2010) Effects of different irrigation methods on water distribution and nitrate nitrogen transport of cucumber in greenhouse. Transactions of the Chinese Society of Agricultural Engineering 26: 67–72. (in Chinese with English abstract)
26. Houba VJG, Temminghoff EJM., Gaikhorst GH (2000) Soil analysis procedures using 0.01M calcium chloride as extraction regent. Commun Soil Sci Plant Anal 31: 1299–1396.
27. Zhang FS,Wang JQ, Zhang WF, Cui ZL, Ma WQ, et al. (2008) Nutrient use efficiencies of major cereal crops in China and measures for improvement. Tu Rang Xue Bao 45: 915–924.
28. Li H (2000) Principles and techniques of plant physiological biochemical experiment. Beijing: Higher Education Press. 246–248 p.
29. Li H (2000) Principles and techniques of plant physiological biochemical experiment. Beijing: Higher Education Press. 195–197 p.
30. Cataldo D, Maroon M, Schrader L, Youngs V (1975) Rapid colorimetric determination of nitrate in plant tissue by nitration of salicylic acid 1. Commun Soil Sci Plan Anal 6: 71–80.
31. De Kroon H, Huber H, Stuefer J F, Van Groenendael J M (2005) A modular concept of phenotypic plasticity in plants. New Phytol 166: 73–82.
32. Hodge A (2006) Plastic plants and patchy soils. J Exp Bot 57: 401–411.
33. Rajput TBS, Patel N (2006) Water and nitrate movement in drip-irrigated onion under fertigation and irrigation treatments. Agricultural Water Management 79: 293–311.

34. Silber A, Xu G, Levkovitch I, Soriano S, Bilu A, et al. (2003) High fertigation frequency: the effects on uptake of nutrients, water and plant growth. Plant Soil 253: 467–477.

35. Silber A, Bruner M, Kenig E, Reshef G, Zohar H, et al. (2005) High fertigation frequency and phosphorus level: Effects on summer-grown bell pepper growth and blossom-end rot incidence. Plant Soil 270: 135–146.

36. Xu G, Levkovitch I, Soriano S, Wallach R, Silber A (2004) Integrated effect of irrigation frequency and phosphorus level on lettuce: P uptake, root growth and yield. Plant Soil 263: 297–309.

37. Belder P, Spiertz J, Bouman B, Lu G, Tuong T (2005) Nitrogen economy and water productivity of lowland rice under water-saving irrigation. Field Crop Res 93: 169–185.

38. Boomsma C, Santini J, Tollenaar M, Vyn T (2009) Maize per-plant and canopy-level morpho-physiological responses to the simultaneous stresses of intense crowding and low nitrogen availability. Agron J 101: 1426–1452.

39. Mi GH, Chen FJ, Wu QP, Lai NW, Yuan LX, et al. (2010) Ideotype root architecture for efficient nitrogen acquisition by maize in intensive cropping systems. Sci China Life Sci 53: 1369–1373.

40. Shen J, Li C, Mi G, Li L, Yuan L, et al. (2012) Maximizing root/rhizosphere efficiency to improve crop productivity and nutrient use efficiency in intensive agriculture of China. J Exp Bot.

41. Guo R, Li X, Christie P, Chen Q, Zhang F (2008) Seasonal temperatures have more influence than nitrogen fertilizer rates on cucumber yield and nitrogen uptake in a double cropping system. Environ pollut 151: 443–451.

42. Wang H, Inukai Y, Yamauchi A (2006) Root development and nutrient uptake. CRC Crit Rev Plant Sci 25: 279–301.

43. Garnett T, Conn V, Kaiser B N (2009) Root based approaches to improving nitrogen use efficiency in plants. Plant Cell Environ 32: 1272–1283.

44. Yoshida C, Iwasaki Y, Makino A, Ikeda H (2011) Effects of Irrigation Management on the Growth and Fruit Yield of Tomato under Drip Fertigation. Horticultural Research (Japan) 10: 325–331.

45. Jackson T, Westermann D, Moore D (1966) The effect of chloride and lime on the manganese uptake by bush beans and sweet corn. Soil Science Society of America Journal 30: 70–73.

46. Wang XZ, Cheng BS, Zhang GZ (1991) Nitrate and its influence factors in vegetables. Chinese Bulletin of Botany 8: 34–37. (in Chinese with English abstract).

47. Koch G, Schulze ED, Percival F, Mooney H, Chu C (1988) The nitrogen balance of Raphanus sativus x raphanistrum plants. II. Growth, nitrogen redistribution and photosynthesis under NO^{3-} deprivation. Plant Cell Environ 11: 755–767.

4

Long-Term Effect of Manure and Fertilizer on Soil Organic Carbon Pools in Dryland Farming in Northwest China

Enke Liu[1,2], Changrong Yan[1,2]*, Xurong Mei[1,2], Yanqing Zhang[1,2]*, Tinglu Fan[3]

1 Institute of Environment and Sustainable Development in Agriculture, Chinese Academy of Agricultural Sciences, Beijing, China, 2 Key Laboratory of Dryland Farming g Agriculture, Ministry of Agriculture of the People's Republic of China (MOA), Beijing, China, 3 Dryland Agricultural Institute, Gansu Academy of Agricultural Sciences, Lanzhou, Gansu, China

Abstract

An understanding of the dynamics of soil organic carbon (SOC) as affected by farming practices is imperative for maintaining soil productivity and mitigating global warming. The objectives of this study were to investigate the effects of long-term fertilization on SOC and SOC fractions for the whole soil profile (0–100 cm) in northwest China. The study was initiated in 1979 in Gansu, China and included six treatments: unfertilized control (CK), nitrogen fertilizer (N), nitrogen and phosphorus (P) fertilizers (NP), straw plus N and P fertilizers (NP+S), farmyard manure (FYM), and farmyard manure plus N and P fertilizers (NP+FYM). Results showed that SOC concentration in the 0–20 cm soil layer increased with time except in the CK and N treatments. Long-term fertilization significantly influenced SOC concentrations and storage to 60 cm depth. Below 60 cm, SOC concentrations and storages were statistically not significant between all treatments. The concentration of SOC at different depths in 0–60 cm soil profile was higher under NP+FYM follow by under NP+S, compared to under CK. The SOC storage in 0–60 cm in NP+FYM, NP+S, FYM and NP treatments were increased by 41.3%, 32.9%, 28.1% and 17.9%, respectively, as compared to the CK treatment. Organic manure plus inorganic fertilizer application also increased labile soil organic carbon pools in 0–60 cm depth. The average concentration of particulate organic carbon (POC), dissolved organic carbon (DOC) and microbial biomass carbon (MBC) in organic manure plus inorganic fertilizer treatments (NP+S and NP+FYM) in 0–60 cm depth were increased by 64.9–91.9%, 42.5–56.9% and 74.7–99.4%, respectively, over the CK treatment. The POC, MBC and DOC concentrations increased linearly with increasing SOC content. These results indicate that long-term additions of organic manure have the most beneficial effects in building carbon pools among the investigated types of fertilization.

Editor: Vishal Shah, Dowling College, United States of America

Funding: This study was supported by the National Basic Research Program of China (973 Program) (No.2012CB955904), the Chinese National Scientific Foundation (No. 31000253, 31170490), the 12th five-year plan of the National Key Technologies R&D Program (No. 2012BAD09B01). The authors declare that no additional external funding was received for this study. The funders had no role in study design, data collection and analysis, decision to publish, or preparation of the manuscript.

Competing Interests: The authors have declared that no competing interests exist.

* E-mail: yancr@ieda.org.cn (CY); zhangyq@ieda.org.cn(YZ)

Introduction

Soil organic matter (SOM) plays a key role in the improvement of soil physical, chemical and biological properties [1]. Conservation of the quantity and quality of soil organic matter (SOM) is considered a central component of sustainable soil management and maintenance of soil quality [2]. Organic manure and inorganic fertilizer are the most common materials applied in agricultural management to improve soil quality and crop productivity [3]. Many studies have shown that balanced application of inorganic fertilizers or organic manure plus inorganic fertilizers can increase SOC and maintain soil productivity [4–7].

However, SOM is not sensitive to short-term changes of soil quality with different soil or crop management practices due to high background levels and natural soil variability [8]. Labile soil organic carbon pools like dissolved organic C (DOC), microbial biomass C (MBC), and particulate organic matter C (POC) are the

fine indicators of soil quality which influence soil function in specific ways (e.g., immobilization–mineralization) and are much more sensitive to change in soil management practices [9,10]. Because these components can respond rapidly to changes in C supply, they have been suggested as early indicators of the effects of land use on SOM quality [11]. Recently, many studies have reported responses of labile SOC pools to management practices [5,12,13], though limited to tillage practices or cropping intensity and rotations management [14]. Few studies have focused on the effect of labile organic C after long-term fertilizer application in northwest China.

In most cases, studies for SOC and SOC fractions have mostly focused on shallow surface soil [15]. The limited information on soil profile SOC and its fractions distribution is a hindrance to conclusive identification of beneficial effects after long-term fertilizer application. Thus, further research is needed to clarify fertilizer application impacts on SOC and SOC fractions for the entire soil profile. Documenting increased SOC levels at deeper depths in the soil profile, however, has been difficult due to a lack of studies where sampling occurred below 30 cm. Nayak et al. [13] found that applications of combined inorganic fertilizers with or without manure can sequester carbon in the 0–60 cm soil layer at the Indian sub-Himalayas. In hot humid subtropical eastern India, Majumder et al. [16] found that after 19 y in a puddle rice-wheat (Triticum aestivum L.) system, NPK+FYM treated plots had 14% larger labile C pools compared with the control plots in the 0–60 cm soil layer.

Northwest China is a vast semi-arid area with average annual precipitation ranging from 300 to 600 mm and more than 90% of the cropland depends on rain fall. Dryland farming has prevailed for several decades in this region. The dry climate and sparse vegetation are mainly responsible for the low SOM. A long period of cultivation and severe erosion in northwest China are likely other potential causes of low SOM [17]. SOC content in the 0–20 cm soil layer of this region is about 11.4 t C ha^{-1} [18]. In recent years, there has been a large increase in the use of inorganic fertilizer with a concomitant decrease in the use of manure. Challenges for dryland farming in Northwest China are low SOC and nutrient retention [19]. However, little is known about the long-term application of inorganic fertilizers either alone or with organic manure on SOC and the distribution of labile organic C fractions at different profile depths. Thus, it is crucial to collect SOC data from long-term experiments in order to understand and estimate the contribution of manure and fertilizer to soil C dynamics. This study provided a unique opportunity to examine the long-term effects of manure and fertilizer on soil organic carbon pools for dryland farming in Northwest China. We hypothesized that long-term fertilizer and manure application would influence the SOC and labile carbon. Moreover, we considered labile organic C fractions would be responsive indicators to SOC change with long histories of fertilizer managements. Our objective was to study the changes of the depth distribution (0–100 cm) in SOC and SOC fractions under a 30-year field experiment in the north of China, and to explain the relationship between different SOC fractions and SOC concentrations. Improved understanding of labile organic matter fractions will provide valuable information for establishing sustainable fertilizer management systems to maintain and enhance soil quality.

Materials and Methods

Experimental Site

The research was based on a long-term fertilizer experiment started in 1979 at the Gaoping Agronomy Farm (35°16′N, 107°30′E, 1254 m altitude), Pingliang, Gansu, China (Fig. 1). Under average climatic conditions, the area has an aridity index (P/PET: precipitation/potential evapotranspiration) of 0.39 and receives 540 mm precipitation, about 60% of which occurs in the summer from July through September. May through June is the driest period for crop growth and little precipitation occurs during the winter months of December and January. The mean annual temperature is 9.8°C. The mean annual sunshine period is 2834 h. The soil is a dark loessial soil classified as calcarid regosols [20]. Analysis of soil samples taken from the experimental area in October 1978 indicated that the surface 15 cm of soil had a pH of 8.2, SOC content of 6.2 g kg^{-1}, total N of 0.95 g kg^{-1}, total P content of 0.57 g kg^{-1}, available P of 7.2 mg kg^{-1} and available K of 165 mg kg^{-1}.

Experimental Design and Treatments

The experiment began in 1979 with a maize crop on land that had been cropped to maize during the previous year, one crop per year. Six fertilization treatments were arranged in a randomized complete block design with three replications. Maize was grown in 1979 and 1980, wheat from 1981 to 1984, maize in 1985 and 1986, wheat from 1987 to 1990, maize in 1991 and 1992, wheat from 1993 to 1998, soybean (Glycine max (L.) Merr.) in 1999, sorghum (Sorghum bicolor (L.) Moench) in 2000, wheat from 2001 to 2004, maize in 2005 and 2006, and wheat in 2006 to 2008.

Winter wheat (Qingxuan 8271, Longyuan 935, and Ping 93-2) was seeded in rows 14.7 cm apart at rates of 165 kg ha^{-1} on about 20 September each year when wheat followed wheat, and in early October when wheat followed maize. Maize was seeded about 20 April each year that maize was grown and Zhongdan 2 was seeded by hand in clumps every 33 cm in rows 66.5 cm apart. About 3 weeks after seeding, maize plants were thinned to one plant per clump. Later, if tillers developed, they were removed to avoid competition. Hand weeding was done to control weeds and plant protection measures were applied when needed. Crops were harvested manually close to the ground and all harvested biomass was removed from the plots. Grain yields were determined by harvesting 20 m^2 for wheat and 40 m^2 for maize at centers of the plots. Grain samples were air-dried on concrete, threshed, and oven-dried at 70°C for 48 hrs, and then weighed.

The experimental area was 0.44 ha. Each plot was 16.7 m×13.3 m with a buffer zone of 1.0 m between each plot. The six treatments were (1) CK, unfertilized control, (2) N, nitrogen fertilizer annually, (3) NP, nitrogen and phosphorus (P) fertilizers annually, (4) NP+S, straw (S) plus N added annually and P fertilizer added every second year, (5) FYM, farmyard manure added annually, and (6) NP+FYM, farmyard manure plus N and P fertilizers added annually. Urea was the N source and was applied to supply 90 kg N ha^{-1} yr^{-1}. Superphosphate was the P source and applied to supply 30 kg P ha^{-1} yr^{-1}. Farmyard manure was added at rate of 75 t ha^{-1} (wet weight). Deep plowing of approximately 23 cm was performed in July after wheat harvest or in October after maize harvest except for the years in which wheat followed maize. In those years, shallow disk tillage was done after maize harvest and wheat was seeded immediately.

Generally, the farmyard manure was a mixture of about 1:5 ratio of wet cattle manure to loess soils and so its nutrient content was quite variable from year to year. The SOC, N, P, and K contents of manure were 11.37, 1.07, 0.69, and 12.3 g kg^{-1} in dry weight, indicating that manure is very low in N, and high in P and K. Although the specific amounts of nutrients added with manure each year were not determined, an application of approximately 75 t ha^{-1} (wet weight) supplied roughly 425 kg C ha^{-1}, 40 kg N ha^{-1}, 26 kg P ha^{-1}, and 460 kg K ha^{-1} in manure annually to crops. For NP+S treatment, mean 5.61 t ha^{-1} y^{-1} of straw (winter wheat and maize) approximately 10 cm in length was returned to the soil prior to plowing, and P fertilizer was added every second year. The straw contained 2.45 t C ha^{-1}, 29.8 kg N ha^{-1}, 7.4 kg P ha^{-1}, and 34.0 kg K ha^{-1}.

Soil Sampling

For SOC trend (0–20 cm layer)study from 1979 to 2008, the composite soil sample (0–20 cm) for each plot was prepared by mixing ten soil cores (4-cm inner diameter) collected randomly after the harvest during 1979 through 1991 and 1996 through 2008 at about 15 d after harvest. The fresh soil was mixed thoroughly, air dried for 7 d, sieved through a 2.0 mm sieve at field moisture content, mixed, and stored in sealed plastic jars for

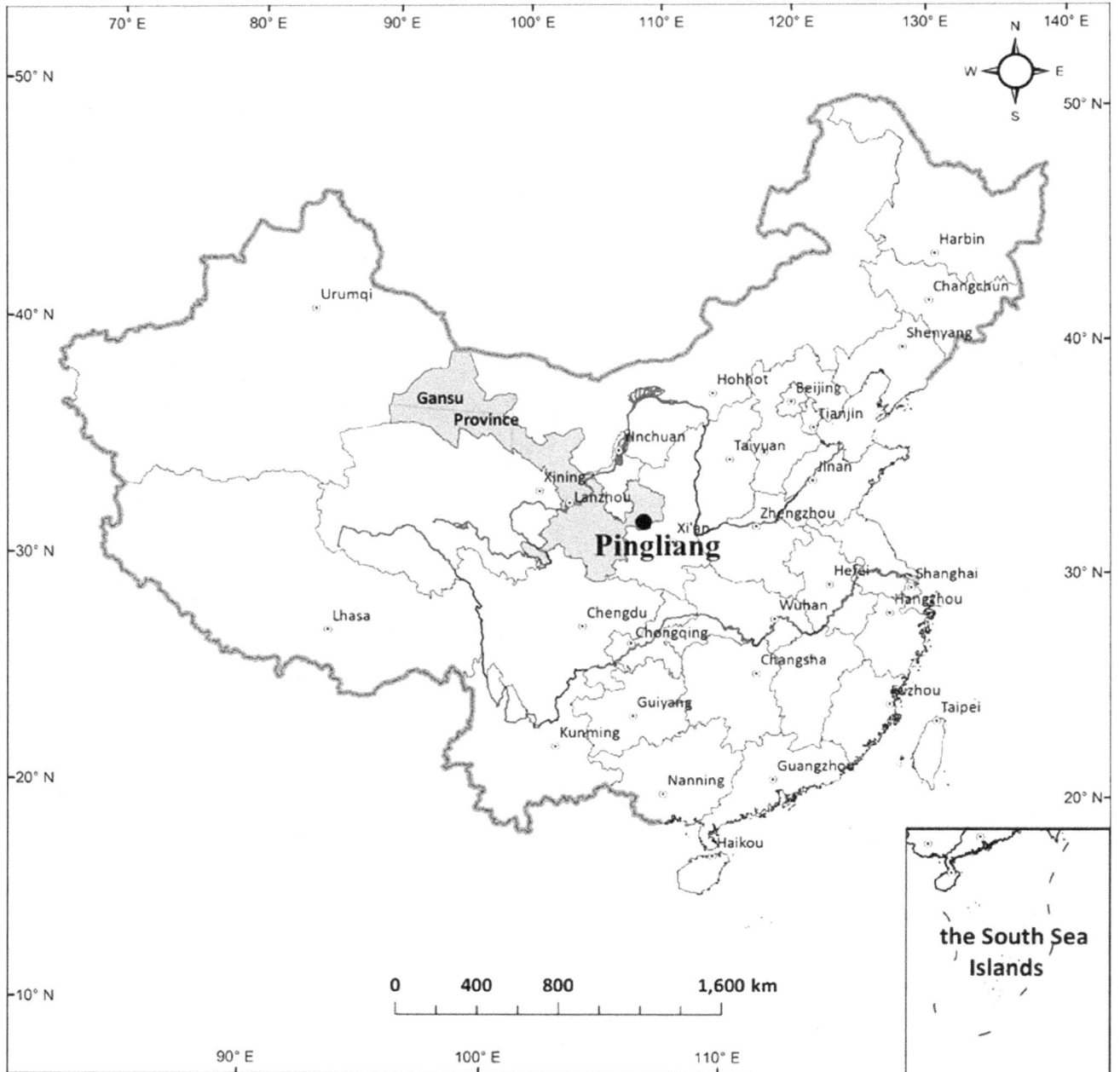

Figure 1. Location of the study site.

analysis. Sub-samples were drawn to determine SOC in the 0–20 cm soil layer. Soil organic C was not determined during 1992 through 1995.

For the distribution of SOC and labile organic C fractions at different profile depths study, soil profile samples (0–20, 20–40, 40–60, 60–80, and 80–100 cm) were collected from six treatments before wheat sowing in September 2008. In each plot the soil was collected from ten points randomly, and mixed into one sample. After carefully removing the surface organic materials and fine roots, each mixed soil sample was divided into two parts. One part of the soil sample was air-dried for the estimation of soil chemical properties and the other part was sieved through a 2 mm wide screen and immediately transferred to the laboratory for bio-

chemical analysis. Soil fresh samples were kept at 4°C in plastic bags for a few days to stabilize the microbiological activity and analyzed within 2 weeks.

Soil Analyses

Soil organic C and bulk densities measured using the method of Blake [21]. DOC was measured using the method of Jiang et al. [22]. POM-C was determined by the method of Cambardella and Elliott [23]. MBC was estimated by fumigation-extraction [24].

Table 1. Estimated mean annual crop biomass and carbon input (Mg C ha^{-1} yr^{-1}) to soil under different fertilizer treatments.

Treatments	Mean annual crop biomass (Mg C ha^{-1} yr^{-1})					Mean annual carbon input (Mg C ha^{-1} yr^{-1})					
	grain yield	straw biomass	root biomass	stubble biomass	Rhizodeposition	Straw-C	Roots-C	Stubble-C	Rhizodeposition-C	FYM-C	Total C
CK	2.11	2.92	0.71	0.29	0.73	0.00	0.29	0.13	0.27	0.00	0.69
N	2.76	3.71	1.09	0.37	0.84	0.00	0.44	0.16	0.35	0.00	0.95
NP	4.59	5.58	1.46	0.56	1.22	0.00	0.59	0.24	0.50	0.00	1.34
FYM	4.33	5.15	1.48	0.52	1.19	0.00	0.60	0.22	0.47	0.43	1.72
NP+S	4.88	5.61	1.68	0.56	1.27	2.45	0.68	0.24	0.52	0.00	3.89
NP+FYM	5.49	6.63	1.96	0.66	1.45	0.00	0.80	0.29	0.60	0.43	2.11

Carbon Inputs

To compute the fraction of added C stabilized cumulative C input was estimated from exogenous supply of C to the soil through straw and FYM and plant input of C through root biomass, stubble and rhizodeposition (Table 1).

The straw and stubble were collected from three 6 m^2 areas for each plot immediately after the harvest of the grains in 1997 and 2005. The straw, stubble and samples were then oven-dried at 60°C for 72 h and weighed. The straw contributed 56.1, 55.6, 54.0, 54.2, 53.0 and 54.6 per cent of total harvestable above ground biomass in winter wheat, and 59.9, 59.5, 56.1, 54.5, 54,2 and 54.8 per cent of total harvestable above ground biomass in maize for CK, N, NP, FYM, NP+S and NP+FYM, respectively. The stubble on an average constituted 10 per cent of the straw.

Root biomasses in winter wheat and maize were calculated using the root: shoot ratio. After harvest, four soil cores (8 cm diameter by 100 cm depth) per plot (two from rows and the other two from between rows) were collected from the 0 to 100 cm soil depths to measure root biomass. The root biomass represented 20.6, 23.6, 19.4, 20.6, 21.6, and 21.8 per cent of the harvestable above ground biomass in winter wheat, and 7.1, 8.0, 7.0, 8.9, 7.7, and 8.8 per cent of the harvestable above ground biomass in maize, respectively in the treatments listed above.

In 1997 and 2005, portions of air-dried straw, stubble and straw were passed through a 0.25 mm sieve for the determinations of the C concentrations. The root biomass, stubble and straw contained 40.4, 42.9 and 42.9 per cent C for winter wheat and 41.5, 44.6 and 44.6 per cent C for maize, respectively. While calculating total rhizodeposition derived from different crops in this study, we used the values mentioned by Bronson et al. [25]. Root exudates therefore represented 15% of aboveground biomass at maturity with a C concentration of 36% in CK and N treatments and 33% in NP, NP+S, FYM and NP+FYM treatments.

Estimation of Soil Organic Carbon Stock

Total SOC stock of profile for each of the five depths (0–20, 20–40, 40–60, 60–80, 80–100 cm) was computed by multiplying the SOC concentration by the bulk density, depth, and factor by 10 (Equation 1).

$$\text{Profile SOC stock} = \text{SOC concentration}(\text{g kg}^{-1})$$
$$\times \text{bulk density}(\text{Mg m}^{-3}) \qquad (1)$$
$$\times \text{depth (m)} \times 10$$

Statistical Analysis

The effects of fertilizer treatments on SOC and labile SOC fractions (MBC, DOC, POC) within each depth were analyzed using one-way ANOVA. Differences were considered significant at P<0.05. Pearsons linear correlation were used to evaluate the relationships between SOC and POC, MBC and DOC. Linear-regression analyses were performed to determine trends using composite soil data from three replicate plots to assess trends of SOC (0–20 cm layer) over the years. For statistical analysis of data, Microsoft Excel (Microsoft Corporation, USA) and SPSS window version 11.0 (SPSS Inc., Chicago, USA) packages were used. Unless otherwise stated, the level of significance referred to in the results is P<0.05.

Results

Soil Organic Carbon (SOC) Trends

The SOC concentrations in the 0–20 cm soil layer for CK, N, NP, FYM, NP+S and NP+FYM treatments at the beginning of the study in 1979 were 5.97, 6.15, 5.92, 6.38, 6.09, 6.03 g kg^{-1}. Above data was the source of the growth rates. Although there were large fluctuations of SOC content with time, the SOC content in CK and N treatments generally constant with time and in NP, slightly increased (Fig. 2). The SOC concentration significantly increased with the lapse of year in the C input treatments (FYM, NP+S and NP+FYM). Across the 30 cropping and fertilization periods, annual SOC concentration rates (slopes of the linear regression vs. time) in Fig. 2 indicated that 0.15, 0.16 and 0.19 g kg^{-1} yr^{-1} were increased each year in NP+S, FYM and NP+FYM treatments, respectively.

Bulk Density

Long-term application of manure and fertilizer significantly affected soil bulk density (BD) to a depth of 40 cm (Fig. 3A). The addition of FYM or straw (FYM, NP+FYM and NP+S) treatments decreased soil bulk density significantly in comparison to that in control plots in all the layers. However, the decrease was more in upper soil layers (0–20 and 20–40 cm) than in the lower layers (40–60, 60–80 and 80–100 cm). Similar was the case with NP treatment, where BD was lower than that in CT treatment at 0–20 and 20–40 cm depths. There were no statistically significant differences in BD among treatments below 40 cm depth.

Depth Distribution of Soil Organic Carbon

The distribution of SOC with depth was dependent on the use of various fertilizers (Fig. 3B). The highest SOC concentration was obtained for 0–20 cm depth and decreased with depth for all

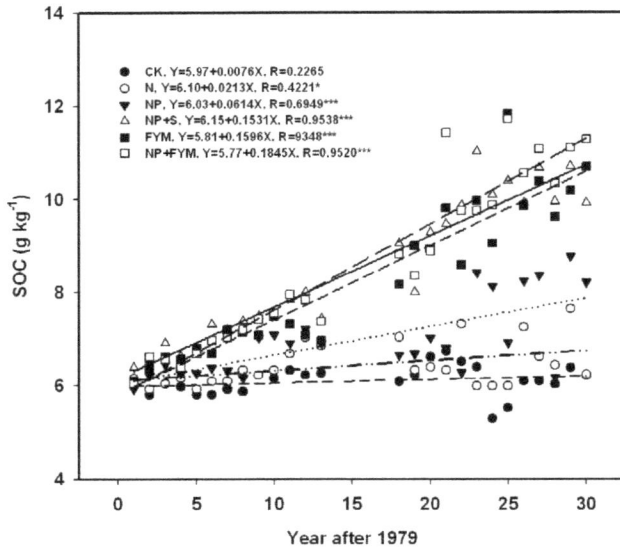

Figure 2. Trend changes of soil organic carbon (SOC) at 0–20 cm top soil layer in a long-term (1979–2008) fertilization experiment in Pingliang, Gansu, China.

treatments. The SOC concentration in 0–20, 20–40 and 40–60 cm depths increased significantly by farmyard manure or straw application. At the 0–20 and 20–40 cm soil depths, SOC was highest in NP+FKM followed by NP+S and FYM treatments and the least in CK treatment. However, the SOC concentration below 60 cm depth was statistically similar among different treatments.

Soil Organic Carbon Storage

The effects of fertilization on SOC storage showed a similar trend to SOC concentration (Fig. 3C). The topsoil (0–20 cm) had the maximum levels of cumulative SOC storage in the 1 m soil depth for the CK, N, NP, FYM, NP+S and NP+FYM treatments, accounting for 24%, 23%, 27%, 30%, 31% and 31%, respectively. At the 20–40 cm and 40–60 cm soil layers, the SOC stocks of the NP, FYM, NP+S and NP+FYM treatments were significantly higher by 17%, 21%, 25% and 37% and 5.3%, 8.1%, 7.3% and

11%, respectively, than that of the CK. The differences of SOC storage between different treatments were not significant in the 60–80 cm and 80–100 cm soil layers. SOC storages were significantly different between fertilization treatments in the 0–100 cm profile. Compared with the CK treatment, SOC storages of the NP+FYM, NP+S, FYM and NP treatments within the 0–100 cm soil depth were increased by nearly 30, 24, 20 and 12%, respectively.

Particulate Organic Carbon

Particulate organic C was found stratified along the soil depth. A higher POC was found in surface soil decreasing with depth (Fig. 4A). At the 0–20 cm, POC content under NP+FYM, NP+S and FYM were 103, 89 and 90% greater than under CK, respectively. In 20–40 cm and 40–60 cm soil layers, NP+FYM had maximum POC which was significantly higher than NP+S and FYM treatments. Even though POC below 60 cm depth was statistically similar among fertilization treatments, the general trend was for increased POC with farmyard manure or straw application down to 100 cm soil depth.

Dissolved Organic Carbon

Irrespective of soil depths, NP+FYM invariably showed higher content of DOC over all other treatments. The CK and N treatments showed lower content of DOC. The DOC concentrations in 0–20 cm, 20–40 cm and 40–60 cm depths were observed highest for NP+FYM followed by NP+S and FYM, and both of them were significant higher than NP (Fig. 4B). However, in the deeper layers (60–80 cm and 80–100 cm), the difference in DOC among the treatments was not significant.

Microbial Biomass Carbon

The MBC differences among treatments not only presented in the surface soil layers, but also presented at deeper depths in the profile. In our study, MBC showed a significant effect at different fertilizer treatments (Fig. 4C). The SOC concentration in 0–80 cm depth increased significantly by farmyard manure or straw application. The mean MBC content in 0–80 cm profile was 82% higher in NP+FYM treatment than in CK treatment.

Comparison of Labile Organic Carbon Pools

POC, MBC and DOC concentrations increased linearly with increasing soil SOC content (Fig. 5), suggesting that total organic

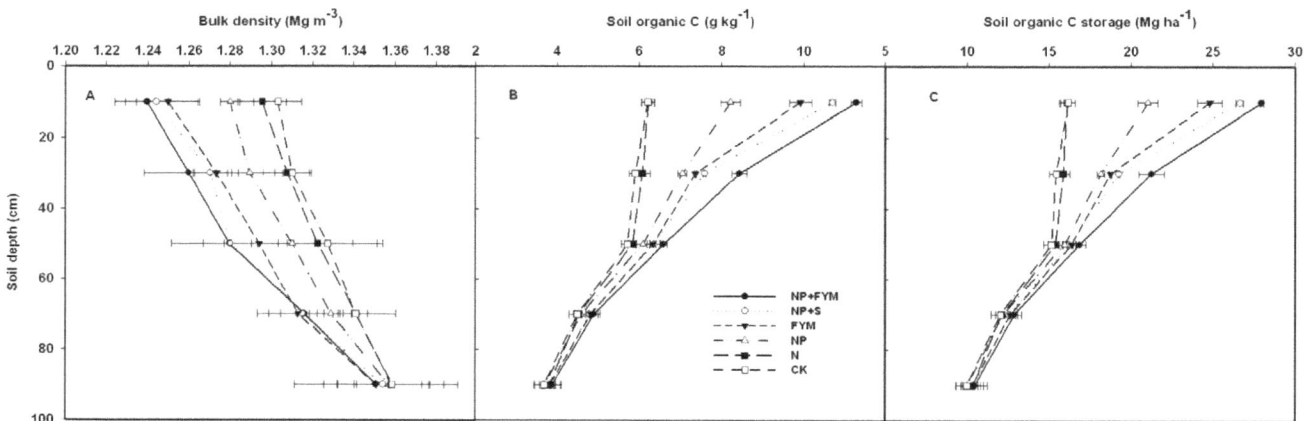

Figure 3. Effect of long-term fertilizer applications on depth distribution of bulk density (A), soil organic C (B) and soil organic C storage (C).

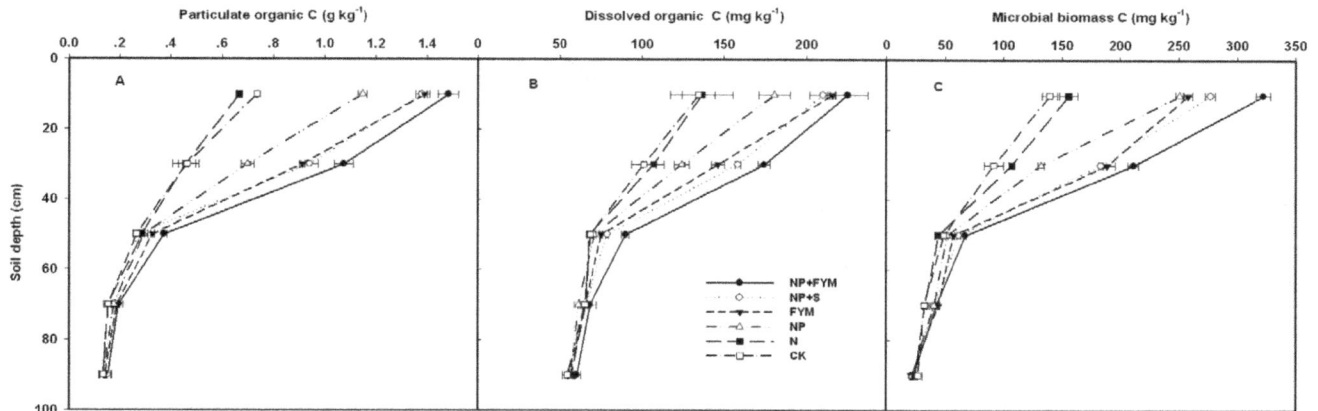

Figure 4. Effect of long-term fertilizer applications on depth distribution of particulate organic C (A), dissolved organic C (B) and microbial biomass C (C).

matter content was a major determinant of the amount of POC, MBC and DOC present. Of the reported C pools, POC was most highly correlated with SOC, followed by MBC then DOC.

Discussion

Soil Total C

From 1979 to 2008, SOC concentrations (0–20 cm layer) increased significantly for all treatments except the CK and N treatments, and the greatest increases occurred for the three plots of FYM, NP+FYM and NP+FYM treatments that received organic materials (Fig. 2). Apparently, application of only N fertilizer did not increase SOC content over long-term cropping. This observation was consistent with that of Goyal et al. [26], who reported that no significant increase in SOC by the addition of only N fertilizer. In the present study, SOC in the CK and N soil were at par, presumably because of lower crop productivity that results in significantly lower accumulation of root biomass. The NP treatment increased SOC concentrations but at a much lower rate of 0.061 g kg^{-1} yr^{-1}. This finding indicated that long-term chemical NP fertilizer alone can increase soil C sequestration, which has been confirmed by other long-term fertilizer experiments in China [27]. A before linear relationship was found between SOC and organic material (i.e., manure and straw) input after 30 y of continuous cropping in northwest China. Most of the soil organic matter (SOM) models assume a linear increase in SOC levels with increasing C input [28].

The SOC concentration differences among treatments not only presented in the surface soil layers, but also presented at deeper depths in the profile. SOC concentration was highest in NP+FYM plots and the least in unfertilized control (CK) at all the sampling depths. Though the average SOC concentration decreased with soil depth, the NP+FYM, NP+S and FYM treatments resulted in a significant increase in organic C even in 40–60 cm soil layer. Manjaiah and Singh [29] and also found that inorganic fertilizers plus organic material increased the SOC content of the soil. The reasons for the higher SOC in manure soils at deeper depths include the following. First, the crop rooting depth between organic manure and inorganic fertilizer soils differ. The organic manure soils can be favorable for the growth of roots into deeper layers due to the relatively loose soil and high soil water content. Second, SOC in organic manure soils can also move to lower depths through earthworm burrows and leaching [30]. Applying straw with N and P fertilizer (NP+S) had the highest total C input

(Table 1), yet decreased SOC concentration over the FYM with NP fertilizer treatment. As NP+FYM treatment significantly increased SOC concentration, this suggests that animal manure is more effective in building soil C than straw, possibly due to the presence of more humified and recalcitrant C forms in animal manure as compared to the straw. For the inorganic fertilizer treatment, the optimum application of inorganic fertilizer NP treatment showed a higher SOC concentration over the application of inorganic fertilizer N treatment at all the sampling depths. The optimum fertilization results in better plant growth including the root biomass (Table 1), which could have added to the SOC particularly as indicated in the lower layers [31].

Similar to the concentration, SOC stock of the profile was also significantly ($P<0.05$) higher in the organic manure treatments (FYM, NP+FYM and NP+S) compared with the only inorganic fertilizer (N, NP) and control treatments. Gami et al. [32] also reported a significant increase in SOC stocks to 60 cm depth under three 23–25-year-old long-term fertility experiments in the Nepal, with application of manure and inorganic fertilizer. Within 1 m soil depth, the cumulative distribution of SOC in the CK, N, NP, FYM, NP+S and NP+FYM treatments were by 50%, 46%, 51%, 53%, 54% and 55% in the 0–40 cm layer, and 68%, 68%, 71%, 72%, 73% and 74% in the 0–60 cm layer, respectively. The SOC storage in the 60–100 cm layer was statistically similar among different treatments. On average the estimate of soil C accumulation to 60 cm depth were 267% and 41% greater than that for soil C accumulated to 20 cm depth and to 40 cm depth, respectively. These findings suggest that the estimate of soil C accumulation to 60 cm depth was more effective than that for soil C accumulated to 40 cm. In this study, C input was increased under the N treatment compared to the CK. However, neither SOC concentration nor C storage was significantly changed under the N treatment. The reason for this is that the N treatment may stimulate soil microbial activity, therefore increasing the C output. The increase in C mineralization might offset the increase in C input. Similar results were also found by Halvorson et al. [33], Su et al. [34], and Lou et al. [35].

Soil C Fractions

The POC fraction has been defined as a labile SOC pool mainly consisting of plant residues partially decomposed and not associated with soil minerals [23,36]. In the present study, the soil amended by FYM or straw contained significantly higher POC in

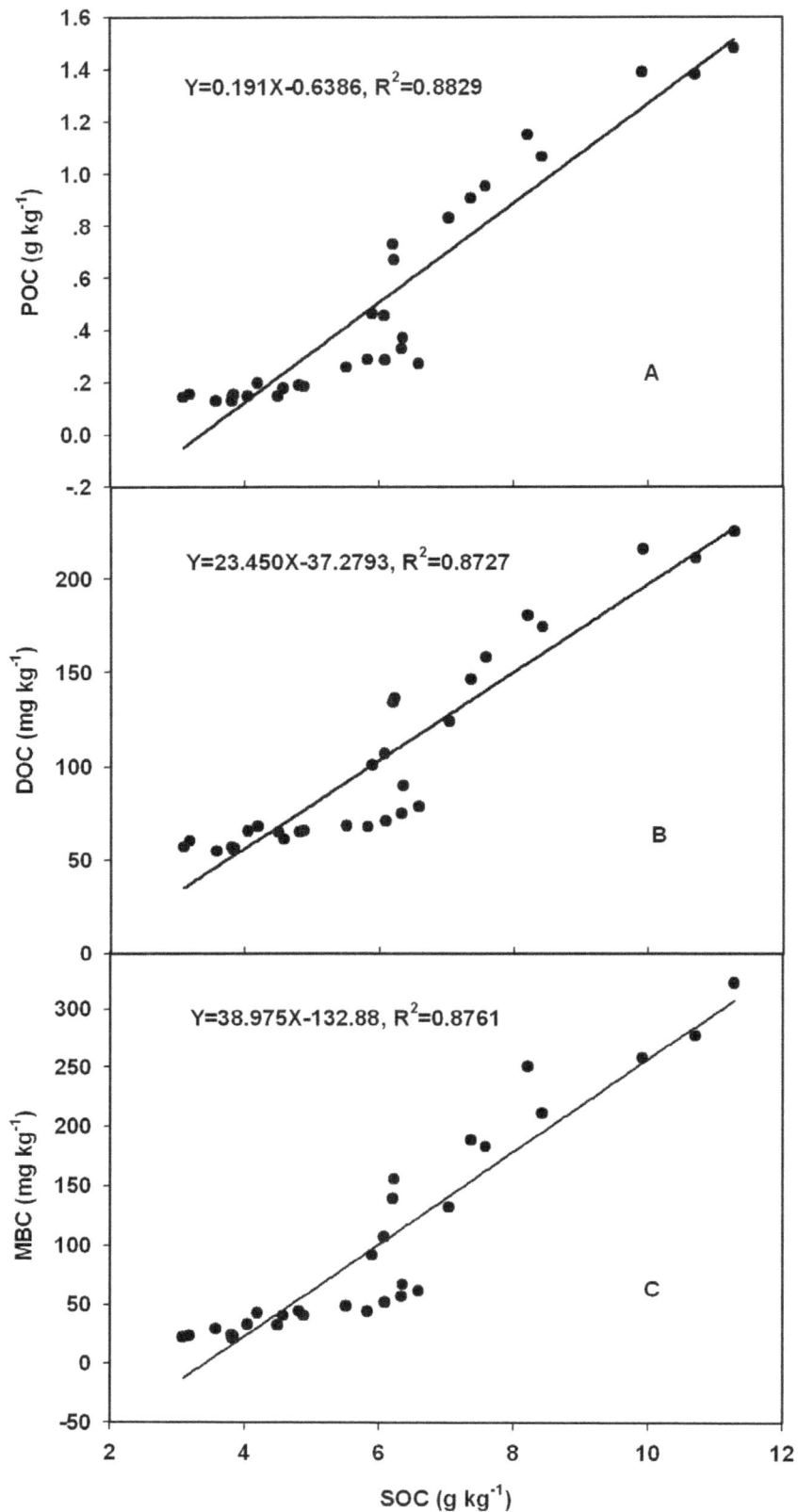

Figure 5. Relationship between soil organic carbon and different labile SOC pools: (A) dissolve organic C (DOC), (B) particulate organic C (POC), and (C) microbial biomass C (MBC).

the 0–60 cm than that in the inorganic fertilizer treatments. Rudrappa et al. [31] reported that the additional organic carbon input could enhance the POC accumulation. Purakayastha et al. [5] concluded that FYM can increase the root biomass and microbial biomass debris which is the main source of POC. It is suggested that the greater biochemical recalcitrance of root litter [37] might have also increased the POC contents in soil depending upon the root biomass produced. The continuous replacement of organic manure on the soil creates a favourable environment for the cycling of C and formation of macroaggregates. Furthermore, POC acts as a cementing agent to stabilise macroaggregates and protect intra-aggregate C in the form of POC [36,38]. Below 60 cm soil layer, the POC declined with increase in soil depth. Chan [39] also found that straw application increased POC in surface soil but not at lower depths.

DOC is believed to be derived from plant roots, litter and soil humus and is a labile substrate for microbial activity [40,41]. The concentration of DOC varied widely among all the treatments and a significant increase was observed in surface soils under different fertilizer treatments compared with CK. In the long-term, the quantity of organic residues are the main factors influencing the amount and composition of DOC. Likewise, in our study, the upper 60 cm soil layer had more DOC concentration than that of lower layer. Below 60 cm, the DOC concentration sharply decreased with soil depth. DOC in subsurface soils may be a result of decomposition of crop residues or translocation from surface soil [42]. Several field studies have sown that concentration and fluxes of DOM in soil solution decrease significantly with soil depth [41].

In our study, MBC was highest in the farmyard manure plus inorganic fertilizer treatment in top soil, an increased MBC content after farmyard application was also reported by Chakraborty et al. [43] and Marschner et al. [44]. This indicated the activation of microorganisms through carbon source inputs consisting of organic residues. Increases in soil organic matter are usually associated with similar increases in microbial biomass because the SOM provides principal substrates for the microorganisms [45]. Among the investigated fertilizer treatments, straw plus inorganic fertilizer had impact on the microbial biomass. This effect is mainly due to the input of straw manure as an organic carbon source. Lynch and Panting [46] reported that eight months after application of straw manure to a loamy arable soil the microbial biomass was almost twice as high as compared to a control. Also, Ocio et al. [47] have demonstrated rapid and significant increases in microbial biomass following straw inputs in field conditions. The MBC was not only correlated with SOC concentration near the surface but also at deeper depths. Though the average MBC decreased with soil depth, the NP+FYM, NP+S and FYM treatments resulted in significant increase in microbial biomass C even in 60–80 cm soil layer. The main source of MBC in deep soil was mainly the left over root biomass and increased microbial biomass debris. It is suggested that the greater biochemical recalcitrance of root litter [37] might have also increased the MBC contents in soil depending upon the root biomass produced. However, the MBC content was lower in the 80–100 cm profile. The reason is that roots may be difficult extend to lower depths. The imbalanced use of fertilizers (CK and N) decreased MBC due to limitation imposed by major nutrients like P and K, which are essential for higher crop production as well as for microbial cell synthesis.

SOC was highly correlated with labile carbon (POC, MBC and DOC). POC, MBC and DOC were significantly and positively correlated with SOC. The correlation coefficient was highest between POC and SOC ($R^2 = 0.883$), followed between MBC and SOC ($R^2 = 0.876$) and DOC and SOC ($R^2 = 0.873$). Such high correlations have also been reported by Rudrappa et al. [31] and

Liang et al. [12], and it is not surprising that the three measures of labile organic matter were closely correlated since they are closely interrelated properties. This result confirms the value of these fractions as sensitive indicators for detecting changes in SOM in the short term, before they are readily measurable in total C. Likewise, these correlations also indicated that SOC was a major determinant of the labile C fractions present.

Conclusions

Fertilizer application has played an important role in improving the total SOC and labile C pools content in the soil after 30 years. Because there was low SOC content in the Northwest of China, the long-term application of organic manure and inorganic fertilizer increased the content of SOC. SOC concentrations and storage were highest in surface soil and depth interval down to 60 cm under NP+FYM and NP+S, below which concentrations did not change with depth. At the same time, on average the estimate of soil C storage to 60 cm depth was higher than that for soil C accumulated to 20 cm depth and to 40 cm depth, respectively. These findings suggest that the estimate of soil C accumulation to 60 cm depth was more effective than that for soil C accumulated to 20 cm depth and to 40 cm depth. NP+FYM was the most efficient management system for sequestering SOC. A large amount of C was also sequestered in soil under NP+S treatment. Soil microbial biomass C, POC and DOC were all significantly greater under organic manure (farmyard manure or straw) plus inorganic fertilizers, especially in the surface. The labile fraction organic C contents decreased significantly with increasing soil depth. These labile pools were highly correlated with each other and SOC, indicating that they were sensitive to changes in SOC.

In Northwest China, the effects of manure and fertilizer application practices on soil C sequestration were studied so that dryland farming soil could contribute to both sustainable food production and mitigation of greenhouse gas emissions through soil C sequestration. Our results have very significant implications for soil C sequestration potential in semiarid agro-ecosystems of northwest China. SOC concentration in surface soil (0–20 cm) and SOC storage of the profile (0–100 cm) were not significantly or slightly increased by the 30 yr of fertilizer treatments (N and NP), but they were sharply increased by the manure and straw amendment (FYM, NP+S and NP+FYM). Thus, returning crop residue to the soil or adding farmyard manure on the soil surface is crucial to improving the SOC level. The large scale implementation of the straw or manure plus inorganic fertilizer amendments will help to enhance the capacity of carbon sequestration and promote food security in the region. Therefore, local government should encourage farmers to manage the nutrients and soil fertility based on integrated nutrient management by combining organic matter with inorganic fertilizer to improve soil carbon pools and increase crop productivity for long-term.

Acknowledgments

We wish to thank Prof. Bing So for his revision of the manuscript. We thank the staff of Pingliang Prefecture Institute of Agricultural Sciences. We are grateful for the support and technical assistance from colleagues at Institute of Environment and Sustainable Development in Agriculture, CAAS.

Author Contributions

Conceived and designed the experiments: EKL CRY YQZ. Performed the experiments: EKL TLF. Analyzed the data: EKL CRY XRM. Contributed reagents/materials/analysis tools: EKL TLF YQZ. Wrote the paper: EKL CRY XRM.

References

1. Ouédraogo E, Mando A, Brussaard L, Stroosnijder L (2007) Tillage and fertility management effects on soil organic matter and sorghum yield in semi-arid West Africa. Soil Till Res 94: 64–74.

2. Doran JW, Sarrantonio M, Liebig MA (1996) Soil health and sustainability. Adv Agron 56: 1–54.

3. Verma S, Sharma PK (2007) Effect of long-term manuring and fertilizers on carbon pools, soil structure, and sustainability under different cropping systems in wet-temperate zone of northwest Himalayas. Biol Fertility Soils 44: 235–240.

4. Blair N, Faulkner RD, Till AR, Korschens M, Schulz E (2006) Long-term management impacts on soil C, N and physical fertility: Part II: Bad Lauchstadt static and extreme FYM experiments. Soil Till Res 91: 39–47.

5. Purakayastha TJ, Rudrappa L, Singh D, Swarup A, Bhadraray S (2008) Long-term impact of fertilizers on soil organic carbon pools and sequestration rates in maize–wheat–cowpea cropping system. Geoderma 144: 370–378.

6. Gong W, Yan X, Wang J, Hu T, Gong Y (2009) Long-term manure and fertilizer effects on soil organic matter fractions and microbes under a wheat–maize cropping system in northern China. Geoderma 149: 318–324.

7. Powlson DS, Bhogal A, Chambers BJ, Coleman K, Macdonald AJ, et al. (2012) The potential to increase soil carbon stocks through reduced tillage or organic material additions in England and Wales: A case study. Agric Ecosyst Environ 146: 23–33.

8. Haynes RJ (2005) Labile organic matter fractions as central components of the quality of agricultural soils: An overview. Adv Agron 85: 221–268.

9. Saviozzi A, Levi-Minzi R, Cardelli R, Riffaldi R (2001) A comparison of soil quality in adjacent cultivated, forest and native grassland soils. Plant Soil 233: 251–259.

10. Xu M, Lou Y, Sun X, Wang W, Baniyamuddin M, et al. (2011) Soil organic carbon active fractions as early indicators for total carbon change under straw incorporation. Biol Fertility Soils 47: 745–752.

11. Gregorich EG, Monreal CM, Carter MR, Angers DA, Ellert BH (1994) Towards a minimum data set to assess soil organic matter quality in agricultural soils. Can J Soil Sci 74: 367–385.

12. Liang Q, Chen H, Gong Y, Fan M, Yang H, et al. (2011) Effects of 15 years of manure and inorganic fertilizers on soil organic carbon fractions in a wheat-maize system in the North China Plain. Nutr Cycl Agroecosyst 92: 1–13.

13. Nayak AK, Gangwar B, Shukla AK, Mazumdar SP, Kumar A, et al. (2012) Long-term effect of different integrated nutrient management on soil organic carbon and its fractions and sustainability of rice–wheat system in Indo Gangetic Plains of India. Field Crop Res 127: 129–139.

14. Dou F, Wright AL, Hons FM (2008) Sensitivity of labile soil organic carbon to tillage in wheat-based cropping systems. Soil Sci Soc Am J 72: 1445–1453.

15. West TO, Post WM (2002) Soil organic carbon sequestration rates by tillage and crop rotation. Soil Sci Soc Am J 66: 1930–1946.

16. Majumder B, Mandal B, Bandyopadhyay PK, Gangopadhyay A, Mani PK, et al. (2008) Organic amendments influence soil organic carbon pools and rice–wheat productivity. Soil Sci Soc Am J 72: 775–785.

17. Janzen HH, Campbell CA, Ellert BH, Bremer E (1997) Soil organic matter dynamics and their relationship to soil quality. In: Gregorich, EG, Carter MR, editors. Soil Quality for Crop Production and Ecosystem Health. Amsterdam: Developments in Soil Science 25, Elsevier, 277–292.

18. Xing NQ, Zhang YQ, Wang LX (2001) The study on dryland agriculture in North China (in Chinese). Beijing: Chinese Agriculture Press.

19. Zhang X, Quine TA, Walling DE (1998) Soil erosion rates on sloping cultivated land on the Loess Plateau near Ansai, Shaanxi Province, China: an investigation using 137Cs and rill measurements. Hydrol Processes 12: 171–189.

20. FAO/UNESCO (1988) Soil map of the world. Paris: UNESCO.

21. Black CA (1965) Methods of soil analysis, Part I and II. Madison, Wis: American Society of Agronomy.

22. Jiang PK, Xu QF, Xu ZH, Cao ZH (2006) Seasonal changes in soil labile organic carbon pools within a Phyllostachys praecox stand under high rate fertilization and winter mulch in subtropical China. Forest Ecol Manag 236: 30–36.

23. Cambardella CA, Elliott ET (1992) Particulate soil organic-matter changes across a grassland cultivation sequence. Soil Sci Soc Am J 56: 777–783.

24. Vance ED, Brookes PC, Jenkinson DS (1987) An extraction method for measuring soil microbial biomass C. Soil Biol Biochem 19: 703–707.

25. Bronson KF, Cassman KG, Wassmann R, Olk DC, Noordwijk MV, et al. (1998) Management of carbon sequestration in soil. In: Lal R, Kimble J, Follet RF, Stewart BA, editors, Soil Carbon Dynamics in Different Cropping Systems in Principal Ecoregions of Asia. Boca Raton: CRC/Press. 35–57.

26. Goyal S, Mishra MM, Hooda IS, Singh R (1992) Organic matter-microbial biomass relationships in field experiments under tropical conditions: Effects of inorganic fertilization and organic amendments. Soil Biol Biochem 24: 1081–1084.

27. Wu T, Schoenau JJ, Li F, Qian P, Malhi SS, et al. (2005) Influence of fertilization and organic amendments on organic-carbon fractions in Heilu soil on the loess plateau of China. J Plant Nutr Soil Sci 168: 100–107.

28. Zhang W, Xu M, Wang B, Wang X (2009) Soil organic carbon, total nitrogen and grain yields under long-term fertilizations in the upland red soil of southern China. Nutr Cycl Agroecosyst 84: 59–69.

29. Kanchikerimath M, Singh D (2001) Soil organic matter and biological properties after 26 years of maize–wheat–cowpea cropping as affected by manure and fertilization in a Cambisol in semiarid region of India. Agric Ecosyst Environ 86: 155–162.

30. Lorenz K, Lal R (2005) The depth distribution of soil organic carbon in relation to land use and management and the potential of carbon sequestration in subsoil horizons. Adv Agron 88: 35–66.

31. Rudrappa L, Purakayastha TJ, Singh D, Bhadraray S (2006) Long-term manuring and fertilization effects on soil organic carbon pools in a Typic Haplustept of semi-arid sub-tropical India. Soil Till Res 88: 180–192.

32. Gami SK, Lauren JG, Duxbury JM (2009) Soil organic carbon and nitrogen stocks in Nepal long-term soil fertility experiments. Soil Till Res 106: 95–103.

33. Halvorson AD, Wienhold BJ, Black AL (2002) Tillage, nitrogen, and cropping system effects on soil carbon sequestration. Soil Sci Soc Am J 66: 906–912.

34. Su YZ, Wang F, Suo DR, Zhang ZH, Du MW (2006) Long-term effect of fertilizer and manure application on soil-carbon sequestration and soil fertility under the wheat–wheat–maize cropping system in northwest China. Nutr Cycl Agroecosyst 75: 285–295.

35. Lou Y, Wang J, Liang W (2011) Impacts of 22-year organic and inorganic N managements on soil organic C fractions in a maize field, northeast China. Catena 87: 386–390.

36. Six J, Conant RT, Paul EA, Paustian K (2002) Stabilization mechanisms of soil organic matter: Implications for C-saturation of soils. Plant Soil 241: 155–176.

37. Puget P, Drinkwater LE (2001) Short-term dynamics of root-and shoot-derived carbon from a leguminous green manure. Soil Sci Soc Am J 65: 771–779.

38. Sá JCM, Lal R (2009) Stratification ratio of soil organic matter pools as an indicator of carbon sequestration in a tillage chronosequence on a Brazilian Oxisol. Soil Till Res 103: 46–56.

39. Chan KY (1997) Consequences of changes in particulate organic carbon in vertisols under pasture and cropping. Soil Sci Soc Am J 61: 1376–1382.

40. Liang BC, Mackenzie AF, Schnitzer M, Monreal CM, Voroney PR, et al. (1997) Management-induced change in labile soil organic matter under continuous corn in eastern Canadian soils. Biol Fertility Soils 26: 88–94.

41. Kalbitz K, Solinger S, Park J.H, Michalzik B, Matzner E (2000) Controls on the dynamics of dissolved organic matter in soils: a review. Soil Sci 165: 277–304.

42. Dou F, Wright AL, Hons FM (2007) Depth distribution of soil organic C and N after long-term soybean cropping in Texas. Soil Till Res 94: 530–536.

43. Chakraborty A, Chakrabarti K, Chakraborty A, Ghosh S (2011) Effect of long-term fertilizers and manure application on microbial biomass and microbial activity of a tropical agricultural soil. Biology and Fertility of Soils 47: 227–233.

44. Marschner P, Kandeler E, Marschner B (2003) Structure and function of the soil microbial community in a long-term fertilizer experiment. Biol Fertility Soils 35: 453–461.

45. Melero S, López-Garrido R, Murillo J.M, Moreno F (2009) Conservation tillage: Short-and long-term effects on soil carbon fractions and enzymatic activities under Mediterranean conditions. Soil Till Res 104: 292–298.

46. Lynch JM, Panting LM (1980) Variations in the size of the soil biomass. Soil Biol Biochem 12: 547–550.

47. Ocio JA, Martinez J, Brookes PC (1991) Contribution of straw-derived N to total microbial biomass N following incorporation of cereal straw to soil. Soil Biol Biochem 23: 655–659.

5

Effects of Different Tillage and Straw Return on Soil Organic Carbon in a Rice-Wheat Rotation System

Liqun Zhu[1]*, **Naijuan Hu**[1], **Minfang Yang**[2], **Xinhua Zhan**[2], **Zhengwen Zhang**[1]

1 College of Agriculture, Nanjing Agricultural University, Nanjing, China, 2 College of Resources and Environmental Science, Nanjing Agricultural University, Nanjing, China

Abstract

Soil management practices, such as tillage method or straw return, could alter soil organic carbon (C) contents. However, the effects of tillage method or straw return on soil organic C (SOC) have showed inconsistent results in different soil/climate/cropping systems. The Yangtze River Delta of China is the main production region of rice and wheat, and rice-wheat rotation is the most important cropping system in this region. However, few studies in this region have been conducted to assess the effects of different tillage methods combined with straw return on soil labile C fractions in the rice-wheat rotation system. In this study, a field experiment was used to evaluate the effects of different tillage methods, straw return and their interaction on soil total organic C (TOC) and labile organic C fractions at three soil depths (0–7, 7–14 and 14–21 cm) for a rice-wheat rotation in Yangzhong of the Yangtze River Delta of China. Soil TOC, easily oxidizable C (EOC), dissolved organic C (DOC) and microbial biomass C (MBC) contents were measured in this study. Soil TOC and labile organic C fractions contents were significantly affected by straw returns, and were higher under straw return treatments than non-straw return at three depths. At 0–7 cm depth, soil MBC was significantly higher under plowing tillage than rotary tillage, but EOC was just opposite. Rotary tillage had significantly higher soil TOC than plowing tillage at 7–14 cm depth. However, at 14–21 cm depth, TOC, DOC and MBC were significantly higher under plowing tillage than rotary tillage except for EOC. Consequently, under short-term condition, rice and wheat straw both return in rice-wheat rotation system could increase SOC content and improve soil quality in the Yangtze River Delta.

Editor: Shuijin Hu, North Carolina State University, United States of America

Funding: This work was supported by Special Fund for Agro-scientific Research in the Public Interest (201103001). The funders had no role in study design, data collection and analysis, decision to publish, or preparation of the manuscript.

Competing Interests: The authors have declared that no competing interests exist.

* E-mail: zhulq@njau.edu.cn

Introduction

Soil organic carbon (C) has profound effects on soil physical, chemical and biological properties [1]. Maintenance of soil organic C (SOC) in cropland is important, not only for improvement of agricultural productivity but also for reduction in C emission [2]. However, short- and medium-term changes of SOC are difficult to detect because of its high temporal and spatial variability [3]. On the contrary, soil labile organic C fractions (i.e., microbial biomass C (MBC), dissolved organic C (DOC), and easily oxidizable C (EOC)) that turn over quickly can respond to soil disturbance more rapidly than total organic C (TOC) [1,3,4]. Therefore, these fractions have been suggested as early sensitive indicators of the effects of land use change on soil quality (e.g. [3,5,6]).

Agricultural practices such as tillage methods are conventionally used for loosening soils to grow crops. But long-term soil disturbance by tillage is believed to be one of the major factors reducing SOC in agriculture [7]. Frequent tillage may destroy soil organic matter (SOM) [8] and speed up the movement of SOM to deep soil layers [9]. As a consequence, agricultural practices that reduce soil degradation are essential to improve soil quality and agricultural sustainability. Crop residue plays an important role in SOC sequestration, increasing crop yield, improving soil organic matter, and reducing the greenhouse gas (e.g. [10–13]). As an important agricultural practice, straw return is often implemented

with tillage in the production process. Although numerous studies have indicated that tillage methods combined with straw return had a significant effect on labile SOC fractions, the results varied under different soil/climate conditions. For example, both no-tillage and shallow tillage with residue cover had significantly higher SOC than conventional tillage without residue cover in Loess Plateau of China [14], while Wang et al. [15] reported that the difference between the treatments of plowing with straw return and no-tillage with straw return on TOC in central China was not significant. Rajan et al. [2] showed that in Chitwan Valley of Nepal, no-tillage with crop residue application at upper soil depth had distinctly higher SOC sequestration than conventional tillage with crop residue. The effects of tillage on soil labile organic C vary with regional climate [16], soil condition (e.g. [17–20]), residue management practice, and crop rotation (e.g. [21,22]). Therefore, the investigation on soil labile organic C for specific soil, climate, and cropping system is necessary to improve the soil quality.

The Yangtze River Delta of China is the main production region of rice and wheat, and rice-wheat rotation is the most important cropping system in this region [23]. The total sown area of rice and wheat in the Yangtze River Delta accounted for about 20.1% of that in China in 2011, and the total yield was 22.1% of the national yield for these two crops [24]. Many field experiments in this region (e.g. [25–27]) about the effects of tillage methods

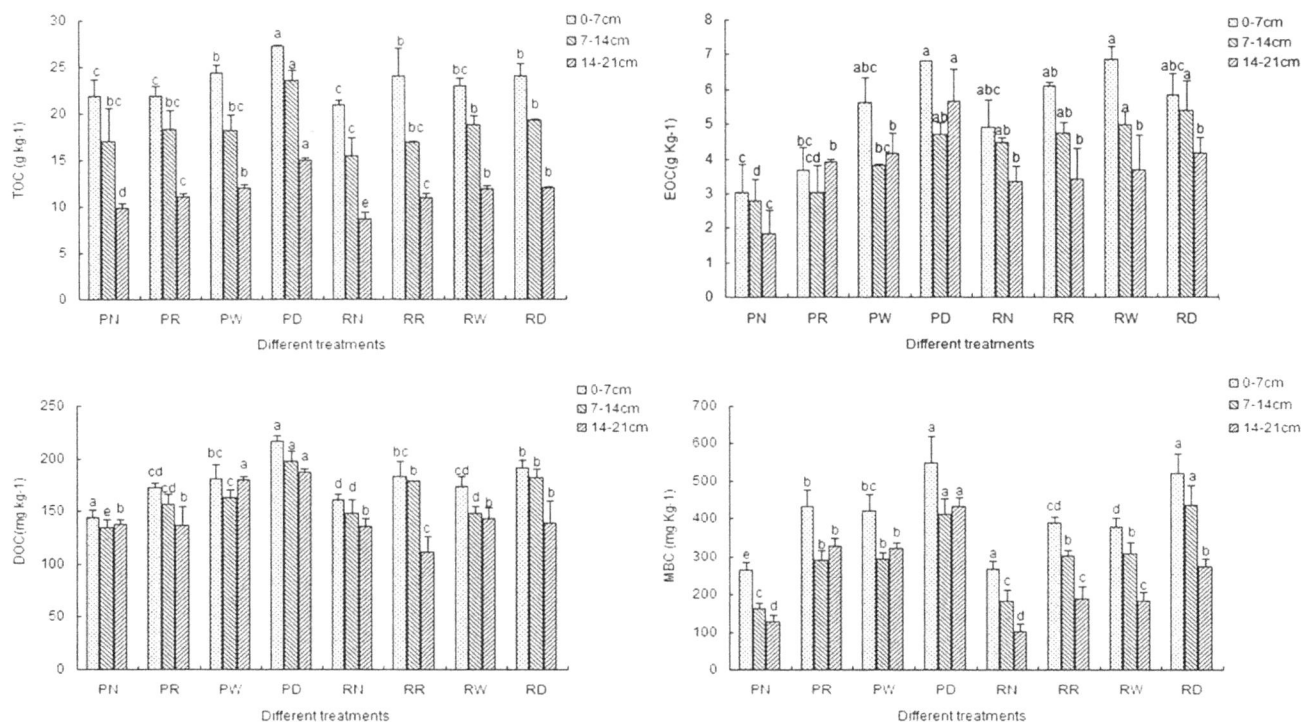

Figure 1. Effects of eight treatments on soil TOC, EOC, DOC and MBC contents at three depths.

combined with straw return on cropland ecosystem in rice-wheat rotation system have been studied during these years. However, most of them are focused on soil physical-chemical properties, soil nutrient and crop yield. To our knowledge, there is little information about the effects of different tillage and straw return on soil labile C fractions in rice-wheat rotation system. Thus, the objectives of this study were (1) to quantify the effects of tillage methods and straw return on soil TOC, MBC, DOC, EOC contents in the rice-wheat rotation system in the Yangtze River Delta, and (2) to explore an optimal management practice combination of tillage and straw return for improving the soil quality and increasing the local crop production.

Table 1. Linear correlations among soil TOC and labile organic C fractions at the 0–21 cm depth.

Index	TOC	EOC	DOC	MBC
TOC	1			
EOC	0.638**	1		
DOC	0.758**	0.684**	1	
MBC	0.741**	0.639**	0.908**	1

TOC: total organic carbon; MBC: microbial biomass carbon; DOC: dissolved organic carbon; EOC: easily oxidizable carbon.
* $P<0.05$.
**$P<0.01$.

Materials and Methods

Site Description

The experiment was conducted at Changwang Country, Youfang Town, Yangzhong City, Jiangsu Province, China (119°42′–119°58′E, 32°–2°19′N, 4–4.5 m above mean sea level) from November, 2009 to June, 2011. Access to the study site was obtained in the form of a rent contract, in which we had to confirm that our study did not involve endangered or protected species.

The experimental site had a subtropical monsoon climate with an average annual precipitation of 1000 mm, an average annual temperature of 15.1°C, and a mean annual sunshine hour of 2135 h. The soil of the experimental site was a loam and classified as an anthrosols. Rice-wheat double cropping system was the most important cropping system in the region. The main properties of soil (0–20 cm depth) sampled in November 2009 were as follows: soil organic matter 29.81 g kg^{-1}; alkali-hydrolyzale nitrogen 194.02 mg kg^{-1}; available phosphorus 13.60 mg kg^{-1}; available potassium 51.45 mg kg^{-1}; and pH 7.34.

The variety of wheat used in this study was Yangmai16 *(Triticum aestivum* L.*)* and rice was Nangeng47 *(Oryza sativa* L.*)*.

Experimental Design and Field Managements

The experiment had a split-plot design with two tillage methods in the main plots and four straw return modes in subplots with three replications (6 m×5 m). Tillage methods included plowing tillage (P) and rotary tillage (R). Straw return modes were as follows: no straw return (N), only rice straw return (R), only wheat straw return (W), and rice and wheat straw both return (D). There were eight treatments in this study: (1) plowing tillage with no straw return (PN: rice with plowing tillage-wheat with plowing tillage); (2) plowing tillage with only rice straw return (PR: rice with

Table 2. Effects of different tillage factor and straw factor on soil TOC, EOC, DOC and MBC at 0–7 cm, 7–14 cm and 14–21 cm depth.

Soil dept(cm)	Factors	Treatments	TOC(g kg^{-1})	labile organic carbon fractions contents		
				EOC (g kg^{-1})	DOC(mg kg^{-1})	MBC(mg kg^{-1})
0–7	Tillage factor	P	23.87±2.67a	4.78±1.23a	178.36±33.74a	417.16±17.20a
		R	23.00±2.38a	5.92±1.48a	177.05±30.96a	389.22±36.56b
	Straw factor	N	21.40±1.87c	3.97±0.92b	152.25±10.76c	266.73±11.68c
		R	23.01±2.97bc	4.88±1.40ab	177.61±33.57b	411.67±27.32b
		W	23.65±1.52b	6.23±1.56a	176.92±30.87b	398.57±27.34b
		D	25.68±2.63a	6.32±1.68a	204.04±22.67a	535.79±20.31a
7–14	Tillage factor	P	19.30±1.23a	3.58±1.76b	163.05±37.30a	311.42±43.89a
		R	17.64±2.48b	4.89±1.52a	164.11±27.23a	306.50±36.09a
	Straw factor	N	16.28±3.20c	3.62±1.76c	141.49±18.49d	172.63±16.58c
		R	17.63±2.86bc	3.90±2.70bc	167.39±21.63b	315.19±20.00b
		W	18.50±1.75b	4.39±0.54ab	155.47±22.41c	313.93±14.31b
		D	21.47±2.02a	5.04±0.76a	189.98±17.71a	434.09±11.65a
14–21	Tillage factor	P	12.00±2.56a	3.89±1.42a	160.06±30.21a	280.68±58.67a
		R	10.90±2.61b	3.64±1.14a	132.00±25.31b	187.08±44.86b
	Straw factor	N	9.28±1.69d	2.58±1.18c	136.34±11.66b	114.99±18.05d
		R	10.99±1.96c	3.66±1.26b	123.67±29.32c	239.95±39.07b
		W	11.97±3.06b	3.91±1.39b	161.19±25.91a	238.92±35.06c
		D	13.56±1.75a	4.90±0.68a	162.93±26.49a	341.67±31.14a

Different letters in a line under a specific influence factor denote significant difference at the 5% level. Different capitals in a column at different soil depths and treatments present significant different at the 0.05 level. TOC: total organic carbon; MBC: microbial biomass carbon; DOC: dissolved organic carbon; EOC: easily oxidizable carbon.

plowing tillage - wheat with plowing tillage+rice straw return); (3) plowing tillage with only wheat straw return (PW: rice with plowing tillage+wheat straw return - wheat with plowing tillage); (4) plowing tillage with rice and wheat straw both return(PD: rice with plowing tillage+wheat straw return -wheat with plowing tillage+rice straw return); (5) rotary tillage with no straw return (RN: rice with rotary tillage - wheat with rotary tillage); (6) rotary tillage with only rice straw return (RR: rice with rotary tillage - wheat with rotary tillage+rice straw return); (7) rotary tillage with only wheat straw return (RW: rice with rotary tillage+wheat straw return - wheat with rotary tillage); (8) rotary tillage with rice and wheat straw both return (RD: rice with rotary tillage+wheat straw return - wheat with rotary tillage+rice straw return).

The experimental site was cultivated with a rice-wheat rotation prior to November 2009, where wheat was planted with plowing tillage from November to the following June, and rice was transplanted by plowing tillage from June to November. In this study, after wheat or rice was harvested, they were cultivated at a depth of 10–15 cm by rotary cultivation in rotary tillage plots while for the plowing tillage plots, cultivation was at a depth of 20–25 cm with a moldboard plough. Before the rice and wheat were sown, the plowing tillage plots were disked and moldboard plowed for weed control and bedding. This was followed by an application of fertilizer. For straw returned plots, the wheat and rice straw were cut into 8–10 cm after air-dried, and placed back on the surface of the soil in June or November of each year, with returned amount of 6000 kg·hm^{-2} for both wheat and rice straw.

In this study, wheat was sown on November 3, 2009 and November 24, 2010, respectively. The seed quantity was 150 kg hm^{-1} by machine. The base fertilizer applied before sowing was 135 kg·hm^{-2} pure N, 67.5 kg·hm^{-2} P$_2$O$_5$, and 67.5 kg·hm^{-2} K$_2$O, and topdressing was at the elongation stage with 135 kg hm^{-1} pure N. For all treatments, N was applied in the form of CO(NH$_2$)$_2$, and the fertilizer in wheat seasons was applied at the same rate. The wheat was harvested on June 3, 2010 and June 8, 2011, respectively. Rice was transplanted at about 3–4 seedlings per hole, 255,000 holes per hectare on June 15, 2010 and June 15, 2011, respectively. The rice was fertilized just before transplanting with a base fertilizer (120 kg hm^{-1} pure N; 60 kg hm^{-1} P$_2$O$_5$; 60 kg hm^{-1} K$_2$O), and at tillering stage and earing stage with a topdressing (180 kg hm^{-1} N (3:1 ratio)). The fertilizer applied in the two rice seasons were the same. The rice was harvested on November 12, 2010 and November 29, 2011, respectively. The pesticide management of both rice and wheat seasons was in accordance with the conventional, and all other management procedures were identical for the eight treatments.

Soil Sampling and Analytical Methods

Soil samples were collected by a geotome (5 cm diameter) on October 29, 2011 (just before the rice was harvested). Five random locations were chosen in each of the 24 observational plots and samples were taken from each location at three soil depths (0–7, 7–14, and 14–21 cm) separately. Soil samples from each depth were about 200 g, fully blended. The collected moist samples were ground and sieved through a 10 mesh screen. Sieved soil samples were divided into two sub-samples. One was air-dried and sieved again through 100 mesh screen for determining soil TOC and EOC. Another was immediately stored in 4°C refrigerators for determining DOC and MBC. During sieving, crop residues, root material and stones were removed.

Table 3. Affecting force analysis of different tillage and straw return and their interaction on soil TOC, EOC, DOC and MBC at 0–7 cm, 7–14 cm and 14–21 cm depth.

Soil depth(cm)	Difference source	Affecting force(%)			
		TOC	EOC	DOC	MBC
0–7	Block	10.35	8.53	6.55	0.26
	Tillage	0.34	28.06**	0.64	2.06**
	Straw return	19.69*	24.79**	38.58**	95.95**
	Straw return × tillage	10.71	4.34	4.86	0.93
	Error	58.92	34.28	49.37	0.79
7–14	Block	19.97**	5.36	11.53	0.29
	Tillage	0.02	13.03*	0.85	0.07
	Straw return	21.41**	13.16	39.21**	97.83**
	Straw return × tillage	26.30**	3.89	11.12*	0.96
	Error	32.32	64.56	37.28	0.86
14–21	Block	10.22	1.61	16.66**	0.27
	Tillage	1.05	3.21	11.77**	23.81**
	Straw return	32.35**	25.06**	35.65**	70.17**
	Straw return × tillage	14.72*	13.95*	9.60*	4.93**
	Error	41.66	56.17	26.33	0.82

The affecting force of tillage = tillage variables (square)/total variables (sum of total squares) ×100%; the affecting force of straw = straw variables (square)/total variable (sum of total squares) ×100%; the affecting force of the interaction = interaction variable (square)/total variables ((total squares of sum) ×100%. TOC: total organic carbon; MBC: microbial biomass carbon; DOC: dissolved organic carbon; EOC: easily oxidizable carbon.
* $P<0.05$.
**$P<0.01$.

Total organic C (TOC) concentration was determined by oxidation with potassium dichromate and titration with ferrous ammonium sulphate [28].

Dissolved organic C (DOC) was extracted from 10 g of moist soil with 1:2.5 ratio of soil to water at 25.8°C [29]. After shaking for 1 h and centrifuging for 10 min at 4500 r min^{-1}, the supernatant was filtered with a 0.45 mm membrane filter. The filtrate was measured by oxidation with potassium dichromate and titration with ferrous ammonium sulphate.

Microbial biomass C (MBC) was analyzed by the fumigation extraction method [30]. Each sample was weighed into two equivalent portions, one was fumigated for 24 h with ethanol-free chloroform and the other was the unfumigated control. Both fumigated and unfumigated soils were shaken for 1 h with 0.5 M K$_2$SO$_4$ (2:5 soil: extraction ratio), centrifuged and filtered.

Easily oxidizable C (EOC) was measured as described by Blair et al. [3]. Finely ground air-dried soil samples were reacted with 333 mmol L^{-1} KMnO$_4$ by shaking at 60 r min^{-1} for 1 h. The suspension was then centrifuged at 2000 r min^{-1} for 5 min. The supernatant was diluted and measured spectrophotometrically at 565 nm. All soil samples were analyzed in triplicate.

Data Analysis

The SPSS 16.0 analytical software package was used for all statistical analyses. A 2-factor analysis of variance (ANOVA) was employed for difference test among eight treatments at $P<0.05$, with separation of means by least significant difference (LSD). Correlation analysis were performed to determine correlations

among soil labile organic C fractions in the 0–21 cm soil depth, and the significant probability levels of the results were given at the $P<0.05$ (*) and $P<0.01$ (**), respectively. Moreover, the affecting force analysis of tillage factor, straw factor and their interaction influence on labile organic C fractions was calculated based on the method of Leng [31]: the affecting force of tillage = tillage variables (square)/total variables (sum of total squares) ×100%; the affecting force of straw return = straw return variables (square)/total variable (sum of total squares) ×100%; the affecting force of interaction = interaction variable (square)/total variables ((total squares of sum) ×100%.

Results

Soil TOC, DOC, MBC and EOC Contents in Different Treatments

As shown in Fig. 1, the different treatments significantly affected the contents of soil TOC and labile organic C fractions, where PD generally had the highest contents of TOC, DOC, MBC and EOC at the three soil depths. Crop straw return treatments (PR, PW, PD, RR, RW, RD) had consistently higher amount of TOC and labile organic C fractions at the three soil depths than without crop straw return treatments (PN, RN). Moreover, PN had significantly lower TOC, DOC, MBC and EOC at 0–7 cm and 7–14 cm, and RN had the lowest TOC and MBC at 14–21 cm compared to other treatments (Fig. 1). Soil TOC and labile organic C fractions generally decreased with an increase in soil depth under all treatments. As expected, soil TOC and labile organic C fractions were significantly and positively correlated with each other (Table 1).

Effects of Different Tillage Methods on Soil TOC and Labile Organic C Fractions

Tillage had a significant effect on MBC at 0–7 cm soil depth, but seldom on soil TOC, DOC and EOC. Soil TOC, DOC and MBC contents were all higher under plowing tillage (P) than rotary tillage (R), while EOC was opposite at 0–7 cm soil depth (Table 2). At 7–14 cm, soil EOC under rotary tillage (R) was significantly higher than plowing tillage (P), but TOC had the contrary results, and there were no significant differences on DOC and MBC (Table 2). At 14–21 cm, soil TOC, DOC and MBC were significantly higher under plowing tillage (P) than rotary tillage (R), except EOC (Table 2).

Effects of Different Straw Return on Soil TOC and Labile Organic C Fractions

Straw return had significant effects on soil TOC and labile organic C at the three depths as shown in Table 2. In general, soil TOC and three labile organic C ranged in the following order: rice and wheat straw both return>only wheat or rice straw return>no straw return at three depths (Table 2). At 7–14 cm depth, only rice straw return in the wheat season had significantly higher DOC than only wheat straw return in the rice season (Table 2). However, at 14–21 cm depth, except for MBC, soil TOC, EOC, DOC under only rice straw return in the wheat season were lower than only wheat straw return in the rice season (Table 2). Moreover, there were significant differences in TOC and MBC among the four straw return at 14–21 cm depth (Table 2).

Affecting Force Analysis of Different Tillage, Straw Return and their Interaction on Soil TOC and Labile Organic C Fractions

Affecting force of different tillage, straw return and their interaction on soil TOC and labile organic C were different with increasing soil depth (Table 3). The affecting force of tillage increased with the increase of soil depth (Table 3). Tillage had significant affecting force on EOC and MBC at 0–7 cm depth, but seldom on TOC and DOC (Table 3). At 7–14 cm depth, the affecting force of tillage on EOC was lower than at 0–7 cm, and there was no significant affecting force on soil TOC and other labile organic C fractions (Table 3). At 14–21 cm depth, tillage had significant affecting force on DOC and MBC. However, there was no significant affecting force on TOC and EOC (Table 3).

Straw return had significant affecting force on soil TOC, DOC and MBC at the three depths, but there was no significant affecting force on EOC at 7–14 cm (Table 3). Among the four indictors, the affecting force of straw return on MBC was the greatest at the three depths, which reached 95.95%, 97.83% and 70.17%, respectively (Table 3).

The affecting force of the interaction generally increased with an increase in soil depth (Table 3). At 0–7 cm soil depth, the interaction had no significant affecting force on soil TOC and labile organic C (Table 3). Soil TOC and DOC were mainly dominated by the interaction at 7–14 cm depth, but there was no significant affecting force on MBC and EOC. At 14–21 cm depth, the interaction had significant affecting force on soil TOC and all labile organic C (Table 3).

Discussion

Suitable soil tillage practice can increase the SOC content, and improve SOC density of the plough layer [32]. The effect size of tillage methods on SOC dynamics depends on the tillage intensity [33]. Compared to conventional tillage (CT), no-tillage and reduced tillage could significantly improve the SOC content in cropland. Frequent tillage under CT easily exacerbate C-rich macroaggregates in soils broken down due to the increase of tillage intensity, then forming a large number of small aggregates with relatively low organic carbon content and free organic matter particles. Free organic matter particles have poor stability and are easy to degradation, thereby causing the loss of SOC [33,34]. In our study, at 0–7 cm soil depth, soil EOC under plowing tillage was lower than rotary tillage (Table 2). The reason could be attributed to the tillage method. Tillage increases the effect of drying–rewetting and freezing-thawing on soil, which increases macroaggregate susceptibility to disruption [21,35,36], and accelerates the labile organic C mineralization and SOM degradation, thus increasing the loss of EOC [14,37]. At 7–14 cm, rotary tillage had higher soil EOC and DOC than plowing tillage, but lower at 14–21 cm soil depth, indicating that tillage affected the vertical distribution of EOC and DOC (Table 2). The difference in soil condition after plowing tillage or rotary tillage affects the rate of straw decomposition, thereby resulting in a difference in the soil nutrient accumulation [38]. Similarly, Liu et al. [39] have found that SOM content under plowing and rotary tillage at deeper soil both were higher than that of the upper soil. The reason might be that rotary tillage and plowing tillage mixed crop straw into the deeper soil layer, making SOM well-distributed at different depths [40].

Carbon input can be increased by adopting straw return in cropland [14]. Fresh residues are C source for microbial activity and nucleation centers for aggregation when returned to cropland. The enhanced microbial activity induces the binding of residue and soil particles into macroaggregates [34,41], which could increase aggregates stability, fix the unstable C, thus improving the concentration of SOC [42] and increasing C sequestration [14]. In our research, straw return had significantly higher soil TOC and labile organic carbon fractions contents at the three soil depths than no crop straw return (Table 2). Soil TOC and labile organic C fractions in both rice and wheat straw return treatments were higher than only wheat or rice straw return (Table 2), indicating that straw return plays an very important role in increasing soil TOC and labile organic C fractions. Similar observations have been reported by other researchers [43–45]. At the three depths, soil TOC in the treatment of only wheat straw return in rice season was higher than only rice straw return in wheat season, moreover, the difference was significant ($p<0.05$) at 7–14 cm (Table 2). This was related to the relatively near-surface higher water content and favorable soil temperature during the rice growing season, resulting in relatively fast straw decomposition [46]. The decomposition of wheat straw provides enough energy and carbon source for soil microorganisms, thus increases the microorganisms' activities. Alternatively, after wheat straw return, the high temperature and humid conditions accelerates the reduction of the C/N ratio of the straw, allowing for sufficient decomposition. More nutrients are released and utilized by the crops, which therefore lightens the pressure of burning straw and improves the soil quality [47]. According to table 1, the study showed that MBC was affected by the straw return factor with an affecting force of 95.95% at 0–7 cm depth and 97.83% at 7–14 cm depth. The probable explanation maybe that crop residue might enter the labile C pool, provide substrate for the soil microorganisms, and contribute to the accumulation of labile C [48].

In our study, PD had the highest content of soil TOC at all the three soil depths (Fig. 1). The reason might be that plowing tillage made the soil and straw in the plow layer turned over quarterly, which increased the stability of the TOC content at each soil layer [49]. In addition, the rice and wheat straw were both returned under PD treatment from 2009, plowing tillage made much SOM enter into the soil and accumulate [45]. However, Tian et al. [50] found that rotary tillage with straw return had higher SOC than plowing tillage with straw return at 0–10 cm soil depth in wheat field. The diverse results might be due to the different regional climate, soil type, crop rotation and the length of study [22]. In this study, at upper soil layer, the interaction effect between tillage and straw return was not significant, but generally increased with an increase in soil depth (Table 3). Rajan et al [2] also found that single effect of residue application was not significant but its significance became apparent after its interaction with tillage system.

In our study, soil labile organic C fractions were significantly and positively correlated with TOC concentrations at 0–21 cm soil depth (Table 1). Such correlations suggested that TOC was a major determinant of soil labile organic C fractions. MBC, DOC and EOC were also significantly and positively correlated with each other in this study (Table 1). The results were consistent with Chen et al. [14], who reported similar correlations between soil TOC, labile organic C fractions (MBC, DOC, particulate organic C, EOC and hot-water extractable C), and macroaggregate C within 0–15 cm depth. Dou et al. [51] also observed the same results. MBC is the living part of SOM, which plays an important role in maintenance of soil fertility [52]. It serves as a sensitive indicator of change and future trends in organic matter level [53]. Dissolved organic C consists of organic compounds present in soil solution, acts as a substrate for microbial activity, and is the primary energy source for soil microorganisms [1]. Easily oxidizable C partly reflects enzymatic decomposition of labile SOC [54]. Therefore, it is not surprising to find the positive correlations among the labile C pools as they have a close association with each other.

Conclusions

In this study, after 2 years of a rice-wheat rotation, soil TOC and labile organic C fractions in PR, PW, PD, and RR, RW, RD were all higher than PN and RN. PD and RD had more significant effects on EOC, DOC and MBC compared to other treatments at 0–21 cm depth. Soil TOC and labile organic C fractions were highly correlated with each other. Under short-term conditions, rice and wheat straw both return in rice-wheat rotation system can increase SOC content and improve soil quality in the Yangtze

River Delta, which is a suitable agricultural practice in this region under rice-wheat cropping system.

Acknowledgments

We thank Mr. Kejun Gu (Jiangsu Province Academy of Agricultural Sciences, China) for supporting the field experiment, Dr. Changqing Chen (Nanjing Agricutural University, China) for his kind help in data analysis,and Mr. Michael Rickaille (University of the West Indies) for his critical reading.

Author Contributions

Conceived and designed the experiments: LQZ. Performed the experiments: MFY NJH ZWZ. Analyzed the data: MFY. Contributed reagents/materials/analysis tools: LQZ. Wrote the paper: LQZ NJH XHZ.

References

1. Haynes RJ (2005) Labile organic matter fractions as central components of the quality of agricultural soils: an overview. Adv. Agron 85: 221–268.
2. Rajan G, Keshav RA, Zueng-Sang C, Shree CS, Khem RD (2012) Soil organic carbon sequestration as affected by tillage, crop residue, and nitrogen application in rice–wheat rotation system. Paddy Water Environ 10: 95–102.
3. Blair GJ, Lefory RDB, Lise L (1995) Soil carbon fractions based on their degree of oxidation and the development of a carbon management index for agricultural system. Aust. J. Agric. Res 46: 1459–1466.
4. Ghani A, Dexter M, Perrott WK (2003) Hot-water extractable carbon in soils: a sensitive measurement for determining impacts of fertilization, grazing and cultivation. Soil Biol. Biochem 35: 1231–1243.
5. Rudrappa L, Purakayastha TJ, Singh D, Bhadraray S (2006) Long-term manuring and fertilization effects on soil organic carbon pools in a Typic Haplustept of semi-arid sub-tropical India. Soil Till. Res 88: 180–192.
6. Yang CM, Yang LZ, Zhu OY (2005) Organic carbon and its fractions in paddy soil as affected by different nutrient and water regimes. Geoderma 124: 133–142.
7. Baker JM, Ochsner TE, Venterea RT, Griffis TJ (2007) Tillage and soil carbon sequestration–what do we really know? Agric Ecosyst Environ 118, 1–5.
8. Hernanz JL, L'opez R, Navarrete L, S'anchez-Gir'on V (2002) Long-term effects of tillage systems and rotations on soil structural stability and organic carbon stratification in semiarid central Spain. Soil Till. Res 66 (2): 129–141.
9. Shan YH, Yang LZ, Yan TM, Wang JG (2005) Downward movement of phosphorus in paddy soil installed in large-scale monolith lysimeters. Agr. Ecosyst. Environ 111 (1–4): 270–278.
10. Zhang ZJ (1998) Effects of long-term wheat-straw returning on yield of crop and soil fertility. Chinese Journal of Soil Science 29(4): 154–155.
11. Sun X, Liu Q, Wang DJ, Zhang B (2007) Effect of long-term application of straw on soil fertility. Chinese Journal of Eco-Agriculture 16(3): 587–592.
12. West TO, Post WM (2002) Soil organic carbon sequestration rates by tillage and crop rotation. Soil Science Society of American Journal 66: 1930–1946.
13. Liu SP, Nie XT, Zhang HC, Dai QG, Huo ZY, et al. (2006) Effects of tillage and straw returning on soil fertility and grain yield in a wheat-rice double cropping system. Transactions of the CSAE 22(7): 48–51.
14. Chen HQ, Hou RX, Gong YS, Li HW, Fan MS, et al. (2009) Effects of 11 years of conservation tillage on soil organic matter fractions in wheat monoculture in Loess Plateau of China. Soil Till. Res 106: 85–94.
15. Wang DD, Zhou L, Huang SQ, Li CF, Cao CG (2013) Short-term Effects of Tillage Practices and Wheat-straw Returned to the Field on Topsoil Labile Organic Carbon Fractions and Yields in Central China. Journal of Agro-Environment Science 32(4): 735–740.
16. Miller AJ, Amundson R, Burke IC, Yonker C (2004) The effect of climate and cultivation on soil organic C and N. Biogeoche mistry 67: 57–72.
17. Diekow J, Mielniczuk J, Knicker H, Bayer C, Dick DP, et al. (2005) Soil C and N stocks as affected by cropping systems and nitrogen fertilisation in a southern Brazil Acrisol managed under no-tillage for 17 years. Soil and Tillage Research 81, 87–95.
18. Galantini JA, Senesi N, Brunetti G, Rosell R (2004) In fluence of texture on organic matter distribution and quality and nitrogen and sulphur status in semiarid Pampean grassland soils of Argentina. Geoderma 123, 143–152.
19. Ouédraogo E, Mando A, Stroosnijder L (2006) Effects of tillage, organic resources and nitrogen fertiliser on soil carbon dynamics and crop nitrogen uptake in semi-arid West Africa. Soil and Tillage Res 91: 57–67.
20. Yamashita T, Feiner H, Bettina J, Helfrich M, Ludwig B (2006) Organic matter in density fractions of water-stable aggregates in silty soils: effect of land use. Soil Biology and Biochemistry 38: 3222–3234.
21. Paustian K, Collins HP, Paul EA (1997) Management controls in soil carbon. In: Paul, E.A., Paustian, K.A., Elliott, E.T., Cole, C.V. (Eds). Soil Organic Matter in Temperate Ecosystems: Long Term Experiments in North America. RC, Boca Raton, FL 15–49.
22. Puget P, Lal R (2005) Soil organic carbon and nitrogen in a Mollisol in central Ohio as affected by tillage and land use. Soil Till. Res 80, 201–213.
23. Ding LL, Cheng H, Liu ZF, Ren WW (2013) Experimental warming on the rice-wheat rotation agro-ecosystem. Plant Science Journal 31(1): 49–56.
24. Editorial Board of China Agriculture Yearbook (2012) China Agriculture Yearbook 2009, Electronic Edition. China Agriculture Press, Beijing, China.
25. Hao JH, Ding YF, Wang QS, Liu ZH, Li GH, et al. (2010) Effect of wheat crop straw application on the quality of rice population and soil properties. Journal of Nanjing Agricultural University 33(3): 13–18.
26. Zhu LQ, Zhang DW, Bian XM (2011) Effects of continuous returning straws to field and shifting different tillage methods on changes of physical-chemical properties of soil and yield components of rice. Chinese Journal of Soil Science 42(1): 81–85.
27. Liu SP, Nie XT, Zhang HC, Dai QG, Huo ZY, et al. (2006) Effects of tillage and straw returning on soil fertility and grain yield in a wheat-rice double cropping system. Transactions of the CSAE 22(7): 48–51.
28. Lu RK (1999) Soil agricultural chemistry analysis. China's agricultural science and technology press, 106–110.
29. Jiang PK, Xu QF, Xu ZH, Cao ZH (2006) Seasonal changes in soil labile organic carbon pools within a Phyllostachys praecox stand under high rate fertilization and winter mulch in subtropical China. For. Ecol. Manage 236: 30–36.
30. Vance F, Brookes P, Jenkinson D (1987) Microbial biomass measurements in forest soils: the use of the chloroform fumigation-incubation method in strongly acid soils. Soil Biochem 19: 697–702.
31. Leng SC (1992) Biological Statistic and Field Experimental Design. Beijing: China Radio& Television Press (in chinese).
32. Duan HP, Niu YZ, Bian XM (2012) Effects of tillage mode and straw return on soil organic carbon and rice yield in direct seeding rice field. Bulletin of Soil and Water Conservation 32(3): 23–27.
33. Yang JC, Han XG, Huang JH, Pan QM (2003) The dynamics of soil organic matter in cropland responding to agricultural practices. Acta Ecologica Sinica 23(4): 787–796.
34. Six J, Elliott ET, Paustian K (1999) Aggregate and soil organic matter dynamics under conventional and no-tillage systems. Soil. Sci. Soc. Am. J. 63, 1350–1358.
35. Beare MH, Hendrix PF, Coleman DC (1994) Water-stable aggregates and organic matter fractions in conventional-tillage and no-tillage soils. Soil Sci. Soc. Am. J. 58, 777–786.
36. Mikha MM, Rice CW (2004) Tillage and manure effects on soil and aggregate-associated carbon and nitrogen. Soil Sci. Soc, AM. J 68: 809–816.
37. Wang J, Zhang RZ, Li AZ (2008) Effect on soil active carbon and C cool management index of different tillages. Agricultural Research in the Arid Areas 26(6): 8–12.
38. Li XJ, Zhang ZG (1999) Influence on soil floods properties of mulching straws and soil-returning straw. Territory and Natural Resources Study 1: 43–45.
39. Liu DY (2009) Physiological and ecological mechanism of stable and high yield of broadcasted rice in paddy field with high standing-stubbles under no-tillage condition. PhD: Sichuan Agricultural University.
40. Gao YJ, Zhu PL, Huang DM, Wang ZM, Li SX (2000) Long-term impact of different soil management on organic matter and total nitrogen in rice-based cropping system. Soil and Environmental Sciences 9(1): 27–30.
41. Jastrow JD (1996) Soil aggregate formation and the accrual of particulate and mineral associated organic matter. Soil Biol. Biochem. 28, 656–676.
42. Govaerts B, Sayre KD, Lichter K, Dendooven L, Deckers J (2007) Influence of permanent raised bed planting and residue management on physical and chemical soil quality in rain fed maize/wheat systems. Plant Soil 291, 39–54.
43. Stockfisch N, Forstreuter T, Ehlers W (1999) Ploughing effects on soil organic matter after twenty years of conservation tillage in Lower Saxony, Germany. Soil Till. Res 52: 91–101.

44. Chen SH, Zhu ZL, Liu DH, Shu L, Wang CQ (2008) Influence of straw mulching with no-till on soil nutrients and carbon pool management index. Plant Nutrition and Fertilizer Science 14(4): 806−809.

45. Song MW, Li AZ, Cai LQ, Zhang RS (2008) Effects of different tillage methods on soil organic carbon pool. Journal of Agro-Environment Science 27(2): 622−626.

46. Zuo YP, Jia ZK (2004) Effect of soil moisture content o n straw decomposing and its dynamic changes. Joural of Northwest Sci-Tech University of Agri. and For. (Nat. Sci. Ed.) 32(5): 61–63.

47. Dai ZG (2009) Study on nutrient release characteristics of crop residue and effect of crop residue returning on crop yield and soil fertility. PhD: Huazhong Agricultural University (in Chinese).

48. Li CF, Yue LX, Kou ZK, Zhang ZS, Wang JP, et al. (2012) Short-term effects of conservation management practices on soil labile organic carbon fractions under a rape–rice rotation in central China. Soil Till, Res 119: 31–37.

49. Zhang P, Li H, Jia ZK, Wang W, Lu WT, et al. (2011) Effects of straw returning on soil organic carbon and carbon mineralization in Semi-arid areas of southern Ningxia, China. Journal of Agro-Environment Science 30(12): 2518−2525.

50. Tian SZ, Ning YT, Wang Y, Li HJ, Zhong WL, et al. (2010) Effects of different tillage methods and straw-returning on soil organic carbon content in a winter wheat field. Chinese Journal of Applied Ecology 21(2): 373–378.

51. Dou FG, Wright AL, Hons FM (2008) Sensitivity of labile soil organic carbon to tillage in wheat-based cropping systems. Soil Sci. Soc. Am. J. 72: 1445–1453.

52. Wu TY, Schoenau JJ, Li FM, Qian PY, Malhi SS, et al. (2004) Influence of cultivation and fertilization on total organic carbon and carbon fractions in soils from the Loess Plateau of China. Soil & Till, Res 77: 59–68.

53. Gregorich EG, Ellert BH, Gregorich EG, Carter MR, Monreal CM, et al. (1994) Towards a minimum data set to assess soil organic matter quality in agricultural soils. Can. J. Soil Sci 74: 367–385.

54. Loginow W, Wisniewski W, Gonet SS, Ciescinska B (1987) Fractionation of organic carbon based on susceptibility to oxidation. Pol. J. Soil Sci 20: 47–52.

6

Prunus persica Crop Management Differentially Promotes Arbuscular Mycorrhizal Fungi Diversity in a Tropical Agro-Ecosystem

Maria del Mar Alguacil[1]*, Emma Torrecillas[1], Zenaida Lozano[2], Maria Pilar Torres[3], Antonio Roldán[1]

1 CSIC-Centro de Edafología y Biología Aplicada del Segura, Department of Soil and Water Conservation, Campus de Espinardo, Murcia, Spain, 2 Universidad Central de Venezuela (UCV), Facultad de Agronomía, Instituto de Edafología, El Limón, Campus Universitario, Maracay, Venezuela, 3 Departamento de Biología Aplicada, Area de Botánica, Universidad Miguel Hernández, Elche, Alicante, Spain

Abstract

Due to the important role of arbuscular mycorrhizal fungi (AMF) in ecosystem functioning, determination of the effect of management practices on the AMF diversity in agricultural soils is essential for the sustainability of these agro-ecosystems. The objective of this study was to compare the AMF diversity in Prunus persica roots under two types of fertilisation (inorganic, with or without manure) combined with integrated or chemical pest management in a Venezuelan agro-ecosystem. The AM fungal small-subunit (SSU) rRNA genes were subjected to PCR, cloning, sequencing and phylogenetic analyses. Twenty-one different phylotypes were identified: 15 belonged to the genus Glomus, one to Claroideoglomus, two to Paraglomus, one to Acaulospora, one to Scutellospora and one to Archaeospora. The distribution of the AMF community composition differed as a consequence of the treatment effects. The treatment combining organic and inorganic fertilisation with chemical pest control had the highest AMF richness and the treatment combining inorganic fertilisation with chemical pest had the lowest. The real causes and effects of these differences in the AMF community are very difficult to establish, since the crop management regimes tested were composed of several interacting factors. In conclusion, the crop management practices can exert a significant influence on the populations of AMF. The treatment combining organic and inorganic fertilisation with chemical pest control appears to be the most suitable agricultural management strategy with respect to improving the AMF diversity in this crop under tropical conditions, and thus for maintaining the agricultural and environmental sustainability of this agro-ecosystem.

Editor: Lee A. Newsom, The Pennsylvania State University, United States of America

Funding: This research was supported by funding from Ramon and Cajal programme (Ministerio de Educación y Ciencia, Spain). MM Alguacil was supported by the Ramon and Cajal programme (Ministerio de Educación y Ciencia, Spain). The funders had no role in study design, data collection and analysis, decision to publish, or preparation of the manuscript.

Competing Interests: The authors have declared that no competing interests exist.

* E-mail: mmalguacil@cebas.csic.es

Introduction

The soil is a complex matrix containing microorganisms that play a key role in the functioning of terrestrial ecosystems. They mediate many processes, including nutrient cycles, organic matter decomposition, soil aggregate formation and plant performance. Arbuscular mycorrhizal fungi (AMF) are among the most-important soil microorganisms, being obligate symbionts in the roots of most land plants in both natural and agricultural ecosystems, where they increase plant uptake of mineral nutrients, especially phosphorus [1]. Other beneficial effects of AMF are plant growth promotion [2], increased tolerance of drought [3], heavy metals [4] and plant protection agents [5]. In fact, in a previous study carried out at the site that is also the subject of the current work [6], it was found that galls produced in Prunus persica roots due to infection with Meloidogyne incognita were extensively colonized by AMF, whose function might be to act as protection agents against opportunistic pathogens. Furthermore, the diversity of AMF influences a number of important ecosystem processes, including plant productivity, plant diversity and soil structure [7,8,9].

Due to the important role of AMF in ecosystem functioning, knowledge of the diversity of the AMF colonising the roots of crop plants in agricultural soils is essential for sustainable management of these agro-ecosystems. Fertilisation is a common practice used to increase the nutrient availability to crops and hence their yields. Studies carried out in recent years have considered the effect of different fertilisation treatments and cropping systems on AMF diversity. Thus, a general decrease in AMF diversity has been found with the use of mineral fertilisers [10,11,12,13,14], although not in all cases [15]. Others studies showed that fertilisation with manures stimulated the AMF populations [13,16,17], but studies on the species composition of the AMF community colonising crop roots in response to other management practices are scarce [18,19,20,21].

Prunus persica (L.) Batsch. (peach) is a fruit tree, native to Asia, introduced into Venezuela. Peach production in Venezuela is an activity that generates steady employment and is aimed primarily at the domestic market [22]. At present, there are 2,500 hectares that can produce more than 15,000 metric tons of fruit. Peach production in Venezuela is a cropping system in which fertilisers and pest control are combined in order to maximise yields while maintaining a suitable soil nutrient content.

Due to the economic importance of this fruit crop, the elucidation of whether there is a fertiliser/pest management combination that can maintain or increase the AMF diversity colonising the roots is an important step towards sustainable soil use and therefore protection of biodiversity. Therefore, the objective of this study was to compare the AMF diversity in *P. persica* roots under two fertilisation treatments (inorganic, with or without manure) combined with integrated or chemical pest management.

Materials and Methods

Ethics Statement

No specific permits were required for the described field studies since these locations are not privately-owned or protected in any way. Field studies did not involve endangered or protected species.

Study site and Sampling

The study was conducted in a *P. persica* orchard located at the "Colonia Tovar" Aragua State, in the north of Venezuela (latitude 10° 29' N, longitude 67° 07' W, 1790 msl). The climate is tropical temperate (mean annual temperature of 16.8 °C, annual average rainfall of 1271 mm). The soil was classified as a sandy loam Inceptisol [23]. The soil characteristics were: pH of 5.18, 5.75% clay, 40.5% silt, 53.75% sand, 6.46 cmol kg^{-1} of cationic exchange capacity, Total N 2.7 g kg^{-1}, Available P 32 $\mu g \ g^{-1}$, 5.9% organic matter and bulk density 1.29 g cm^{-3}

The plants used in this survey were 13-year old peach (*Prunus persica* (L). Batsch cv. Criollo Amarillo). The experimental sampling was a randomised block design with two factors and four replication blocks (100 m^2 each) in an experimental area of approximately 1000 m^2. The first factor consisted of two types of fertilization and the second factor was two different procedures of pest control. Four treatments were established in the sampling design. The treatments were selected in order to provide more sustainable practices to producers, since the regular management practices are limited to exclusively use of inorganic fertilization and an excessive use of pesticides

T1: Combination of organic and inorganic fertilization (ComFert) and integrated pest management (IntM).

T2: Inorganic fertilization (InorgFert) and integrated pest management (IntM).

T3: Inorganic fertilization (InorgFert) and chemical pest control (ChemM).

T4: Combination of organic and inorganic fertilization (ComFert) and chemical pest control (ChemM).

-ComFert consisted of application of chicken manure (1400 kg ha^{-1}), urea (140 kg ha^{-1}), complex fertilizer (NPK) 12-12-17 (280 kg ha^{-1}), and potassium sulfate (40 kg ha^{-1}).

-InorgFert consisted of application of urea (140 kg ha^{-1}), complex fertilizer (NPK) 12-12-17 (400 kg ha^{-1}) and potassium sulfate (70 kgha^{-1}).

-IntM consisted of weekly applications of *Beauveria bassiana* (300 g spores ha^{-1}) for one month, subsequently weekly applications for two months of *Trichoderma harzianum* (300 g spores ha^{-1}) and lastly applications every 15 days for two more months of *Trichoderma harzianum* (300 g spores ha^{-1}). We applied these products as biocontrol agents against fungal diseases.

-ChemM consisted of applications of different chemicals from the beginning of flowering aimed at insect pests control and subsequently the incidence of foliar diseases. Thus, weekly applications for six weeks of Profenofos 0.6 kg a.i. ha^{-1} (Curacron

®) + Mancozeb 8 kg i.a. ha^{-1} (Dithane ®) were made. For control of *Oidium leucoconium*, was applied twice the mixture Urea 10 kg ha^{-1} + Flusilazol 0.4 kg a.i. ha^{-1} (Punch ®) + Mancozeb 4 kg a.i. ha^{-1} (Dithane ®), then weekly applications of Mancozeb 4 kg a.i. ha^{-1} (Dithane ®) + Profenofos 0.6 kg a.i. ha^{-1} (Curacron ®) + Endosulfuran 2.8 kg a.i. ha^{-1} (Thionil). For control of *Monilia cinerea* were performed four weekly applications of Carbendazin 2 kg a.i. ha^{-1} (Bavistin ®).

The treatments were applied for one year and sampling was conducted after fruit harvest (February 2011). Four plants (one per block) of each treatment were sampled providing 16 samples in total. The roots were sampled using three soil cores from three points/single tree/block.

Root DNA extraction and PCR

All PCR experiments were run using DNA preparations consisting of pooled roots of individual plants. DNA extractions from 16 root samples were carried out.

For each sample (total 16), total DNA was extracted from (0.1 g) fine root material using a DNeasy plant mini Kit following the manufacturer's recommendations (Qiagen). The roots samples were placed into a 2-ml screw-cap propylene tube together with two tungsten carbide balls (3 mm) and beaten (3 min, 13000 r.p.m.) using a mixer mill (MM 400, Retsch, Haan, Germany). The extracted DNA was resuspended in 20 μl of water. Several dilutions of extracted DNA (1/10, 1/50, 1/100) were prepared and 2 μl were used as template. Partial small-subunit (SSU) ribosomal RNA gene fragments were amplified using nested PCR with the universal eukaryotic primers NS1 and NS4 [24]. PCR was carried out in a final volume of 25 ml using the "ready to go" PCR beads (Amersham Pharmacia Biotech, Piscataway, N.J.), 0.2 mM dNTPs and 0.5 mM of each primer (PCR conditions: 94 °C for 3 min, then 30 cycles at 94 °C for 30 s, 40 °C for 1 min, 72 °C for 1 min, followed by a final extension period at 72 °C for 10 min). Two μl of several dilutions (1/10, 1/20, 1/50 and 1/100) from the first PCR were used as template DNA in a second PCR reaction performed using the specific primers AML1 and AML2 [44]. PCR reactions were carried out in a final volume of 25 ml using the "ready to go" PCR beads (Amersham Pharmacia Biotech, Piscataway, N.J.), 0.2 mM dNTPs and 0.5 mM of each primer (PCR conditions: 94 °C for 3 min, then 30 cycles of 1 min denaturation at 94 °C, 1 min primer annealing at 50 °C and 1 min extension at 72 °C, followed by a final extension period of 10 min at 72 °C). Positive and negative controls using PCR positive products and sterile water respectively were also included in all amplifications. All the PCR reactions were run on a Perkin Elmer Cetus DNA Thermal Cycler. Reactions yields were estimated by using a 1.2% agarose gel containing GelRed^{TM} (Biotium).

Cloning and sequencing

The PCR products were purified using a Gel extraction Kit (Qiagen) cloned into pGEM-T Easy (Promega) and transformed into *Escherichia coli* (XL2-Blue). Thirty two positive transformants were screened in each resulting SSU rRNA gene library, using 0.8 units of RedTaq DNA polymerase (Sigma) and a re-amplification with AML1 and AML2 primers with the same conditions described above. Product quality and size were checked in agarose gels as described above. All clones having inserts of the correct size in each library were sequenced.

Clones were grown in liquid culture and the plasmid extracted using the QIAprep Spin Miniprep Kit (Qiagen). The sequencing was done by Laboratory of Sistemas Genómicos (Valencia, Spain) using the universal primers SP6 and T7. Sequence editing was done using the program Sequencher version 4.1.4 (Gene Codes

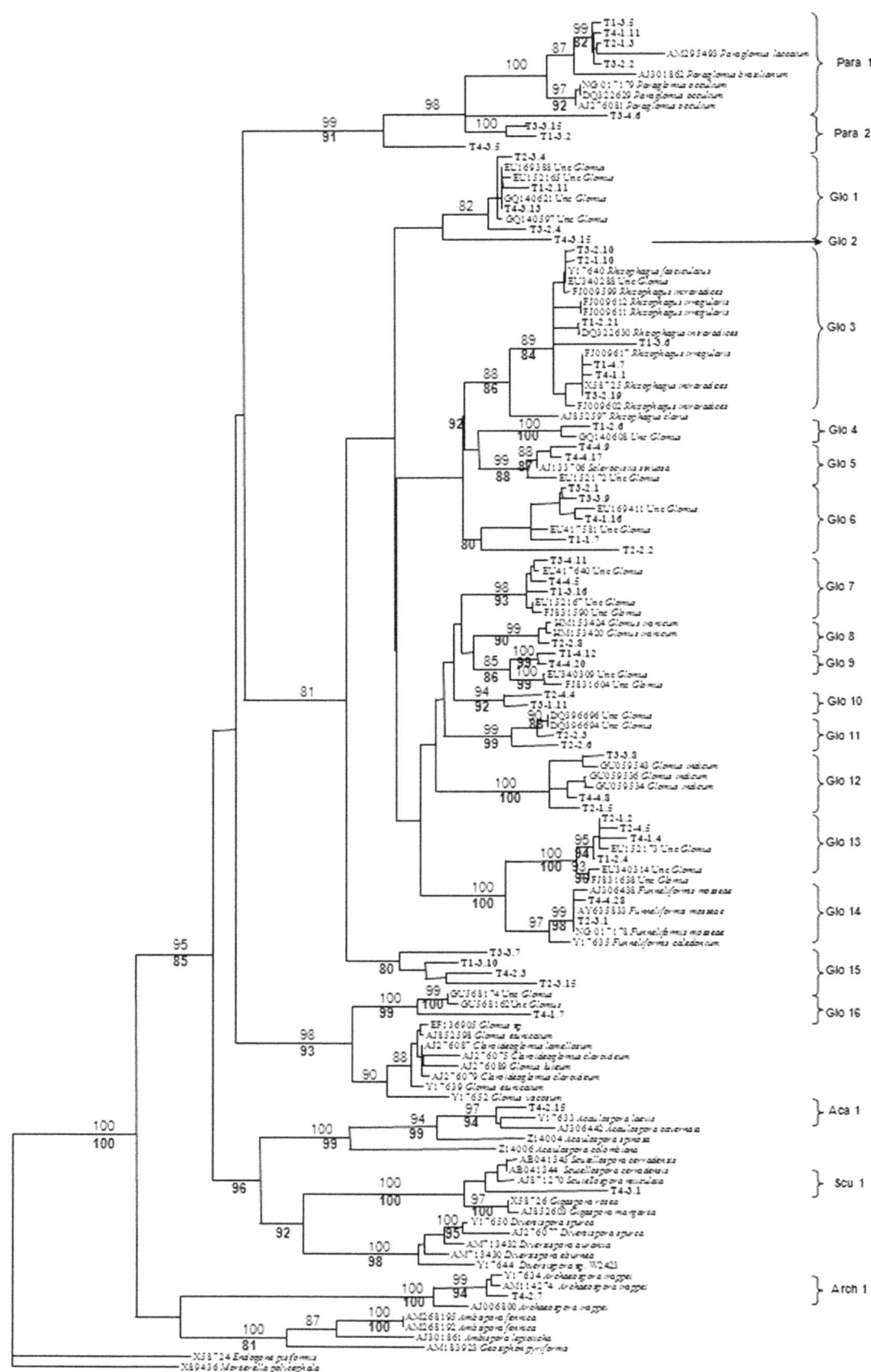

Figure 1. Phylogenetic tree of AMF sequences isolated from the *Prunus persica* roots under different treatments (T1: Combination of organic and inorganic fertilization and integrated pest management; T2: Inorganic fertilization and integrated pest management; T3: Inorganic fertilization and chemical pest control; T4: Combination of organic and inorganic fertilization and chemical pest control), reference sequences corresponding to the closest matches from GeneBank as well as sequences from cultured AMF taxa including representatives of the major taxonomical groups. Numbers above branches indicate the bootstrap values determined for Neighbour-Joining (NJ) analysis; bold numbers below branches indicate the bootstrap values of the maximum likelihood analysis. Sequences are labelled with the number of treatment from which they were obtained (T1, T2, T3, T4) and the clone identity number, Group identifiers (for example Glo 1) are AM fungal sequences types found in our study. Since identical sequences were detected, the clones producing the same sequence for each treatment were represented once in the alignment for clarity (Table S1 in File S1 material show a detailed description of the total number of clones of each AMF phylotype that were recovered from each treatment).

Corporation). Unique sequences of the clones generated in this study have been deposited at the National Centre for Biotechnology Information (NCBI) GenBank (http://www.ncbi.nlm.nih.gov) under the accession numbers HE613450 to HE613504.

Phylogenetical analysis

Sequence similarities were determined using the Basic Local Alignment Search Tool (BLASTn) sequence similarity search tool [25] provided by GenBank.

Phylogenetic analysis was carried out on the sequences obtained in this study and those corresponding to the closest matches from GenBank as well as sequences from cultured AMF taxa including representatives of the major taxonomical groups described by Schüßler et al. [26]. Sequences were aligned using the program ClustalX [27] and the alignment was adjusted manually in GeneDoc [28]. Neighbour-joining (NJ) [29] and maximum likelihood (ML) phylogenetic analyses were performed with the programs PAUP4.08b [30] and RAxML v.7.0.4 [31], respectively. The evolutionary distances for the NJ tree were computed using the maximum composite likelihood method with 1000 bootstrap replicates. For the ML analysis, a GTR-GAMMA model of evolution was used. The ML bootstrap values were calculated with

1000 replicates using the same substitution model. *Endogone pisiformis* Link and *Mortierella polycephala* Coem, were used as the out-groups.

Statistical analysis

Treatments effects on the number of phylotypes per root sample were compared using analysis of variance and comparisons among means were made using the Duncan's test calculated at $P<0.05$. The effect of two factors: types of fertilizer and pest management on AMF community composition were tested using a two-way analysis of variance, The statistical procedures were carried out with the software package SPSS 19.0 for Windows.

Canonical-correspondence analysis (CCA) with the relative abundance of clones per AMF sequence types found in *P. persica* roots under different treatments was performed. The results were summarized in an ordination diagram conducted in CANOCO for Windows v. 4.5 [32]. CCA is a multivariate statistical method that allows comparisons of AM fungal community compositions between four treatments.

The Shannon-Weaver (H') index was calculated as an additional measure of diversity, as it combines two components of diversity, i.e., species richness and evenness. It is calculated from

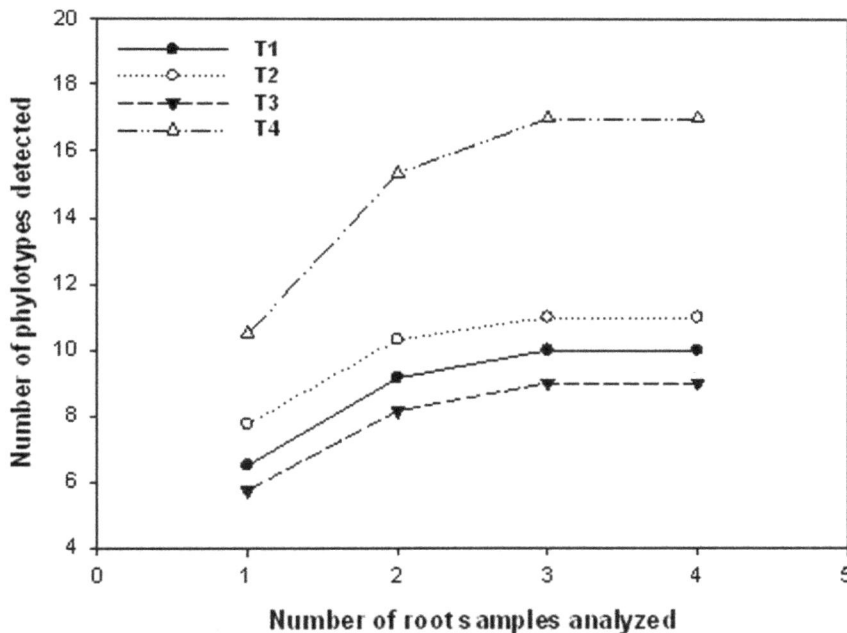

Figure 2. Sampling effort curves for *Prunus persica* roots under different treatments analysed. T1: Combination of organic and inorganic fertilization and integrated pest management; T2: Inorganic fertilization and integrated pest management; T3: Inorganic fertilization and chemical pest control; T4: Combination of organic and inorganic fertilization and chemical pest control. The number of clones for each AMF phylotypes in each root sample was used to construct the sampling effort curves (with 95% confidence intervals) using the software EstimateS 8.00 (Colwell, [33]). The sample order was randomized by 100 replications.

the equation H' = $-\sum p_i(\ln p_i)$, where p_i is the proportion of individuals found in the ith species (in a sample, the true value of p_i is unknown but is estimated as n_i/N, [here and throughout, n_i is the number of individuals in the ith species; N is the total number of individuals of all species]).

The number of clones for each AMF phylotypes in each soil sample was used to construct the sampling effort curves (with 95% confidence intervals) using the software EstimateS 8.00 [33]. The sample order was randomized by 100 replications.

Results

Sequence identity and phylogenetic analysis

In our study, 16 root samples (four repetitions per treatment) that were subjected to DNA extraction produced clonable PCR products of the expected size (about 795 bp). Overall, 512 clones from 16 clone libraries were screened by PCR; out of these, 449 contained the small-subunit rRNA gene fragment and an average of 15 clones per root sample were sequenced (in total, 242 sequences). According to the BLAST search in GenBank, sequences with a high degree of similarity (98−99%) to taxa belonging to the phylum Glomeromycota were produced by 227 clones. The remaining 15 clones produced incomplete sequences.

The 227 sequences were grouped in 21 different AMF sequence types or phylotypes, with sequence similarities varying from 97 to 100% and bootstrap values ≥ 80% (Fig. 1). These phylotypes were grouped in five families: the Glomeraceae, Paraglomeraceae, Acaulosporaceae, Gigasporaceae and Archaeosporaceae. Sixteen of these sequence groups belonged to the genus *Glomus*, two to the genus *Paraglomus*, one to the genus *Acaulospora*, one to the genus *Scutellospora* and one to the genus *Archaeospora*.

For the number of clones sequenced, the sampling effort curves showed a decreasing rate of accumulation of phylotypes, reaching the asymptote (Fig. 2). This pattern indicates that the clones analysed covered the AMF diversity colonising the *P. persica* roots under the four treatments. Therefore, no more clones were sequenced.

Nine phylotypes corresponded to morphologically-defined species: six were related to sequences from single, morphological-ly-described species (Para 1 to *Paraglomus laccatum*, Glo 5 to *Sclerocystis sinuosa*, Glo 8 to *Glomus iranicum*, Glo G12 to *Glomus indicum*, Glo 14 to *Funneliformis mosseae* and Arch 1 to *Archaeospora trappei*) and three were related to sequences belonging to two or three different, morphologically-described species (Glo 3 was related to a species group including *Rhizophagus intraradices/ irregularis/fasciculatus*, Aca 1 to an *Acaulospora cavernata/laevis* group and Scu 1 to a *Scutellospora cerradensis/reticulata* group). Eight phylotypes were related to uncultured glomalean species that have not been characterised morphologically (Glo 1, Glo 4, Glo 6, Glo 7, Glo 9, Glo 11, Glo 13 and Glo 16) and the remaining four phylotypes were not related to any sequences of AMF in the database (Glo 2, Glo 10, Glo 15 and Para 2) (Fig. 1).

The AMF community composition

There were significant differences in the AMF taxon richness. The ComFert+ChemM treatment harboured the highest mean number of AMF phylotypes per root sample (8.00), which was significantly different from the value for the InorgFert+ChemM treatment (4.25) according to Duncans multiple-comparison test. The mean number of AMF phylotypes detected in the tree roots receiving the ComFert+IntM or InorgFert+IntM treatments was the same (5.75) and no significant differences between these and either of the treatments mentioned above were found.

The factorial analysis showed that both factors: fertilization and pest management had a significant effect on the AMF community

composition ($p = 0.002$, $F = 9.734$ and $p = 0.007$, $F = 7.412$, respectively) The interaction between these factors was not significant ($p = 0.906$, $F = 0.014$).

The AMF communities of tree roots in the InorgFert+ChemM treatment had the lowest diversity ($H = 1.78$), with the lowest total number of AMF sequence types (9). The trees from the ComFert+ IntM and InorgFert+IntM treatments had similar AMF diversity ($H \approx 2.0$), while the treatment ComFert+ChemM yielded the highest number of different AMF sequence types (17) and showed the highest diversity index ($H = 2.69$). In the CCA diagram (Fig. 3), the different distributions of the AMF phylotypes, as a consequence of the treatments, can be observed. The symbols representing different treatments are distant to each other, which demonstrates that the treatments had a significant effect on the AMF community composition, with the different treatments hosting distinct phylotypes. This diagram also shows the AMF phylotypes found exclusively in each treatment.

Discussion

In this study we compared the diversity of AMF in *Prunus persica* roots under two types of fertilisation (inorganic, with or without manure) combined with integrated or chemical pest management in a tropical agro-ecosystem.

It is worth noting the high number of phylotypes found in this study (twenty-one) in comparison with other studies carried out also in agricultural soils. Daniell et al. [34] and Helgason et al. [35] found 10 phylotypes in arable soils around North Yorkshire, U.K. Toljander et al. [36] found eight phylotypes, all belonging to the

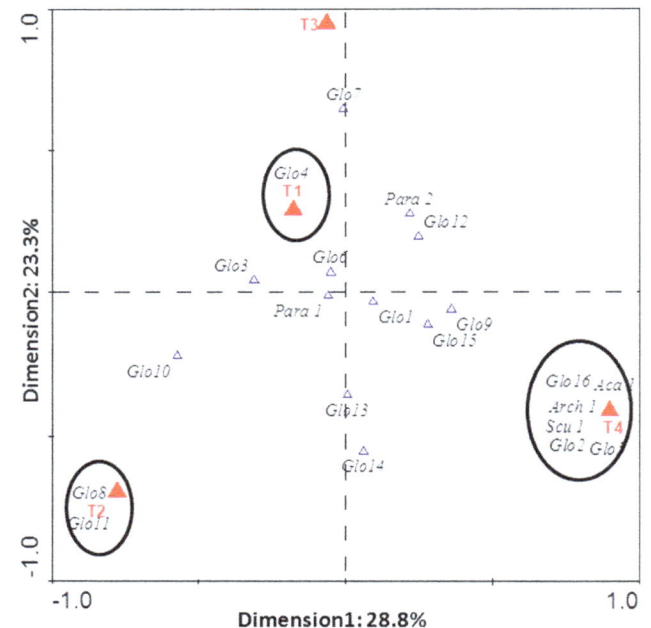

Figure 3. Canonical Correspondence analysis (CCA) of the AM fungal community composition found in the roots of *P. persica* under different treatments (T1: Combination of organic and inorganic fertilization and integrated pest management; T2: Inorganic fertilization and integrated pest management; T3: Inorganic fertilization and chemical pest control; T4: Combination of organic and inorganic fertilization and chemical pest control). Full triangles represent the treatments and the open triangles the AMF phylotypes. Open circles represent the AMF phylotypes exclusively found in individual treatments.

Table 1. Relative abundance of the different AMF sequence types observed in *Prunus persica* roots under the different treatments analysed.

Phylotypes	Treatments			
	ComFert+IntM (n = 60)	InorgFert+IntM(n = 55)	InorgFert+ChemM (n = 49)	ComFert+ChemM (n = 63)
Para 1	13.3	14.6	12.2	11.1
Para 2	3.3	0	4.1	3.2
Glo1	23.3	10.9	8.2	19.1
Glo2	0	0	0	3.2
Glo3	26.7	25.5	16.3	3.2
Glo4	3.3	0	0	0
Glo5	0	0	0	4.8
Glo6	8.3	5.5	6.1	4.8
Glo7	10.0	0	28.6	3.2
Glo8	0	3.6	0	0
Glo9	3.3	0	0	3.2
Glo10	0	9.1	4.1	0
Glo11	0	10.9	0	0
Glo12	0	3.6	16.3	11.1
Glo13	5.0	9.1	0	7.9
Glo14	0	3.6	0	3.2
Glo15	3.3	3.6	4.1	9.5
Glo16	0	0	0	3.2
Aca 1	0	0	0	3.2
Scu 1	0	0	0	3.2
Arch 1	0	0	0	3.2

ComFert: Combination of organic and inorganic fertilization; IntM: Integrated pest management; InorgFert: Inorganic fertilization; ChemM: Chemical pest control.

genus *Glomus*, in a field experiment using different organic and mineral fertilisers, while Hijri et al. [37] found 10 phylotypes in a conventional maize field in Germany. Also Verbruggen et al. [38] found that arbuscular mycorrhizal fungi richness varied from one to 11 phylotypes among organically and conventionally managed fields in the Netherlands. Alguacil et al. [39] found nine phylotypes in a study carried out in tropical savanna soils planted with leguminous forage under different doses of phosphorus fertiliser.

We observed different AMF community composition between treatments. Some studies have pointed out that soil physical-chemical properties, such as pH and nutrient content, are the factors influencing the structure of AMF communities in agricultural systems [17,36,40,41,42,43,44]. In our study, we did not find significant differences in soil characteristics among treatments (Table S2 in File S1), so the observed differences in the number of AMF phylotypes or AMF richness can be attributed to the different treatments applied. It has been reported that the AMF diversity is higher in soils amended with different organic substrates (or that the AMF diversity increases in organically-managed soils) [13,17,36,45,46]. Moreover, the addition of chemical pest-control products could have prevented the colonisation by non-AM fungi inside the roots, favouring the most-tolerant AMF species.

In contrast to some studies which indicated that species of the *Paraglomeraceae* appear to be rare or poor in agricultural soils [37,47,48], we found that Para 1 was one of the most-abundant groups in our study (12.8% of the AMF clones analysed), together with Glo 1 (15.9%) and Glo 3 (17.6%) (Table 1); Glo 6 and Glo 15

were found in all treatments, but their occurrence frequency was low (<6.2% of clones). The presence of the genus *Paraglomus* has been observed also in other agricultural management studies using group-specific primers [37,49] or the same pair of primers as ourselves [50]. In the case of Glo 3 (related to the *R. intraradices/irregularis* species complex group), our results are in accordance with several studies carried out in agricultural soils where this AMF taxon showed the highest representation in the clone libraries [36,37,50,51,52]. On the other hand, this species also showed the highest abundance of clones in *P. persica* roots when integrated pest management treatments (ComFert+IntM and InorgFert+IntM) were applied (Table 1). Several studies have shown that the presence of *Trichoderma harzianum* significantly increases root colonisation by *Rhizophagus intraradices* in melon crops [53,54], *R. intraradices* being the only taxon that increased the *T. harzianum* populations [53]. Therefore, a synergistic relationship between *T. harzianum* and *R. intraradices* could have existed in *P. persica* roots under the ComFert+IntM and InorgFert+IntM treatments, increasing their respective abundances.

Interestingly, in contrast to Lee et al. [55] who reported that the AML1/AML2 primer pairs do not amplify sequences belonging to the family *Archaeosporaceae*, we detected the Arch 1 phylotype, which showed 99% homology with sequences related to *Archaeospora trappei*.

There were phylotypes that occurred exclusively in some treatments; for example, Glo 2, Glo 5, Glo 16, Aca 1, Scu 1 and Arch 1 seemed to be specific for trees treated with ComFert+ChemM. The low abundance and specificity found for these phylotypes could be attributable to different colonisation strategies by these different phylogenetic groups. For example, it has been reported that mycelia of *Acaulospora* species have low root and soil colonisation levels [56]

Phylotypes Glo 8 and Glo 11 only appeared in the roots of trees receiving the InorgFert+IntM treatment and Glo 4 occurred exclusively with the ComFert+IntM treatment. The remaining phylotypes did not have a clear distribution, occurring haphazardly in two or three treatments, as in the case of Glo 14, related to sequences belonging to *Funneliformis mosseae*. Although in our study this phylotype was of low abundance (6.81%), this taxon appears to be one of the most-typical and dominant taxa in many agricultural fields [34,35,37,47,57]. Together with *R. intraradices*, these taxa are sometimes called the "typical AMF of arable lands" [47].

Conclusions

The real causes and effects of these differences in the AMF community composition observed among treatments are very difficult to establish, bearing in mind that the different crop management regimes studied consist of several influencing parameters. In fact, the ComFert+ChemM - which produced the highest diversity of AMF - included the highest number of parameters. Further investigation of AMF diversity, including

analysis of each factor separately and subsequently their interactions, could help to ascertain the cause of the effects reported here.

Different crop management strategies can exert a clear influence on the populations of AMF. The treatment including a combination of organic and inorganic fertilisation together with chemical pest control appears to be the most-suitable with respect to improve of the AMF diversity in this crop under tropical conditions, thus improving the agricultural and environmental sustainability of this agro-ecosystem.

Supporting Information

File S1 Supporting tables. Table S1. AMF sequences obtained in the present study for each phylotype. **Table S2.** Chemical properties of soil in response to different treatments analysed at the time of sampling (n = 4).

Author Contributions

Conceived and designed the experiments: MMA ZL AR. Performed the experiments: MMA ZL AR. Analyzed the data: MMA ET MPT. Contributed reagents/materials/analysis tools: MMA ET AR. Wrote the paper: MMA AR.

References

1. Smith SE, Read DJ (2008) Mycorrhizal Symbiosis. New York: Academic Press. 787 p.
2. Caravaca F, Alguacil MM, Figueroa D, Barea JM, Roldán A (2003) Re-establishment of *Retama sphaerocarpa* as a target species for reclamation of soil physical and biological properties in a semi-arid Mediterranean area. Forest Ecol Manag 182: 49−58.
3. Querejeta JI, Allen MF, Alguacil MM, Roldán A (2007) Plant isotopic composition provides insight into mechanisms underlying growth stimulation by AM fungi in a semiarid environment. Funct Plant Biol 34: 683−691.
4. Hildebrandt U, Regvar M, Bothe H (2007) Arbuscular mycorrhiza and heavy metal tolerance. Phytochemistry 68: 139−146.
5. Pozo MJ, Azcón-Aguilar C (2007) Unraveling mycorrhiza-induced resistance. Curr Opin Plant Biol 10: 393−398.
6. Alguacil MM, Torrecillas E, Lozano Z, Roldán A (2011) Evidence of differences between the communities of arbuscular mycorrhizal fungi colonizing galls and roots of *Prunus persica* infected by the root-knot nematode *Meloidogyne incognita*. Appl Environ Microb 77: 8656−8661.
7. Van Der Heijden MGA, Klironomos JN, Ursic M, Moutoglis P, Streitwolf-Engel R, et al. (1998) Mycorrhizal fungal diversity determines plant biodiversity, ecosystem variability and productivity. Nature 396: 69−72.
8. Rillig MC (2004) Arbuscular mycorrhizae and terrestrial ecosystem processes. Ecol Lett 7: 740−754.
9. Wilson GWT, Rice CW, Rillig MC, Springer A, Hartnett DC (2009) Soil aggregation and carbon sequestration are tightly correlated with the abundance of arbuscular mycorrhizal fungi: Results from long-term field experiments. Ecol Lett 12: 452−461.
10. Joner EJ (2000) The effect of long-term fertilization with organic or inorganic fertilizers on mycorrhiza-mediated phosphorus uptake in subterranean clover. Biol Fertil Soils 32: 435−440.
11. Na Bhadalung N, Suwanarit A, Dell B, Nopamornbodi O, Thamchaipenet A, et al. (2005) Effects of long-term NP-fertilization on abundance and diversity of arbuscular mycorrhizal fungi under a maize cropping system. Plant Soil 270: 371−382.
12. Bradley K, Drijber RA, Knops J (2006) Increased N availability in grassland soils modifies their microbial communities and decreases the abundance of arbuscular mycorrhizal fungi. Soil Biol Biochem 38: 1583−1595.
13. Gryndler M, Larsen J, Hrselová H, Rezáčová V, Gryndlerová H, et al. (2006) Organic and mineral fertilization, respectively, increase and decrease the development of external mycelium of arbuscular mycorrhizal fungi in a long-term field experiment. Mycorrhiza 16: 159−166.
14. Sasvári Z, Hornok L, Posta K (2011) The community structure of arbuscular mycorrhizal fungi in roots of maize grown in a 50-year monoculture. Biol Fertil Soils 47: 167−176.
15. Beauregard MS, Hamel C, Atul N, St-Arnaud M (2010) Long-term phosphorus fertilization impacts soil fungal and bacterial diversity but not AM fungal community in alfalfa. Microb Ecol 59: 379−389.
16. Mäder P, Edenhofer S, Boller T, Wiemken A, Niggli U (2000) Arbuscular mycorrhizae in a long-term field trial comparing low-input (organic, biological) and high-input (conventional) farming systems in a crop rotation. Biol Fertil Soils 31: 150−156.
17. Alguacil MM, Díaz-Pereira E, Caravaca F, Fernández DA, Roldán A (2009) Increased diversity of arbuscular mycorrhizal fungi in a long-term field experiment via application of organic amendments to a semiarid degraded soil. Appl Environ Microb 75: 4254−4263.
18. Alguacil MM, Lumini E, Roldán A, Salinas-García JR, Bonfante P, et al. (2008) The impact of tillage practices on arbuscular mycorrhizal fungal diversity in subtropical crops. Ecol Appl 18: 527−536.
19. Van Der Gast CJ, Gosling P, Tiwari B, Bending GD (2011) Spatial scaling of arbuscular mycorrhizal fungal diversity is affected by farming practice. Environ Microbiol 13: 241−249.
20. Borriello R, Lumini E, Girlanda M, Bonfante P, Bianciotto V (2012) Effects of different management practices on arbuscular mycorrhizal fungal diversity in maize fields by a molecular approach. Biol Fertil Soils 48: 911−922
21. Lumini E, Vallino M, Alguacil MM, Romani M, Bianciotto V (2011) Different farming and water regimes in Italian rice fields affect arbuscular mycorrhizal fungal soil communities. Ecol Appl 21: 1696−1707.
22. Avilán L, Leal F (1990) Suelos, fertilizantes y encalado para frutales.; América E, editor. Caracas, Venezuela.
23. SSS (2010) Keys to Soil Taxonomy. Washington DC: USDA. Natural Resources Conservation Service. 338 p.
24. White TJ, Bruns T, Lee S, Taylor J (1990) Amplification and direct sequencing of fungal ribosomal RNA genes for phylogenetics. In: Innis MA, Gelfand, D.H., Sninsky, J.J., and White, T.J., editor. PCR protocols A guide to methods and applications. San Diego. pp. 315−322.
25. Altschul SF, Madden TL, Schäffer AA, Zhang J, Zhang Z, et al. (1997) Gapped BLAST and PSI-BLAST: a new generation of protein database search programs. Nucleic Acids Res 25: 3389−3402.
26. Schüßler A, Schwarzott D, Walker C (2001) A new fungal phylum, the Glomeromycota: Phylogeny and evolution. Mycol Res 105: 1413−1421.
27. Thompson JD, Gibson TJ, Plewniak F, Jeanmougin F, Higgins DG (1997) The ClustalX windows interface: flexible strategies for multiple sequence alignment aided by quality analysis tools. Nucleic Acids Res 24: 4876−4882.
28. Nicholas KB, Jr. NHB (1997) GeneDoc: a tool for editing and annotating multiple sequence alignments.
29. Saitou N, Nei M (1987) The neighbor-joining method: a new method for reconstructing phylogenetic trees. Mol Biol Evol 4: 406−425.
30. Swofford DL (2002) PAUP: Phylogenetic Analysis Using Parsimony (and other methods). Version 4.08b10 for Machintosh ed. Sunderland, MA.: Sinauer Associates Inc.
31. Stamatakis A (2006) RaxML-VI-HPC: maximum likelihoodbased phylogenetic analyses with thousands of taxa and mixed models. Bioinformatics 22: 2688−2690.
32. ter Braak CFJ, Smilauer P (2004) CANOCO reference manual and CanoDraw for Windows. version 4.5 ed. Wageningen, Netherlands: Biometris. pp. Software for canonical community ordination.

33. Colwell R (2005) EstimateS: statistical estimation of species richness and shared species from samples, version 8.0.

34. Daniell TJ, Husband R, Fitter AH, Young JPW (2001) Molecular diversity of arbuscular mycorrhizal fungi colonising arable crops. FEMS Microbiol Ecol 36: 203−209.

35. Helgason T, Daniell TJ, Husband R, Fitter AH, Young JPW (1998) Ploughing up the wood-wide web? [4]. Nature 394: 431.

36. Toljander JF, Santos-González JC, Tehler A, Finlay RD (2008) Community analysis of arbuscular mycorrhizal fungi and bacteria in the maize mycorrhizo-sphere in a long-term fertilization trial. FEMS Microbiol Ecol 65: 323−338.

37. Hijri I, Sýkorová Z, Oehl F, Ineichen K, Mäder P, et al. (2006) Communities of arbuscular mycorrhizal fungi in arable soils are not necessarily low in diversity. Mol Ecol 15: 2277−2289.

38. Verbruggen E, Van Der Heijden MGA, Weedon JT, Kowalchuk GA, Röling WFM (2012) Community assembly, species richness and nestedness of arbuscular mycorrhizal fungi in agricultural soils. Mol Ecol 21: 2341−2353.

39. Alguacil MM, Lozano Z, Campoy M, Roldán A (2010) Phosphorus fertilisation management modifies the biodiversity of AM fungi in a tropical savanna forage system. Soil Biol Biochem 42: 1114−1122.

40. Gosling P, Hodge A, Goodlass G, Bending GD (2006) Arbuscular mycorrhizal fungi and organic farming. Agric Ecosyst Environ 113: 17−35.

41. Helgason T, Fitter AH (2009) Natural selection and the evolutionary ecology of the arbuscular mycorrhizal fungi (Phylum Glomeromycota). J Exp Bot 60: 2465−2480.

42. Schreiner RP, Mihara KL (2009) The diversity of arbuscular mycorrhizal fungi amplified from grapevine roots (Vitis vinifera L.) in Oregon vineyards is seasonally stable and influenced by soil and vine age. Mycologia 101: 599−611.

43. Balestrini R, Magurno F, Walker C, Lumini E, Bianciotto V (2010) Cohorts of arbuscular mycorrhizal fungi (AMF) in Vitis vinifera, a typical Mediterranean fruit crop. Env Microbiol Rep 2: 594−604.

44. Oehl F, Laczko E, Bogenrieder A, Stahr K, Bösch R, et al. (2010) Soil type and land use intensity determine the composition of arbuscular mycorrhizal fungal communities. Soil Biol Biochem 42: 724−738.

45. Verbruggen E, Röling WFM, Gamper HA, Kowalchuk GA, Verhoef HA, et al. (2010) Positive effects of organic farming on below-ground mutualists: Large-scale comparison of mycorrhizal fungal communities in agricultural soils. New Phytol 186: 968−979.

46. Gosling P, Ozaki A, Jones J, Turner M, Rayns F, et al. (2010) Organic management of tilled agricultural soils results in a rapid increase in colonisation potential and spore populations of arbuscular mycorrhizal fungi. Agric Ecosyst Environ 139: 273−279.

47. Oehl F, Sieverding E, Ineichen K, Mäder P, Boller T, et al. (2003) Impact of land use intensity on the species diversity of arbuscular mycorrhizal fungi in agroecosystems of Central Europe. Appl Environ Microb 69: 2816−2824.

48. Oehl F, Sieverding E, Ineichen K, Ris EA, Boller T, et al. (2005) Community structure of arbuscular mycorrhizal fungi at different soil depths in extensively and intensively managed agroecosystems. New Phytol 165: 273−283.

49. Galván GA, Parádi I, Burger K, Baar J, Kuyper TW, et al. (2009) Molecular diversity of arbuscular mycorrhizal fungi in onion roots from organic and conventional farming systems in the Netherlands. Mycorrhiza 19: 317−328.

50. Santos-González JC, Nallanchakravarthula S, Alström S, Finlay RD (2011) Soil, but not cultivar, shapes the structure of arbuscular mycorrhizal fungal assemblages associated with strawberry. Microb Ecol 62: 25−35.

51. Mathimaran N, Ruh R, Jama B, Verchot L, Frossard E, et al. (2007) Impact of agricultural management on arbuscular mycorrhizal fungal communities in Kenyan ferralsol. Agric Ecosyst Environ 119: 22−32.

52. Cesaro P, Van Tuinen D, Copetta A, Chatagnier O, Berta G, et al. (2008) Preferential colonization of Solanum tuberosum L. roots by the fungus Glomus intraradices in arable soil of a potato farming area. Appl Environ Microb 74: 5776−5783.

53. Martínez-Medina A, Pascual JA, Lloret E, Roldán A (2009) Interactions between arbuscular mycorrhizal fungi and Trichoderma harzianum and their effects on Fusarium wilt in melon plants grown in seedling nurseries. J Sci Food Agric 89: 1843−1850.

54. Martínez-Medina A, Roldán A, Pascual JA (2011) Interaction between arbuscular mycorrhizal fungi and Trichoderma harzianum under conventional and low input fertilization field condition in melon crops: Growth response and Fusarium wilt biocontrol. Appl Soil Ecol 47: 98−105.

55. Lee J, Lee S, Young JPW (2008) Improved PCR primers for the detection and identification of arbuscular mycorrhizal fungi. FEMS Microbiol Ecol 65: 339−349.

56. Hart MM, Reader RJ (2002) Taxonomic basis for variation in the colonization strategy of arbuscular mycorrhizal fungi. New Phytol 153: 335−344.

57. Bharadwaj DP, Lundquist PO, Alström S (2007) Impact of plant species grown as monocultures on sporulation and root colonization by native arbuscular mycorrhizal fungi in potato. Appl Soil Ecol 35: 213−225.

7

Burkholderia ambifaria and *B. caribensis* Promote Growth and Increase Yield in Grain Amaranth (*Amaranthus cruentus* and *A. hypochondriacus*) by Improving Plant Nitrogen Uptake

Fannie I. Parra-Cota[1], Juan J. Peña-Cabriales[1], Sergio de los Santos-Villalobos[2], Norma A. Martínez-Gallardo[1], John P. Délano-Frier[1]*

[1] Centro de Investigación y de Estudios Avanzados-Unidad Irapuato, Irapuato, Guanajuato, México, [2] Departamento de Ciencias del Agua y del Medio Ambiente, Instituto Tecnológico de Sonora, Ciudad Obregón, Sonora, México

Abstract

Grain amaranth is an emerging crop that produces seeds having high quality protein with balanced amino-acid content. However, production is restricted by agronomic limitations that result in yields that are lower than those normally produced by cereals. In this work, the use of five different rhizobacteria were explored as a strategy to promote growth and yields in *Amaranthus hypochondriacus* cv. Nutrisol and *A. cruentus* cv. Candil, two commercially important grain amaranth cultivars. The plants were grown in a rich substrate, high in organic matter, nitrogen (N), and phosphorus (P) and under greenhouse conditions. *Burkholderia ambifaria* Mex-5 and *B. caribensis* XV proved to be the most efficient strains and significantly promoted growth in both grain amaranth species tested. Increased grain yield and harvest index occurred in combination with chemical fertilization when tested in *A. cruentus*. Growth-promotion and improved yields correlated with increased N content in all tissues examined. Positive effects on growth also occurred in *A. cruentus* plants grown in a poor soil, even after N and P fertilization. No correlation between non-structural carbohydrate levels in roots of inoculated plants and growth promotion was observed. Conversely, gene expression assays performed at 3-, 5- and 7-weeks after seed inoculation in plants inoculated with *B. caribensis* XV identified a tissue-specific induction of several genes involved in photosynthesis, sugar- and N- metabolism and transport. It is concluded that strains of *Burkholderia* effectively promote growth and increase seed yields in grain amaranth. Growth promotion was particularly noticeable in plants grown in an infertile soil but also occurred in a well fertilized rich substrate. The positive effects observed may be attributed to a bio-fertilization effect that led to increased N levels in roots and shoots. The latter effect correlated with the differential induction of several genes involved in carbon and N metabolism and transport.

Editor: Nancy E. Freitag, University of Illinois at Chicago College of Medicine, United States of America

Funding: This work was largely supported financially by the European Commission 6th Framework Programme, AMARANTH: FUTURE-FOOD, Contract No. 032263. Financial support was provided by México Tierra de Amaranto A. C. and The Deborah Presser-Velder Foundation. FIPC and SSV were supported by postgraduate scholarships (codes 10506 and 10512, respectively) granted by The National Council for Science and Technology (CONACyT, México). The funders had no role in study design, data collection and analysis, decision to publish, or preparation of the manuscript.

Competing Interests: The authors have declared that no competing interests exist.

* E-mail: jdelano@ira.cinvestav.mx

Introduction

The genus *Amaranthus* L. (Caryophyllales: Amaranthaceae) comprises C4 dicotyledonous herbaceous plants classified into approximately 70 species having a relatively high level of genetic variability. They have a worldwide distribution, although most species predominate in the warm temperate and tropical regions of the world [1]. Many amaranth species are cultivated as ornamentals or as a source of highly nutritious pseudocereals (e.g. *grain amaranths*) and/or of vitamin- and mineral-rich leaf vegetables [2–4]. Others are notoriously aggressive weeds of commercial crops [5,6]. The grain amaranths (predominantly *Amaranthus hypochondriacus* L., *A. cruentus* L., and *A. caudatus* L.) offer attractive nutritional and health-related traits (recently reviewed in [7] and [8]), in addition to many desirable agronomic characteristics. Thus, amaranth seeds are notable for their high contents of gluten-free protein possessing a nutritionally balanced amino-acid composition, the ability to release bioactive peptides when digested and relatively high levels of squalene-rich oil. Moreover, grain amaranths offer a viable alternative to cereals and other crops in agricultural settings where soil moisture conditions vary considerably between growing seasons [1]. Their ability to withstand drought and salt stress has been attributed to their superior water use efficiency [9–11], which is higher than other crops, including wheat, corn, cotton and sorghum [12]. Other contributing factors to abiotic stress resistance are the use the C4 pathway for CO_2 fixation, an indeterminate flowering habit and the capacity to grow long taproots and develop an extensive lateral root system in response to water shortage in the soil [11,13]. Osmolyte accumulation and the activation of stress-related genes are also associated with stress tolerance in grain amaranth [14,15].

Augmented and consistent yields, increased pest resistance, and improved harvestability are important breeding goals that grain amaranth shares with all grain crops. The typical yields of current amaranth cultivars oscillate around ~1000 kg/ha, although the potential exists for producing significantly higher yields that surpass 3000 kg/ha [1,16]. If such potential could be more uniformly expressed, it should be possible to improve grain amaranth yields substantially [1].

Nowadays, several bio-fertilizers of bacterial or fungal origin are commercially available and may be utilized to improve productivity. In addition, they offer great ecological benefits associated with a number of properties that impinge positively on both the soil and the plants growing in it. These include the ability to fix atmospheric nitrogen, degrade organic compounds, including pesticides, and suppress various soil-borne pathogens via the synthesis of antibiotics, hydrogen cyanide and/or siderophores. Nitrogen fixation is of paramount importance considering that the natural supply of soil N usually limits plant yields in most agricultural cropping systems [17]. For this reason N fertilizer application is predicted to greatly increase in the next decades [18] unless N use efficiency (NUE) is significantly increased. NUE is defined as the total biomass or grain yield produced per unit of applied fertilizer N, and it integrates both Nitrogen uptake efficiency (the capacity of plant roots to acquire N from the soil) and Nitrogen utilization efficiency (the fraction of plant-acquired N to be converted to total plant biomass or grain yield) ([19] and references therein). Its importance is underlined by the deleterious effects that excess N compounds released from agricultural systems can have on the quality of air, water, and soil [17,20].

Beneficial soil bacteria and fungi can also confer immunity against a wide range of foliar diseases and insects via the long-distance activation of plant defenses [21]. Growth promotion is believed to be tightly associated with the synthesis of bacterial auxins, giberellins and cytokinins, volatile compounds and/or vitamins, the induction of 1-aminocyclopropane-1-carbocylate (ACC) deaminase, which coupled to an increased superficial root area acting together with the secretion of siderophores, facilitate the absorption of limiting nutrients, such as iron, phosphorus and other minerals [22–28]. The bio-fertilizers of bacterial origin are commonly part of what are known as plant growth promoting rhizobacteria (PGPR). They constitute a large group on non-pathogenic soil bacteria that promote growth and/or control soil pathogens or insect pests when grown in a non-symbiotic association with plants [29–35]. Illustrative examples are the capacity to promote growth and protect against Fusarium wilts, in maize [36] and anthracnose, in mango [37] observed in plants inoculated with Burkholderia cepacia, and the induced systemic response against whitefly (Bemisia tabaci) pests detected in tomato plants inoculated with a Bacillus subtilis strain [38]. Other PGPR species belong to the Rhizobium, Mesorhizobium and Bradyrhizobium [39,40], Azospirillum [41], Agrobacterium, Azotobacter [42], Arthrobacter, Alcaligenes, Pseudomonas [43,44], Serratia, Enterobacter, Beijerinckia, Klebsiella, Clostridium, Variovovax, Xanthomonas, and Phyllobacterium genera [45–49]. Bio-fertilizers also include a group of phosphate-solubilizing microorganisms [50,51] and certain mycoparasitic filamentous fungi of the genus Trichoderma [52]. Their use is considered to be innocuous both to man and the environment. They usually are more efficient in low-fertility soils and are economical and easily transported although care must be exercised to maintain their biological activity [47].

Information regarding the use of PGPR in Amaranthus is limited to reports that focused on growth promotion in leafy species [53,54] and germination inhibition of weedy A. hybridus [55]. In addition, Nair and Anith [56] explored the use of PGPR for the control of leaf blight in A. tricolor. The comparative study herewith reported describes the effect that the inoculation of selected PGPR had on the growth and productivity of grain amaranth. Morphologic, metabolic and molecular studies were concomitantly performed in an effort to understand the possible mechanisms by means of which PGPR promote growth and increase yield and total biomass in grain amaranth. The information presented here has the potential to be employed to enhance the agronomic performance of grain amaranths in the field, while limiting N fertilizer application and thereby ameliorating the ecological damage associated to N pollution of the environment [57].

Results

Growth promotion experiments

The main objective of this work was to determine whether the utilization of PGPR with demonstrated agronomic potential was effective in promoting growth and increasing grain yield and total biomass in two species of grain amaranth. Biochemical and molecular tests were concomitantly performed to determine the possible mechanisms responsible for the changes observed.

The main characteristics of the five PGPR initially tested are shown in Table 1. They all showed at least one trait usually associated with plant growth promotion such as auxin production, ACC deaminase activity and siderophore production. The presence of acetylene reduction activity in the Burkholderia strains was indicative of the possible presence of nitrogenase activity required for nitrogen fixation.

Initial exploratory experiments, performed in the commercially important A. hypochondriacus cv. Nutrisol and A. cruentus cv. Candil genotypes, showed that bacterial inoculation via direct seed-soaking produced better results than root drenching of seedlings. Growth promotion was determined 8 weeks post inoculation (wpi) by measuring total biomass, in general, and also separately in leaves, stems and roots. Plant height and stem diameter were two additional parameters determined. A positive effect was produced on both species and was observed with all PGPR tested, with the exception of B. subtilus BEB-DN (results not shown). Additional experiments were performed with those strains showing the best growth promoting efficiency, namely B. caribensis XV and B. ambifaria Mex-5, and to a lesser degree, B. cepacia XXVI. Significantly positive effects (Tukey test; P≤0.05, n = 5) on total biomass and leaf area, plant height and stem diameter (Figure 1), and leaf, stem and root biomass (Figure 2) were observed in both species at 8 wpi. The effect was still noticeable in plants that received chemical N and P fertilization, in particular in plants inoculated with B. caribensis XV and B. ambifaria Mex-5. It also tended to be more evident in A. cruentus than in A. hypochondriacus (Figures 1 and 2). Foliar nitrogen levels were also significantly higher in PGPR-treated A. cruentus and A. hypochondriacus plants (Figures 2C and D). For A. cruentus, this effect was evident even in plants subjected to N fertilization (Figure 2D).

All further experiments were performed with A. cruentus cv. Candil which was inoculated with the best performing PGPR, i.e. B. caribensis XV and B. ambifaria Mex-5. The choice of A. cruentus was based on its relative insensitivity to the photo-period, a useful characteristic which allowed extended experimentation during early and late periods of the year, which are unsuitable for A. hypochondriacus [58]. All growth parameters tested were significantly increased in PGPR-inoculated A. cruentus plants (Tukey test; P≤ 0.05, n = 6), as shown for, stem diameter (average 23.6% [XV] and 18.5% [Mex5]) and plant height (average 17.6% [XV] and

Table 1. Properties of the rhizobacteria used in this study.

	Rhizobium sp. XXV[3]	Bacillus subtilis BEB-DN[2]	Burkholderia ambifaria Mex5[3]	Burkholderia cepacia XXVI[1]	Burkholderia caribensis XV[3]
Auxin production	+	+	+	−	−
ACC deaminase activity	−	+	+	−	+
Siderophore production	−	ND	+	+	+
Acetylene Reduction Activity	−	ND	+	+	+
Nitrogenase gene, nifH 2	ND	ND	ND	ND	ND

ND = Not determined.
[1] = Described in Reference [37];
[2] = Described in reference [91];
[3] = Described in Reference [92].

16.9% [Mex5]) (Figure 3A). Increased leaf area (average 38.6%, [XV] and 23.1% [Mex5]) (Figure 3B) as well as total biomass in dry (average 55% and 29.3% increase for XV and Mex5, respectively) and fresh weight basis (Figures 3C and 3D), were also observed. The latter data were corroborated by significantly increased leaf (average 58.3% [XV] and 24.7% [Mex5]), stem (average 65.8% [XV] and 42.1% [Mex5]) and root (average 33.1% [XV] and 19.5% [Mex5]) biomass, which occurred during the entire duration of the experiment (Figure 3F). Total nitrogen content in the latter tissues was also significantly higher in PGPR-inoculated plants (Tukey test; $P \leq 0.05$, n = 6) (Figure 3G). The

increments ranged from 27.7% (in leaves, [Mex5]) to 79.8% (in stems, [XV]). In general, growth promotion and NUE were more effective in plants inoculated with *B. caribensis* XV, as shown by the above data (see also Figure 3E).

After 7 weeks of growth, 1×10^6 cfu per g substrate/soil were detected. This demonstrated the efficient long term colonization of the substrate/soil by *B. ambifaria* Mex5 and *B. caribensis* XV.

However, contrasting effects on the non-structural carbohydrates (NSC) levels in the different tissues examined were observed between the two PGPR tested. Inoculation with *B. caribensis* XV had a negative to neutral effect on all NSC tested and in all tissues

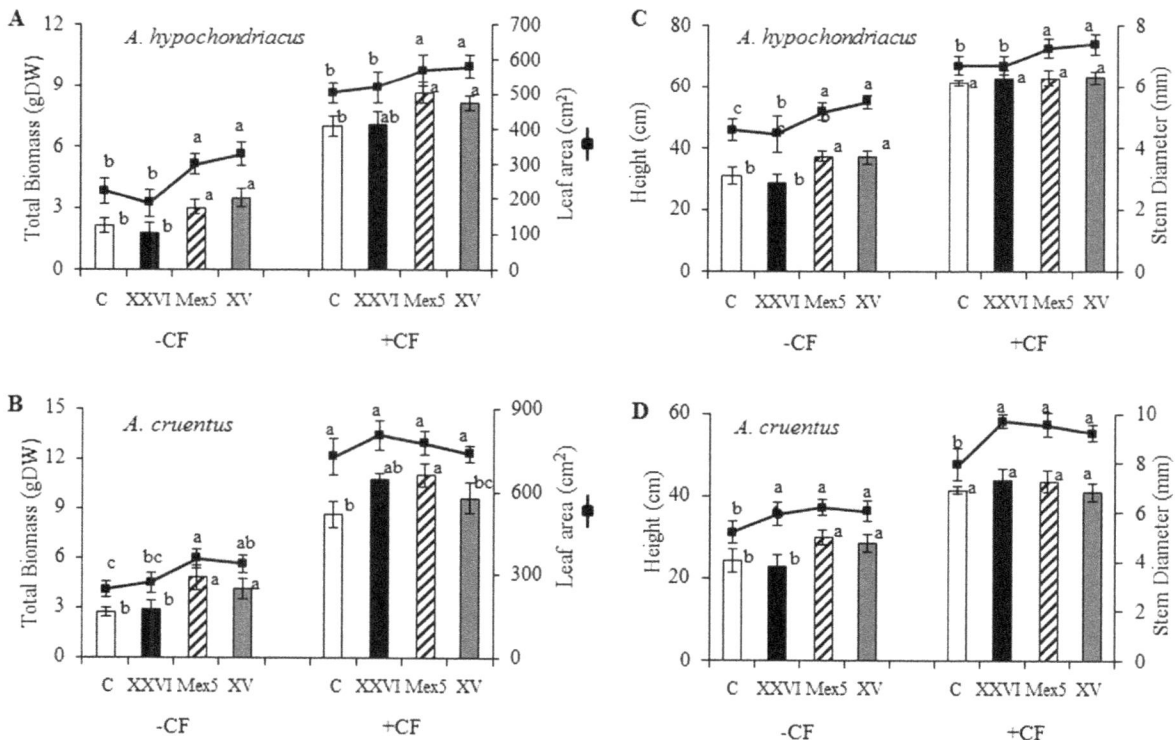

Figure 1. PGPR positively affect growth of grain amaranth plants. The effect of PGPR on different growth parameters, produced 8 weeks after inoculation with three strains of *Burkholderia* (*B. cepacia* XXVI, *B. ambifaria* Mex5 and *B. caribensis* XV), was determined in (**A**) and (**B**) *Amaranthus hypochondriacus* and (**C**) and (**D**) *A. cruentus* plants grown in a rich substrate, with (+CF) or without (−CF) chemical fertilization. Parameters measured were: total biomass, leaf area, plant height and stem diameter. Mean values ± SE are presented. Different letters over the bars and lines represent statistically different values at $P \leq 0.05$. Experiments were performed twice, and representative results are shown.

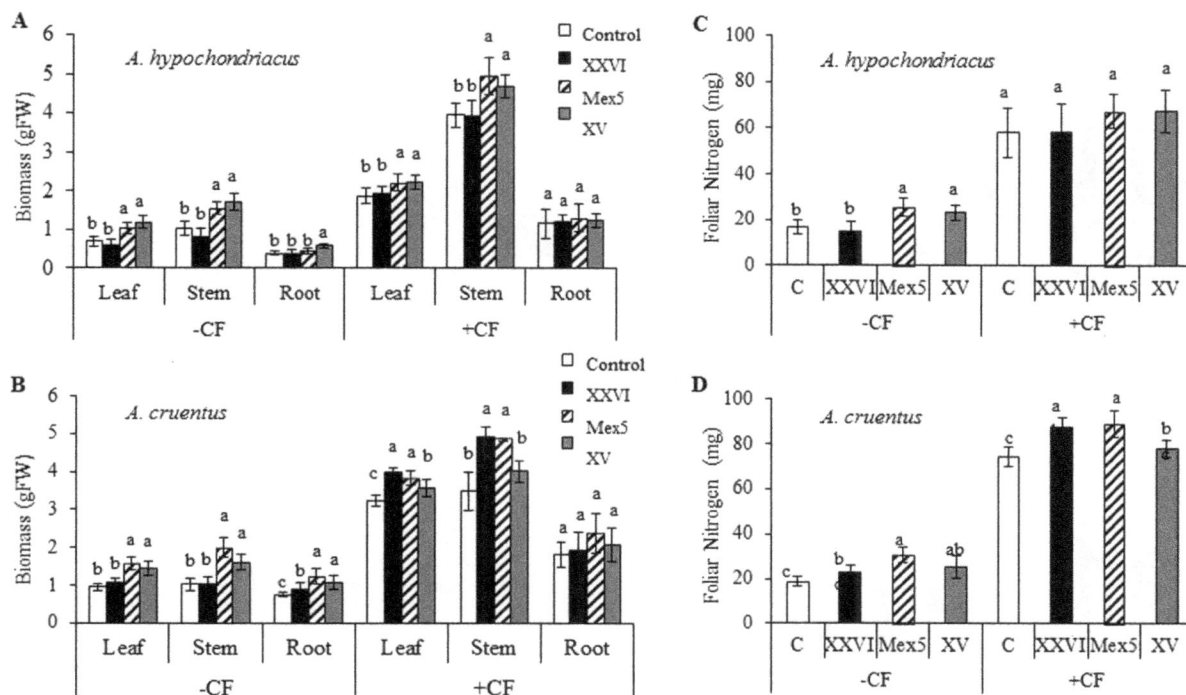

Figure 2. PGPR positively affect growth and nitrogen content of grain amaranth plants. The effect on the total biomass of leaves, stems and roots, (**A**) and (**B**), and on foliar nitrogen, (**C**) and (**D**), was measured 8 weeks after inoculation with three strains of *Burkholderia* (*B. cepacia* XXVI, *B. ambifaria* Mex5 and *B. caribensis* XV) in *Amaranthus hypochondriacus* and *A. cruentus* grown in a rich substrate, with (+CF) or without (−CF) chemical fertilization. Mean values ± SE are presented. Different letters over the bars and lines represent statistically different values at *P*≤0.05. Experiments were performed twice, and representative results are shown.

examined, except for the late increase of hexoses in stem (GLC and FRC) and leaves (FRC) observed at 7 wpi (Table 2). On the other hand, and except for a few cases (i.e. starch in leaves at 3 wpi and FRC in stems, at 3 and 5 wpi), *A. cruentus* plants inoculated with *B. ambifaria* Mex-5 showed a biphasic oscillation in NSC levels, being neutral to negative at 3 and 5 wpi, and becoming predominantly positive at 7 wpi (Table 2). Thus, only growth promotion by *B. ambifaria* Mex-5 was associated with a gradual increase of NSC levels in roots and shoots.

The growth of *A. cruentus* plants was drastically reduced when they were grown in a poor soil, which contrasted with the rich substrate used in all other experiments by the much lower levels of fertility shown, predominantly in terms of organic matter and available N and P (Table S1). Inoculation with *B. caribensis* XV and *B. ambifaria* Mex5 significantly enhanced growth of *A. cruentus* plants as observed 7 wpi: stem diameter and plant height were increased between 29% (e.g stem diameter in Mex5) to 42% (e.g. plant height in XV) (Tukey test; *P*≤0.05, n = 7; Figure 4A).Leaf area and total biomass (in fresh and dry weight basis, respectively) were increased more than two-fold (Figures 4B to 5D). This was mirrored by the measurement of the leaf, stem and root biomasses, which were also significantly increased in inoculated plants. The effect was particularly evident in plants inoculated with *B. caribensis* XV (Figure 4E), in which the biomass of each tissue examined was more than doubled. In roots, for example, dry weigh biomass was augmented 5.6-fold (Tukey test; *P*≤0.05, n = 7; Figures 4F and G). Nitrogen content of inoculated plants grown in poor soil was also greatly increased. Increments ranged between 2- and 4- fold in leaves and roots of plants treated *B. caribensis* XV (Figure 4H).

Similar results were obtained with *B. ambifaria* Mex5 (data not shown).

Taken together, the results demonstrated that the growth promotion induced by these two bacterial strains was increased on different substrates with different fertility levels, and was particularly striking in plants grown in a low fertility sandy soil, sampled in the Bajío region of central Mexico.

Effects on yield, harvest index, and seed size

The results shown in Figure 5 indicate the effect that both *Burkholderia* strains employed had on the different production parameters tested. In contrast to the growth promotion experiments, *B. ambifaria* Mex5 was found to have a similar effect on production parameters as *B caribensis* XV. Yields were increased by chemical fertilization and the effect was significant in one repetition of the experiment (Tukey test; *P*≤0.05, n = 11; Figure 5A). However, in both cases, the combination with a bacterial partner significantly increased yields (41.4% for XV and 155.4% for Mex5, respectively). A similar tendency was observed when measuring the harvest index, although this parameter was increased solely by chemical fertilization and only marginal increases of 39.5% and 22.6% were detected when fertilized plants were treated together with *B. caribensis* XV and *B. ambifaria* Mex5, respectively (Tukey test; *P*≤0.05, n = 11; Figure 5B). Seed size was not affected by chemical fertilization; however, it was significantly increased (in a range of 7.2% to 15.5%) in the presence of the bacterial inoculates (Figure 5C).

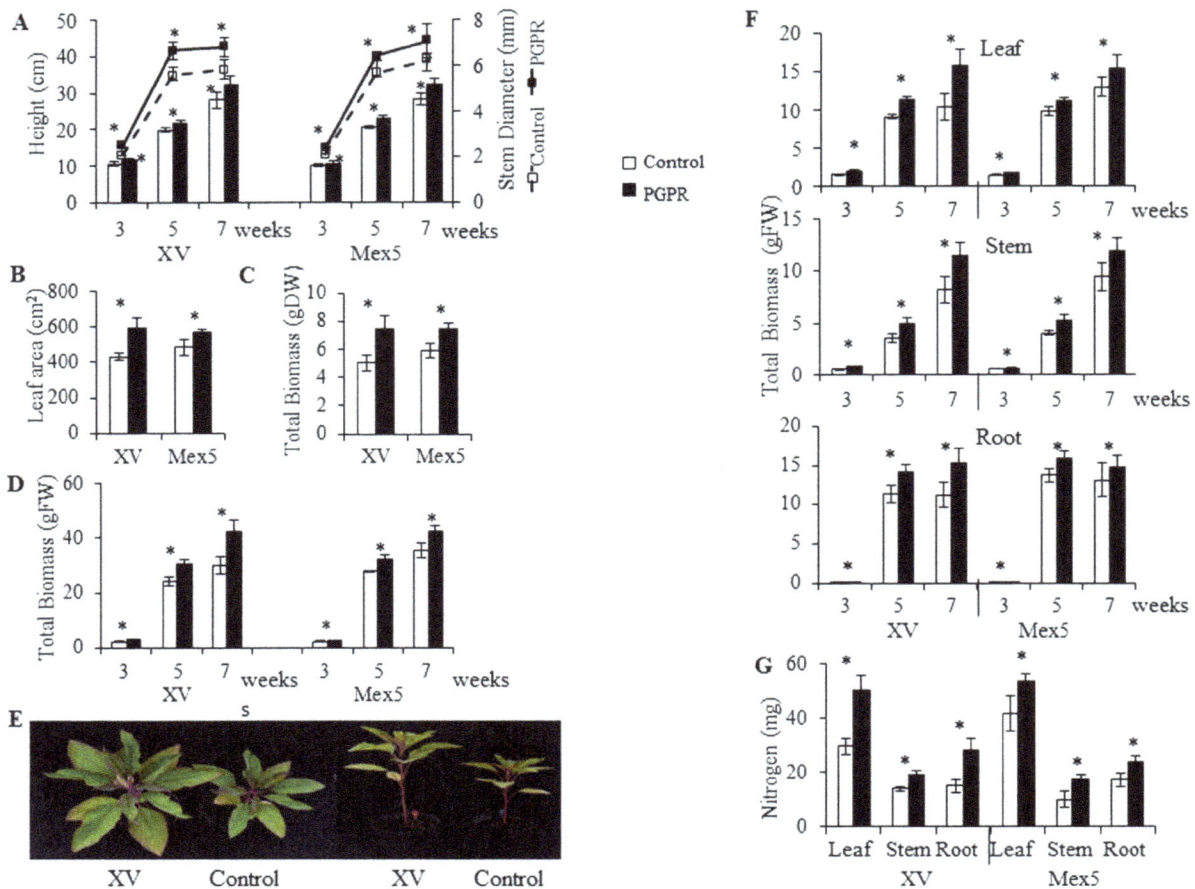

Figure 3. Time-course effect on different growth parameters produced in PGPR-inoculated amaranth plants. *A. cruentus* plants grown in a rich substrate were inoculated with two strains of *Burkholderia* (*B. ambifaria* Mex5 and *B. caribensis* XV) and growth-related parameters were measured at 3, 5 and 7 weeks after inoculation. These were (**A**) plant height and stem diameter; (**B**) leaf area; (**C**) and (**D**) total biomass in fresh and dry weight basis and (**F**) total leaf, stem and root biomass in a FW basis and (**G**) total foliar Nitrogen levels. Differences in plant height and leaf area produced between controls and plants inoculated with *B. caribensis* XV are shown in (**E**). Mean values ± SE are presented. Asterisks over the bars and lines represent statistically different values at $P \le 0.05$. The results presented were obtained from a typical experiment that was repeated three times.

Gene expression assays

The results shown in Figure 6A showed that two genes remained up-regulated in roots of inoculated plants at the three sampling stages analyzed (3, 5 and 7 wpi). These coded for *AhBAMY* a ß-amylase and *AhSUT1*, a sucrose transporter. Others, such as *AhDRM3*, an auxin responsive gene, *AhSUS2*, sucrose synthase 2 and *AhNRT.1.1*, a nitrate transporter type 1.1 were up-regulated at 3 and 5 wpi but returned to basal levels at 7 wpi, whereas the pyruvate orthophosphate dikinase gene, *AhPPDK*, was induced at the latter stages of the process first appearing at 5 wpi and remaining up-regulated at 7 wpi. Other genes, including various associated with nitrogen metabolism were up-regulated at definite stages of the process. Such was the case of an alanine aminotransferase, *AhAlaAT* (up-regulated at 3 wpi), and NADH dependent glutamate synthase, *AhNADH GOGAT*, the *DOF1* transcription factor, *AhDOF1*, and cytosolic glutamine synthase 1, *AhGS1* (up-regulated at 5 wpi). A hexose transporter 1, *AhHT1* was also up-regulated early in the process (at 3 wpi), whereas a neutral cytosolic invertase1, *AhA/NI-1*, showed a contrasting pattern of expression, being repressed at 3 wpi but induced at 5 wpi.

The results obtained in leaves are shown in Figure 6B. The pattern differed amply from the gene expression patterns observed in roots. No gene was found to be expressed at all three stages examined. Genes that were up-regulated at only one time point were *AhDOF1*, *AhNADH GOGAT*, *AhSUT1* and a nitrate transporter 3 (*AhNRT.3*), that were induced at 3 wpi; *AhBAMY*, and two photosynthesis-related genes, namely phosphoenolenol-pyruvate carboxylase, *AhPEPC*, and *AhPPDK*, were induced at 5 wpi, and α-expansin 3 (*AhEXP3α*), glutamine-dependent aspar-agine synthase 1 (*AhASN1*), *AhNRT.1.1* and *AhSUS2*, were induced at 7 wpi. All other genes were either expressed at 3 and 5 wpi (i.e. *AhA/NI-1*), 3 and 7 wpi (i.e. *Ah TPT*, a triose phosphate/phosphate transporter) and more frequently at 5 and 7 wpi (i.e. a NADPH-dependent malic enzyme malic enzyme, *AhME*, glutamate dehyrdrogenase 2, *AhGDH2*, *AhGS1*, *AhDRM3*, and *AhHT1*).

Discussion

The growth promoting effects of rhizobacteria have the potential to be widely applied in agriculture, mainly as biofertilization, biocontrol and phytoremediation agents [59–61]. Several mechanisms are believed to be acting to permit these

Table 2. Time-course changes in non-structural carbohydrate levels in different tissues of PGPR-inoculated *A. cruentus* plants.

μmol/gFW

Strain[1]	CHO	Tissue	3 weeks		5 weeks		7 weeks	
			Control	PGPR	Control	PGPR	Control	PGPR
XV	Starch	Leaf	20.97±0.85	19.37±1.42	43.19±2.93	41.51±1.63	51.76±2.55	46.22±1.09
		Stem	8.07±0.45	*5.44±0.65	33.85±3.27	31.51±2.80	52.09±3.28	47.28±2.91
		Root	ND	ND	1.25±0.14	1.24±0.09	2.51±0.21	2.56±0.20
	Sucrose	Leaf	1.67±0.09	1.51±0.08	3.91±0.47	3.54±0.24	2.93±0.11	3.00±0.21
		Stem	1.35±0.58	0.71±0.02	6.62±1.13	7.42±0.83	16.99±0.75	15.69±0.67
		Root	ND	ND	3.87±0.36	4.49±0.06	6.64±0.13	6.25±0.33
	Glucose	Leaf	7.49±0.51	*5.01±0.30	11.20±0.80	11.23±0.22	12.31±0.34	11.83±0.51
		Stem	5.57±1.04	4.51±0.05	30.62±4.01	26.77±1.88	22.50±0.69	*26.41±0.50
		Root	ND	ND	2.32±0.12	*0.99±0.14	1.94±0.19	*0.96±0.08
	Fructose	Leaf	2.00±0.13	*1.51±0.16	1.58±0.14	1.35±0.08	0.71±0.06	*1.13±0.15
		Stem	2.99±0.18	2.76±0.05	6.62±0.78	6.44±0.51	3.68±0.28	4.50±0.10
		Root	ND	ND	0.59±0.06	*0.45±0.02	0.44±0.04	*0.24±0.01
Mex5	Starch	Leaf	17.90±1.48	*23.53±1.45	41.49±2.45	*34.97±1.34	35.88±1.71	*48.99±3.50
		Stem	6.19±0.27	6.47±0.45	36.68±1.00	*32.32±1.46	42.83±3.72	*63.81±2.20
		Root	ND	ND	1.85±0.06	*1.03±0.05	2.60±0.17	*3.60±0.31
	Sucrose	Leaf	1.99±0.16	1.81±0.17	3.39±0.19	*2.24±0.09	2.13±0.06	*2.47±0.14
		Stem	0.61±0.04	*0.74±0.03	7.81±0.18	*6.82±0.24	15.86±0.63	14.98±0.99
		Root	ND	ND	3.67±0.04	*2.89±0.03	6.14±0.24	*7.76±0.39
	Glucose	Leaf	4.71±0.15	4.11±0.24	9.70±0.33	9.46±0.35	9.58±0.28	*11.35±0.39
		Stem	4.02±0.14	*4.91±0.14	31.32±0.42	*32.54±0.35	20.84±0.91	*28.41±0.88
		Root	ND	ND	2.52±0.24	*1.34±0.13	1.45±0.24	*2.97±0.23
	Fructose	Leaf	1.32±0.07	1.21±0.08	1.10±0.11	0.89±0.09	1.43±0.23	*0.71±0.09
		Stem	2.62±0.07	*3.19±0.05	7.54±0.08	*8.16±0.19	4.12±0.25	*5.50±0.50
		Root	ND	ND	0.63±0.04	*0.30±0.02	0.32±0.05	*0.61±0.07

[1] = Burkholderia caribensis XV and B. ambifaria Mex5;
ND = Not determined;
* = Significant difference with controls at $P<0.05$.

benefits, such as an enhanced nutrient uptake efficiency [19], hormone production or transformation [62], or improved defense against pathogens [60]. However, the molecular events underlying plant growth promotion by PGPR are still poorly understood. In this study, the long-term effect of diverse growth promoting rhizobacteria, including three potentially diazotrophic bacteria, on growth promotion, plant biomass accumulation and seed yield in grain amaranth were examined. In addition, changes in gene transcription and in sugar and nitrogen levels were also analyzed.

Grain amaranth is a marginal crop that has consistently attracted interest worldwide. This is mostly because grain amaranths can be utilized for the production of high quality grain in conditions that are unsuitable for cereal crops. However, there are many agronomic characteristics that must be improved (see [1]) in order to increase yields, which are much lower than those reported for cereal crops. Thus, it is imperative to develop appropriate agronomic practices for the cultivation of grain amaranth if higher yields are to be achieved.

Chemical fertilization is, together with optimal plant density, one key factor to boost grain amaranth yields, particularly when grown in poor or degraded soils. Up to a certain limit (≥90 Kg N/ha), amaranth grain yield is known to respond positively to nitrogen fertilizer, without increasing its tendency to lodge [63–65]. Also, nitrogen fertilization has been found to augment grain weight, biomass, grain yield and harvest index, although a negative effect was observed as nitrogen fertilizer rates were increased [66]. A similar effect was obtained in this study, since productivity was generally increased in fertilized *A. cruentus* plants. Interestingly, the beneficial effect observed was potentiated by inoculation with either *B. caribensis* XV or *B. ambifaria* Mex5 (see Figure 5).

The positive effect of N and P fertilization on grain amaranth's growth and biomass accumulation were also corroborated in this work, as shown by Figures 1 to 4, in which all parameters examined were significantly increased by N and P fertilization, including N uptake. Interestingly, and similarly to the productivity experiment, these parameters were also improved by the inoculation of the PGPR tested, even in chemically fertilized plants. Once again, this effect was particularly evident with two PGPR strains, namely *B. caribensis* XV and *B. ambifaria* Mex5. These are free-living and presumably diazotrophic *Burkholderia* strains that show promise for agro-biotechnological applications.

Growth promotion by inoculation with the *Burkholderia* strains was accompanied always by increased N levels in all plant tissues tested (i.e. roots, stems and leaves). It is therefore valid to propose that growth promotion effects were the result, at least partly, of an enhanced N uptake efficiency. This 'biofertilization' effect was

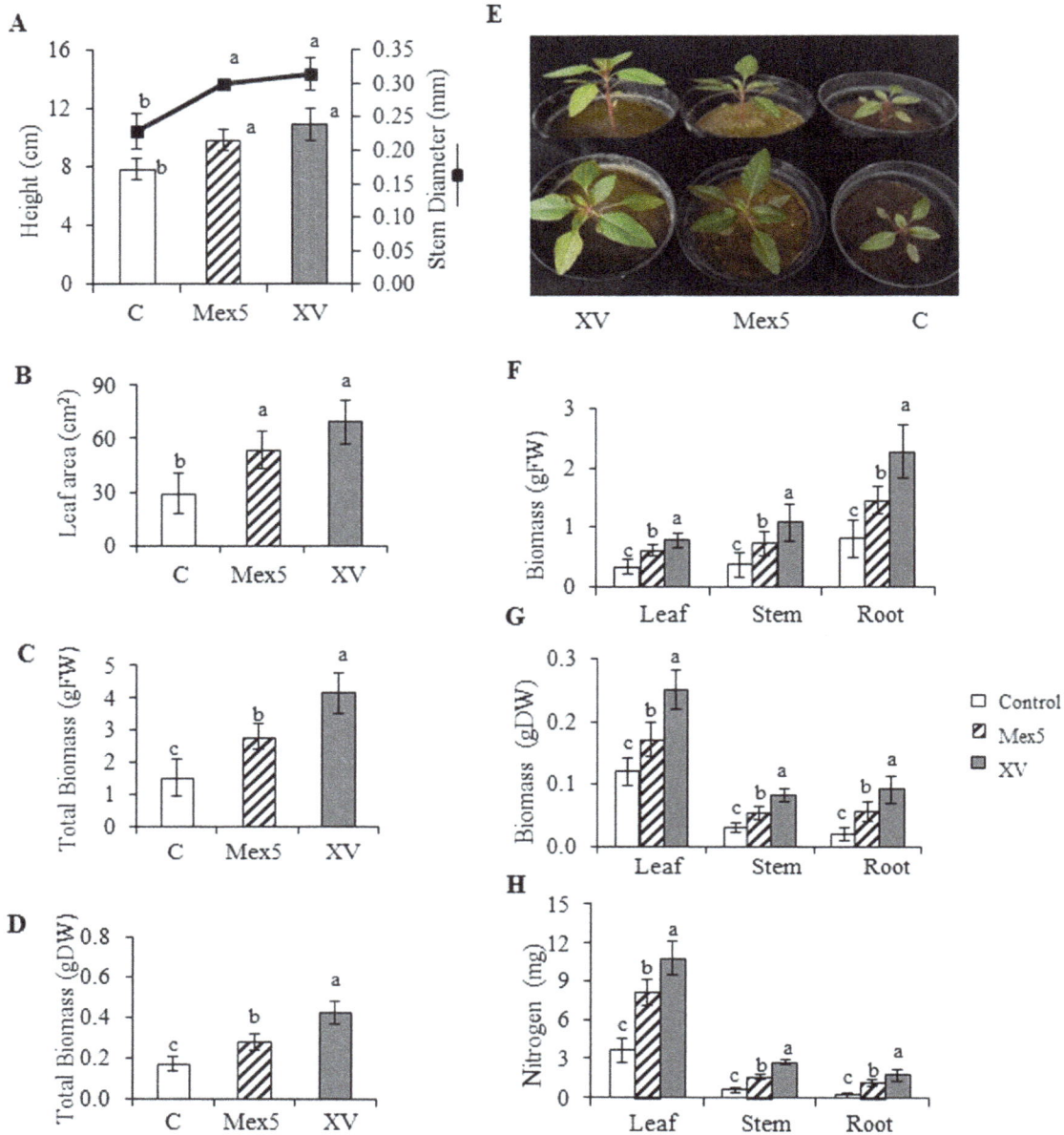

Figure 4. Growth promoting effect of PGPR inoculation on grain amaranth plants maintained in a low-fertility soil. The effect on different growth parameters were determined in *A. cruentus* plants grown in a low fertility soil and inoculated with two strains of *Burkholderia* (*B. ambifaria* Mex5 and *B. caribensis* XV). The parameters measured 7 weeks after inoculation were the following: **A**) plant height and stem diameter; **B**) leaf area; **C** and **D**) total biomass in fresh and dry weight basis, respectively, and **F** and **G**) total leaf, stem and root biomass in a FW and DW basis, respectively. Differences in plant height and leaf area between controls and plants inoculated with *B. caribensis XV* or *B. ambifaria* Mex5 are shown in (**E**). The effect on total nitrogen levels in leaves, stems and roots produced in plants inoculated with *B. caribensis* XV is shown in (**H**). Mean values $\pm$ SE are presented. Asterisks over the bars and lines represent statistically different values at $P \leq 0.05$. Experiments were performed twice, and representative results are shown.

consistent with the fact that, similarly to most crop plants, N availability is the main yield-limiting factor in grain amaranth (see above). The implied ability that these bacteria have to convert molecular nitrogen into ammonia by virtue of the nitrogenase enzyme complex ([67]; see Table 1), raised the possibility that the positive effects on plant growth and yield observed in grain amaranth were associated with an increased N provision occurring as a result of its fixation by the rhizobacterial partners. However, most experimental evidence gathered so far indicates that growth promotion by diazotrophic PGPR does not rely on the N_2-fixation

process, most probably because of its high energetic cost (see [59]). Other, more probable, scenarios have been raised in which stimulated plant growth is proposed to be the result of improved N nutrition occurring as a consequence of increased N uptake in the form of NO_3^-. This was in agreement with the results shown here, since growth promotion and grain yield were consistently shown to increase concomitantly with the N status of the substrate/soil in which grain amaranth was grown, being lowest in a low fertility soil deficient in N and NO_3^- contents (see Table S1 and Figure 4) and highest in the rich substrate supplemented with chemical N

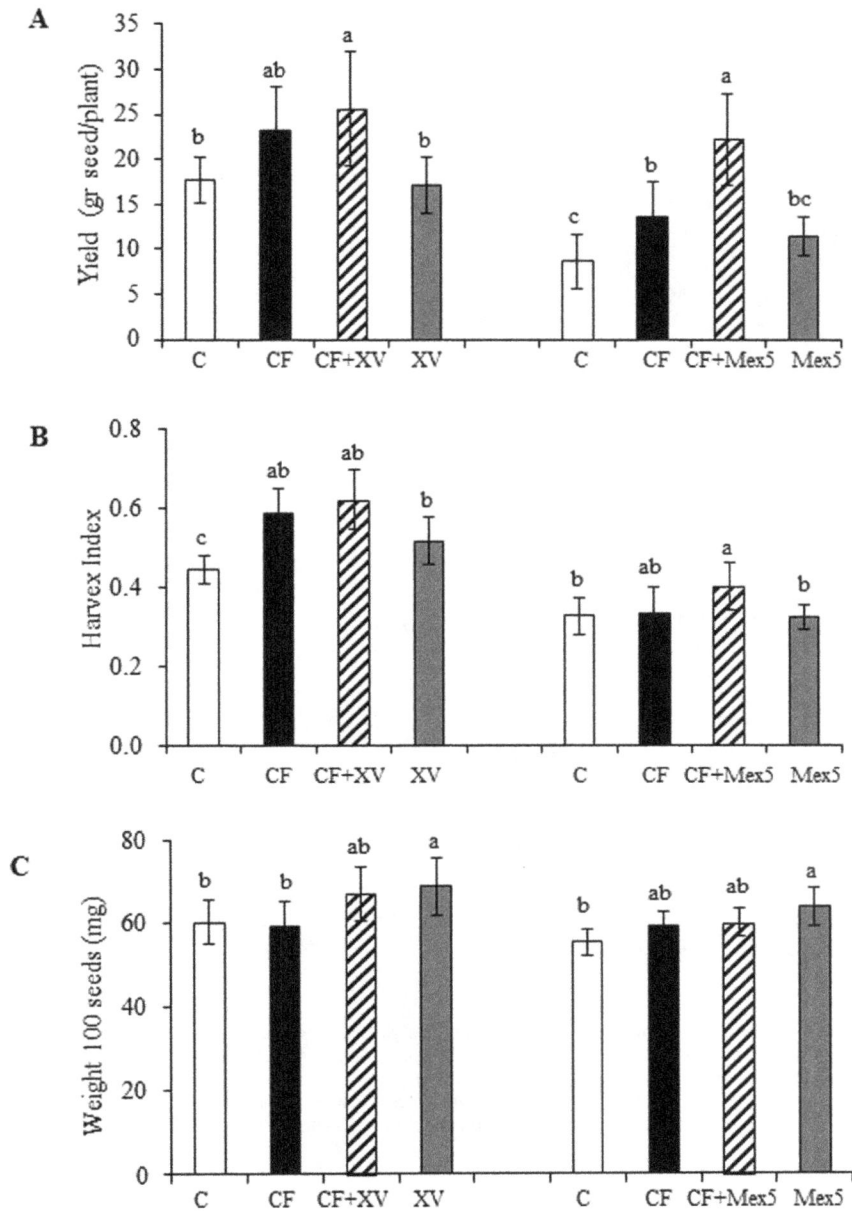

Figure 5. Effect on production parameters measured in *A. cruentus* plants inoculated with different PGPR. (A) Seed yield, **(B)** harvest index and **(C)** weight of 100 seeds were determined in *A. cruentus* plants inoculated with two strains of *Burkholderia* (*B. ambifaria* Mex5 or *B. caribensis* XV) and grown to maturity in a rich substrate. Inoculated plants ± chemical fertilization (CF) were compared with un-inoculated plants ± CF. Mean values ± SE are presented. Different letters over the bars represent statistically different values at *P*≤0.05. The results of a representative experiment that was performed in duplicate are shown.

and P (see Figures 1 to 3). The increased expression of the two nitrate transporters examined was in accordance with this possibility, considering the results of various expression studies that suggest that NO_3^- uptake is primarily regulated at the transcriptional level (see below).

The experiments performed with *A. cruentus* cv. Candil (Figure 3) consistently showed that *B. caribensis* XV produced the best results in terms of growth promotion and grain yield. Curiously, plants inoculated with this strain showed no increase of sucrose (SUC) (in stem and roots) and starch (in leaves and stem) at 7 wpi (Table 2). These results have some similarity with a recent report in *Arabidopsis thaliana* showing that growth promotion resembled a sugar starvation-like transcriptional phenotype that was somehow induced by an unidentified signal from the associated bacterium [68]. These workers speculated that such response could be indicative of an increased metabolic demand for sugars and energy. Likewise, it could be proposed that the best gains in growth promotion and yield observed in *A. cruentus*, which were presumably caused by improved N uptake, occurred at the cost of a higher investment in C resources for the maintenance of the bacterial partner in the rhizosphere. More investigations are needed to prove this hypothesis.

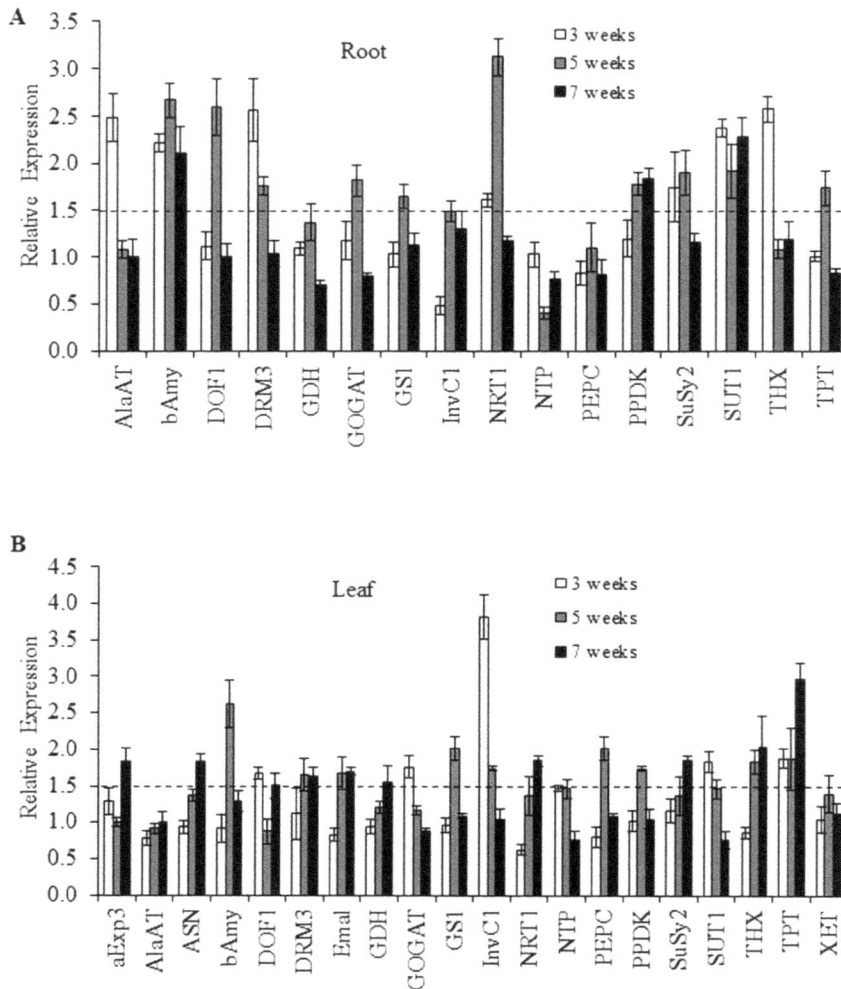

Figure 6. Real-time PCR analysis of gene expression in different tissues of PGPR-inoculated *A. cruentus* **plants.** The expression levels of a battery of genes involved in C and N metabolism and transport were measured in roots (**A**), and leaves (**B**), of *A. cruentus* plants inoculated with *Burkholderia caribensis* XV. The relative expression levels were determined by qPCR at 3, 5 and 7 weeks after seed inoculation, using the $2^{-\Delta\Delta Ct}$ method, as described in [94]. The bars represent mean values ± SE. Dashed lines indicate upper and lower limits beyond which genes were considered to be up- and down-regulated, respectively. Experiments were performed thrice, and results from a representative experiment are shown.

However, the gene expression analysis was accordance with the above possibility. It showed that many genes involved in sugar transport and metabolism were up-regulated in response to the inoculation with *B. caribensis* XV in at least one sampling time point during the seven week period of experimentation. In roots, the sucrose transporter *AhSUT1* remained constantly up-regulated during this period, as well as *AhBAMY1*. Also relevant was the expression of a hexose transporter as well as a cytosolic invertase and an *AhSUS2* gene within the first five weeks after inoculation. The latter genes were also expressed in leaves, in addition to the *AhTPT* gene.

In the context of growth promotion, a previous report showed that the expression of *AtSUT1* and *AhBAMY1* were associated with the high tolerance to defoliation observed in grain amaranth. This report proposed that the up-regulation of these genes facilitated SUC transport and starch degradation in the early stages of plant recovery [69]. Additionally, the constitutive overexpression of a hexose transporter, STP13, in Arabidopsis, was shown to increase the expression of a high affinity nitrate transporter and total N

uptake with the concomitant promotion of plant growth [70]. Moreover, the increased expression of cytosolic invertase1 and *AhSUS2* probably contributed to increase the hexose levels in order to fuel the observed growth promotion in roots and leaves. Importantly, the increased root surface area produced by the association with PGPR most probably enabled the plant to forage a larger volume of soil, which may have led to an enhanced nutrient uptake and consequent promotion of plant growth [59]. It could also be argued that increased transport and metabolism of sugars was probably supporting the augmented flow of C to the root-associated microorganisms present in the rhizosphere (reviewed in [71] and [72]).

In addition, the major role predicted for auxins in rhizobacterial growth promotion [62,73] was supported by the expression in both leaves and roots of the *AhDRM3* gene. This gene was also found to be up-regulated in Arabidopsis plants inoculated with a naturally associated rhizobacterium [68]. The induction of the *AhDof1* gene in both roots and leaves of grain amaranth was in agreement with findings obtained from Arabidopsis and rice plants genetically engineered with a *Dof1* transcription factor, which showed better growth under N-limiting conditions and an enhanced net N assimilation, which was closely associated with the up-regulation of *PEPC, PPDK* (also induced in leaves, i.e. *AhPEPC*, and both leaves and roots, i.e. *AhPPDK*, of inoculated amaranth plants), and other genes coding for enzymes responsible for building the C skeletons used as platforms for inorganic N uptake [74,75].

It is considered that despite their ability to fix atmospheric N_2, diazotrophic PGPR are unlikely to provide large amounts of this form of N to the plants. However, they may greatly influence N nutrition by increasing NO_3^- uptake capacity. One of the proposed mechanisms is by direct stimulation of NO_3^- transport systems. The possibility that this mechanism was also responsible for the growth promotion observed in grain amaranth plants is supported by the expression of the two nitrate transporter genes examined in this study, most predominantly in roots. Such proposal is supported by numerous studies showing that NO_3^- uptake is primarily regulated at the transcriptional level [76–79]. In addition, it was found that the constitutive expression of a high affinity nitrate transporter in rice led to the enhancement of vegetative growth under low nitrogen conditions [19].

The induction of genes involved in N assimilation was in accordance with the results obtained in a recent study in soybean whose aim was to identify genes associated with an enhanced nitrogen use efficiency [80]. Thus, similarly to this study, the gene expression analysis performed in inoculated grain amaranth showed that, in addition to genes involved in nitrate transport (see above), several other genes involved in N assimilation were induced in roots and/or leaves of grain amaranth inoculated with *B. caribensis XV*. These included genes coding for a glutamate dehydrogenase, an NADH GOGAT precursor, and an asparagine synthetase, which is known to be regulated by the carbon (C)/ nitrogen (N) status of the plant. The expression of other genes involved in N assimilation, such as *AhGS1* and *AhAlaAT*, were also in accordance with several other related studies that have shown a positive correlation between the overexpression of cytosolic *GS1* and enhanced growth and/or yields in several plants species (reviewed in [19]) and with reports that demonstrated that the expression of a barley alanine aminotransferase gene in rice, led to significantly increased nitrogen uptake efficiency, biomass, and/or grain yields ([81,82].

An increased expression of C4 photosynthesis-related genes, (*AhNADPH-ME, AhPEPC* and *AhPPDK*), mostly expressed in leaves, may have also indicated a need to increase CO_2 uptake in order to sustain the enhanced plant growth produced by the association with the different PGPR tested. In this respect, various plant-microbe interactions have been previously described as having a strong effect on plant C metabolism [83,84]. This may presumably represent an attempt by the bacteria to manipulate plant metabolism in order to gain access to nutrients, but may also be a manifestation of the positive growth effects of PGPR on plants.

Conclusions

Grain amaranth is a highly tolerant species to adverse environmental conditions, including poor soils, lack of water and severe defoliation. However, grain amaranth production world-wide is hindered by relatively low yields. These are the consequence of several agronomic characteristics that negatively affect productivity [1]. This study demonstrated that both yield and biomass were significantly increased when grain amaranth plants were inoculated with free-living diazotrophic PGPR, which proved to be superior to other PGPR such as *B. subtilis* and

Rhizobium spp. The effect was evident in both a rich substrate with high fertility and in an unfertile soil low in organic matter and primary nutrients, and was still relevant after chemical fertilization of the plants. Growth promotion appeared to be more evident in *A. cruentus* plants, particularly when inoculated with *B. caribensis XV*, a PGPR isolated from the rhizosphere of mango trees. An analysis of gene expression in *A. cruentus* plants inoculated with *B. caribensis XV* revealed that growth promotion was associated with the up-regulation of genes involved in C and N transport and metabolism. Thus, the application of PGPR to grain amaranth could be a strategy to improve productivity, particularly in poor soils with low fertility and could be also be employed to reduce chemical fertilization with the consequential reduction of the environmental pollution problems associated with excessive nitrogen fertilization.

Methods

Plant material and growth conditions

Seeds of *Amaranthus hypochondriacus* cultivar *Nutrisol* and *A. cruentus* cv. *Candil* were provided by Eduardo Espitia (INIFAP, México) and Universidad Nacional de La Pampa, Facultad de Agronomía (Argentina), respectively. The materials were chosen due to their commercial and agronomic importance in these countries. All experiments were performed in the greenhouse, under natural conditions of light and temperature, from mid-February to the end of November, which is the suitable growth season for grain amaranth in central Mexico.

Bacterial growth conditions

The *Burkholderia* spp. strains were cultivated in LB medium [85]; *Bacillus subtilis* BEB-DN was cultivated in Potato Dextrose Broth as described previously [38], whereas *Rhizobium* sp. XVI was cultivated in LGI medium [86]. For inocula preparation, the bacteria were grown aerobically in 1.0-L to 1.5-L of the respective media (initial $A_{600} = 0.1$) on a rotary shaker (145 rpm) using 72 h incubation at 28°C to obtain bacteria in the exponential phase. The culture of bacterial cells was pelleted by centrifugation (5000×g, 7 min, 10°C), washed twice and re-suspended in sterile distilled-deionized water. To obtain 1×10^9 colony forming units (cfu) per ml in the inoculum, the volume was adjusted based upon a correspondence established between the absorbance measured at 600 nm and the bacterial concentration. The density of bacteria was further estimated by plating dilutions of inoculum in Petri dishes containing 1.5% agar plus the respective medium (w/v). Bacteria were inoculated at a density of 1×10^6 cfu/gr of substrate/soil.

Bacterial re-isolation

Samples of 1 g of rhizospheric soil were collected 7 weeks after inoculation to determine the bacterial population of *B. ambifaria* Mex5 and *B. caribensis* XV. This was done following the methodology described by Constantino et al. [87]. The 16S rRNA gene sequences were determined by PCR amplification [88] and direct sequencing. For the phylogenetic analyses, related 16S rRNA gene sequences within the genus *Burkholderia* were included. 16S rDNA sequences were aligned by using the ClustalX program. The phylogenetic tree for the datasets was inferred from the neighbor-joining method described by Saitou and Nei [89] by using the Molecular Evolutionary Genetics Analysis (MEGA) software, version 5 [90] (data not shown).

In planta screening for growth promotion

Briefly, in order to determine inoculation effects, two initial growth promotion pot experiments (GPPE) with two amaranth cultivars, five bacterial strains and two inoculation procedures were followed by three final GPPEs with one amaranth cultivar, two bacterial strains and one inoculation method. These experiments were performed in the years 2011 and 2012. In addition, a yield pot experiment (YPE) was performed with one amaranth cultivar, two bacterial strains and a mixed inoculation procedure in the summer/fall of 2012. In both the preliminary GPPEs and the YPE, the effect of chemical fertilization on PGPR bio-fertilization was evaluated. An additional comparative experiment was performed in the fall of 2012 with plants grown in a poor soil (GPPE-PS) collected from a field located in the town of San Juan de la Vega in the municipality of Celaya in the state of Guanajuato, Mexico (Table S1). No chemical fertilization was applied in this experiment.

The initial GPPE was performed (February to April, 2011) with both grain amaranth species and with five prospective growth promoting rhizobacterial strains having biocontrol properties. These were the following: *Bacillus subtilis* BEB-DN, originally isolated from the rhizosphere of field-cultivated potato plants in the municipality of León, state of Guanajuato, México [91] and known to confer resistance against whitefly infestation in tomato [38]; *Rhizobium spp.* XXV, *Burkholderia caribensis* XV, and *B. cepacia* XXVI, shown to be an effective biocontrol agent against anthracnose in mango fruits and isolated from the rhizosphere of mango trees growing in orchards located in the municipality of Apatzingán, State of Michoacán, México and Chauites, Oaxaca, México [37,92]. *B. ambifaria* Mex-5 was isolated from teosinte plants (*Zea perennis*) growing in a natural reserve ("Reserva de la biósfera, Sierra de Manantlán") located in the municipality of Autlán in the state of Jalisco, México. Other salient characteristics of these bacterial strains are shown in Table 1. Two inoculation procedures were tested: 1) seed soaking with bacterial cultures, for 30 min, when sowing in 2.5-L plastic pots and 2) soil application by drenching the base of the seedlings, three weeks after germination and at the moment of their transfer to 2.5-L pots. All inoculations were done with bacterial suspensions containing the equivalent of 1×10^6 colony-forming units (cfu)/g of substrate. Inoculated seedlings had been previously germinated in 60-space germinating trays as described elsewhere [93]. The pots were filled with a sterile substrate composed of 3 parts Sunshine Mix 3^{TM} (SunGro Horticulture, Bellevue, WA), 1 part loam, 2 parts mulch, 1 part vermiculite (SunGro Horticulture) and 1 part perlite (Termolita S.A., Nuevo León, México). The physicochemical characteristics of this rich substrate are shown in Table S1. All experiments were performed in greenhouses located at Cinvestav, Irapuato, México (20°40′18″N 101°20′48″W) under natural conditions of light and temperature. Several morphometric traits were measured in five plantlets per treatment at 8 weeks (soil drenching of 3 week-old seedlings) or 7 weeks (seed soaking at sowing) after inoculation. These were the following: plant height, stem diameter, total biomass (leaf, stem and roots, in both a dry [DWB] and fresh weight basis [FWB]) and leaf surface area. The latter was measured using a Portable Area Meter LI-3000 (Li-COR; Lincoln, NE, USA). The results of the first set of experiments, established the basis of a second one performed in September to November 2011, with both *A. cruentus* and *A. hypochondriacus*, in which only three bacterial strains (i.e. *B. caribensis* XV, *B. cepacia* XXVI and *B. ambifaria* Mex-5) were inoculated by seed soaking at sowing. In this second experiment, the performance of these PGPR was tested 8 weeks after germination in groups of five plants that included un-inoculated controls (± chemical fertilization) and inoculated controls (± chemical fertilization). Chemical fertilization was done by adding 1.25 g of N as $(NH_4)_2SO_4$ and 0.857 g of P as P_2O_5 to the 2.5-L pots at the start of the experiments. The fertilization regime was based on the amount of N: P: K (180: 40: 00 Kg/ha) recommended for irrigated grain amaranth cultivation in Mexico (E Espitia-Rangel, personal communication).

Pot experiments for growth promotion, nitrogen and carbohydrate content levels and variations in gene expression

Based on the above data, three additional GPPEs were performed in the late spring and summer of 2012 (May 7 to August 27). These experiments were performed under greenhouse conditions, as described above, and as follows: seeds of *A. cruentus* cv. Candil were soak-inoculated at sowing in 2.5-L plastic pots with 1×10^6 CFU/g substrate of *B. caribensis* XV or *B. ambifaria* Mex-5. Plant height, stem diameter, leaf surface area, total biomass in both a FWB and DWB and leaf, stem and root biomass, in a FWB, were measured at 3, 5 and 7 weeks after sowing. Tissue sampling of six plants per treatment was performed at the same time points. The tissues sampled were leaf, stems and roots. They were stored at $-80°C$ until required for the determination of total nitrogen content, non-structural carbohydrates (NSC) (starch, sucrose, glucose and fructose) levels and for gene expression analysis (see below).

Pot experiments for seed yield, harvest index and weight of 100 seeds

Seeds of *A. cruentus* cv. Candil were soak-inoculated at sowing in 16-L plastic pots with 1×10^6 CFU/g substrate of *B. caribensis* XV and *B. ambifaria* Mex-5. A second inoculation was performed 8 weeks after sowing by direct application to the substrate $(1 \times 10^6$ CFU/g) surrounding the roots. These experiments included groups of eleven plants comprising un-inoculated controls (± chemical fertilization), and inoculated plants (± chemical fertilization). This experiment was performed in the greenhouse under the above conditions, from May to November 2012. Step-wise harvest of the plants was started in late October and terminated in mid-November. Two replicates of the experiment were performed simultaneously. Colonization by *B. caribensis* XV and *B. ambifaria* Mex5 was corroborated in all experiments performed by collecting roots and isolating associated bacterial, as described above.

Extraction of total RNA and cDNA preparation

Total RNA was extracted from 100–200 mg of frozen tissue with the Trizol reagent (Invitrogen, Carlsbad, CA, USA), according to the manufacturer's instructions, with modifications. These consisted of the addition of a salt solution (sodium citrate 0.8 M+1.2 M NaCl) during precipitation in a 1:1 v/v ratio with isopropanol and further purification with LiCl (8 M) for one hour at 4°C. All RNA samples were analyzed by formaldehyde agarose gel electrophoresis and visual inspection of the ribosomal RNA bands upon ethidium bromide staining. Total RNA samples (1 μg for leaf and 3 μg for root) were reverse-transcribed to generate the first-strand cDNA using an oligo dT_{20} primer and 200 units of SuperScript II reverse transcriptase (Invitrogen).

Gene expression analysis by quantitative real-time RT-PCR (qRT-PCR)

The cDNA employed for the qRT-PCR assays was initially prepared from 4 μg total RNA. It was then diluted ten-fold in sterile deionized-distilled (dd) water prior to qRT-PCR. Amplifications were performed using SYBR Green detection chemistry and run in triplicate in 96-well reaction plates with the CFX96

Real Time System (Bio-Rad, Hercules, CA, USA). Reactions were prepared in a total volume of 20 μl containing: 2 μl of template, 2 μl of each amplification primer (2 μM), 8 μl of IQ SYBR SuperMix (Bio-Rad) and 6 μl of sterile dd water. Quantitative real-time PCR was performed in triplicate for each sample using the primers listed in Table S2. Primers were designed for each gene, based on partial cDNA sequences derived from the transcriptomic analysis of *A. hypochondriacus* [93] or from complete cDNAs generated in a related study [69]. Primer design was performed using DNA calculator software (Sigma-Aldrich St. Louis, MO, USA) and included, when possible, part of unique 3' non-coding regions to ensure specificity.

The following protocol was followed for all qRT-PCR runs: 15 min at 95°C to activate the *Taq* Polymerase, followed by 40 cycles of denaturation at 95°C for 15 s and annealing at 60°C for 1 min. Slow amplifications requiring an excess of 32 cycles were not considered for analysis. The specificity of the amplicons was verified by melting curve analysis after 40 cycles and agarose gel electrophoresis. Baseline and threshold cycles (Ct) were automatically determined using Real-Time PCR System software. PCR efficiencies for all genes tested were greater than 95%. Relative expression was calculated using the comparative cycle threshold method [94], where delta (Δ) cycle threshold of cDNA from un-inoculated controls was defined as 100% transcript presence.

The selection of genes was partly based on a recent report describing that the natural association of *A. thaliana* seedlings with growth promoting *Pseudomonas. sp.* G62 rhizobacteria induced a rapid and stable starvation-like transcriptional response which included genes involved in cell wall modification, C- and N-metabolism and auxin signaling [68]. These were *AhXET*, (xyloglucan endo-transglycosylase-related, isotig 04370), *AhEXP3α* (α-expansin 3, isotig 07296), *AhASN1* (Glutamine-dependent asparagine synthetase 1, isotig 11850), *AhGDH2* (Glutamate Dehydrogenase 2, isotig 09281), and *AhNRT.3* (Nitrate transporter 3, isotig 03624) and *DRM3*, an auxin responive gene (isotig 02637). Genes were also selected from a group of carbohydrate metabolism and C4 photosynthesis-related genes used to monitor changes in leaf gene expression in response to source-sink perturbation caused by partial shading of 12-month-old sugar cane plants [95]. These included the following: *AhME* (NADP-dependent malic enzyme, isotig 05148), *AhPEPC* (phosphoenole-nolpyruvate carboxylase, isotig 16713), *Ah TPT* (triose phosphate/phosphate transporter, isotig 12255), and *AhHT* (hexose transporter, isotig 11515). Finally, genes involved in C mobilization and whose expression was positively correlated with defoliation tolerance in grain amaranth [69], were analyzed too. These included the following: *AhBAmy1* (*β-amilase1*, isotig 03918); *AhA/NI-1* (*cytosolic invertase 1*; accession No. JQ012920), *AhSUT1* (sucrose transporter1, isotig 00313), and *AhSus2* (Sucrose synthase2, accession No. JQ012919). Genes were also selected on the basis of results obtained from transgenic approaches designed to improve plant nitrogen use efficiency (NUE) (reviewed in [19]). These included the following: *AhNRT.1.1* (nitrate transporter1.1; isotig 05430); *AhAlaAT* (alanine aminotransferase; contig 19731); *AhGS1* (cytosolic glutamine synthetase 1; isotig 04849); *AhNADH GOGAT* (NADH dependent glutamate synthase; isotig 12310); *AhDof1* (Dof1 transcription factor; isotig 15733), and *AhPPDK* (pyruvate orthophosphate dikinase; isotig 00544).

Transcript abundance data were normalized against the average transcript abundance of two reference genes: *actin* (isotig 10321) and *β-tubulin* (isotig 05486). These were obtained from the above transcriptomic study. The fold change in expression of the target genes in each treatment was calculated using the following equation: $2^{-\Delta\Delta Ct}$, where $\Delta\Delta Ct = $ (Ct target gene - average Ct reference genes)$_{treatment} - $ (Ct target gene - average Ct reference genes)$_{control}$ [94]. Values reported are the mean of three repetitions $\pm$ SE of one representative experiment. The qRT-PCR expression analysis was validated in three independent experiments.

Determination of non-structural carbohydrate and nitrogen levels

All tissues (leaves, stems, roots and panicles) were collected at the beginning of the dark period (~6:30 p.m.) and flash frozen in liquid nitrogen. Frozen ground tissue (200 mg) was extracted with 500 μl 80% aqueous ethanol (v/v) and incubated at 4°C for 10 min with stirring. After refrigerated centrifugation at 10,000 rpm (4°C for 10 min), the cleared supernatants were transferred into new tubes and concentrated by centrifugation (Heto Maxi Dry Lyo, Heto-Holten, Denmark). The residue was re-dissolved in 500 μl of 100 mM Hepes buffer, pH 7.4, and 5 mM $MgCl_2$, and used for the determination of soluble sugars. The pellet derived from the centrifugation step was used for the determination of starch. To this end, it was homogenized with 500 μl of 10 mM KOH and incubated at 99°C for 2 h. Sucrose (SUC), glucose (GLC), fructose (FRC) and starch contents were measured using enzyme-based methods as instructed (Boehringer Mannheim/R-Biopharm, Darmstadt, Germany), except that the final reaction volume was reduced to fit a micro-plate format (250 μl per reaction).

Leaf N was determined by the micro-Kjeldahl method [96].

Statistical analysis

All statistical analyses of the physiological and biochemical data were done using JPM8 at the $\alpha = 0.05$ level (SAS Institute Inc., Cary, NC). Data were analyzed using an ANOVA. A Tukey test was performed with each ANOVA. In all figures, mean values and vertical bars representing standard errors (SE) are shown. In Table 2, standard errors are also listed beside mean values.

Supporting Information

Table S1 Characteristics of the rich substrate and of the sandy, infertile soil used in the growth promotion experiments.

Table S2 Primers used for gene expression analysis by qRT PCR.

Author Contributions

Conceived and designed the experiments: JPDF FIPC. Performed the experiments: FIPC NAMG. Analyzed the data: JPDF FIPC SSV JJPC. Contributed reagents/materials/analysis tools: SSV JJPC. Wrote the paper: JPDF FIPC.

References

1. Brenner DM, Baltensperger DD, Kulakow PA, Lehmann JW, Myers RL, et al. (2000) Genetic resources and breeding of *Amaranthus*. In Janick J, editor. Plant Breeding Reviews. vol. 19. New York: John Wiley & Sons, Inc. pp 227–285.

2. Hill RM, Rawate PD (1982) Evaluation of food potential, some toxicological aspects, and preparation of a protein isolate from the aerial part of amaranth (pigweed). J Agric Food Chem 30: 465–469.

3. Shukla S, Bhargava A, Chatterjee A, Srivastava J, Singh N, et al. (2006) Mineral profile and variability in vegetable amaranth (*Amaranthus tricolor*). Plant Foods Hum Nutr 61: 23–28.

4. Akubugwo IE, Obasi NA, Chinyere GC, Ugbogu AE (2007) Nutritional and chemical value of *Amaranthus hybridus* L. leaves from Afikpo, Nigeria. Afr J Biotechnol 6: 2833–2839.

5. Weaver SE, McWilliams EL (1980) The biology of canadian weeds: 44. *Amaranthus retroflexus* L., *A. powellii* S. Wats. and *A. hybridus* L. Can J Plant Sci 60: 1215–1234.

6. Steckel LE (2007) The dioecious *Amaranthus* spp.: here to stay. Weed Technol 21: 567–570.

7. Huerta-Ocampo J, Barba de la Rosa A (2011) Amaranth: a pseudo-cereal with nutraceutical properties. Curr Nutr Food Sci 7: 1–9.

8. Caselato-Sousa VM, Amaya-Farfán J (2012) State of knowledge on amaranth grain: a comprehensive review. J Food Sci 77: R93–R104.

9. Li J, Wang S, Liu X, Li X, Gou J (1989) An observation of the root system growth of grain amaranth and its drought resistance. Agric Res Arid Areas 3: 34–41.

10. Johnson BL, Henderson TL (2002) Water use patterns of grain amaranth in the northern Great Plains. Agron J 94: 1437–1443.

11. Omami EN, Hammes PS, Robbertse PJ (2006) Differences in salinity tolerance for growth and water-use efficiency in some amaranth (*Amaranthus* spp.) genotypes. New Zeal J Crop Hort Sci 34: 11–22.

12. Weber LE (1990) Amaranth grain production guide. New Crops Department, Rodale Research Center, Rodale Press, Emmaus, PA.

13. Kadereit G, Borsch T, Weising K, Freitag H (2003) Phylogeny of Amaranthaceae and Chenopodiaceae and the evolution of C-4 photosynthesis. Int J Plant Sci 164: 959–986.

14. Huerta-Ocampo JA, Briones-Cerecero EP, Mendoza-Hernandez G, De Leon-Rodriguez A, Barba de la Rosa AP (2009) Proteomic analysis of amaranth (*Amaranthus hypochondriacus* L.) leaves under drought stress. Int J Plant Sci 170: 990–998.

15. Huerta-Ocampo JA, Leon-Galvan MF, Ortega-Cruz LB, Barrera-Pacheco A, De Leon-Rodriguez A, et al. (2011) Water stress induces up-regulation of DOF1 and MIF1 transcription factors and down-regulation of proteins involved in secondary metabolism in amaranth roots (*Amaranthus hypochondriacus* L.). Plant Biol (Stuttg) 13: 472–482.

16. Myers R (1996) Amaranth: New crop opportunity. In: Janick J, editor. Progress in new crops. ASHS Press, Alexandria, VA. pp. 207–220.

17. Robertson GP, Vitousek PM (2009) Nitrogen in agriculture: balancing the cost of an essential resource. Annu Rev Environ Resour 34: 97–125.

18. Good AG, Shrawat AK, Muench DG (2004) Can less yield more? Is reducing nutrient input into the environment compatible with maintaining crop production? Trends Plant Sci 9: 597–605.

19. Xu GH, Fan XR, Miller AJ (2012) Plant nitrogen assimilation and use efficiency. Annu Rev Plant Biol 63: 153–182.

20. Guo JH, Liu XJ, Zhang Y, Shen JL, Han WX, et al. (2010) Significant acidification in major chinese croplands. Science 327: 1008–1010.

21. van Loon LC, Bakker PAHM, Pieterse CMJ (1998) Systemic resistance induced by rhizosphere bacteria. Annu Rev Phytopathol 36: 453–483.

22. Al-Taweil HI, Osman MB, Hamid AA, Wan Yussof WM (2009) Development of microbial inoculants and the impact of soil application on rice seedlings growth. Am J Agric Biol Sci 4: 79–82.

23. Gamalero E, Glick BR (2011) Mechanisms used by plant growth-promoting bacteria. In: Maheshwari DK, editor. Bacteria in Agrobiology: Plant Nutrient Management. Springer-Verlag, Berlin Heidelberg. pp. 17–46.

24. Kloepper JW (2003) A review of mechanisms for plant growth promotion by PGPR. In: Reddy MS, Anandaraj M, Eapen SJ, Sarma YR, Kloepper JW, editors. Abstracts and short papers. 6th International PGPR workshop, 5–10 october 2003. Indian Institute of Spices Research, Calicut, India. pp. 81–92.

25. Suneja P, Dudeja SS, Narula N (2007) Development of multiple co-inoculants of different biofertilizers and their interaction with plants. Arch Agron Soil Sci 53: 221–230.

26. Rengel Z, Marschner P (2005) Nutrient availability and management in the rhizosphere: exploiting genotypic differences. New Phytol 168: 305–312.

27. Ryu CM, Farag MA, Hu CH, Reddy MS, Wei HX, et al. (2003) Bacterial volatiles promote growth in *Arabidopsis*. Proc Natl Acad Sci USA 100: 4927–4932.

28. Zhang H, Kim MS, Krishnamachari V, Payton P, Sun Y, et al. (2007) Rhizobacterial volatile emissions regulate auxin homeostasis and cell expansion in *Arabidopsis*. Planta 226: 839–851.

29. Bashan Y, Holguin G (1998) Proposal for the division of plant growth-promoting rhizobacteria into two classifications: Biocontrol-PGPB (Plant Growth-Promoting Bacteria) and PGPB. Soil Biol Biochem 30: 1225–1228.

30. Belimov AA, Safronova VI, Sergeyeva TA, Egorova TN, Matveyeva VA, et al. (2001) Characterization of plant growth promoting rhizobacteria isolated from polluted soils and containing 1-aminocyclopropane-1-carboxylate deaminase. Can J Microbiol 47: 642–652.

31. Kiely PD, Haynes JM, Higgins CH, Franks A, Mark GL, et al. (2006) Exploiting new systems-based strategies to elucidate plant-bacterial interactions in the rhizosphere. Microb Ecol 51: 257–266.

32. Pineda A, Zheng SJ, van Loon JJ, Pieterse CM, Dicke M (2010) Helping plants to deal with insects: the role of beneficial soil-borne microbes. Trends Plant Sci 15: 507–514.

33. Vessey JK (2003) Plant growth promoting rhizobacteria as biofertilizers. Plant Soil 255: 571–586.

34. Whipps JM (2001) Microbial interactions and biocontrol in the rhizosphere. J Exp Bot 52: 487–511.

35. Zehnder GW, Murphy JF, Sikora EJ, Kloepper JW (2001) Application of rhizobacteria for induced resistance. Eur J Plant Pathol 107: 39–50.

36. Bevivino A, Sarrocco S, Dalmastri C, Tabacchioni S, Cantale C, et al. (1998) Characterization of a free-living maize-rhizosphere population of *Burkholderia cepacia*: effect of seed treatment on disease suppression and growth promotion of maize. FEMS Microbiol Ecol 27: 225–237.

37. de los Santos-Villalobos S, Barrera-Galicia GC, Miranda-Salcedo MA, Pena-Cabriales JJ (2012) *Burkholderia cepacia* XXVI siderophore with biocontrol capacity against *Colletotrichum gloeosporioides*. World J Microbiol Biotechnol 28: 2615–2623.

38. Valenzuela-Soto JH, Estrada-Hernandez MG, Ibarra-Laclette E, Delano-Frier JP (2010) Inoculation of tomato plants (*Solanum lycopersicum*) with growth-promoting *Bacillus subtilis* retards whitefly *Bemisia tabaci* development. Planta 231: 397–410.

39. Khurana AL, Namdeo SL, Dudeja SS (1997) On-farm experiments on rhizobial inoculants: problems and possible solutions. In: Rupela OP, Johansen C, Herridge DF, editors. Managing legume nitrogen fixation in cropping systems of Asia: 20–24 Aug 1996; ICRISAT Asia Center. pp. 217–226.

40. Khurana AL, Dudeja SS (1997) Biological nitrogen fixation technology for pulses interaction in India. Technical bulletin, Indian Institute of Pulses Research, Kanpur. pp. 1–18.

41. Caballero-Mellado J, Carcaño Montiel M, Mascarua-Esparza M (1992) Field inoculation of wheat (*Triticum aestivum*) with *Azospirillum brasilense* under temperate climate. Symbiosis 13: 243–253.

42. Narula N, Yadav KS (1989) Nitrogen fixation research in India with *Azotobacter*. In: Dadarwal KR, Yadav KS, editors. Biological nitrogen fixation research status in India. The Society for Plant Physiology and Biochemistry, New Delhi. pp. 87–124.

43. Derylo M, Skorupska A (1993) Enhancement of symbiotic nitrogen-fixation by vitamin-secreting fluorescent *Pseudomonas*. Plant Soil 154: 211–217.

44. Dudeja SS, Duhan JS (2005) Biological nitrogen fixation research in pulses with special reference to mungbean and urdbean. Indian J Pulses Res 18: 107–118.

45. Defreitas JR, Germida JJ (1990) Plant-growth promoting rhizobacteria for winter-wheat. Can J Microbiol 36: 265–272.

46. de Silva A, Patterson K, Rothrock C, Moore J (2000) Growth promotion of highbush blueberry by fungal and bacterial inoculants. Hort Sci 35: 1228–1230.

47. Kloepper JW, Lifshitz R, Zablotowicz RM (1989) Free-living bacterial inocula for enhancing crop productivity. Trends Biotechnol 7: 39–44.

48. Lucy M, Reed E, Glick B (2004) Applications of free living plant growth-promoting rhizobacteria. Antonie van Leeuwenhoek 86: 1–25.

49. Lugtenberg BJJ, Chin-A-Woeng TFC, Bloemberg GV (2002) Microbe-plant interactions: principles and mechanisms. Antonie Van Leeuwenhoek 81: 373–383.

50. Tiwari V, Pathak A, Lehri L (1993) Rock phosphate super-phosphate in wheat in relation to inoculation with phosphate solubilizing organisms and organic waste. Indian J Agric Res 27: 137–145.

51. Toro M, Azcon R, Barea JM (1998) The use of isotopic dilution techniques to evaluate the interactive effects of *Rhizobium* genotype, mycorrhizal fungi, phosphate-solubilizing rhizobacteria and rock phosphate on nitrogen and phosphorus acquisition by *Medicago sativa*. New Phytol 138: 265–273.

52. Harman G, Howell C, Viterbo A, Chet I, Lorito M (2004) *Trichoderma* species-oportunistic, biological nitrogen fixation research in pulses. Nat Rev Microbiol 2: 43–56.

53. Adesemoye AO, Torbert HA, Kloepper JW (2008) Enhanced plant nutrient use efficiency with PGPR and AMF in an integrated nutrient management system. Can J Microbiol 54: 876–886.

54. Nair CB, Anith KN, Sreekumar J (2007) Mitigation of growth retardation effect of plant defense activator, acibenzolar-S-methyl, in *Amaranthus* plants by plant growth-promoting rhizobacteria. World J Microbiol Biotech 23: 1183–1187.

55. Martinez-Mendoza EK, Mena-Violante HG, Mendez-Inocencio C, Oyoque-Salcedo G, Cortez-Madrigal H, et al. (2012) Effects of *Bacillus subtilis* extracts on weed seed germination of *Sorghum halepense* and *Amaranthus hybridus*. Afric J Microbiol Res 6: 1887–1892.

56. Nair C, Anith K (2009) Efficacy of acibenzolar-S-methyl and rhizobacteria for the management of foliar blight disease of amaranth. J Trop Agric 47: 43–47.

57. Kraiser T, Gras DE, Gutierrez AG, Gonzalez B, Gutierrez RA (2011) A holistic view of nitrogen acquisition in plants. J Exp Bot 62: 1455–1466.

58. Espitia-Rangel E, Mapes-Sánchez C, Escobedo-López D, de la O-Olán M, Rivas-Valencia P, et al. (2010) Conservación y uso de los recursos genéticos de amaranto en México. SINAREFI-INIFAP-UNAM, Centro de Investigación Regional Centro, Celaya, Guanajuato, , México. 201 p.

59. Mantelin S, Touraine B (2004) Plant growth-promoting bacteria and nitrate availability: impacts on root development and nitrate uptake. J Exp Bot 55: 27–34.

60. van de Mortel JE, de Vos RCH, Dekkers E, Pineda A, Guillod L, et al. (2012) Metabolic and transcriptomic changes induced in *Arabidopsis* by the rhizobacterium *Pseudomonas fluorescens* SS101. Plant Physiol 160: 2173–2188.

61. Sheng XF, Xia JJ, Jiang CY, He LY, Qian M (2008) Characterization of heavy metal-resistant endophytic bacteria from rape (*Brassica napus*) roots and their

potential in promoting the growth and lead accumulation of rape. Environ Pollut 156: 1164–1170.

62. Persello-Cartieaux F, Nussaume L, Robaglia C (2003) Tales from the underground: molecular plant-rhizobacteria interactions. Plant Cell Environ 26: 189–199.

63. Elbehri A, Putnam DH, Schmitt M (1993) Nitrogen-fertilizer and cultivar effects on yield and nitrogen-use efficiency of grain amaranth. Agron J 85: 120–128.

64. Myers RL (1998) Nitrogen fertilizer effect on grain amaranth. Agron J 90: 597–602.

65. Olaniyi JO, Adelasoye KA, Jegede CO (2008) Influence of nitrogen fertilizer on the growth, yield and quality of grain amaranth varieties. World J Agric Sci 4: 506–513.

66. Thanapornpoonpong S (2004) Effect of nitrogen fertilizer on nitrogen assimilation and seed quality of amaranth (Amaranthus spp.) and quinoa (Chenopodium quinoa Willd). PhD Thesis. Georg-August-University of Göttingen, Sweden.

67. Postgate JR (1982) Biological Nitrogen-Fixation: Fundamentals. Phil Trans R Soc Lond 296: 375–385.

68. Schwachtje J, Karojet S, Thormählen I, Bernholz C, Kunz S, et al. (2011) A naturally associated rhizobacterium of Arabidopsis thaliana induces a starvation-like transcriptional response while promoting growth. PLoS ONE 6 (12): e29382.

69. Castrillón-Arbeláez PA, Martínez-Gallardo N, Avilés-Arnaut H, Tiessen A, Délano-Frier JP (2012) Metabolic and enzymatic changes associated with carbon mobilization, utilization and replenishment triggered in grain amaranth (Amaranthus cruentus) in response to partial defoliation by mechanical injury or insect herbivory. BMC Plant Biol 12: 163.

70. Schofield RA, Bi YM, Kant S, Rothstein SJ (2009) Over-expression of STP13, a hexose transporter, improves plant growth and nitrogen use in Arabidopsis thaliana seedlings. Plant Cell Environ 32: 271–285.

71. Jones DL, Nguyen C, Finlay RD (2009) Carbon flow in the rhizosphere: carbon trading at the soil-root interface. Plant Soil 321: 5–33.

72. Dennis PG, Miller AJ, Hirsch PR (2010) Are root exudates more important than other sources of rhizodeposits in structuring rhizosphere bacterial communities? FEMS Microbiol Ecol 72: 313–327.

73. Persello-Cartieaux F, David P, Sarrobert C, Thibaud MC, Achouak W, et al. (2001) Utilization of mutants to analyze the interaction between Arabidopsis thaliana and its naturally root-associated Pseudomonas. Planta 212: 190–198.

74. Yanagisawa S, Akiyama A, Kisaka H, Uchimiya H, Miwa T (2004) Metabolic engineering with Dof1 transcription factor in plants: Improved nitrogen assimilation and growth under low-nitrogen conditions. Proc Natl Acad Sci USA 101: 7833–7838.

75. Kurai T, Wakayama M, Abiko T, Yanagisawa S, Aoki N, et al. (2011) Introduction of the ZmDof1 gene into rice enhances carbon and nitrogen assimilation under low-nitrogen conditions. Plant Biotechnol J 9: 826–837.

76. Forde BG (2000) Nitrate transporters in plants: structure, function and regulation. Biochim Biophys Acta 1465: 219–235.

77. Vidmar JJ, Zhuo D, Siddiqi MY, Schjoerring JK, Touraine B, et al. (2000) Regulation of high-affinity nitrate transporter genes and high-affinity nitrate influx by nitrogen pools in roots of barley. Plant Physiol 123: 307–318.

78. Glass ADM, Britto DT, Kaiser BN, Kinghorn JR, Kronzucker HJ, et al. (2002) The regulation of nitrate and ammonium transport systems in plants. J Exp Bot 53: 855–864.

79. Nazoa P, Vidmar JJ, Tranbarger TJ, Mouline K, Damiani I, et al. (2003) Regulation of the nitrate transporter gene AtNRT2.1 in Arabidopsis thaliana: responses to nitrate, amino acids and developmental stage. Plant Mol Biol 52: 689–703.

80. Hao QN, Zhou XA, Sha AH, Wang C, Zhou R, et al. (2011) Identification of genes associated with nitrogen-use efficiency by genome-wide transcriptional analysis of two soybean genotypes. BMC Genomics 12: 525.

81. Good AG, Johnson SJ, De Pauw M, Carroll RT, Savidov N (2007) Engineering nitrogen use efficiency with alanine aminotransferase. Can J Bot 85: 252–262.

82. Shrawat AK, Carroll RT, DePauw M, Taylor GJ, Good AG (2008) Genetic engineering of improved nitrogen use efficiency in rice by the tissue-specific expression of alanine aminotransferase. Plant Biotechnol J 6: 722–732.

83. Biemelt S, Sonnewald U (2006) Plant-microbe interactions to probe regulation of plant carbon metabolism. J Plant Physiol 163: 307–318.

84. Chen LQ, Hou BH, Lalonde S, Takanaga H, Hartung ML, et al. (2010) Sugar transporters for intercellular exchange and nutrition of pathogens. Nature 468: 527–532.

85. Weaver VB, Kolter R (2004) Burkholderia spp. alter Pseudomonas aeruginosa physiology through iron sequestration. J Bacteriol 186: 2376–2384.

86. Cavalcante VA, Dobereiner J (1988) A new acid-tolerant nitrogen-fixing bacterium associated with sugarcane. Plant Soil 108: 23–31.

87. Constantino M, Gomez-Alvarez R, Alvarez-Solis JD, Geissen V, Huerta E, et al. (2008) Effect of inoculation with rhizobacteria and arbuscular mycorrhizal fungi on growth and yield of Capsicum chinense Jacquin. J Agric Rural Dev Trop 109: 169–180.

88. Lane DJ (1991) 16S/23S rRNA sequencing. In: Stackebrandt E, Goodfellow M, editors. Nucleic acid techniques in bacterial systematics. John Wiley and Sons, New York. pp. 115–175.

89. Saitou N, Nei M (1987) The neighbor-joining method: a new method for reconstructing phylogenetic trees. Mol Biol Evol 4: 406–425.

90. Tamura K, Peterson D, Peterson N, Stecher G, Nei M, et al. (2011) MEGA5: Molecular Evolutionary Genetics Analysis using maximum likelihood, evolutionary distance, and maximum parsimony methods. Mol Biol Evol 28: 2731–2739.

91. Jiménez-Delgadillo M (1999) Evaluación y caracterización fisiológica de rizobacterias empleadas como posibles agentes de biocontrol. MSc Thesis. Cinvestav I.P.N., Unidad Irapuato; Irapuato, México.

92. de los Santos-Villalobos S, Folter S, Délano-Frier J, Gómez-Lim M, Guzmán-Ortiz D, et al. (2013) Growth promotion and flowering induction in mango (Mangifera indica L. cv "Ataulfo") trees by Burkholderia and Rhizobium inoculation: morphometric, biochemical, and molecular events. J Plant Growth Regul 32: 615–627.

93. Délano-Frier JP, Avilés-Arnaut H, Casarrubias-Castillo K, Casique-Arroyo G, Castrillón-Arbeláez PA, et al. (2011) Transcriptomic analysis of grain amaranth (Amaranthus hypochondriacus) using 454 pyrosequencing: comparison with A. tuberculatus, expression profiling in stems and in response to biotic and abiotic stress. BMC Genomics 12: 363.

94. Livak KJ, Schmittgen TD (2001) Analysis of relative gene expression data using real-time quantitative PCR and the $2^{-\Delta\Delta Ct}$ method. Methods 25: 402–408.

95. McCormick AJ, Cramer MD, Watt DA (2008) Changes in photosynthetic rates and gene expression of leaves during a source-sink perturbation in sugarcane. Ann Bot 101: 89–102.

96. Humphries E (1956) Mineral components and ash analysis. In: Peach K, Tracy M, editors. Modern methods of plant analysis. Springer Verlag, Berlin. pp. 468–502.

8

Nitrogen Deposition Enhances Carbon Sequestration by Plantations in Northern China

Zhihong Du[1,2], Wei Wang[1]*, Wenjing Zeng[1], Hui Zeng[2]

1 Department of Ecology, College of Urban and Environmental Sciences, and Key Laboratory for Earth Surface Processes of the Ministry of Education, Peking University, Beijing, China, **2** Key Laboratory for Urban Habitat Environmental Science and Technology, Peking University Shenzhen Graduate School, Shenzhen, China

Abstract

Nitrogen (N) deposition and its ecological effects on forest ecosystems have received global attention. Plantations play an important role in mitigating climate change through assimilating atmospheric CO_2. However, the mechanisms by which increasing N additions affect net ecosystem production (NEP) of plantations remain poorly understood. A field experiment was initialized in May 2009, which incorporated additions of four rates of N (control (no N addition), low-N (5 g N m^{-2} yr^{-1}), medium-N (10 g N m^{-2} yr^{-1}), and high-N (15 g N m^{-2} yr^{-1})) at the Saihanba Forestry Center, Hebei Province, northern China, a locality that contains the largest area of plantations in China. Net primary production (NPP), soil respiration, and its autotrophic and heterotrophic components were measured. Plant tissue carbon (C) and N concentrations (including foliage, litter, and fine roots), microbial biomass, microbial community composition, extracellular enzyme activities, and soil pH were also measured. N addition significantly increased NPP, which was associated with increased litter N concentrations. Autotrophic respiration (AR) increased but heterotrophic respiration (HR) decreased in the high N compared with the medium N plots, although the HR in high and medium N plots did not significantly differ from that in the control. The increased AR may derive from mycorrhizal respiration and rhizospheric microbial respiration, not live root respiration, because fine root biomass and N concentrations showed no significant differences. Although the HR was significantly suppressed in the high-N plots, soil microbial biomass, composition, or activity of extracellular enzymes were not significantly changed. Reduced pH with fertilization also could not explain the pattern of HR. The reduction of HR may be related to altered microbial C use efficiency. NEP was significantly enhanced by N addition, from 149 to 426.6 g C m^{-2} yr^{-1}. Short-term N addition may significantly enhance the role of plantations as an important C sink.

Editor: Shuijin Hu, North Carolina State University, United States of America

Funding: This research was supported by the National Basic Research Program of China (No. 2010CB950600 and 2013CB956303), Projects of the National Natural Science Foundation of China (No. 31222011, 31270363, and 31070428), Projects of Innovative Research Groups of the National Natural Science Foundation of China (No. 31021001) and University Construction Projects from Central Authorities in Beijing. The funders had no role in study design, data collection and analysis, decision to publish, or preparation of the manuscript.

Competing Interests: The authors have declared that no competing interests exist.

* E-mail: wangw@urban.pku.edu.cn

Introduction

Terrestrial ecosystems sequester nearly 30% of anthropogenic carbon (C) emissions, offering the most effective, yet natural, means to mitigate climate change [1]. Nitrogen (N) is a major limiting nutrient to plant growth in most terrestrial ecosystems [2] and thus affects C sequestration in terrestrial ecosystems [3]. Human activity has led to a significant increase in N deposition owing to industrialization, agricultural practices, and the combustion of fossil fuels [4–6]. Numerous studies have shown that N deposition can increase net ecosystem production (NEP), as an indicator of ecosystem C sequestration [7–9]. However, the magnitude of the increased NEP following N addition varied greatly from 24.5 to 225 kg C per kg N [10–12]. Therefore, there is an urgent need to explore the mechanisms underlying this effect.

NEP is determined by the difference between net primary production (NPP) and soil heterotrophic respiration (HR). One important issue that needs addressing is how additional N affects the process of plant growth and thus enhances NPP. Many studies have attributed the increased tree growth to significantly higher foliar N concentrations in fertilized plots [8,13–15]. The increased

foliar N concentrations could improve biomass production through the following three pathways: by increasing the uptake of CO_2 [16–18], by increasing water-use efficiency of foliage via altering CO_2 assimilation and stomatal conductance [19], and by reducing the thermally dissipated light [20]. At the same time, N addition may also decrease leaf N resorption [21], and thus increase litter N concentrations. Consequently, more available N is released via decomposition to supply plant growth [22]. However, little research has been conducted to comprehensively analyze the mechanism of plant biomass growth caused by N addition.

How soil respiration (SR) responds to N addition is also relevant. SR consists of autotrophic respiration (AR, respiration by live roots, rhizospheric microorganism, and mycorrhizal fungi) and HR, which mainly originates from microbial decomposition of soil organic matter. With N addition, AR was either inhibited by decreasing the below-ground C allocation and fine root biomass of trees [23] or promoted by increasing the N concentration in fine roots [24–28]. At the same time, the enhanced tree growth caused by N addition is also likely to lead to more plant photosynthate being transported from above ground to below ground, thus increasing AR. HR is also commonly considered to be related to microbial biomass and activity [29–31]. For instance, decreased HR was observed along with a consistent decrease in microbial biomass and extracellular enzyme activity [32]. Soil acidification caused by N deposition [33] is also a potential factor that could contribute to decreased HR. However, the inherent reasons concerning the responses of AR and HR to N addition are poorly understood.

Although there have been numerous studies investigating the effects of N deposition on ecosystem C sequestration [27,34,35], most of them focused on natural forests. Plantations are becoming a key component of world forest resources and play important roles in the context of overall sustainable forest management. Well-designed, multi-purpose plantations can reduce pressure on natural forests, restore some ecological services provided by natural forests, and mitigate climate changes through direct C sequestration [36]. However, there remain great uncertainties in the potential of plantations to sequestrate C [37]. Compared with natural forests, plantations appear to have lower NPP, root biomass, and soil microbial biomass [37]. Whether plantations have the same ecosystem C sequestration capacity as natural forests remains to be confirmed [38–40]. Among a few studies, increased ecosystem C sequestration with N deposition has been observed [41,42]. However the underlying mechanisms by which N increases the plant C accumulation and affects SR and its autotrophic and heterotrophic components are still poorly understood.

In China, the total plantation area reached 5.33×10^7 ha in 1998, accounting for 30% of the total forest area of China and 29% of the world's total plantation area [43]. C accumulation in China is mainly ascribable to its extensive afforestation efforts, as 80% of the observed increase in tree C stocks in China occurred on its 213,106 ha of plantations [43]. These reforestation and afforestation programs are considered to influence C storage in China. Thus, to assess the C sequestration capacity of plantations and optimize their role as C sinks, it is necessary to systematically explore ways in which N deposition affects C sequestration. Consequently, a 3-year field N addition experiment was conducted in the Saihanba Forestry Center, Hebei Province, northern China, which contains the largest area of plantations in China, with the dominant species being *Pinus sylvestris* var. *mongolica* (Mongolia pine). NPP, SR, and its autotrophic and heterotrophic components were measured. Relevant influential factors were also measured, including plant tissue C and N concentrations (foliage, litter, and fine roots), microbial biomass C, microbial community composition, potential extracellular enzyme activities (EEAs), and soil pH values. The study aimed to address three questions: (1) how does the N addition affect NPP? (2) What are the responses of SR and its autotrophic and heterotrophic components to N addition? (3) What is the effect of N addition on NEP? We hypothesized that: (1) N addition would increase NPP via increasing foliage or litter N concentrations; (2) AR would remain stable because of the contrasting effects from decreasing below-ground C allocation and fine root biomass and increased fine root N concentrations and photosynthate transport from above ground to below ground; HR would be reduced because of decreased microbial biomass, inhibited microbial activity, and reduced pH values; and (3) NEP would be enhanced because of the increasing NPP and decreased HR.

Materials and Methods

Ethics Statement

The administration of the Saihanba Forestry Center gave permission for the use of their plantation for our study site. We confirm that the field studies did not involve endangered or protected species.

Site description

The study was conducted at the Saihanba Forestry Center in Hebei Province, northern China (117°12′–117°30′ E, 42°10′–42°50′ N, 1400 m a.s.l.). The study area belongs to a typical forest-steppe ecotone of the temperate area. The climate is semi-arid and semi-humid, with a long and cold winter (November to March), and a short spring and summer. Annual mean air temperature and precipitation over the period from 1964 to 2004 were −1.4°C and 450.1 mm, respectively. The soils are predominantly sandy. The study site is located in the largest area of plantations in China, with the dominant species being *Pinus sylvestris* var. *mongolica*. The herbaceous layer is dominated by *Carex rigescens*, *Thalictrum aquilegifolium*, *Galium verum*, *Geum aleppicum*, *Artemisia tanacetifolia*, and *Agrimonia pilosa*.

Experimental treatments

The N addition experiment was initiated in May 2009. Urea solution was evenly sprayed once a month from May to September with the same dose each year. Four N addition treatments (in three replicates) were established, including a control (without N added), low N (5 g N m^{-2} yr^{-1}), medium N (10 g N m^{-2} yr^{-1}), and high N (15 g N m^{-2} yr^{-1}). Twelve plots, each of 20 m × 20 m dimensions were established, each surrounded by a 10-m wide buffer strip. All plots and treatments were randomly laid out. During each application, the fertilizer was weighed, mixed with 10 L of water, and applied to each plot below the canopy using a backpack sprayer. The control plot received 10 L of water without N.

Field measurements

Biomass production and accumulation. An allometric method was used to estimate biomass production through establishing the relationship between component biomass (foliage, branches, stem, and roots) and diameter at breast height (DBH) and tree height (H) [44]. In July 2010, stems were cut at the soil surface in the area near our experimental plots. Total tree heights, length of live crown, DBH, and diameter at the base of the live crown were measured and recorded. All foliage on each live branch was collected and weighed. All live and dead branches from each canopy position were cut and weighed separately. The stems were cut into 1-m sections and weighed. Litter from each deforested tree was carefully collected and weighed. The entire root system of the sample trees was excavated using a combination of a pulley device and manual digging, and cleaned of adhering soil. The fresh mass of each component was determined to the nearest 1 g using an electronic balance. All of these procedures were conducted in the field immediately after the tree was felled. The total biomass was calculated as the sum of foliage, branch, litter, stem, and root biomass.

An allometric equation was established as:

$$Biomass production(BP) = a(D^2 H)^b (R^2 = 0.96, P < 0.01)$$

where H is the height of trees (m), D is DBH (cm), and *a* and *b* are regression constants (*b* = 0.70, *a* = 107.01). DBH was recorded on all living stems in each plot in July 2010 and July 2012. The height of each living tree was measured using a DME (Haglöf Vertex IV, Sweden). Because biomass production (BP) constitutes the largest fraction of NPP, BP is commonly used as a proxy for NPP [45–48]. It is important to note that NPP includes numerous

C-consuming processes such as plant growth, root exudation, and C allocation to symbionts [49]. The NPP in our study was thus underestimated. We used 50% as the C concentration in plant tissue [28]. The net primary production was calculated by the following equation:

$$NPP = (BP_{2012} - BP_{2010})/2,$$

where NPP is the net primary productivity (g C m^{-2} yr^{-1}) and BP is the estimated biomass production (g C m^{-2} yr^{-1}) in 2012 and 2010. The annual net ecosystem production (NEP, g C m^{-1} yr^{-1}) was calculated as the annual NPP minus annual soil HR.

Fine root biomass. In July 2011 and 2012, five soil core samples were taken randomly using a 5.8-cm-diameter soil corer around the trees in each replicate plot, causing as little disturbance as possible to the surrounding soil. The roots were transported to the laboratory where they were washed free of soil, dried at 70°C, and weighed.

SR and its autotrophic and heterotrophic components. SR was measured using a Li-8100 soil CO$_2$ flux system (LI-COR Inc. Lincoln, NE, USA). Measurements were conducted at least once per month from May to October in 2011 and 2012. There were five subsamples (i.e., SR collars) in each plot. We used two kinds of soil collars in each plot to measure total SR and HR. A shallow surface collar (10 cm inside diameter, 6 cm height) that penetrated 3 cm into the soil was used to measure SR. The other kind of collar (10 cm inside diameter, 35 cm height), which was used for HR measurement, was inserted 30 cm into the soil with a 5-cm height above the soil surface. Because the majority of roots are found within the upper 30 cm of the soil profile in this forest (data not shown), these deeper collars should eliminate the majority of live roots and their contributions to respiration. All the polyvinyl chloride (PVC) collars were installed 6 months prior to the first measurements to minimize any disturbance of the soil environment. SR in the growing season was obtained from the monthly data directly measured in the field experiment using linear extrapolation methods. Winter SR was obtained from the data of Wang et al. (2010) [50] from the same study site.

Laboratory analyses

Plant chemical analyses. Five subsamples were collected in each plot for chemical analyses. Green foliage was sampled from vigorously growing trees in late July 2012 using a pole pruner and a steel ladder. Foliar litter was collected from litter traps. Fine root samples were selected after the soil was passed through a 2-mm sieve. All the green foliage, foliar litters, and fine roots were dried at 60°C to constant mass, and ground using an intermediate mill (0.5-mm mesh screen) to generate homogeneous samples for chemical analysis. The C and N concentrations were measured using an element analyzer (Vario EL III, Elementar, Hanau, Germany).

Soil chemical analyses. In July 2012, mineral soils were sampled at 0–10 cm depth from five random locations per plot using 5.8-cm-diameter soil corers. Once collected, the soils were immediately placed in a cooler and transported to the nearby laboratory (less than 30 min travel time per site). The cores from each plot were then combined and frozen for later processing. Within 24 h, frozen soils were allowed to thaw at room temperature. Plant litter in the upper layer, as well as all the coarse and fine roots, was carefully removed. The soils were then separated into four sub-samples for laboratory analysis, including pH, microbial biomass C and N, microbial community composition (PLFAs), and potential EEAs.

Air-dried soil had any roots removed, and was passed through a 2-mm sieve. Soil pH was determined using a 1:5 soil:water ratio with a pH meter (Model PHS-2; INESA Instrument, Shanghai, China).

MBC and MBN were measured using the chloroform fumigation extraction technique [51,52]. Two replicate samples, one unfumigated and one fumigated with alcohol-free CHCl$_3$ for 24 h, were pre-incubated at 25°C for 7 days and then extracted with 0.5 mol/L K$_2$SO$_4$ (1:2.5 w/v). The extracts were analyzed for total dissolved C and N using a total C analyzer (TOC-500; Shimadzu, Kyoto, Japan). The microbial biomass was calculated as the difference in extractable C and N between the fumigated and unfumigated soils. The efficiency factors used to calculate the respective MBC and MBN were $K_C = 0.379$ [52] and $K_N = 0.54$ [51].

Phospholipid fatty acids (PLFAs) analysis was used to assess microbial community composition. PLFAs were extracted and analyzed using a procedure described by [53]. Briefly, the soil was extracted in a single-phase mixture of chloroform: methanol: citrate buffer (1: 2: 0.8) [54]. After extraction, the lipids were separated into neutral lipids, glycolipids, and polar lipids (phospholipids) on a silicic acid column. The phospholipids were methylated and separated on a gas chromatograph equipped with a flame ionization detector. Peak areas were quantified by adding methyl nonadecanoate fatty acid (19:0) as the internal standard before the methylation step. Peaks were identified by chromatographic retention time and a standard qualitative mix in the range of C9–C30 using a microbial identification system (Microbial ID Inc., Newark, DE, USA). The fatty acid 18: 2ω6, 9 was recognized as the fungal biomarker [55]. The sum of the following PLFAs was used a measure of the bacterial biomass: i14:0, i15:0, a15:0, 15:0, i16:0, 10Me16:0, i17:0, a17:0, cy17:0, 17:0, br18, 10Me17:0, 18:1ω7, 10Me18:0, and cy19:0 [56].

Seven EEAs were measured, including five enzymes involving C metabolism (α-glucosidase (AG), β-1,4-glucosidase (BG), leucine aminopeptidase (LAP), β-D-cellobiosidase (CB), and xylosidase (XS)), one involving N metabolism (N-acetyl-glucosaminidase (NAG), and one involving phosphorus (P) metabolism (acid phosphatase (AP)). The measurements were conducted following the method of Saiya-Cork et al. (2002) [57]. Briefly, sample suspensions were prepared by adding 2 g of fresh soil to 90 ml of 50 mmol/L, pH 6.0 acetate buffer and homogenizing for 1 min.

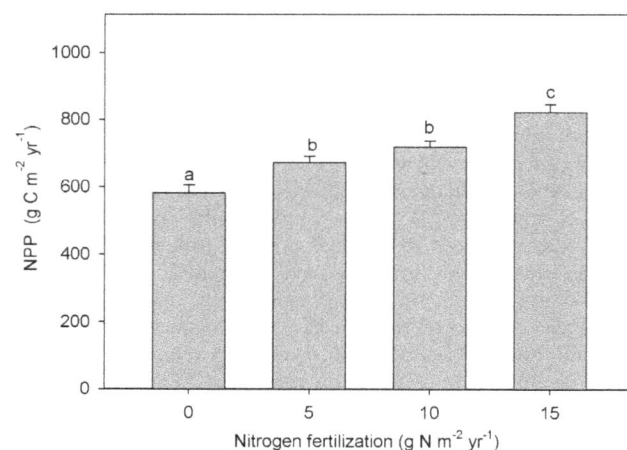

Figure 1. Average net primary productivity (NPP) in control and nitrogen (N) fertilized treatments. Significant differences among N treatments are indicated by different letters.

Continuously, 200-μl suspensions were combined with the corresponding substrate in a 96-well microplate. There were six replicate wells per sample per assay. The micro-plates were incubated at 25°C for up to 3 h. Fluorescence was then measured using a microplate reader with 365-nm excitation and 450-nm emission filters (Tecan Infinite M200, Salzburg, Austria). Finally, the concentration was divided by incubation time and dry weight soil to estimate potential enzyme activity.

Statistical analysis

All statistical analyses were performed using SPSS statistical software (SPSS 18.0 for Windows; SPSS Inc., Chicago, IL, USA). One-way analysis of variance with Duncan's test was used to test the differences among the different N addition treatments in NPP, SR, and its AR and HR components, and in NEP, as well as plant and soil chemical parameters. Significant effects were determined

at $P < 0.05$ unless otherwise stated. Data was expressed as mean values ± S.E. (standard error).

Results

Biomass production and accumulation

No significant differences were observed in DBH for both 2010 and 2012 among the different N addition treatments (Fig. S1a). However, in 2012, the tree height significantly increased with fertilization (Fig. S1b). The averaged NPP was 582.3±22.8 g C m^{-2} yr^{-1} in the control plots. N addition increased NPP by 15.45%, 23.51%, and 41.21%, respectively, in the low-, medium-, and high-N plots (Fig. 1). Fine root biomass showed a decreasing trend with fertilization although no significant differences were observed (Fig. S2).

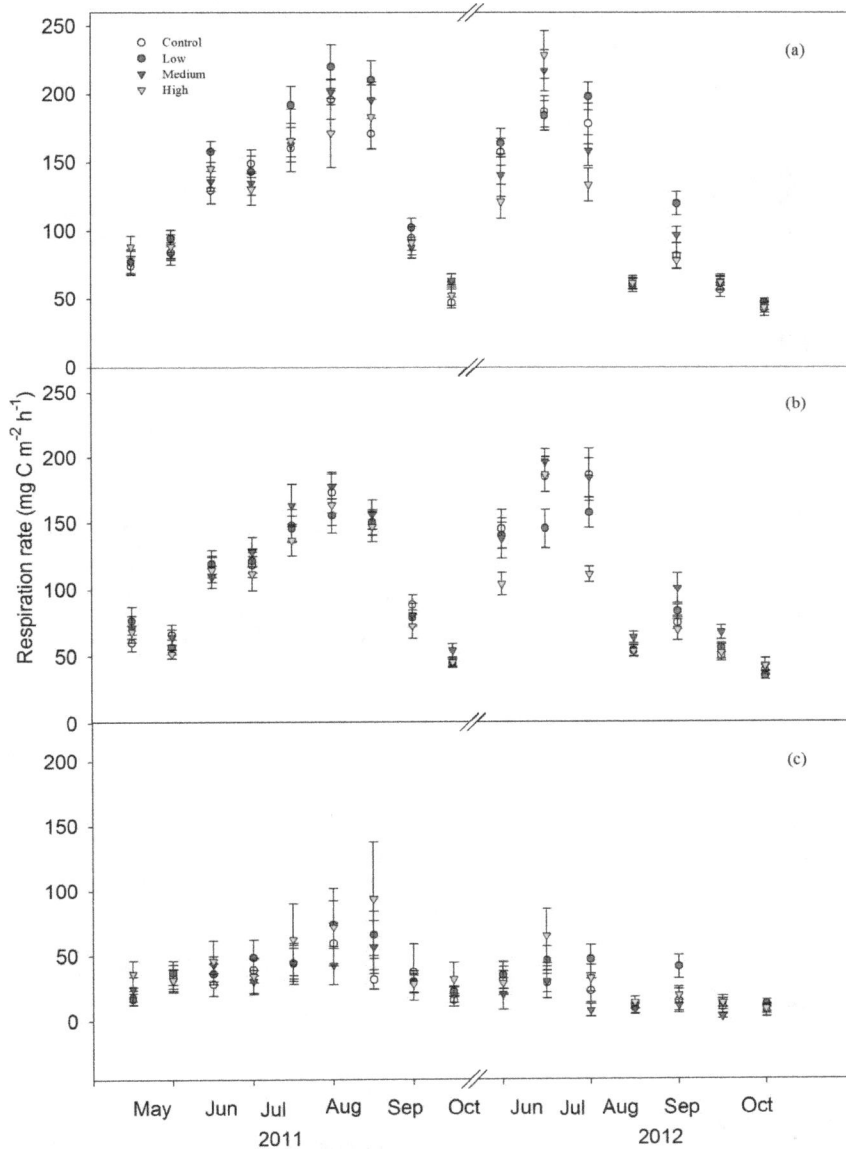

Figure 2. Soil respiration (a), heterotrophic respiration (b), and autotrophic respiration (c) in control (yellow circle), low-nitrogen (N) (red circle), medium-N (blue triangle), and high-N (green triangle) plots during the growing seasons of 2011 and 2012.

SR and its autotrophic and heterotrophic components

Both total SR and HR followed a clear seasonal pattern with the highest rates in June–August and the lowest rates in spring and autumn for all the treatments (Fig. 2). SR was not significantly different among control and fertilized treatment plots in both 2011 ($P = 0.43$) and 2012 ($P = 0.36$) (Fig. 3a). There was no significant variance between the control and low-N treatment for AR and HR in both 2011 and 2012 ($P > 0.05$) (Fig. 3b, Fig. 3c). With the fertilization gradient increasing, significantly higher AR (Fig. 3c) and lower HR (Fig. 3b) occurred in the high-N plots compared with the medium-N treatment in 2012.

Net ecosystem productivity (NEP)

With a decrease in HR and increased NPP, NEP significantly increased with fertilization, from 149 to 426.6 g C m^{-2} yr^{-1} (Fig. 4). The amount of C (kg) fixed by 1 kg N/ha N addition was in the range of 116.6–209.8 kg C per kg N/ha.

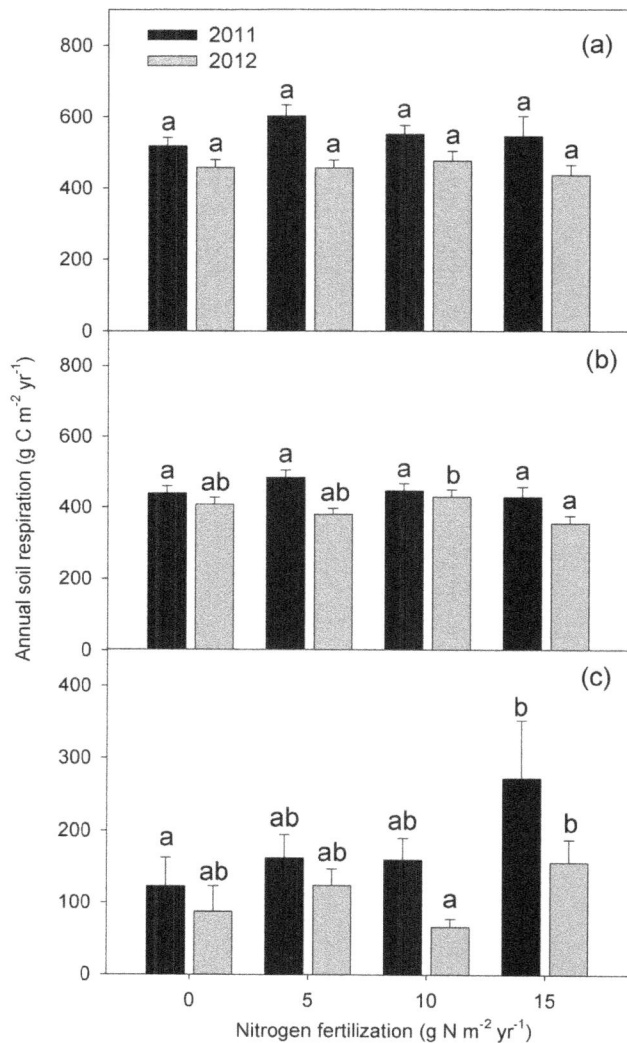

Figure 4. Net ecosystem productivity (NEP) in control and nitrogen (N) fertilized treatments. Significant differences among N treatments are indicated by different letters.

Plant chemical parameters

N addition significantly increased foliar N concentrations in the herbaceous layer (Fig. S3). However, N concentrations of the tree foliage showed no significant differences among the control and fertilization treatments (Fig. 5). C and N concentrations of foliar litter were significantly higher in the high-N and medium-N plots than in the control (Fig. 5). Fine root N concentrations showed no

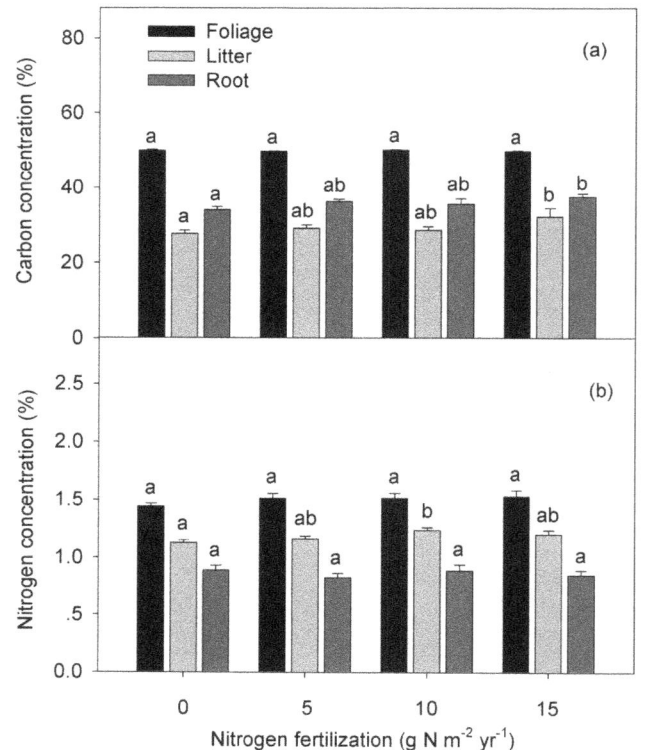

Figure 3. Comparison of annual soil respiration (a), autotrophic respiration (b), and heterotrophic respiration (c) among control, low-, medium-, and high-nitrogen (N) plots in 2011 (black) and 2012 (gray). Significant differences among N treatments are indicated by different letters.

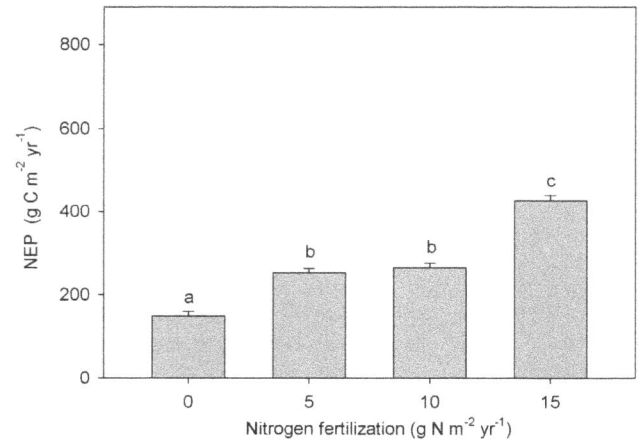

Figure 5. Carbon (a) and nitrogen (b) concentrations of foliage, litter, and fine roots in control and nitrogen (N) fertilized treatments. Significant differences among N treatments are indicated by different letters.

Table 1. Effects of nitrogen (N) addition on microbial biomass carbon (MBC), microbial biomass N (MBN), bacterial biomass (BB), and fungal biomass (FB).

Microbial properties	N treatment			
	Control	Low N	Medium N	High N
MBC	132.67±22.39[a]	157.55±18.31[a]	162.14±22.51[a]	145.68±21.05[a]
MBN	48.9±82.46[a]	51.88±4.81[a]	55.55±66.62[a]	60.59±5.98[a]
BB	17.0±1.52[a]	18.94±2.73[a]	16.19±2.53[a]	12.91±1.58[a]
FB	7.08±0.75[a]	9.62±2.15[a]	7.41±1.21[a]	6.19±0.11[a]

Data are expressed as mean ± S.E. (standard error). Different superscript letters indicated significant differences among N treatment plots ($P<0.05$).

significant differences while fine root C concentrations increased by 10.28% in the high-N plots compared with the control (Fig. 5).

Soil chemical parameters

Soil pH significantly decreased with fertilization (Fig. S4). MBC and MBN were 132.67–145.68 mg C kg^{-1} dry soil and 48.98–60.59 mg C kg^{-1} dry soil, respectively. Although there were no significant differences among the control and fertilized treatments, MBC and MBN generally increased along the fertilization gradient (Table 1). Neither bacterial nor fungal biomass varied significantly with N addition (Table 1). The activities of all seven enzymes involving C, N, and P metabolism showed no significant differences between the control and fertilized plots (Fig. 6).

Discussion

Effects of N addition on NPP

N fertilization significantly increased NPP by 15.4–41.2% (Fig. 1) through the vertical growth of trees (Fig. S1). This maximum rate of increase of NPP is more than twice the average level of temperate forests (19.5%) [58]. The increased NPP could

not be related to fresh foliar N concentrations, because no significant differences occurred between the control and fertilized plots (Fig. 5). This is inconsistent with commonly observed increases in foliar N with N addition [59,13–15]. For instance, May et al. (2005) [15] found that foliar N concentrations averaged 11% higher in fertilization treatments than in the control in a mixed-deciduous forest. In this study, litter N concentrations significantly increased in the fertilized plots relative to the control ($P<0.05$) (Fig. 5b), suggesting a likely decrease in leaf N resorption by fertilization [15,60,61]. Litter with higher N concentration would be easily decomposed by microbes and release large amounts of available N for plant growth, thus potentially increasing forest productivity [62–65].

Foliage N concentrations of trees showed no significant differences among control and fertilized plots (Fig. 5), while foliage N concentrations in plants in the herbaceous layer significantly increased (Fig. S3). This may be because of differences in leaf shape. Compared with coniferous trees, the broad-leaved herbaceous plants may invest more N in foliage to produce enzymes and proteins associated with photosynthetic processes or increase their foliage area to improve photosynthesis. Increased

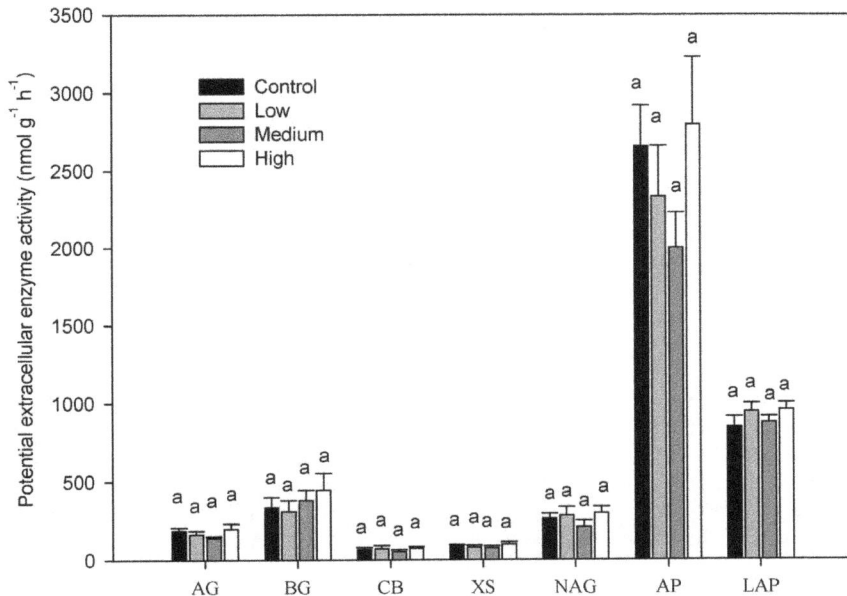

Figure 6. Potential extracellular enzyme activity at a soil depth of 0–10 cm in control and fertilized plots measured in 2012. Significant differences among nitrogen (N) treatments are indicated by different letters. AG = α-glucosidase; BG = β-1,4-glucosidase; CB = β-D-cellobiosidase; XS = xylosidase; NAG = N-acetyl-glucosaminidase; AP = acid phosphatase; LAP = leucine aminopeptidase.

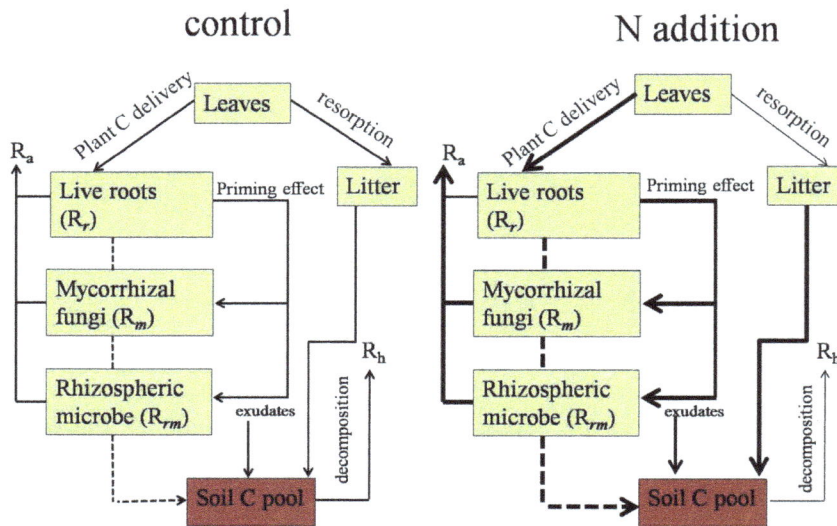

Figure 7. Carbon sequestration and its response to nitrogen (N) addition in plantations. R_a and R_h are autotrophic and heterotrophic respiration, respectively; R_r is live root respiration, R_m is respiration of mycorrhizal fungi, and R_{rm} is rhizospheric microbial respiration. Thick arrows represent the enhanced process and thin arrows represent the declined progress in the N addition treatment compared with the control.

foliage N concentrations following N addition have been commonly observed in previous studies of broadleaf species (i.e., *Acer rubrum, Liriodendron tulipifera, Prunus serotina, Acer saccharum,* and *Betula alleghaniensis*) [13–15]. Therefore, the nutrient use strategy of plants may be closely related to their foliage shape. Thus, foliage shape should be taken into consideration in the future when it comes to assessing the response of ecosystem C sequestration to N deposition.

SR and its autotrophic and heterotrophic components

No significant differences were observed in total SR among the control and fertilized treatments (Fig. 3a), which is inconsistent with a commonly reported reduction in SR following N addition [9,66–68]. Significantly higher AR in high-N plots than in the medium-N treatment was observed, although neither treatment showed any significant differences to the control (Fig. 3c). No significant differences in fine root biomass (Fig. S2) and N concentrations were observed among the different treatments (Fig. 5b), implying that live root respiration may not change with fertilization. Instead, fine root C concentrations significantly increased by 4.89%, 6.72%, and 10.28%, respectively, in the low-, medium-, and high-N addition foliage plots compared with the control treatment (Fig. 5a). This suggests an increased supply of photosynthetic products from above ground to below ground following N addition. Consequently, plant C may prime the growth and activity of mycorrhizal fungi [29] and rhizospheric microbes [23,69–72], thus promoting AR. However, fertilization may also suppress rhizospheric microbial respiration to a greater extent than that in the bulk soil because of the decreased C allocation to root symbionts and exudation [68]. Therefore, there is a need to distinguish different components of AR in the future to accurately explore the internal mechanism underlying the increased AR.

HR in the high-N treatment was significantly lower than that in the medium-N plot after 3 years' fertilization, although it did not significantly differ from the control and low-N plots (Fig. 3c). Decreased HR is believed to be mainly driven by a decreased microbial biomass [29,30] and depressed phenol oxidase activity (a

lignin-degrading enzyme) [31,73]. In contrast to most previous studies, we did not observe significant variation in the soil microbial biomass (Table 1), extracellular enzyme activity (Fig. 6), and microbial community composition (Table 1). Although soil pH was significantly lower in the high-N plots than in the control plots, it did not significantly differ from values in the medium- and low-N plots (Fig. S4). Hence, the decreased HR in the high-N plot could not be attributable to changes in the microbial biomass, extracellular enzyme activity, microbial community composition, or soil pH. The variation in HR may depend on the C-use efficiency of the decomposers (defined as the ratio of C employed in the new biomass relative to C consumed for respiration) [74]. When N availability is high, microbes may increase their efficiency leading to an efficient increase in biomass and a relatively low release of C to the atmosphere [75]. Because microbial biomass was measured only in July 2012, a greater frequency of measurement of microbial biomass should be conducted to better explore the reasons for the decrease in HR.

Effect of N fertilization on NEP

With the increasing NPP and decreasing HR, NEP greatly increased from the control to the high-N addition plots (149 versus 426.6 g C m^{-2} yr^{-1}). NEP in our control plot fell within the published range for boreal forests (40–180 g C m^{-2} yr^{-1}) [76]. The amount of C fixed per unit of added N fertilizer was in the range of 116.6–209.8 kg C/ha per kg N in this study, which is broadly similar to the range proposed by Magnani et al. (2008) (175–225 kg C per kg N) [11]. Thus, short-term N fertilization can greatly increase the NEP of plantations in northern China and enhance the role of plantations as an important C sink.

However, N fertilization may also induce alterations in the availability of other nutrients such as P, potassium, and calcium, because of the intrinsic stoichemical constraints of plant growth. This could have an important influence on NEP. N deposition is also likely to be accompanied by other environmental changes including rising atmospheric CO_2 concentrations, global warming, and soil acidification and these changes will interact with N availability in complex ways. The complexity of these interacting controls (i.e., temperature and nutrient availability) further restricts our ability to forecast future C sequestration capacities. In addition, we should be careful when extrapolating our results and mechanisms to systems with long-term N inputs. Finally, the stand age of plantations may also be a potentially influential factor in evaluating their sequestration capacity. This suggests a need for future studies in stands of various ages and incorporating long-term multi-factorial experiments.

Conclusion

This study has comprehensively analyzed the effects of N addition on biomass accumulation, SR, and its autotrophic and heterotrophic components in plantations of northern China. N addition might alter C sequestration capacity through the following possible pathways (Fig. 7): (1) increased litter N concentration because of decreased N resorption by foliage; (2) enhancement of the amount of photosynthetic products transported downward; (3) increased AR through the priming effect of plant C on rhizospheric microbial and mycorrhizal fungi activity; and (4) suppressed HR through increased microbial C use efficiency. Increasing N deposition is likely to stimulate NEP and slow the accumulation of atmospheric CO_2. In the context of global atmospheric N deposition, we highlighted that plantations might offer an important role to mitigate the future climate change.

References

1. Le Quéré C, Raupach MR, Canadell JG, Marland G (2009) Trends in the sources and sinks of carbon dioxide. Nature Geoscience 2: 831–836.
2. Vitousek PM, Cassman K, Cleveland C, Crews T, Field CB, et al. (2002) Towards an ecological understanding of biological nitrogen fixation. Biogeochemistry 57: 1–45.
3. Thomas DC, Zak DR, Filley TR (2012) Chronic N deposition does not apparently alter the biochemical composition of forest floor and soil organic matter. Soil Biology and Biochemistry 54: 7–13.
4. Vitousek PM, Aber JD, Howarth RW, Likens GE, Matson PA, et al. (1997) Human alteration of the global nitrogen cycle: sources and consequences. Ecological Applications 7: 737–750.
5. Aber JD, Goodale CL, Ollinger SV, Smith ML, Magill AH, et al. (2003) Is nitrogen deposition altering the nitrogen status of northeastern forests? BioScience 53: 375–389.
6. Galloway JN, Townsend AR, Erisman JW, Bekunda M, Cai Z, et al. (2008) Transformation of the nitrogen cycle: recent trends, questions, and potential solutions. Science 320: 889–892.
7. Magnani F, Mencuccini M, Borghetti M, Berbigier P, Berninger F, et al. (2007) The human footprint in the carbon cycle of temperate and boreal forests. Nature 447: 849–851.
8. Pregitzer KS, Burton AJ, Zak DR, Talhelm AF (2008) Simulated chronic nitrogen deposition increases carbon storage in Northern Temperate forests. Global Change Biology 14: 142–153.
9. Janssens I, Dieleman W, Luyssaert S, Subke JA, Reichstein M, et al. (2010) Reduction of forest soil respiration in response to nitrogen deposition. Nature Geoscience 3: 315–322.
10. de Vries W, Solberg S, Dobbertin M, Sterba H, Laubhahn D, et al. (2008) Ecologically implausible carbon response? Nature 451: E1–E3.
11. Magnani F, Mencuccini M, Borghetti M, Berninger F, Delzon S, et al. (2008) Ecologically implausible carbon response? Reply. Nature 451: E3–E4.
12. Liu L, Greaver TL (2009) A review of nitrogen enrichment effects on three biogenic GHGs: the CO_2 sink may be largely offset by stimulated N_2O and CH_4 emission. Ecology letters 12: 1103–1117.
13. Boggs JL, McNulty SG, Gavazzi MJ, Myers JM (2005) Tree growth, foliar chemistry, and nitrogen cycling across a nitrogen deposition gradient in southern Appalachian deciduous forests. Canadian Journal of Forest Research 35: 1901–1913.
14. Elvir JA, Rustad L, Wiersma GB, Fernandez I, White AS, et al. (2005) Eleven-year response of foliar chemistry to chronic nitrogen and sulfur additions at the Bear Brook Watershed in Maine. Canadian Journal of Forest Research 35: 1402–1410.
15. May JD, Burdette SB, Gilliam FS, Adams MB (2005) Interspecific divergence in foliar nutrient dynamics and stem growth in a temperate forest in response to chronic nitrogen inputs. Canadian Journal of Forest Research-Revue Canadienne De Recherche Forestiere 35: 1023–1030.
16. Aber JD, Nadelhoffer KJ, Steudler P, Melillo JM (1989) Nitrogen saturation in northern forest ecosystems. Bioscience 39: 378–286.
17. Aber J, McDowell W, Nadelhoffer K, Magill A, Berntson G, et al. (1998) Nitrogen saturation in temperate forest ecosystems. Bioscience 48: 921–934.
18. Driscoll CT, Lawrence GB, Bulger AJ, Butler TJ, Cronan CS, et al. (2001) Acidic deposition in the northeastern United States: sources and inputs, ecosystem effects, and management strategies. Bioscience 51: 180–198.
19. Guerrieri R, Mencuccini M, Sheppard L, Saurer M, Perks M, et al. (2011) The legacy of enhanced N and S deposition as revealed by the combined analysis of $\delta^{13}C$, $\delta^{18}O$ and $\delta^{15}N$ in tree rings. Global Change Biology 17: 1946–1962.
20. Tomaszewski T, Sievering H (2007) Canopy uptake of atmospheric N deposition at a conifer forest: Part II-response of chlorophyll fluorescence and gas exchange parameters. Tellus B 59: 493–501.
21. van Heerwaarden LM, Toet S, Aerts R (2003) Nitrogen and phosphorus resorption efficiency and proficiency in six sub-arctic bog species after 4 years of nitrogen fertilization. Journal of Ecology 91: 1060–1070.
22. Sullivan PF, Sommerkorn M, Rueth HM, Nadelhoffer KJ, Shaver GR, et al. (2007) Climate and species affect fine root production with long-term fertilization in acidic tussock tundra near Toolik Lake, Alaska. Oecologia 153: 643–652.
23. Högberg MN, Briones MJ, Keel SG, Metcalfe DB, Campbell C, et al. (2010) Quantification of effects of season and nitrogen supply on tree below-ground carbon transfer to ectomycorrhizal fungi and other soil organisms in a boreal pine forest. New Phytologist 187: 485–493.
24. Janssens IA, Crookshanks M, Taylor G, Ceulemans R (1998) Elevated atmospheric CO_2 increases fine root production, respiration, rhizosphere respiration and soil CO_2 efflux in Scots pine seedlings. Global Change Biology 4: 871–878.
25. Pregitzer KS, Laskowski MJ, Burton AJ, Lessard VC, Zak DR (1998) Variation in sugar maple root respiration with root diameter and soil depth. Tree Physiology 18: 665–670.
26. Jia SX, Wang ZQ, Li XP, Sun Y, Zhang XP, et al. (2010) N fertilization affects on soil respiration, microbial biomass and root respiration in Larix gmelinii and Fraxinus mandshurica plantations in China. Plant and Soil 333: 325–336.
27. Burton AJ, Jarvey JC, Jarvi MP, Zak DR, Pregitzer KS (2011) Chronic N deposition alters root respiration-tissue N relationship in northern hardwood forests. Global Change Biology 18: 258–266.
28. Tu LH, Hu TX, Zhang J, Li RH, Dai HZ, et al. (2011) Short-term simulated nitrogen deposition increases carbon sequestration in a Pleioblastus amarus plantation. Plant and Soil 340: 383–396.
29. Craine JM, Morrow C, Fierer N (2007) Microbial nitrogen limitation increases decomposition. Ecology 88: 2105–2113.
30. Fierer N, Lauber CL, Ramirez KS, Zaneveld J, Bradford MA, et al. (2011) Comparative metagenomic, phylogenetic and physiological analyses of soil microbial communities across nitrogen gradients. The ISME journal 6: 1007–1017.
31. Zak DR, Pregitzer KS, Burton AJ, Edwards IP, Kellner H (2011) Microbial responses to a changing environment: implications for the future functioning of terrestrial ecosystems. Fungal Ecology 4: 386–395.

Supporting Information

Figure S1 Diameter at breast height (DBH) and height of trees with nitrogen (N) fertilization in 2010 (black) and 2012 (gray).

Figure S2 Fine root biomass among different nitrogen (N) fertilization gradients in 2012. Significant differences among N treatments are indicated by different letters.

Figure S3 Effects of nitrogen (N) addition on foliar N concentrations of herbaceous layer plants. Significant differences among N treatments are indicated by different letters.

Figure S4 Soil pH in control and nitrogen (N) treatments plots after 3 years fertilization. Significant differences among N treatments are indicated by different letters.

Author Contributions

Conceived and designed the experiments: WW. Performed the experiments: ZHD WJZ. Analyzed the data: ZHD. Contributed reagents/materials/analysis tools: HZ. Wrote the paper: ZHD.

32. Ramirez KS, Craine JM, Fierer N (2010) Nitrogen fertilization inhibits soil microbial respiration regardless of the form of nitrogen applied. Soil Biology and Biochemistry 42: 2336–2338.

33. Phoenix GK, Emmett BA, Britton AJ, Caporn SJ, Dise NB, et al. (2012) Impacts of atmospheric nitrogen deposition: responses of multiple plant and soil parameters across contrasting ecosystems in long-term field experiments. Global Change Biology 18: 1197–1215.

34. Hagedorn F, Kammer A, Schmidt MW, Goodale CL (2012) Nitrogen addition alters mineralization dynamics of ^{13}C-depleted leaf and twig litter and reduces leaching of older DOC from mineral soil. Global Change Biology 18: 1412–1427.

35. Hasselquist NJ, Metcalfe DB, Högberg P (2012) Contrasting effects of low and high nitrogen additions on soil CO_2 flux components and ectomycorrhizal fungal sporocarp production in a boreal forest. Global Change Biology 18: 3596–3605.

36. Paquette A, Messier C (2010) The role of plantations in managing the world's forests in the Anthropocene. Frontiers in Ecology and the Environment 8: 27–34.

37. Liao C, Luo Y, Fang C, Li B (2010) Ecosystem carbon stock influenced by plantation practice: implications for planting forests as a measure of climate change mitigation. Plos one 5: e10867.

38. Harmon ME, Ferrell WK, Franklin JF (1990) Effects on carbon storage of conversion of old-growth forests to young forests. Science 247: 699–702.

39. Chen GS, Yang YS, Xie JS, Guo JF, Gao R, et al. (2005) Conversion of a natural broad-leafed evergreen forest into pure plantation forests in a subtropical area: effects on carbon storage. Annals of forest science 62: 659–668.

40. Yang YS, Guo J, Chen G, Xie J, Gao R, et al. (2005) Carbon and nitrogen pools in Chinese fir and evergreen broadleaved forests and changes associated with felling and burning in mid-subtropical China. Forest Ecology and Management 216: 216–226.

41. Hyvönen R, Ågren GI, Linder S, Persson T, Cotrufo MF, et al. (2007) The likely impact of elevated CO_2, nitrogen deposition, increased temperature and management on carbon sequestration in temperate and boreal forest ecosystems: a literature review. New Phytologist 173: 463–480.

42. Zhao M, Xiang W, Tian D, Deng X, Huang Z, et al. (2013) Effects of Increased Nitrogen Deposition and Rotation Length on Long-Term Productivity of Cunninghamia lanceolata Plantation in Southern China. Plos one 8: e55376.

43. Fang J, Chen A, Peng C, Zhao S, Ci L (2001) Changes in forest biomass carbon storage in China between 1949 and 1998. Science 292: 2320–2322.

44. Wang C (2006) Biomass allometric equations for 10 co-occurring tree species in Chinese temperate forests. Forest Ecology and Management 222: 9–16.

45. Waring R, Landsberg J, Williams M (1998) Net primary production of forests: a constant fraction of gross primary production? Tree Physiology 18: 129–134.

46. DeLUCIA E, Drake JE, Thomas RB, Gonzalez-Meler M (2007) Forest carbon use efficiency: is respiration a constant fraction of gross primary production? Global Change Biology 13: 1157–1167.

47. Drake J, Davis S, Raetz L, DeLucia E (2011) Mechanisms of age-related changes in forest production: the influence of physiological and successional changes. Global Change Biology 17: 1522–1535.

48. Goulden ML, McMillan A, Winston G, Rocha A, Manies K, et al. (2011) Patterns of NPP, GPP, respiration, and NEP during boreal forest succession. Global Change Biology 17: 855–871.

49. Vicca S, Luyssaert S, Penuelas J, Campioli M, Chapin F, et al. (2012) Fertile forests produce biomass more efficiently. Ecology letters 15: 520–526.

50. Wang W, Peng S, Wang T, Fang J (2010) Winter soil CO_2 efflux and its contribution to annual soil respiration in different ecosystems of a forest-steppe ecotone, north China. Soil Biology and Biochemistry 42: 451–458.

51. Brookes P, Landman A, Pruden G, Jenkinson D (1985) Chloroform fumigation and the release of soil nitrogen: a rapid direct extraction method to measure microbial biomass nitrogen in soil. Soil Biology and Biochemistry 17: 837–842.

52. Vance E, Brookes P, Jenkinson D (1987) An extraction method for measuring soil microbial biomass C. Soil Biology and Biochemistry 19: 703–707.

53. Frostegard A, Tunlid A, Baath E (1993) Phospholipid fatty acid composition, biomass, and activity of microbial communities from two soil types experimentally exposed to different heavy metals. Applied and Environmental Microbiology 59: 3605–3617.

54. Bossio D, Scow K (1998) Impacts of carbon and flooding on soil microbial communities: phospholipid fatty acid profiles and substrate utilization patterns. Microbial Ecology 35: 265–278.

55. Zelles L (1997) Phospholipid fatty acid profiles in selected members of soil microbial communities. Chemosphere 35: 275–294.

56. Frostegard A, Baath E (1996) The use of phospholipid fatty acid analysis to estimate bacterial and fungal biomass in soil. Biology and Fertility of Soils 22: 59–65.

57. Saiya-Cork K, Sinsabaugh R, Zak D (2002) The effects of long term nitrogen deposition on extracellular enzyme activity in an Acer saccharum forest soil. Soil Biology and Biochemistry 34: 1309–1315.

58. LeBauer DS, Treseder KK (2008) Nitrogen limitation of net primary productivity in terrestrial ecosystems is globally distributed. Ecology 89: 371–379.

59. Reich PB, Tjoelker MG, Pregitzer KS, Wright IJ, Oleksyn J, et al. (2008) Scaling of respiration to nitrogen in leaves, stems and roots of higher land plants. Ecology letters 11: 793–801.

60. Li X, Hu Y, Han S, Liu Y, Zhang Y (2010) Litterfall and litter chemistry change over time in an old-growth temperate forest, northeastern China. Annals of forest science 67: 206.

61. Vergutz L, Manzoni S, Porporato A, Novais RF, Jackson RB (2012) Global resorption efficiencies and concentrations of carbon and nutrients in leaves of terrestrial plants. Ecological Monographs 82: 205–220.

62. Vitousek P (1982) Nutrient cycling and nutrient use efficiency. American Naturalist 119: 553–572.

63. Wieder WR, Cleveland CC, Townsend AR (2009) Controls over leaf litter decomposition in wet tropical forests. Ecology 90: 3333–3341.

64. Wood TE, Lawrence D, Clark DA, Chazdon RL (2009) Rain forest nutrient cycling and productivity in response to large-scale litter manipulation. Ecology 90: 109–121.

65. He H, Bleby TM, Veneklaas EJ, Lambers H (2011) Dinitrogen-fixing Acacia species from phosphorus-impoverished soils resorb leaf phosphorus efficiently. Plant, cell and environment 34: 2060–2070.

66. Bowden RD, Davidson E, Savage K, Arabia C, Steudler P (2004) Chronic nitrogen additions reduce total soil respiration and microbial respiration in temperate forest soils at the Harvard Forest. Forest Ecology and Management 196: 43–56.

67. Burton AJ, Pregitzer KS, Crawford JN, Zogg GP, Zak DR (2004) Simulated chronic NO_3^- deposition reduces soil respiration in northern hardwood forests. Global Change Biology 10: 1080–1091.

68. Phillips RP, Fahey TJ (2007) Fertilization effects on fineroot biomass, rhizosphere microbes and respiratory fluxes in hardwood forest soils. New Phytologist 176: 655–664.

69. De Nobili M, Contin M, Mondini C, Brookes P (2001) Soil microbial biomass is triggered into activity by trace amounts of substrate. Soil Biology and Biochemistry 33: 1163–1170.

70. Paterson E (2003) Importance of rhizodeposition in the coupling of plant and microbial productivity. European Journal of Soil Science 54: 741–750.

71. Marschner P, Crowley D, Yang CH (2004) Development of specific rhizosphere bacterial communities in relation to plant species, nutrition and soil type. Plant and Soil 261: 199–208.

72. Talbot JM, Allison SD, Treseder KK (2008) Decomposers in disguise: mycorrhizal fungi as regulators of soil C dynamics in ecosystems under global change. Functional Ecology 22: 955–963.

73. Edwards IP, Zak DR (2011) Fungal community composition and function after long-term exposure of northern forests to elevated atmospheric CO_2 and tropospheric O_3. Global Change Biology 17: 2184–2195.

74. Del Giorgio PA, Cole JJ (1998) Bacterial growth efficiency in natural aquatic systems. Annual Review of Ecology and Systematics 21: 503–541.

75. Manzoni S, Taylor P, Richter A, Porporato A, Agren GI (2012) Environmental and stoichiometric controls on microbial carbon-use efficiency in soils. New Phytologist 196: 79–91.

76. Bonan GB (2008) Forests and climate change: Forcings, feedbacks, and the climate benefits of forests. Science 320: 1444–1449.

9

Energy Potential and Greenhouse Gas Emissions from Bioenergy Cropping Systems on Marginally Productive Cropland

Marty R. Schmer[1]*, Kenneth P. Vogel[2], Gary E. Varvel[1], Ronald F. Follett[3], Robert B. Mitchell[2], Virginia L. Jin[1]

1 Agroecosystem Management Research Unit, United States Department of Agriculture-Agricultural Research Service (USDA-ARS), Lincoln, Nebraska, United States of America, 2 Grain, Forage and Bioenergy Research Unit, United States Department of Agriculture-Agricultural Research Service (USDA-ARS), Lincoln, Nebraska, United States of America, 3 Soil-Plant Nutrient Research Unit, United States Department of Agriculture-Agricultural Research Service (USDA-ARS), Ft. Collins, Colorado, United States of America

Abstract

Low-carbon biofuel sources are being developed and evaluated in the United States and Europe to partially offset petroleum transport fuels. Current and potential biofuel production systems were evaluated from a long-term continuous no-tillage corn (*Zea mays* L.) and switchgrass (*Panicum virgatum* L.) field trial under differing harvest strategies and nitrogen (N) fertilizer intensities to determine overall environmental sustainability. Corn and switchgrass grown for bioenergy resulted in near-term net greenhouse gas (GHG) reductions of −29 to −396 grams of CO_2 equivalent emissions per megajoule of ethanol per year as a result of direct soil carbon sequestration and from the adoption of integrated biofuel conversion pathways. Management practices in switchgrass and corn resulted in large variation in petroleum offset potential. Switchgrass, using best management practices produced 3919±117 liters of ethanol per hectare and had 74±2.2 gigajoules of petroleum offsets per hectare which was similar to intensified corn systems (grain and 50% residue harvest under optimal N rates). Co-locating and integrating cellulosic biorefineries with existing dry mill corn grain ethanol facilities improved net energy yields (GJ ha^{-1}) of corn grain ethanol by >70%. A multi-feedstock, landscape approach coupled with an integrated biorefinery would be a viable option to meet growing renewable transportation fuel demands while improving the energy efficiency of first generation biofuels.

Editor: Shuijin Hu, North Carolina State University, United States of America

Funding: This research was funded by the U.S. Department of Agriculture (USDA), Agricultural Research Service (ARS) including funds from the USDA-ARS GRACEnet effort, and partly by the USDA Natural Resources Conservation Service. The funders had no role in study design, data collection and analysis, decision to publish, or preparation of the manuscript.

Competing Interests: The authors have declared that no competing interests exist.

* E-mail: marty.schmer@ars.usda.gov

Introduction

Reduction in greenhouse gas (GHG) emissions from transportation fuels can result in near- and long-term climate benefits [1]. Biofuels are seen as a near-term solution to reduce GHG emissions, reduce U.S. petroleum import requirements, and diversify rural economies. Depending on feedstock source and management practices, greater reliance on biofuels may improve or worsen long-term sustainability of arable land. U.S. farmers have increased corn (*Zea mays* L.) production to meet growing biofuel demand through land expansion, improved management and genetics, increased corn plantings, or by increased continuous corn monocultures [2–4]. Productive cropland is finite, and corn expansion on marginally-productive cropland may lead to increased land degradation, including losses in biodiversity and other desirable ecosystem functions [4–6]. We define marginal cropland as fields whose crop yields are 25% below the regional average. The use of improved corn hybrids and management practices have increased U.S. grain yields by 50% since the early 1980's [7] with an equivalent increase in non-grain biomass or stover yields. Corn stover availability and expected low feedstock

costs make it a likely source for cellulosic biofuel. However, excessive corn stover removal can lead to increased soil erosion and decreased soil organic carbon (SOC) [8] which can negatively affect future grain yields and sustainability. Biofuels from cellulosic feedstocks (e.g. corn stover, dedicated perennial energy grasses) are expected to have lower GHG emissions than conventional gasoline or corn grain ethanol [9–13]. Furthermore, dedicated perennial bioenergy crop systems such as switchgrass (*Panicum virgatum* L.) have the ability to significantly increase SOC [14–16] while providing substantial biomass quantities for conversion into biofuels under proper management [17,18].

Long-term evaluations of feedstock production systems and management practices are needed to validate current and projected GHG emissions and energy efficiencies from the transportation sector. In a replicated, multi-year field study located 50 km west of Omaha, NE, we evaluated the potential to produce ethanol on marginal cropland from continuously-grown no-tillage corn with or without corn residue removal (50% stover removal) and from switchgrass harvested at flowering (August) versus a post-killing frost harvest. Our objectives were to compare the effects of long-term management practices including harvest strategies and N fertilizer input intensity on continuous corn grain and switchgrass to determine ethanol production, potential petroleum offsets, and net energy yields. We also present measured SOC changes (0 to 1.5 m) over a nine year period from our biofuel cropping systems to determine how direct SOC changes impact net GHG emissions from biofuels. Furthermore, we evaluate the potential efficiency advantages of co-locating and integrating cellulosic conversion capacity with existing dry mill corn grain ethanol plants.

Materials and Methods

This study is located on the University of Nebraska Agricultural Research and Development Center, Ithaca, Nebraska, USA on a marginal cropland field with Yutan silty clay loam (fine-silty, mixed, superactive, mesic Mollic Hapludalf) and a Tomek silt loam (fine, smectitic, mesic Pachic Argiudoll) soil. Switchgrass plots were established in 1998 and continuous corn plots were initiated in 1999. The study is a randomized complete block design (replications = 3) with split-split plot treatments. Main treatments are two cultivars of switchgrass, 'Trailblazer' and 'Cave-in-Rock', and a glyphosate tolerant corn hybrid. Main treatment plots are 0.3 ha which enables the use of commercial farm equipment. Switchgrass is managed as a bioenergy crop, and corn is managed under no-tillage conditions (no-till farming since 1999). Split-plot treatments are nitrogen (N) fertilizer levels and split-split plots are harvest treatments. Annual N fertilizer rates (2000–2007) were 0 kg N ha^{-1}, 60 kg N ha^{-1}, 120 kg N ha^{-1}, and 180 kg N ha^{-1} as NH$_4$NO$_3$, broadcast on the plots at the start of the growing season. The 0 kg N ha^{-1}, 60 kg N ha^{-1}, 120 kg N ha^{-1} fertilizer rates were used on switchgrass [19] while the 60 kg N ha^{-1}, 120 kg N ha^{-1}, and 180 kg N ha^{-1} fertilizer rates were used for corn. Switchgrass harvest treatments were initiated in 2000 and consist of a one-cut harvest either in early August or after a killing frost. Corn stover treatments were initiated in 2000 and are either no stover harvest or stover removal, where the amount of stover removed approximates 50% of the aboveground biomass after corn grain is harvested.

Baseline soil samples were taken in 1998 at the center of each subplot and re-sampled in 2007 at increments of 0–5, 5–10, 10–30, 30–60, 60–90, 90–120, and 120–150 cm depths [15]. Average changes in total SOC (0–1.5 m) from 1998–2007 were used to estimate direct soil C changes. Further management practices and detailed soil property values from this study have been previously reported [15,20]. Summary of petroleum offsets (GJ ha^{-1}), ethanol production (L ha^{-1}), greenhouse gas (GHG) emissions (g CO$_2$e MJ^{-1}), net GHG emissions (Mg CO$_2$e ha^{-1}), and GHG reductions (%) for corn grain, corn grain with stover removal, and switchgrass are presented in Table S1 in File S1.

Statistical Analyses

Yield data analyzed were from 2000 to 2007, where 2000 was the initiation of harvest treatments for continuous corn and switchgrass and 2007 was the last year that SOC was measured for this study. Data from switchgrass cultivars were pooled together based on their similar aboveground biomass yields over years and similar changes in SOC [15]. Data were analyzed using a linear mixed model approach with replications considered a random effect. Mean separation tests were conducted using the Tukey-Kramer method. Significance was set at P≤0.05.

Life-cycle assessment

For energy requirements in the production, conversion, and distribution of corn grain ethanol and cellulosic ethanol, values from the Greenhouse Gases, Regulated Emissions, and Energy Use in Transportation (GREET v. 1.8) [21], Energy and Resources Group Biofuel Analysis Meta-Model (EBAMM) [22], and Biofuel Energy Systems Simulator (BESS) [23] life cycle assessment models were used as well as previous agricultural energy estimates for switchgrass [12]. Energy use in the agricultural phase consisted of agricultural inputs (seed, herbicides, fertilizers, packaging), machinery energy use requirements, material transport, and diesel requirements used in this study. Stover energy requirements from the production phase were from the diesel requirements to bale, load, and stack corn stover and the embodied energy of the farm machinery used. A proportion of the N fertilizer and herbicide requirements were allocated to the amount of stover harvested.

Multiple biorefinery configurations are presented to evaluate different conversion scenarios and how this affects GHG emissions, petroleum offset credits, and net energy yield (NEY) values. Biorefinery scenarios evaluated in this study are: (i) a natural gas (NG) dry mill corn grain ethanol plant with dry distillers grain (DDGS) as a co-product for the corn grain-only harvests [23–25], (ii) a co-located dry mill corn grain and cellulosic ethanol plant with combined heat and power (CHP) and DDGS co-product, where corn stover is primarily used to displace dry mill ethanol plant natural gas requirements [25,26], (iii) and a standalone cellulosic (switchgrass or corn stover) ethanol plant (sequential hydrolysis and fermentation) with CHP capability and electricity export [22,27–29]. Chemical and enzyme production costs and related GHG emissions for corn grain and cellulosic conversion to ethanol were also incorporated [28]. Ethanol recovery for corn grain was estimated to be 0.419 L kg^{-1} [23]. Ethanol recovery for corn stover and switchgrass were based on cell wall composition from harvested biomass samples. Ground aboveground switchgrass samples were scanned using a near-infrared spectrometer to predict cell wall and soluble carbohydrate biomass composition [30]. Ground corn stover samples were analyzed using a near-infrared spectrometer-based calibration equation developed by the National Renewable Energy Laboratory to predict corn stover cell wall composition [31]. Switchgrass and corn stover cell wall conversion to ethanol was based on composition components of glucan, xylose and arabinose [30,31]. Glucan to ethanol conversion was assumed to be 85.5%, and xylose and arabinose was estimated to have 85% ethanol recovery efficiency [29]. Estimated ethanol recovery for corn stover was 327 L Mg^{-1} which was similar to other findings [29]. For switchgrass, ethanol recovery based on glucan, xylose, and arabinose concentrations was estimated to be 311 L Mg^{-1} and 344 L Mg^{-1} for an August harvest and a post-frost harvest, respectively.

Ethanol plant size capacity was estimated to be 189 million L yr^{-1} for the corn grain-only and cellulosic-only scenarios. For the co-located facility, total plant size was assumed to be 378 million L yr^{-1} capacity. Fossil fuel energy requirements for the conventional corn grain ethanol plant is assumed to be 7.69 MJ L^{-1} for natural gas to power the plant and to dry DGS, 0.59 MJ L^{-1} for corn grain transportation from farm to ethanol plant, 0.67 MJ L^{-1} for electricity purposes, 0.13 MJ L^{-1} to capital depreciation costs, and 0.58 MJ L^{-1} for wastewater processing and effluent restoration [10,22]. Fossil fuel requirements for the corn grain/cellulosic ethanol plant are feedstock transportation 0.63 MJ L^{-1} for corn stover, 0.59 MJ L^{-1} for corn grain transportation from farm to ethanol plant, 0.44 MJ L^{-1} to capital depreciation costs, and 0.58 MJ L^{-1} for wastewater treatment and processing (Table S2 in File S1). Cellulosic ethanol plant fossil fuel requirements are 0.63 MJ L^{-1} for switchgrass transportation from field to ethanol plant, 0.06 MJ L^{-1} diesel requirements for biomass transport within the ethanol plant grounds, 0.44 MJ L^{-1} to capital depreciation costs, and 0.58 MJ L^{-1} for wastewater processing, effluent restoration, and recovery (Table S2 in File S1).

For the co-located corn grain and cellulosic facility, we assumed (i) power and electrical utilities were shared [26]; (ii) power requirements were supplied mainly from the lignin portion of stover with combined ethanol purification from the starch and cellulosic ethanol conversion pathways [26]; and (iii) extra stover biomass would be required in addition to the lignin to meet steam requirements. A co-location facility would require additional unprocessed bales to be used in addition to lignin which lowered the amount of ethanol being generated from stover at a co-located facility compared to a standalone cellulosic facility that uses stover as their primary feedstock (Table S1 in File S1). Electricity would be imported from the grid in this scenario and DDGS exported as the only co-product. Recent analysis [29] of converting cellulose to ethanol has estimated a higher internal electrical demand than previously assumed [26]; suggesting electricity export under this configuration would be unlikely. The value of DDGS as animal feed would likely preclude its use in meeting power requirement in a co-located facility. We based our total biomass energy requirement on the lignin concentration in stover and the expected biomass energy use requirements to power a co-located ethanol plant [25]. Estimated biomass requirements were 11 MJ L^{-1} ethanol and embodied energy value of 16.5 MJ kg^{-1} (low heating value) for stover biomass.

Net energy yield (NEY) values (renewable output energy – fossil fuel input energy) were calculated for each feedstock and conversion scenario. Output energy was calculated from ethanol output plus co-product credits. Co-product credit for DDGS is 4.13 MJ L^{-1} for the corn grain-only ethanol plant and the co-located corn grain/cellulosic ethanol plant [32]. Electricity co-product credit for standalone cellulosic ethanol was estimated to be 1.68 MJ L^{-1} [29]. Petroleum offsets (GJ ha^{-1}) were calculated in a similar fashion as NEY with total ethanol production (MJ ha^{-1}) along with petroleum displacement from co-products minus petroleum inputs consumed in the production, conversion, and distribution phase (Tables S1 and S3 in File S1). Petroleum offsets were calculated as the difference between ethanol output and petroleum inputs from the agricultural, conversion, and distribution phase (Table S1 in File S1). Petroleum requirements for each cropping system were calculated from input requirements from this study and derived values from the EBAMM model [22]. For input requirements without defined petroleum usage, we used the default parameter in EBAMM that estimates U.S. average petroleum consumption at 40% for input source. Petroleum offset credits associated with corn grain ethanol co-products were estimated to be 0.71 MJ L^{-1} while credits for corn stover and switchgrass cellulosic ethanol co-products (standalone facility) were 0.12 MJ L^{-1} (Table S3 in File S1). Petroleum offset credits were calculated from GREET (v 1.8).

Greenhouse gas emissions

Greenhouse gas offsets associated with the production of corn grain and cellulosic ethanol were modeled from the EBAMM and BESS models [22,23]. Agricultural GHG emissions were based on fuel use, fertilizer use, herbicide use, farm machinery requirements, and changes in SOC. Direct land use change by treatment plot can either be a GHG source or a GHG sink depending on

SOC changes from this study [15]. Co-product GHG credits for DDGS or electricity export were derived from the BESS [23] and GREET (v. 1.8) models [21]. Co-product GHG credits for DDGS was -347 g CO_2e L^{-1} ethanol and -304 g CO_2e L^{-1} ethanol for cellulosic electricity export (Table S4 in File S1). Indirect land use changes for corn grain ethanol or switchgrass were not estimated in this analysis. GHG offsets were calculated on both an energy and areal basis (Table S1 in File S1).

Greenhouse gas emissions from N fertilizer were evaluated from the embodied energy requirements and subsequent nitrous oxide (N_2O) emissions (Table S4 in File S1). Direct and indirect nitrous oxide emissions were calculated in this study using Tier 1 Intergovernmental Panel on Climate Change calculations. Greenhouse gas emission values for the agricultural phase are included in Table S4 in File S1 and for the conversion and distribution phase in Table S5 in File S1. For the agricultural phase, total GHG emissions were calculated from the production of fertilizers, herbicides, diesel requirements, drying costs for corn grain, and the embodied energy in farm machinery minus direct soil C changes occurring for the study period (Table S4 in File S1). GHG emissions were reported on an energy basis, areal basis, and the difference between ethanol and conventional gasoline (Table S1 in File S1). For net GHG emissions (Mg CO_2e ha^{-1}), calculations were based on GHG intensity values (g CO_2e MJ^{-1}) multiplied by biofuel production (MJ ha^{-1}) for each cropping system. GHG reductions (Table S1 in File S1) were calculated as the percent difference from conventional gasoline as reported by the California Air Resource Board (99.1 g CO_2 MJ^{-1}) [33].

Results and Discussion

Harvest and N fertilizer management treatments affected grain and biomass yields in both crops over eight growing seasons (Fig. 1A). Switchgrass harvested after a killing frost had 27% to 60% greater biomass yields compared with an August harvest under similar fertilization rates. Highest harvested biomass yields (mean = 11.5 Mg ha^{-1} yr^{-1}) were from fertilized (120 kg N ha^{-1}) switchgrass harvested after a killing frost while continuous corn showed similar grain and stover yields [factorial analysis of variance (ANOVA), P = 0.72] under the highest N fertilizer levels (180 kg N ha^{-1}) (Fig. 1A).

Potential ethanol yields varied from 2050 to 2774 L ha^{-1} yr^{-1} for corn grain-only harvests while those for corn grain with stover removal ranged from 2862 to 3826 L ethanol ha^{-1} yr^{-1} (Fig. 1B). Ethanol contribution from corn stover ranged from 820 to 998 L ha^{-1} yr^{-1} when stover is converted at a standalone cellulosic plant (Fig. 1B). Separate ethanol facilities showed slightly higher potential ethanol yields (L ha^{-1}) than at a co-located facility (Table S1 in File S1) because a larger portion of corn stover biomass was required to meet thermal power requirements at a co-located facility (SI text in File S1). Unfertilized switchgrass had potential ethanol yield values similar to corn stover. Switchgrass under optimal management practices had 17% higher biomass yields than the highest yielding corn with stover removal treatment. Potential ethanol yield for switchgrass, however, was similar (factorial ANOVA, P>0.05) to corn with stover removal (Fig. 1B) due to lower cellulosic ethanol recovery efficiency than exists for corn grain ethanol conversion efficiency. Switchgrass ethanol conversion efficiency from this study was based on updated biochemical conversion processes [29] using known cell wall characteristics [30] that result in lower conversion rates than previous estimates [12,18].

Net energy yield (NEY) (renewable output energy minus fossil fuel input energy) and GHG emission intensity (grams of CO_2

A

B

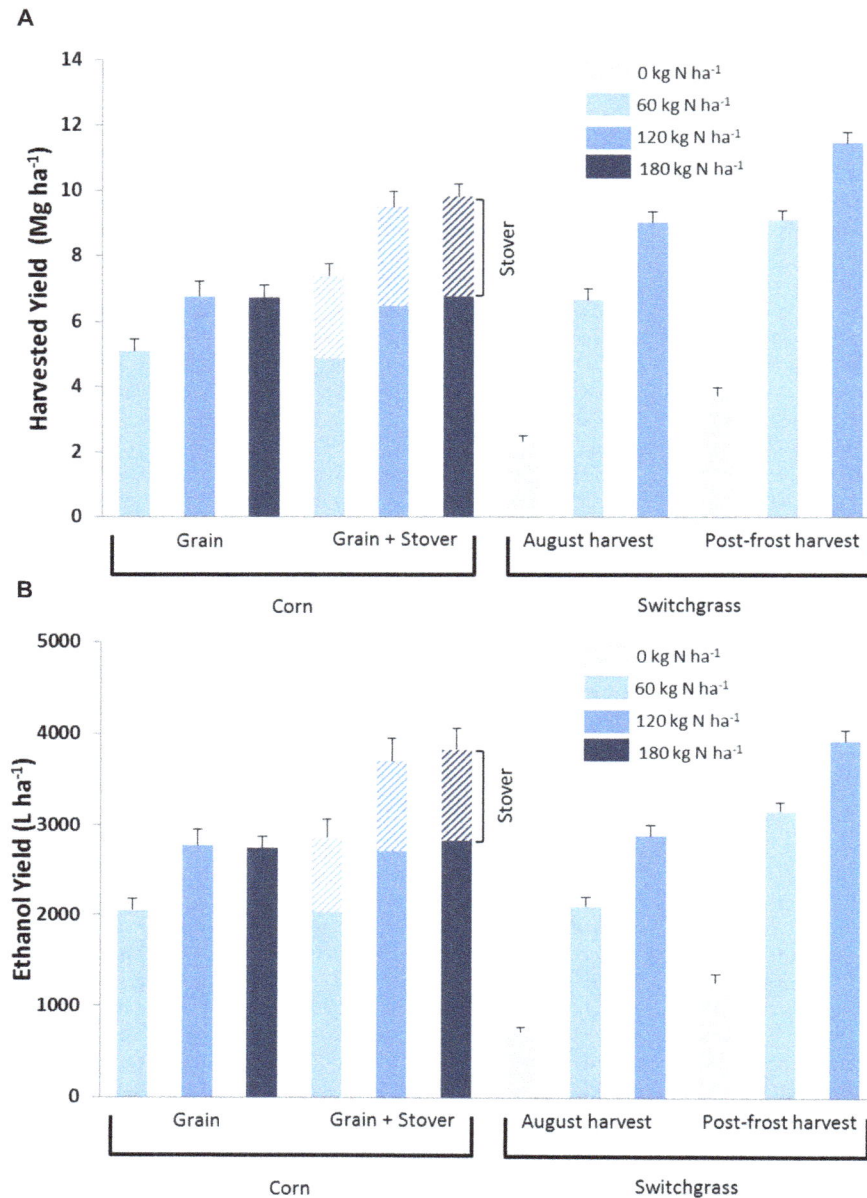

Figure 1. Harvested mean annual yield ± standard error (A) and ethanol energy ± SEM (B) for no-till continuous corn (grain-only harvest or grain and stover harvest) and switchgrass (August harvest or Post-frost harvest) under variable nitrogen rates on marginally-productive rainfed cropland for 2000–2007 (n = 3 replicate corn system plots and 6 replicate switchgrass plots).

equivalents per megajoule of fuel, or g CO_2e MJ^{-1}) are considered the two most important metrics in estimating fossil fuel replacement and GHG mitigation for biofuels [34]. Switchgrass harvested after a killing frost (120 kg N ha^{-1}) and the co-located grain and stover conversion pathway (120 kg N ha^{-1} and 180 kg N ha^{-1} treatments) had the highest overall NEY values (Fig. 2). Net energy yields for continuous corn were higher at a co-located facility because stover biomass and lignin replaced natural gas for thermal energy (Fig. 2). Ethanol conversion of corn grain and stover at separate facilities was intermediate in NEY while traditional corn grain-only natural gas (NG) dry mill ethanol plants had the lowest NEY values for the continuous corn systems. Delaying switchgrass

harvest from late summer to after a killing frost resulted in significant improvement in NEY and potential ethanol output under similar N rates. Unfertilized switchgrass had similar NEY values compared with corn grain processed at a NG dry mill ethanol plant (factorial ANOVA, P = 0.12) while fertilized switchgrass harvested after a killing frost had higher NEY values (factorial ANOVA, P<0.0001) than NG dry mill corn grain ethanol plants (Fig. 2).

Both the continuous corn and switchgrass systems showed significant petroleum offset (ethanol output minus petroleum inputs) capability, with the intensified bioenergy cropping systems having the highest petroleum offsets (Fig. 3). Petroleum use varied

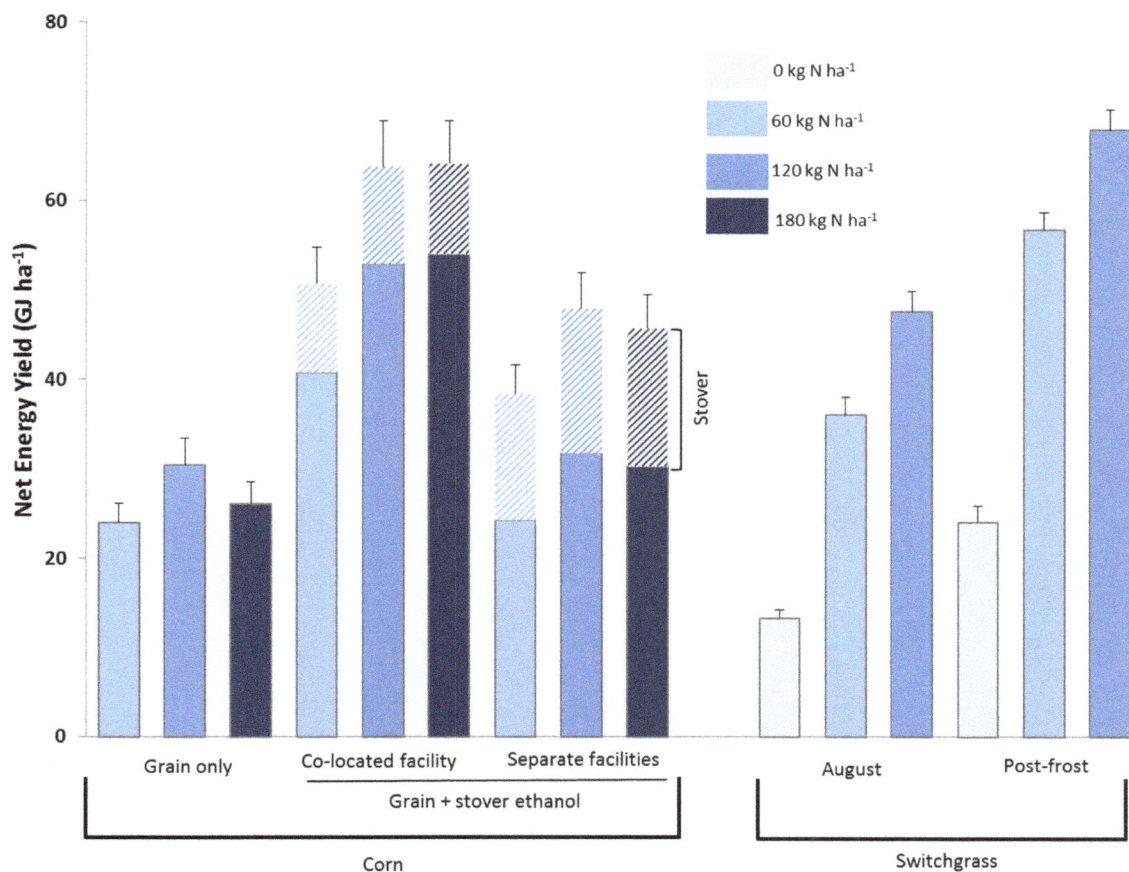

Figure 2. Net energy yield ± standard error for no-till continuous corn (grain-only or grain and stover harvest) and switchgrass (August harvest or post-frost harvest) under variable nitrogen rates on marginally-productive cropland (n = 3 replicate corn system plots and 6 replicate switchgrass plots). Conversion processes evaluated include corn grain-only harvest at a natural gas (NG) dry mill, corn grain with stover harvest at a co-located facility (lignin portion of stover used as primary energy source for grain and cellulose conversion), corn grain with stover harvest at separate ethanol facilities (NG dry mill and a cellulosic ethanol plant), and switchgrass (cellulosic ethanol plant).

by cropping system in the agricultural phase with continuous corn systems having higher overall petroleum requirements than switchgrass. Petroleum requirements (mainly diesel fuel) to harvest corn stover are small relative to corn grain harvest as a result of low harvested stover yields. Lowest petroleum offsets for continuous corn systems were from stover harvests at a separate dedicated cellulosic facility (Table S1 in File S1). Corn grain-only harvests offset less petroleum compared with grain and stover at separate ethanol facilities under similar fertilizer rates (factorial ANOVA, P<0.01). Management practices in switchgrass resulted in the largest variation in petroleum offset credits (Fig. 3B). Petroleum offsets (GJ ha^{-1}) were positively associated with NEY values [$-1.81+0.84$ (Petroleum offset); (P<0.0001); (R^2 = 0.76)], indicating that bioenergy cropping systems with large NEY values will likely result in higher petroleum displacement.

All bioenergy cropping systems evaluated in our study had SOC sequestration rates exceeding 7.3 Mg CO$_2$ yr^{-1} (Table S4 in File S1), with over 50% of SOC sequestration occurring below the 0.3 m soil depth [15]. Soil organic C increased even with corn stover removal, indicating that removal rates were sustainable in terms of SOC and grain yield for this time period. No-tillage continuous corn systems have lower stover retention requirements to maintain SOC than continuous corn with tillage or corn-soybean (*Glycine max* (L.) Merr.) rotations [8]. Consequently, all

conversion pathways had negative GHG emission values as a result of SOC sequestration offsetting GHG emissions from the production, harvest, conversion and distribution phases for corn grain ethanol and cellulosic ethanol. For switchgrass, SOC storage values were similar to other findings within the same ecoregion [16] and a long-term Conservation Reserve Program grassland [35]. Measured SOC storage from the continuous corn systems (Table S4 in File S1) were significantly higher than modeled SOC storage estimates from this region [36]. Corn grain grown with low N rates (60 kg ha^{-1}) had GHG intensity values similar to continuous corn under optimum N rates (120 kg ha^{-1}) but resulted in lower ethanol yields and lower petroleum offset potential (Fig. 3A). Lowest GHG emission intensity values on an energy basis (g CO$_2$e MJ^{-1}) were from unfertilized switchgrass (Table S1 in File S1) due to lower ethanol yields, lower agricultural energy emissions, and similar SOC storage compared with the other biofuel cropping systems. For switchgrass, management practices that resulted in the lowest GHG emission on an energy basis resulted in the lowest petroleum offset potential (Fig. 3B). Direct N$_2$O emissions (Table S4 in File S1) were estimated using Intergovernmental Panel on Climate Change methodology and are in agreement with study site N$_2$O flux measurements from a later time series which indicated N rate as the major contributor to N$_2$O emissions [37]. When evaluating GHG emissions on a per

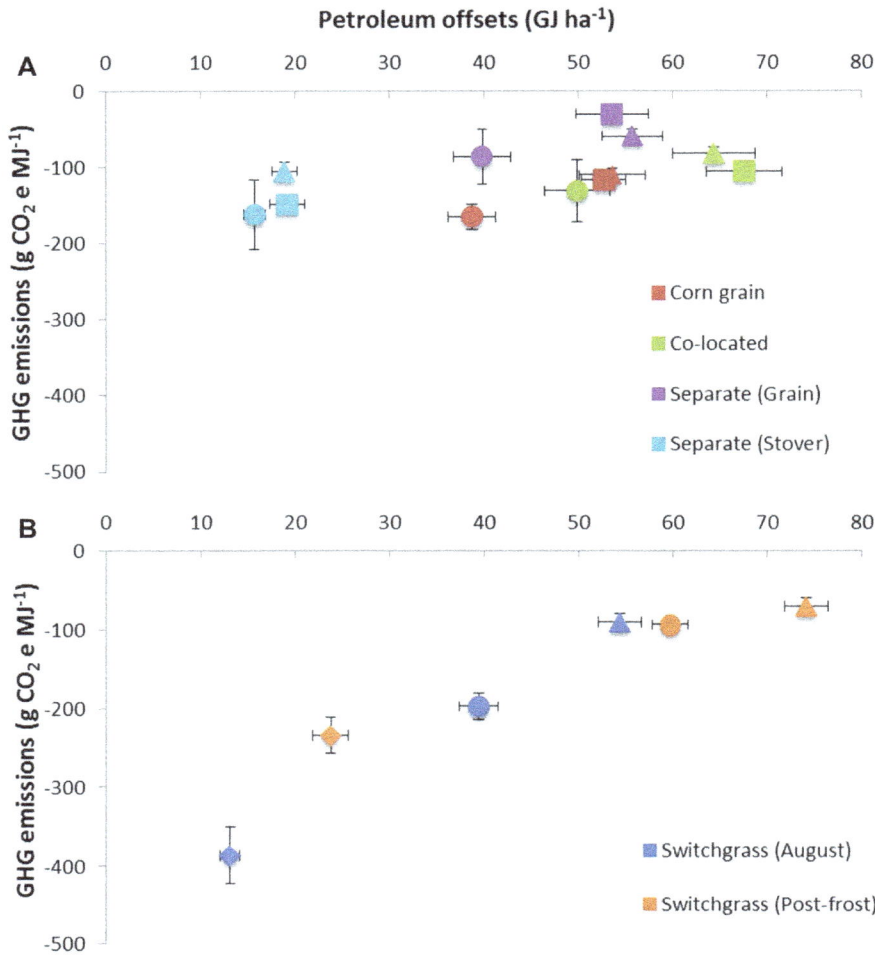

Figure 3. Petroleum offsets compared with GHG emissions (g CO₂e MJ⁻¹ ethanol) for continuous corn and switchgrass grown on marginally-productive cropland (n=3 replicate corn system plots and 6 replicate switchgrass plots). (A) Continuous corn values represent harvest method (stover harvested or retained) and ethanol conversion pathway (co-located facility or at a separate ethanol facilities). (B) Switchgrass values are based on harvest date and N fertilizer rate. Fertilizer rates are 0 kg N ha⁻¹ (♦), 60 kg N ha⁻¹ (●), 120 kg N ha⁻¹ (▲), and 180 kg N ha⁻¹ (■). Error bars indicate standard errors of the mean.

unit area basis (g CO_2e ha^{-1}), unfertilized switchgrass and corn grain-only systems showed similar results with the more intensified cropping systems (Table S1 in File S1).

Both switchgrass and continuous corn with stover removal produced similar ethanol potential, NEY values, petroleum offsets, and GHG emissions but overall values and metric efficiencies were dependent on management practices and downstream conversion scenarios. Dedicated perennial grass systems used for bioenergy will need to have similar or greater yield potential than existing annual crops for widespread adoption to meet renewable energy demands and provide similar economic returns to producers. We have previously shown that switchgrass ethanol yields were comparable with regional corn grain ethanol yields [12]. Here we demonstrate that when switchgrass is optimally managed, ethanol potential is similar to a continuous corn cropping system with stover removal and exceeds ethanol yield for corn grain-only systems on marginally-productive cropland. Furthermore, breeding improvements for bioenergy specific switchgrass cultivars have shown higher yield potential than cultivars evaluated here [38].

Coupling sustainable agricultural residue harvests with dedicated energy crops improves land-use efficiency and reduces biomass

constraints for a mature cellulosic biofuel industry. Recent analysis has shown that sufficient land exists in the U.S. Corn Belt to support a cellulosic ethanol industry without impacting productive cropland [18,39,40]. The effect of dedicated energy crops and corn grain on indirect land use change varies significantly based on the assumptions and models used [13,41,42] but bioenergy crops grown on marginally-productive cropland will have less impact on indirect land use change than bioenergy crops grown on more productive cropland. Likewise, model assumptions underlying direct SOC sequestration will impact system evaluations of GHG emissions and mitigation. Measured SOC sequestration values presented here were based on production years evaluated and were not extrapolated beyond this time-frame. Extrapolating SOC values from this time-frame to a 30-yr time horizon or 100-yr time horizon is still larger than current life cycle assessment assumptions on SOC sequestration potential of switchgrass or no-till corn [12,42,43]. This highlights the importance of accounting for direct SOC changes at depth to accurately estimate GHG emissions for biofuels under both marginal and productive cropland. Further long term evaluation of management practices (e.g. tillage, stover removal) on SOC sequestration potential for corn grain systems under irrigated conditions on productive cropland is warranted [44].

A multi-feedstock, landscape approach minimizes economic and environmental risks in meeting feedstock demands for cellulosic ethanol production by providing sufficient feedstock availability while maintaining ecosystem services. A co-located cellulosic biorefinery is expected to have economic advantages by reducing capital costs requirements for cellulosic conversion and through sharing of infrastructure costs. In this study, we used corn stover as the feedstock for the co-located cellulosic biorefinery but the benefits will apply to other cellulosic feedstocks. A co-located facility can increase NEY values by decreasing natural gas use for thermal energy, but current and forecasted U.S. natural gas prices [45] may affect large scale adoption of co-location unless there are incentives for displacing fossil energy in existing NG dry mill ethanol plants [46]. Integrating cellulosic refining capacity with existing corn grain ethanol plants can improve the sustainability of first generation biofuels and enable the implementation of cellulosic biofuels into the U.S. transportation sector.

Acknowledgments

Authors would also like to acknowledge the contribution of Edward J. Wolfrum and his staff at the National Renewable Energy Laboratory (Golden, CO) in conducting the corn stover analysis. Mention of trade names or commercial products in this publication does not imply recommendation or endorsement by the U.S. Department of Agriculture. USDA is an equal opportunity provider and employer.

Author Contributions

Conceived and designed the experiments: KV GV RF. Performed the experiments: KV GV RF RM MS VJ. Analyzed the data: MS. Contributed reagents/materials/analysis tools: MS KV RM RF. Wrote the paper: MS KV GV RF RM VJ.

References

1. Unger N, Bond TC, Wang JS, Koch DM, Menon S, et al. (2010) Attribution of climate forcing to economic sectors. Proc Natl Acad Sci USA 107(8): 3382–3387.
2. Wallander S, Claassen R, Nickerson C (2011) The ethanol decade: An expansion of U.S. corn production, 2000–09. *USDA-ERS* Economic Information Bulletin No. (EIB-79).
3. Claassen R, Carriazo F, Cooper J, Hellerstein D, Ueda K (2011) Grassland to cropland conversion in the northern plains: The role of crop insurance, commodity, and disaster programs Economic Research Report No. (ERR-120).
4. Wright CK, Wimberly MC (2013) Recent land use change in the western corn belt threatens grasslands and wetlands. Proc Natl Acad Sci USA 110(10): 4134–4139.
5. Wiens J, Fargione J, Hill J (2011) Biofuels and biodiversity. Ecol Appl 21(4): 1085–1095.
6. Donner SD, Kucharik CJ (2008) Corn-based ethanol production compromises goal of reducing nitrogen export by the Mississippi river. Proc Natl Acad Sci USA 105(11): 4513–4518.
7. USDA-NASS. 2013 U.S. Corn Yields. Available: http://www.nass.usda.gov/Charts_and_Maps/Field_Crops/cornyld.asp.
8. Wilhelm WW, Johnson JMF, Karlen DL, Lightle DT (2007) Corn stover to sustain soil organic carbon further constrains biomass supply. Agron J 99(6): 1665–1667.
9. Adler PR, Del Grosso SJ, Parton WJ (2007) Life-cycle assessment of net greenhouse-gas flux for bioenergy cropping systems. Ecol Appl 17(3): 675–691.
10. Farrell AE, Plevin RJ, Turner BT, Jones AD, O'Hare M, et al. (2006) Ethanol can contribute to energy and environmental goals. Science 312(5781): 506–508.
11. Sheehan J, Aden A, Paustian K, Killian K, Brenner J, et al. (2003) Energy and environmental aspects of using corn stover for fuel ethanol. J Indust Ecol 7(3–4): 117–146.
12. Schmer MR, Vogel KP, Mitchell RB, Perrin RK (2008) Net energy of cellulosic ethanol from switchgrass. Proc Natl Acad Sci USA 105(2): 464–469.
13. Wang M, Han J, Dunn JB, Cai H, Elgowainy A (2012) Well-to-wheels energy use and greenhouse gas emissions of ethanol from corn, sugarcane and cellulosic biomass for US use. Environ Res Letters 7(4): 045905.
14. Frank AB, Berdahl JD, Hanson JD, Liebig MA, Johnson HA (2004) Biomass and carbon partitioning in switchgrass. Crop Sci 44: 1391–1396.
15. Follett RF, Vogel KP, Varvel GE, Mitchell RB, Kimble J (2012) Soil carbon sequestration by switchgrass and no-till maize grown for bioenergy. BioEnerg Res 5: 866–875.
16. Liebig MA, Schmer MR, Vogel KP, Mitchell RB (2008) Soil carbon storage by switchgrass grown for bioenergy. BioEnerg Res 1(3–4): 215–222.
17. DOE US (2011) U.S. billion-ton update: Biomass supply for bioenergy and bioproducts industry. ORNL/TM-2011/224.
18. Gelfand I, Sahajpal R, Zhang X, Izaurralde R, Gross KL, et al. (2013) Sustainable bioenergy production from marginal lands in the US Midwest. Nature 493: 514–517.
19. Vogel KP, Brejda JJ, Walters DT, Buxton DR (2002) Switchgrass biomass production in the Midwest USA: Harvest and nitrogen management. Agron J 94: 413–420.
20. Varvel GE, Vogel KP, Mitchell RB, Follett RF, Kimble J (2008) Comparison of corn and switchgrass on marginal soils for bioenergy. Biomass Bioenergy 32(1): 18–21.
21. Greenhouse Gases, Regulated Emissions, and Energy Use in Transportation (GREET) (2012) Argonne National Laboratory. Available: http://greet.es.anl.gov/.
22. Renewable and Applicable Energy Laboratory (2007) Energy and resources group biofuel analysis meta-model. Available: http://rael.berkeley.edu/sites/default/files/EBAMM/.
23. Liska AJ, Yang HS, Bremer VR, Erickson G, Klopfenstein T, et al. (2008) BESS: Biofuel energy systems simulator; life-cycle energy and emissions analysis model for corn-ethanol (v. 2008.3.0).
24. Liska AJ, Yang HS, Bremer VR, Klopfenstein TJ, Walters DT, et al. (2009) Improvements in life cycle energy efficiency and greenhouse gas emissions of corn-ethanol. J Ind Ecol 13(1): 58–74.
25. Wang M, Wu M, Huo H (2007) Life-cycle energy and greenhouse gas emission impacts of different corn ethanol plant types. Environ Res Letters 2(2): 024001.
26. Wallace R, Ibsen K, McAloon A, Yee W (2005) Feasibility study for co-locating and integrating ethanol production plants from corn starch and lignocellulosic feedstocks. USDOE Rep No. NREP/TP-510-37092.
27. Spatari S, Bagley DM, MacLean HL (2010) Life cycle evaluation of emerging lignocellulosic ethanol conversion technologies. Bioresource Technol 101(2): 654–667.
28. MacLean HL, Spatari S (2009) The contribution of enzymes and process chemicals to the life cycle of ethanol. Environ Res Letters 4(1): 014001.
29. Humbird D, Davis R, Tao L, Kinchin C, Hsu D, et al. (2011) Process Design and Economics for Biochemical Conversion of Lignocellulosic Biomass to Ethanol. USDOE (Department of Energy NREL/TP-5100-47764, Golden, CO), pp 136.
30. Vogel K, Dien BS, Jung HG, Casler MD, Masterson SD, et al. (2011) Quantifying actual and theoretical ethanol yields for switchgrass strains using NIRS analyses. BioEnerg Res 4(2): 96–110.
31. Templeton DW, Sluiter AD, Hayward TK, Hames BR, Thomas SR (2009) Assessing corn stover composition and sources of variability via NIRS. Cellulose 16:621–639.
32. Graboski MS (2002) Fossil energy use in the manufacture of corn ethanol. National Corn Growers Association.
33. California Air Resources Board (2013) Low carbon fuel standard. Available: http://www.arb.ca.gov/Fuels/Lcfs/Lcfs.htm.
34. Liska AJ, Cassman KG (2008) Towards standardization of life-cycle metrics for biofuels: Greenhouse gas emissions mitigation and net energy yield. J Biobased Materials and Bioenergy 2(3): 187–203.
35. Gelfand I, Zenone T, Jasrotia P, Chen J, Hamilton SK, et al. (2011) Carbon debt of conservation reserve program (CRP) grasslands converted to bioenergy production. Proc Natl Acad Sci 108(33): 13864–13869.
36. Davis SC, Parton WJ, Del Grosso SJ, Keough C, Marx E, et al. (2012) Impact of second-generation biofuel agriculture on greenhouse-gas emissions in the corn-growing regions of the US. Front Ecol Environ 10:69–74.
37. Jin VL, Varvel GE, Wienhold BJ, Schmer MR, Mitchell RB, et al. (2011) Field emissions of greenhouse gases from contrasting biofuel feedstock production systems under different N fertilization rates. ASA-CSSA-SSSA International Annual Meetings. Oct 16–19, San Antonio TX.
38. Vogel KP, Mitchell RB (2008) Heterosis in switchgrass: Biomass yield in swards. Crop Sci 48(6): 2159–2164.
39. Mitchell RB, Vogel KP, Uden DR (2012) The feasibility of switchgrass for biofuel production. Biofuels 3(1): 47–59.
40. Uden DR, Mitchell RB, Allen CR, Guan Q, McCoy T (2013) The feasibility of producing adequate feedstock for year-round cellulosic ethanol production in an intensive agricultural fuelshed. BioEnerg Res 6(3): 930–938.
41. Searchinger T, Heimlich R, Houghton RA, Dong F, Elobeid A, et al. (2008) Use of U.S. croplands for biofuels increases greenhouse gases through emissions from land-use change. Science 319(5867): 1238–1240.

42. Dunn J, Mueller S, Kwon H, Wang M (2013) Land-use change and greenhouse gas emissions from corn and cellulosic ethanol. Biotechnology for Biofuels 6(1): 51.

43. Fargione J, Hill J, Tilman D, Polasky S, Hawthorne P (2008) Land clearing and the biofuel carbon debt. Science 319(5867): 1235–1238.

44. Follett RF, Jantalia CP, Halvorson AD (2013) Soil carbon dynamics for irrigated corn under two tillage systems. Soil Sci Soc Am J 77: 951–963.

45. Energy Information Agency (2013) Annual energy outlook for 2013 with projections to 2040. DOE/EIA-0383.

46. Plevin RJ, Mueller S (2008) The effect of CO_2 regulations on the cost of corn ethanol production. Environ Res Letters 3(2): 024003.

10

Effects of Soil Data and Simulation Unit Resolution on Quantifying Changes of Soil Organic Carbon at Regional Scale with a Biogeochemical Process Model

Liming Zhang[1,2], Dongsheng Yu[2]*, Xuezheng Shi[2], Shengxiang Xu[2], Shihe Xing[1]*, Yongcong Zhao[2]

1 College of Resource and Environment, Fujian Agriculture and Forestry University, Fuzhou, China, 2 State Key Laboratory of Soil and Sustainable Agriculture, Institute of Soil Science, Chinese Academy of Sciences, Nanjing, China

Abstract

Soil organic carbon (SOC) models were often applied to regions with high heterogeneity, but limited spatially differentiated soil information and simulation unit resolution. This study, carried out in the Tai-Lake region of China, defined the uncertainty derived from application of the DeNitrification-DeComposition (DNDC) biogeochemical model in an area with heterogeneous soil properties and different simulation units. Three different resolution soil attribute databases, a polygonal capture of mapping units at 1:50,000 (P5), a county-based database of 1:50,000 (C5) and county-based database of 1:14,000,000 (C14), were used as inputs for regional DNDC simulation. The P5 and C5 databases were combined with the 1:50,000 digital soil map, which is the most detailed soil database for the Tai-Lake region. The C14 database was combined with 1:14,000,000 digital soil map, which is a coarse database and is often used for modeling at a national or regional scale in China. The soil polygons of P5 database and county boundaries of C5 and C14 databases were used as basic simulation units. Results project that from 1982 to 2000, total SOC change in the top layer (0–30 cm) of the 2.3 M ha of paddy soil in the Tai-Lake region was +1.48 Tg C, −3.99 Tg C and −15.38 Tg C based on P5, C5 and C14 databases, respectively. With the total SOC change as modeled with P5 inputs as the baseline, which is the advantages of using detailed, polygon-based soil dataset, the relative deviation of C5 and C14 were 368% and 1126%, respectively. The comparison illustrates that DNDC simulation is strongly influenced by choice of fundamental geographic resolution as well as input soil attribute detail. The results also indicate that improving the framework of DNDC is essential in creating accurate models of the soil carbon cycle.

Editor: Jose Luis Balcazar, Catalan Institute for Water Research (ICRA), Spain

Funding: The work was funded by National Natural Science Foundation of China (No. 41001126), and the National Basic Research Program of China (973 Program) (2010CB950702). The funders had no role in study design, data collection and analysis, decision to publish, or preparation of the manuscript.

Competing Interests: The authors have declared that no competing interests exist.

* E-mail: dshyu@issas.ac.cn (DY); fafuxsh@126.com (SX)

Introduction

An estimated 1500 Pg of C is held in the form of soil organic carbon (SOC), representing 2/3 of the global terrestrial organic carbon pool [1–3]. SOC plays a vital role in the global carbon cycle, where a slight alteration of the soil carbon pool can cause profound changes in atmospheric CO_2 concentrations. Agro-ecosystems, accounting for 10% of the total terrestrial area, are one of the most sensitive terrestrial ecosystems subject to heavy human activity [3]. Increasing agricultural soil C sequestration is recognized as one strategy for achieving food security and improving soil quality.

Paddy soil is a major cultivated soil in China, and a unique type of anthropogenic soil recognized by Chinese Soil Taxonomy [3–5]. The total area of paddy soils is 45.7 M ha, which accounts for 34% of the total cultivated land in China [6]. This area also accounts for 22% of the total waterlogged farming area worldwide and produces about 44% of all grain in China [4]. Therefore, accurate estimation of paddy soil SOC change in China is vitally important for a comprehensive understanding of SOC dynamics and agro-ecosystem sustainability.

Recently, scientists have applied modeling to estimate SOC change in cropping systems [7–14]. The DeNitrification-DeComposition (DNDC) model, developed by Li et al. [15,16], is a process-based model focused on agrosystem carbon and nitrogen cycling and has been widely used for regional studies in the USA [17], China [11], India [18] and Europe [19]. Recently the DNDC model was determined to be one of the well performing models based on seven long-term experiments selected by the Global Change and Terrestrial Ecosystems Soil Organic Matter Network (GCTE SOMNET), which evaluated model performance using three different land uses, a range of climatic conditions within the temperate region, and different treatments [11,14].

In China, scientists have studied SOC change using the DNDC model for many years. At the regional scale, Tang et al. [11] simulated SOC changes for cropland in China for 1998 using the DNDC model, and they found that SOC would be lost at a rate of 78.89 Tg C $year^{-1}$. Zhang et al. [20] linked the DNDC model and 1: 14,000,000 soil database to estimate SOC stock changes for the year 2000 in Northwest China, revealing a decline in SOC stock. At the field scale, Wang et al. [21] tested DNDC uncertainty based on six long-term (10–20 year) SOC datasets from the Northeast, North, Northwest, Central South, East, and Southwest China. Results from the six validation tests supported the previous

Figure 1. Geographical location of the study area in China.

conclusions that the DNDC model was capable of quantifying SOC change in the agroecosystems across the entire area of China.

To date, the county boundary was used as the basic simulation unit in most DNDC simulations conducted at regional scale [11,20]. As a result, these simulations are often subject to great uncertainties since the soil property data were averaged for the area, which greatly ignore the impacts of soil heterogeneity therein [18,22]. Moreover, many researchers used coarse soil attribute data obtained from the books such as Soil in China (Vol. 1–6) and 1: 14,000,000 soil maps at national or a regional scale in China [11,20]. However, studies have already pointed out that the effect of soil heterogeneity on SOC change estimation is a major source of uncertainty when using the DNDC model at the regional scale [18,22,23].

This study, which was carried out in the rice-dominated Tai-Lake Region of China, provides a chance to test the uncertainty of the DNDC model caused by different precisions of soil data and basic simulation unit. The goals of this study were to: (1) compare SOC changes modeled with different resolutions of soil databases and varied basic simulation units, (2) assess the uncertainty derived from these soil databases with different resolutions and basic simulation units, and (3) give some suggestions for improving the performance of the biogeochemical DNDC model applied at the regional scale.

Materials and Methods

Study area

The Tai-Lake region (118°50′-121°54′E, 29°56′-32°16′N), an area of intensive rice cultivation, is located in the middle and lower reaches of the Yangtze River paddy soil region of China. The region includes the entire Shanghai City administrative area and a part of Jiangsu and Zhejiang provinces, and covers a total area of 36,500 km^2 (**Fig. 1**) [4]. The Tai-Lake region mainly consists of plains formed on deltas with numerous rivers and lakes. The climate is warm and moist with abundant sunshine and a long growing season. Annual rainfall is 1,100–1,400 mm, with a mean temperature of 16°C, and average annual sunshine of 1,870–2,225 hours. The frost-free period is over 230 days. The study area is one of the oldest agricultural regions in China, with a long history of rice cultivation spanning several centuries. Most

cropland in the region is managed as a rice and winter wheat rotation. Rice is planted in June and harvested in October and wheat is planted in November and harvested in May [24].

Approximately 66% of the total land area is covered with paddy soils [24]. Paddy soils in the Tai Lake area are derived mostly from loess, alluvium, and lacustrine deposits, and are classified into 6 soil subgroups according to the Genetic Soil Classification of China (GSCC) system which are represented in the 1:50,000 digital soil map (**Table 1**). As map scale decreased, the soil subgroups of submergenic, bleached, percogenic and degleyed on the 1:50,000 soil map was eliminated and emerged into the soil subgroups of degleyed and hydromorphic in the 1:14,000,000 soil map. Therefore, those were only two paddy soil subgroups of degleyed and hydromorphic in the 1:14,000,000 soil map. The GSCC nomenclature as well as the subgroup's reference name in US Soil Taxonomy (ST) include; Hydromorphic (Typic Epiaquepts), Submergenic (Typic Endoaquepts), Bleached (Typic Epiaquepts), Gleyed (Typic Endoaquepts), Percogenic (Typic Epiaquepts), and Degleyed (Typic Endoaquepts) [25,26].

Description of the DNDC model

The DNDC model (Version 9.1) is a process-based soil biogeochemical research tool that was developed to estimate the impact of management strategies on the fate of nitrogen (N) and carbon (C) in agroecosystems. It integrates crop growth and soil biogeochemical processes on a daily time step and simulates N and C cycles in plant-soil systems.

The model contains six interacting sub-models which describe the generation, decomposition, and transformation of organic matter, and outputs the dynamic components of SOC and greenhouse gas fluxes. The six sub-models include: 1) a soil climate component which use soil physical properties, air temperature, and precipitation data to calculate soil temperature, moisture, and redox potential (Eh) profiles and soil water fluxes through time. The results of the calculation are then fed to the other sub-models; 2) a nitrification component; 3) a denitrification module, which calculates hourly denitrification rates and N_2O, NO, and N_2 production during periods when the soil Eh decreases due to rainfall, irrigation, flooding, or soil freezing; 4) simulation of SOC decomposition and CO_2 production through soil microbial respiration; 5) a plant growth component, which calculates daily

Table 1. The subgroups of paddy soil in the Tai-Lake region, China.

Subgroups	Horizonation*	Descriptions
Bleached	A-P-E-C	Mainly distributed in foothills, usually no underground water, impervious layer at 60 cm depth, soil reaction close to neutral or slightly acid.
Gleyed	Aa-Ap-G-C	Mainly distributed in depressional areas, high underground water level, poorly drained, distinct gleyization, soil reaction was slightly acid.
Percogenic	Aa-Ap-C	Mainly distributed on gentle hill slopes, no underground water, associated with rain-fed paddy fields, soil reaction was neutral to slightly acid.
Degleyed	Aa-Ap-Gw-G	Same distribution area as Gleyed paddy soils, after man-made drainage the underground water level decreases leading to degley processes, soil reaction was slightly acid.
Submergenic	A-Ap-P-C	Mainly distributed in alluvial plain or low flat ground, moderate drainage, underground water level was below 60 cm, soil reaction was neutral.
Hydromophic	Aa-Ap-P-W-G-C	Mainly distributed in floodplain, long cultivation history, well-drained, underground water level was below 90 cm, soil reaction was neutral.

*According to GSCC (Genetic Soil Classification of China), Aa means arable layer, Ap plow pan, C undeveloped parent material, Ds fragmental deposit horizon, E bleached horizon, G gley horizon, Gw degley horizon, P percogenic horizon, W waterlogogenic horizon.

root respiration, water, and N uptake by plants, and plant growth; and 6) a fermentation module, which calculates daily methane (CH_4) production and oxidation. The DNDC model can simulate C and N biogeochemical cycles in paddy rice ecosystems, as the model has been modified by adding a series of anaerobic processes [15,16,22,23,27,28,29,30].

At present, the DNDC model has been utilized by scientists in many countries, for example, the model is applied to simulate the carbon cycle in paddy field in Italy, China and Germany, in wheat fields in Canada, and it has been used to simulate the dynamics of soil organic matter in a 100 year experimental field in Rothamsted Experimental Station in England [14,31]. At the international conference on global change in Asia-Pacific areas in 2000, the DNDC model was recommended as the primary method for SOC studies in the in the Asia-Pacific region [31].

Database development

A major challenge for using an ecosystem model at regional scale is to assemble adequate datasets required to initialize and run the model. We examined the influence of database choices by executing simulation runs with different input sets using individual or combinations of databases. The geographic resolution or fundamental simulation unit could be represented by any of three assessment unit format datasets, polygon-based database of 1:50,000 (P5), county-based database of 1:50,000 (C5), and county-based database of 1:14,000,000 (C14). The three soil datasets covered 37 counties in Tai-Lake region.

The polygon-based database of 1:50,000 (P5) was linked a digital soil map (1:50,000), the most detailed of the three databases, in the Tai-Lake region contains 52,034 paddy soil polygons (**Table 2**). The polygons were derived from 1,107 soil profiles extracted from the latest national soil map (1:50,000), the Second National Soil Survey of China in the 1980s-1990s, with attribute assignment using the Pedological Knowledge Based (PKB) method based on GSCC [32]. The 1:50,000 digital soil database consists of many soil attributes, such as soil name, horizon thickness, bulk density, organic carbon content, clay content, pH, etc.

Soil parameters in C5 were derived from the 1:50,000 digital soil map (**Fig. 2 and Table 2**). However the attributes for C14 were derived from different sources than C5, primarily the 1:14,000,000 national soil map [33,34] (**Fig. 2**). C14 was widely used when the DNDC model was applied to national or regional scale in China [11,20]. The C14 in the Tai-Lake region contained 8 polygons of paddy soils representing 49 paddy soil profiles, and was also compiled via the Pedological Knowledge Based (PKB) method based on GSCC [32].

The C5 and C14 were built from the default method developed for DNDC, in which the maximum and minimum values of soil texture, pH, bulk density, and organic carbon content were recorded for each county (**Fig. 2**). So, the DNDC modeling of C5 and C14 methods conducted have used counties as the basic simulation unit in the Tai-Lake region (**Fig. 2**). After regional runs with C5 and C14 database, the DNDC model produced two SOC

Table 2. Characteristics of different resolution soil attribute databases of paddy soils in GSCC in the Tai-Lake region, China.

Soil database	Map scale	Source of soil maps	Source of soil data	Basic map units	Number of soil profiles	Number of polygons	Simulation unit
P5	1:50,000	Soil Survey Office of County in Jiangsu Province, Zhejiang Province and Shanghai City	Soil Series of County in Jiangsu Province, Zhejiang Province and Shanghai City	Soil Species	1,107	52,034	polygon
C5	1:50,000	Soil Survey Office of County in Jiangsu Province, Zhejiang Province and Shanghai City	Soil Series of County in Jiangsu Province, Zhejiang Province and Shanghai City	Soil Species	1,107	52,034	county
C14	1:14,000,000	Institute of Soil Science, Chinese Academy of Sciences	Soil Series of China	Subgroups	49	8	county

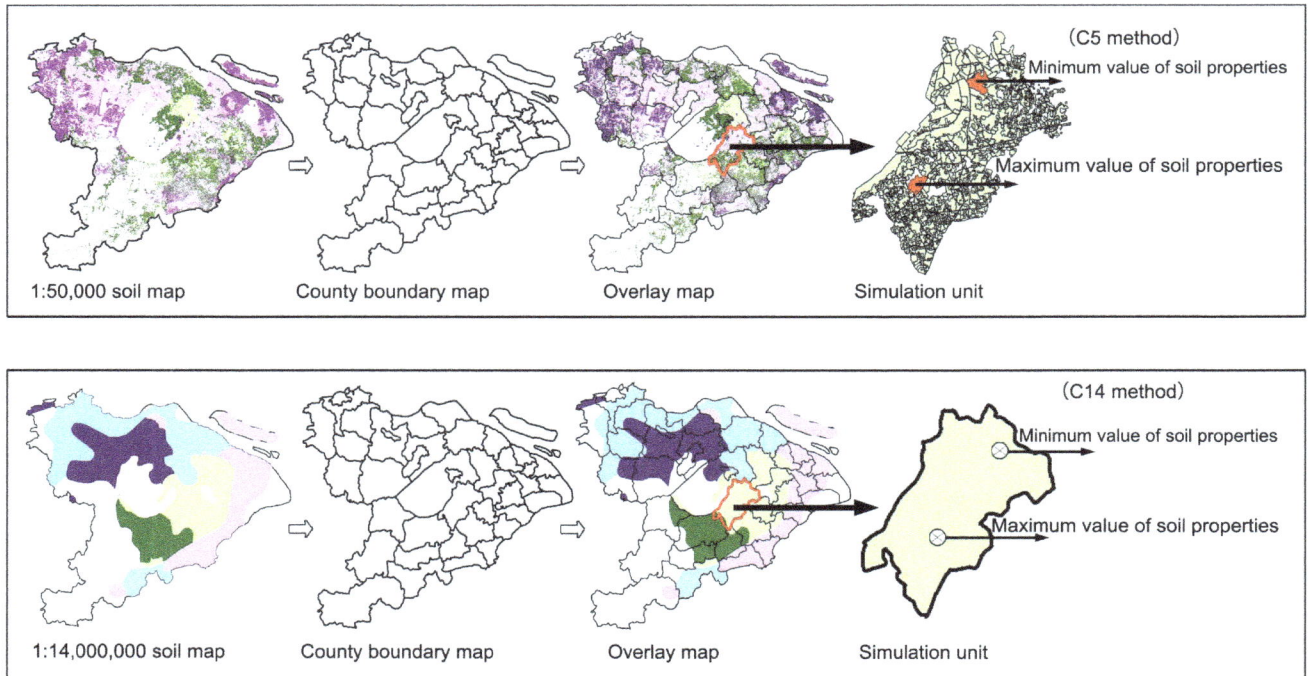

Figure 2. Description of C5 and C14 methods in the Tai-Lake region of China.

change (0–30 cm) resulting from two runs with the maximum and minimum soil values in each county. In this paper we present the mean results (average of maximum and minimum estimates) [11]. The DNDC modeling of P5 method conducted has used polygon as the basic simulation unit in the Tai-Lake region (**Table 2**). Therefore, the DNDC model runs with P5 database produced a single annual SOC change (0–30 cm) for each polygon. The total

SOC change of each county in the P5 was calculated by summing the SOC change of all polygons in a county. For a more complete description of P5 method see Zhang et al [35,36] and Xu et al [37].

For comparison in this study, both the polygon-based (P5) and county-based (C5 and C14) soil databases in the Tai-Lake region were run concurrently so the DNDC model could generalize regional SOC change from 1982 to 2000. The results simulated by DNDC with the two types of databases were compared to assess the advantages of using detailed, polygon-based 1:50,000 soil dataset (P5) [35,36,38,39].

In this study, the crop dataset included physiological data for summer rice and winter wheat in the Tai-Lake region. The crop parameters were obtained from thorough testing with that reflected the typical conditions of Tai-Lake region, which were founded on a wide range of information form Chinese literature published during the past decade and a publication of Gou et al [40,41].

Daily meteorological data (precipitation, maximum and minimum air temperature) for 1982–2000 from 13 weather stations across and near the Tai-Lake region were acquired from the National Meteorological Information Center, China Meteorological Administration (**Fig. 3**) [42]. Each county in the simulation was assigned to the nearest weather station [11,20,31].

The agricultural management dataset included sowing acreage, nitrogen fertilizer application rates, livestock, planting and harvest dates, and agricultural population at the county level from 1982 to 2000 in three resolution databases. The crop management practices of different counties were almost the same because the Tai-Lake region was a plain in topography. The main measures of farming management in the study area included: (1) fertilizer application: nitrogen synthetic fertilizer was applied for 6 times in the basal, tillering and heading stage for rice, and in the basal, jointing and heading stage for wheat; and organic manure (20% of

Figure 3. Geographical location of weather stations across or near the Tai-Lake region, China.

livestock wastes and 10% of human wastes) was applied twice as base fertilizer for rice and wheat at the rates calculated based on the local livestock numbers (866, 44, 95, and 23 kg C head^{-1} yr^{-1} for cattle, sheep, swine and human, respectively); and N concentration in rainfall was 2.07 ppm; (2) crop residue management: 15% of aboveground crop residue was returned to the soil; (3) water management: one time of midseason and 5 time of shallow flooding (from June 17 to July 23, from July 28 to August 12, from August 24 to September 11, from September 18 to September 25, and from September 27 to October 2, respectively) were applied at summer rice; (4) tillage: twice at the 20 cm tilling depth for rice and 10 cm for wheat on the planting dates before 1990; and no-till applied for wheat after 1990; (5) growing period: rice is planted in June and harvested in October and wheat is planted in November and harvested in May; (6) optimum yield: rice is 7500 kg dry matter ha^{-1} and wheat is 3750 kg dry matter ha^{-1} [11,14,24,35,41,43]. All simulation methods within a certain county have the same feature input value such as crops, agricultural management, and climate, except soil feature [38,39].

Evaluation of simulation accuracy in three resolution databases

In order to evaluate the accuracy in three resolution databases, the simulated results of DNDC model were tested against measured data from paddy soils of the Tai-Lake region, which is the same area examined here.

From the perspective of previous studies, most dynamic models were only tested or validated with static long-term field-scale observations due to a lack of available soil data with temporal and spatial variation. Since these models have not yet been validated by regional scale data, uncertainty concerning their accuracy exists when they were applied to larger area dynamic SOC simulation [3].This study compared simulation results with the spatial distribution of SOC measurements from 1033 paddy soil sampling sites acquired in 2000, to validate and assess model performance in different simulation methods (P5, C5, and C14). The bias in the total difference between simulation and measurement were determined by calculating the correlation coefficient (r), the

relative error (E), the mean absolute error (MAE) and the root mean square error (RMSE), as follows: [8,44].

$$r = \frac{\sum_{i=1}^{n} \left(V_{oi} - \overline{V_{oi}} \right) \left(V_{Si} - \overline{V_{Si}} \right)}{\left[\sum_{i=1}^{n} \left(V_{oi} - \overline{V_{oi}} \right)^2 \right]^{1/2} \left[\sum_{i=1}^{n} \left(V_{Si} - \overline{V_{Si}} \right)^2 \right]^{1/2}} \quad (1)$$

$$E = \frac{100}{n} \times \sum_{i=1}^{n} \frac{V_{oi} - V_{Si}}{V_{oi}} \quad (2)$$

$$RMSE = \sqrt{\frac{1}{n} \sum_{i=1}^{n} (V_{oi} - V_{Si})^2} \quad (3)$$

$$MAE = \frac{1}{n} \sum_{i=1}^{n} ABS(V_{oi} - V_{Si}) \quad (4)$$

Where V_{oi} are the observed values, $\overline{V_{oi}}$ is the mean of the observed data, V_{Si} is the simulated value, $\overline{V_{Si}}$ is the mean of the simulated value, $Si \in (P5, C5, C14)$, and n is the number in the sequence of the data pairs. If E is less than 5% or between 5% and 10%, the simulation is satisfactory or acceptable, respectively; otherwise, it is unacceptable [44]. The greater r value is and the smaller RMSE or MAE value is, the greater prediction accuracy is. Conversely, a lower r value and more elevated RMSE or MAE value, the lower prediction accuracy is.

Data comparison and analysis

SOC change as quantified by DNDC modeling with the P5 assessment unit data set are recognized as a benchmark for comparison with the results of the DNDC model runs with the other two assessment unit data sets as input. The P5 are thought

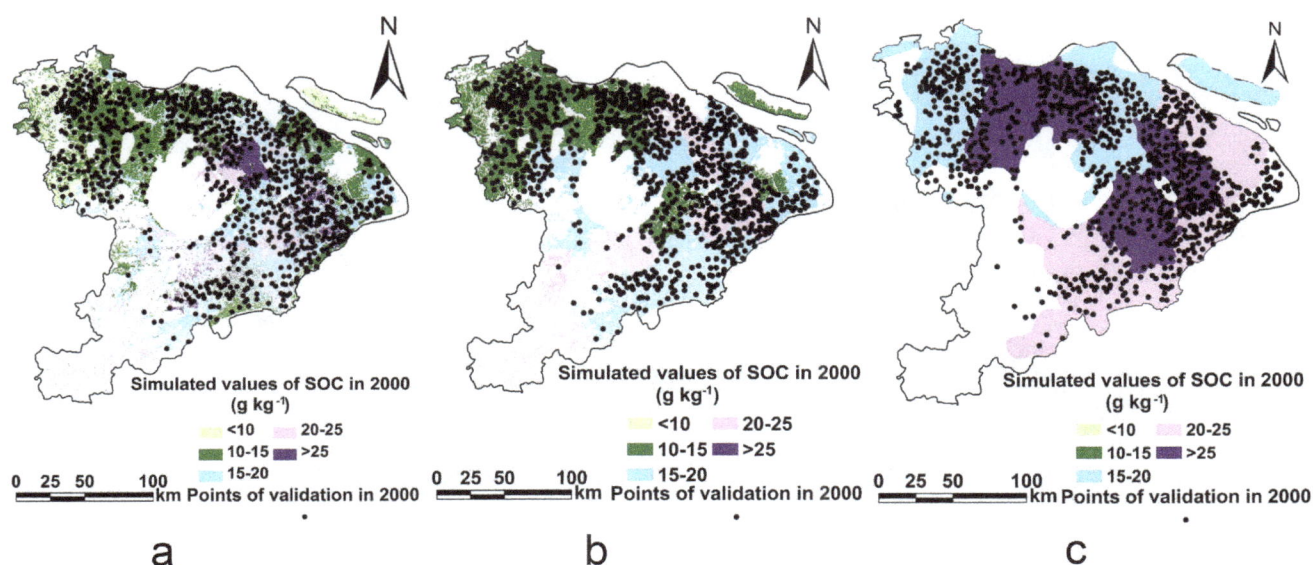

Figure 4. Spatial distribution of validation points and simulated SOC values from different simulation methods for the Tai-Lake region for 2000 (a: P5, b: C5, and c: C14).

theoretically to be more accurate than the C5 and C14 because of their relative greater detail and accuracy [35,36,38,39]. Relative variation of an index value (VIV, %) of C5 and C14 methods is calculated as the formula (1). The index values (IV) were quantified from the P5 data set (IV_{P5}) and other data sets (IV_{Ci}) to support data set comparison [23,38,39].

$$VIV(\%) = ABS\left(100 \times (IV_{P5} - - IV_{Ci})/IV_{P5}\right) \qquad (5)$$

Where ABS is the absolute value function, IV_{P5} is the total SOC change with P5, and IV_{Ci} is the total SOC change produced by C5 (or C14).

Previous results from the sensitivity tests of the DNDC model indicated that the spatial heterogeneity of soil properties (e.g. texture, SOC content, bulk density, and pH) are the major sources of uncertainty for simulating SOC changes under specific management conditions at regional scale [18,20,22,23]. In order to test the most sensitive soil properties factor, the correlation of soil properties and average annual SOC changes were determined by step-wise regression analysis by using SPSS statistical software [37,45]. The step-wise regression is useful in checking how entering each variable affects the overall regression model, which begins by entering the variable with the largest partial statistic and checking the importance of the coefficient of the variable [45,46]. This method keeps adding more variables, each time recalculating the coefficients. During the incorporation of a variable into the model, the partial statistic of the already entered variable changes and might cause it to be unimportant. The operation stops when the model has incorporated the variables with the most significant contribution and discarded the least significant ones [47].

Results and Discussion

Difference of simulation accuracy in three resolution databases

Three maps of average SOC content for paddy soils at surface layers (0–15 cm) in the study area in 2000 were constructed on the basis of simulated data in different simulation methods (P5, C5, and C14) (**Fig. 4**). Also, corresponding SOC validation points were constructed from measurements of the surface layer (0–15 cm) of 1033 paddy soil samples taken in the study area in 2000. Fig. 4 demonstrates that the observed SOC in 2000 varied from 1.9 g kg^{-1} to 36 g kg^{-1}. By comparison, Fig. 4 also illustrates that

simulated SOC in 2000 varied from 5.1 g kg^{-1} to 34 g kg^{-1} in P5, from 11 g kg^{-1} to 24 g kg^{-1} in C5, and from 17 g kg^{-1} to 28 g kg^{-1} in C14; where 99.6%, 84.1% and 57.1% of simulated paddy soil samples in P5, C5 and C14 were within the ranges produced by the observed SOC data. Furthermore, the relative errors (E) of P5 and C5 were 6.4% and 5.0%, respectively; and within the range of 5%–10%, demonstrating that the DNDC model in P5 and C5 were acceptable for modeling SOC of paddy soils in the Tai-Lake region according to the evaluation criteria described earlier (**Fig. 5a and b**) [8,44]. Moreover, the small values of MAE (4.0 g kg^{-1}) and RMSE (5.0 g kg^{-1}) in P5 and C5 also indicated that the modeled results were encouragingly consistent with observations in the Tai-Lake region (**Fig. 5a and b**). However, the E, MAE and RMSE of C14 reached −33%, 6.0 g kg^{-1} and 7.0 g kg^{-1}, respectively, suggesting that the simulated results of C14 were not suitable for simulating paddy soils in the Tai-Lake region (**Fig. 5c**).

Overall, though the values of E, MAE and RMSE between P5 and C5 had no significant differences, P5 was recognized better due to high correlation coefficient (0.5) and accurate simulation range (99.6%) (**Fig. 5a and b**). Furthermore, the simulation of P5 can differentiate the difference of paddy soil type within a county. Some studies showed that SOC content spatial variability was correlated with soil type spatial variability(**Fig. 5a**) [32,48,49]. Compared to the SOC validation of DNDC model in cropland by other scientists, accurate simulation of P5 (r = 0.50**; E = 6.4%; MAE = 4.0 g kg^{-1}; RMSE = 5.0 g kg^{-1} and n = 1033) and C5 (r = 0.40**; E = 5.0%; MAE = 4.0 g kg^{-1}; RMSE = 5.0 g kg^{-1} and n = 1033) are higher than those of Liu et al [50] (r = 0.25–0.66 and n = 68), Liu et al [51] (E = 27.6% and n = 49), and Xu et al [52] (r = 0.22**; RMSE = 4.4 g kg^{-1} and n = 243); and are almost similar to that of the Xu et al [52] (r = 0.52**; RMSE = 4.1 g kg^{-1} and n = 1385). However, the SOC accurate simulation of DNDC model in P5, C5 and C14 are lower than those of Studdert et al [53] (r = 0.73** and n = 286) by using the RothC model and Yu et al [54] (r = 0.98** and n = 349) by using the Agro-C model. Therefore, the results mentioned above suggest that modification of the DNDC model is necessary to better simulate SOC change from cropping systems. With continued modification, DNDC model could become a powerful tool for estimating SOC change at regional and national scales.

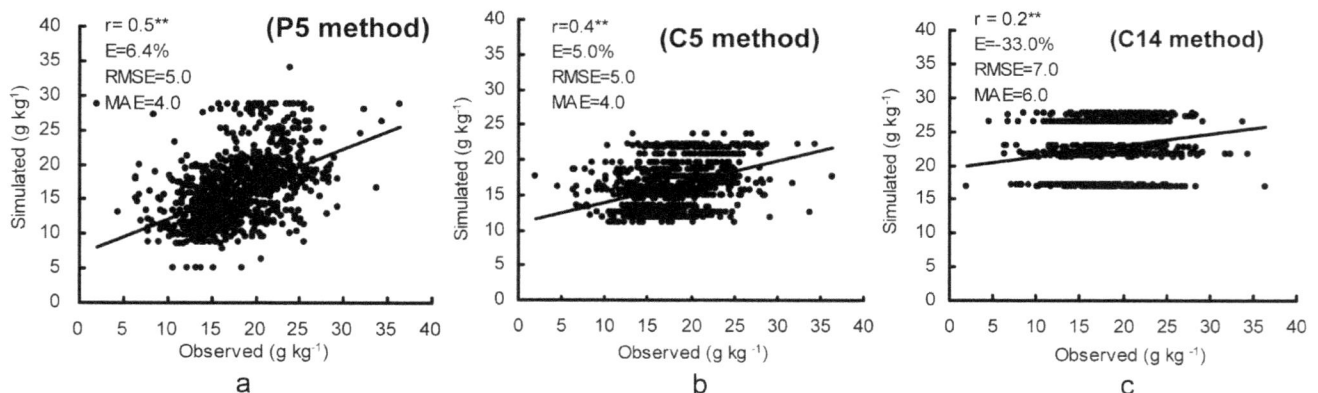

Figure 5. Comparison between simulated and observed SOC values from different simulation methods of the Tai-Lake region for 2000 (a: P5, b: C5, and c: C14).

Table 3. Soil properties at three resolution soil attribute databases contributing to the variability of average annual SOC change in Tai-Lake region paddy soils from 1982 to 2000.

Soil database	Number of simulation units	$\triangle R^{2a}$				Adjusted R^2
		Initial SOC (g kg^{-1})	Clay(%)	pH	Bulk density (g cm^{-3})	
P5	52,034	0.778***	0.066***	0.009***	0.025***	0.878***
C5	37	0.881***			0.062***	0.939***
C14	37	0.185***		0.757***		0.938***

***significant at 0.001 probability levels, respectively.
[a]The change in the R^2 statistic is produced by adding a soil property into stepwise multiple regressions.

Variation of soil properties derived as input for DNDC modeling in three resolution databases in Tai-Lake region

Results of the contribution of soil properties to the variability of average annual SOC change are given in Table 3. All variables (i.e., initial SOC content, pH, bulk density, and clay content) were included in the step-wise regression analysis. For the P5 and C5 resolution databases, initial SOC content accounted for 77.8%–88.1% of the difference of average annual SOC change for paddy soils from 1982 to 2000, while other soil parameters only accounted for less than 6.6% of the difference. For the C14 resolution database, initial SOC content accounted for 18.5% of the difference of average annual SOC change for paddy soils from 1982 to 2000, and soil pH accounted for 75.7% of the difference. Therefore, it could be inferred that the differences in SOC change modeled with the three resolution databases were primarily due to the differences in initial SOC content and pH.

Table 4 shows the initial SOC content (0–5 cm), clay content (0–10 cm), pH (0–10 cm), and bulk density (0–10 cm) derived as input for DNDC modeling, from P5, C5 and C14 for the Tai-Lake region. As for the entire Tai-Lake region, the average initial SOC values sourced from P5 was lower than that from C5 and C14. Another difference is that the average values of clay content and pH sourced from C14 were also higher than those from P5 and C5. The average bulk density sourced from C5 was higher than that from P5 and C14.

The differentiation of soil properties was also shown at the county scale in the Tai-Lake region (**Table 4**). The average values of initial SOC content and bulk density sourced from C5 for 24 counties were higher than those from P5; the other was that the average values of clay content for 24 counties and pH for 20 counties in C5 were lower than those from P5. Although the average clay content sourced from P5 for 25 counties was slightly lower than that from C14, but the average initial SOC content sourced from C14 for 34 counties was obviously higher than that of P5. According to statistics describing the 1:50,000 digital soil database of the Tai-Lake region, initial SOC content of six paddy soil subgroups, namely submergenic, bleached, percogenic, hydromorphic, degleyed and gleyed, were 10 g kg^{-1}, 10 g kg^{-1}, 11 g kg^{-1}, 15 g kg^{-1}, 19 g kg^{-1}, and 25 g kg^{-1}, respectively. As map scale decreased from 1:50,000 to 1:14,000,000, the submergenic, bleached, percogenic and degleyed subgroups on the 1:50,000 digital soil map were eliminated and merged into the hydromorphic and degleyed subgroups in the 1:14,000,000 digital soil map [32,39]. The initial SOC content of the hydromorphic and gleyed subgroups in the 1:14,000,000 digital soil database were 17 g kg^{-1} and 28 g kg^{-1}, respectively, which were higher than most paddy soil subgroups in the 1:50,000 digital soil

database. Therefore, the average initial SOC content of most counties in C14 was significantly higher than that from P5, while the average values of bulk density for 20 counties and pH for 24 counties in C14 was lower than those from P5. The results demonstrated that the soil properties (i.e., texture, SOC content, bulk density, and pH) in three resolution databases methods had large differences in the Tai-Lake region. Many studies have showed that SOC spatial variability is expressed by map delineations and map unit composition which varied with scales, resulting in the assignment of different soil properties at each scale of aggregation [32,48,49]. As such, an improper of soil map scales and simulation unit may lead to SOC estimation inaccuracy.

Variation of the average annual-, total SOC change modeled with the three resolution databases in Tai-Lake region

Similar trends can be observed in estimates of average annual-, total SOC change over the 19 year study period for three resolution databases decreased from P5 to C14 (**Fig. 6**). Simulation results demonstrate that total SOC change of P5 in the top layer (0–30 cm) of the 2.3 M ha of paddy rice fields in the Tai-Lake region was +1.48 Tg C from 1982 to 2000, with the annual SOC change ranging from -45 kg C ha^{-1} yr^{-1} to 92 kg C ha^{-1} yr^{-1} (**Fig. 6**). From 1982 to 1988, the SOC change modeled with P5 inputs was almost negative with annual changes ranging from -3.2 kg C ha^{-1} yr^{-1} to -45 kg C ha^{-1} yr^{-1}. According to agricultural statistical data, chemical fertilizer application rate ranged from 180 kg N ha^{-1} yr^{-1} to 350 kg N ha^{-1} yr^{-1}, which is a relatively low value. Low fertilizer application rates often result in reduced SOC sequestration [31,55]. From 1989 to 2000, rural economic development led to increased fertilizer application from 350 kg N ha^{-1} yr^{-1} to 400 kg N ha^{-1} yr^{-1}. Increasing fertilizer application results in enhanced crop production and residue accumulation, and the latter leads to an increase of SOC. Further, much of the region has been utilizing no-tillage practices in planting wheat since 1991, which contribute to reduced SOC decomposition [35].

Although three resolution databases within a certain county have the same feature input value such as crops, agricultural management, and climate; SOC balance of C5 (or C14) in the Tai-Lake region was almost negative with annual changes ranging from 86 kg C ha^{-1} yr^{-1} to -205 kg C ha^{-1} yr^{-1} (or -185 kg C ha^{-1} yr^{-1} to -693 kg C ha^{-1} yr^{-1}) from 1982 to 2000 (**Fig. 6**). The total SOC changes of C5 and C14 in the Tai-Lake region were -3.99 Tg C and -15.38 Tg C, respectively, from 1982 to 2000. With the total SOC change as modeled with P5 inputs as the baseline, the relative deviation of C5 and C14 were 368% and 1126%, respectively.

Table 4. Statistics for soil properties derived as input for DNDC modeling in different counties, from P5, C5 and C14 for the Tai-Lake region.

County	P5 SOC	Clay	BD	pH	C5 SOC Range	Ave	Clay Range	Ave	BD Range	Ave	pH Range	Ave	C14 SOC Range	Ave	Clay Range	Ave	BD Range	Ave	pH Range	Ave
	-----WA-----																			
Zhangjiagang	14	28	1.22	7.7	10–17	14	5–32	19	1.16–1.33	1.25	7.4–8.0	7.7	12–21	17	24–31	28	1.19–1.23	1.21	6.0–7.4	6.7
Changshu	17	25	1.23	7.0	9–38	24	9–34	22	1.05–1.46	1.26	5.5–8.1	6.8	12–21	17	24–31	28	1.19–1.23	1.21	6.0–7.4	6.7
Taicnang	14	33	1.20	7.7	9–20	15	23–42	33	1.11–1.38	1.25	7.4–8.6	8.0	11–31	21	18–48	33	1.12–1.27	1.20	5.5–7.4	6.5
Kunshan	19	35	1.15	7.1	11–34	18	22–44	33	0.94–1.40	1.17	6.4–7.6	7.0	23–33	28	32–56	44	1.14–1.14	1.14	6.2–6.9	6.6
Wuxian	24	41	1.08	6.6	6–30	23	26–47	37	0.97–1.47	1.22	3.4–7.4	5.6	12–21	17	24–31	28	1.19–1.23	1.21	6.0–7.4	6.7
Wujiang	17	36	1.06	5.9	3–26	15	17–58	38	0.89–1.67	1.28	4.9–6.9	5.9	23–33	28	32–56	44	1.14–1.14	1.14	6.2–6.9	6.6
Wuxi	14	28	1.16	6.7	4–17	11	14–34	24	1.09–1.39	1.24	5.3–7.2	6.3	21–33	27	25–31	28	1.13–1.23	1.18	6.0–6.3	6.1
Jiangyin	13	13	1.28	6.2	6–17	12	8–25	17	0.99–1.54	1.27	5.4–8.0	6.7	21–33	27	25–31	28	1.13–1.23	1.18	6.0–6.3	6.1
Wujin	12	9	1.22	6.8	7–18	13	4–13	9	1.08–1.51	1.30	6.2–7.9	7.1	21–33	27	25–31	28	1.13–1.23	1.18	6.0–6.3	6.1
Jintan	10	9	1.32	6.8	7–14	11	4–13	9	1.17–1.58	1.38	5.5–7.6	6.6	12–21	17	24–31	28	1.19–1.23	1.21	6.0–7.4	6.7
Liyang	10	10	1.23	6.2	6–17	12	7–12	10	1.05–1.37	1.21	6.0–7.2	6.6	12–21	17	24–31	28	1.19–1.23	1.21	6.0–7.4	6.7
Yixing	13	27	1.17	6.0	3–30	17	10–53	32	1.11–1.58	1.35	4.4–8.5	6.5	21–33	27	25–31	28	1.13–1.23	1.18	6.0–6.3	6.1
Dantu	7	36	1.25	6.6	2–19	11	12–49	31	1.07–1.39	1.23	5.8–8.0	6.9	12–21	17	24–31	28	1.19–1.23	1.21	6.0–7.4	6.7
Jurong	10	30	1.23	5.4	6–13	10	15–38	27	1.10–1.29	1.20	5.1–7.4	6.3	12–21	17	24–31	28	1.19–1.23	1.21	6.0–7.4	6.7
Danyang	12	29	1.24	6.7	8–16	12	16–53	35	1.07–1.36	1.22	5.8–7.8	6.8	12–21	17	24–31	28	1.19–1.23	1.21	6.0–7.4	6.7
Jiaxing	19	34	1.19	6.5	10–26	18	20–56	38	0.98–1.34	1.16	5.8–7.6	6.7	23–33	28	32–56	44	1.14–1.14	1.14	6.2–6.9	6.6
Jiashan	21	39	1.23	6.2	15–27	21	22–44	33	1.03–1.34	1.19	5.7–7.0	6.4	23–33	28	32–56	44	1.14–1.14	1.14	6.2–6.9	6.6
Pinghu	15	35	1.10	6.6	9–24	17	22–43	33	0.92–1.48	1.20	6.3–7.2	6.8	11–31	21	18–48	33	1.12–1.27	1.20	5.5–7.4	6.5
Haiyan	17	40	1.17	6.7	7–25	16	22–52	37	0.92–1.51	1.22	5.7–7.3	6.5	11–31	21	18–48	33	1.12–1.27	1.20	5.5–7.4	6.5
Haining	14	36	1.19	6.6	7–24	16	19–52	36	0.92–1.51	1.22	6.0–7.5	6.8	11–31	21	18–48	33	1.12–1.27	1.20	5.5–7.4	6.5
Tongxiang	14	30	1.05	6.5	7–29	18	20–52	36	0.91–1.34	1.13	6.0–7.4	6.7	14–33	24	25–34	30	1.12–1.13	1.13	6.3–6.6	6.5
Huzhou	23	30	1.10	6.2	13–37	25	7–42	25	0.99–1.37	1.18	5.6–6.7	6.2	14–33	24	25–34	30	1.12–1.13	1.13	6.3–6.6	6.5
Changxing	17	31	1.14	5.8	6–31	19	9–47	28	0.84–1.53	1.19	3.6–7.1	5.4	14–33	24	25–34	30	1.12–1.13	1.13	6.3–6.6	6.5
Anji	18	22	1.16	6.0	12–34	23	12–42	27	0.84–1.44	1.14	5.4–6.7	6.1	14–33	24	24–35	30	1.12–1.13	1.13	6.3–6.6	6.5
Deqing	19	32	1.12	6.3	7–26	17	18–38	28	0.87–1.53	1.20	5.2–7.2	6.2	14–33	24	24–35	30	1.12–1.13	1.13	6.3–6.6	6.5
Yuhang	15	5	1.16	6.6	9–21	15	16–48	32	0.95–1.34	1.15	5.9–7.3	6.6	11–31	21	18–48	33	1.12–1.27	1.20	5.5–7.4	6.5
Linan	22	22	1.09	6.2	18–27	23	8–29	19	0.91–1.14	1.03	5.5–7.8	6.7	11–31	21	18–48	33	1.12–1.27	1.20	5.5–7.4	6.5
Minhang	13	26	1.18	7.6	10–18	14	17–46	32	1.11–1.30	1.21	6.4–8.0	7.2	11–31	21	18–48	33	1.12–1.27	1.20	5.5–7.4	6.5
Jiading	13	28	1.10	7.6	9–20	15	13–44	29	0.94–1.24	1.09	6.5–8.1	7.3	11–31	21	18–48	33	1.12–1.27	1.20	5.5–7.4	6.5
Chuangsha	12	29	1.15	7.6	9–20	15	17–36	27	1.06–1.33	1.20	7.3–8.0	7.7	11–31	21	18–48	33	1.12–1.27	1.20	5.5–7.4	6.5
Nanhui	16	31	1.18	7.4	13–22	18	8–35	22	1.11–1.21	1.16	6.5–8.1	7.3	11–31	21	18–48	33	1.12–1.27	1.20	5.5–7.4	6.5
Qingpu	21	27	1.15	7.1	7–33	20	11–36	24	0.94–1.53	1.24	5.6–8.3	7.0	23–33	28	32–56	44	1.14–1.14	1.14	6.2–6.9	6.6
Songjiang	23	26	1.20	6.8	10–33	22	8–37	23	1.03–1.47	1.25	5.6–8.1	6.9	23–33	28	32–56	44	1.14–1.14	1.14	6.2–6.9	6.6
Jinshan	20	29	1.22	7.0	11–37	24	18–36	27	1.11–1.47	1.29	4.6–8.3	6.5	11–31	21	18–48	33	1.12–1.27	1.20	5.5–7.4	6.5
Fengxian	15	25	1.20	7.4	12–18	15	19–39	29	1.11–1.49	1.30	6.9–8.1	7.5	11–31	21	18–48	33	1.12–1.27	1.20	5.5–7.4	6.5
Baoshan	11	23	1.21	7.9	9–19	14	8–44	26	1.11–1.28	1.20	7.2–8.2	7.7	11–31	21	18–48	33	1.12–1.27	1.20	5.5–7.4	6.5
Chongming	10	17	1.11	8.1	9–13	11	15–29	22	1.11–1.21	1.12	7.8–8.1	8.0	12–16	14	24–39	31	1.17–1.27	1.22	7.3–7.4	7.5
Tai-Lake region	**15**	**26**	**1.18**	**6.7**	**2–38**	**16**	**4–58**	**27**	**0.84–1.54**	**1.23**	**3.4–8.3**	**6.7**	**11–33**	**22**	**18–56**	**32**	**1.12–1.27**	**1.18**	**5.5–7.4**	**6.5**

WA = Weighted average of soil properties by the area of each polygon; SOC = Initial SOC content (g kg^{-1}); Clay = Clay content (%); BD = Bulk Density (g cm^{-3}); Range = Range of maximum and minimum soil properties; Ave = Average of maximum and minimum soil properties.

As Table 3 illustrated, initial SOC content was the most sensitive parameter controlling SOC change among all soil factors in P5 and C5 [20,22]. The average initial SOC value of P5 and C5 were 15 g kg^{-1} and 16 g kg^{-1} for the entire Tai-Lake region, respectively. Furthermore, the average initial SOC content sourced from P5 for 24 counties was lower than that from C5, while the average clay content sourced from P5 for 24 counties was also higher than that from C5. Many previous studies showed that soils with lower initial organic carbon and higher clay content tended to sequester C [20,22,35]. The high SOC sequestration

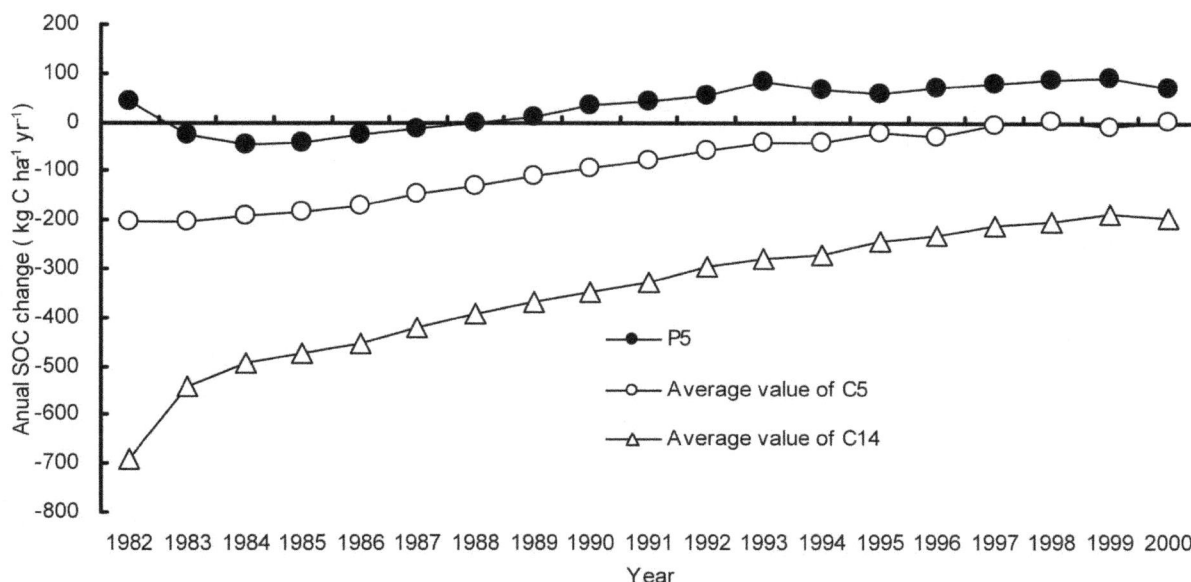

Figure 6. Temporal distribution of average annual SOC change modeled with P5, C5 and C14 from 1982 to 2000 in the Tai-Lake region, China.

rate (34 kg C ha^{-1} yr^{-1}) was thus associated with P5 (**Fig. 7a**). Conversely, the high SOC losses rate (-91 kg C ha^{-1} yr^{-1}) was associated with C5 (**Fig. 7b**). The SOC losses rate (-349 kg C ha^{-1} yr^{-1}) in C14 was the highest in the three resolution databases (**Fig. 7c**). Table 3 demonstrates that pH and initial SOC content are the most sensitive parameters controlling SOC change among all soil factors in C14. The average initial SOC value (22 g kg^{-1}) of C14 was significantly higher than that of P5 (15 g kg^{-1}) and C5 (16 g kg^{-1}) for the entire Tai-Lake region. In addition, the average pH value of C14 for 34 counties was close to neutral (6.5–7.5), and the average initial SOC contents of C14 for 28 counties were higher than 20 g kg^{-1}. Some studies showed that soils with neutral pH value and higher organic carbon content were favorable for CO_2 production by providing more substrates and better living environment for microbes [22,56].

The comparison illustrates that using different basic simulation units and soil data sources will produce different conclusions as to C sequestration or C liberation in the same study area. The implication is that more precise soil data and high resolution simulation units were necessary for better simulating regional scale SOC dynamics. The simulation outcome can be attributed to how the databases represent soil types and spatial heterogeneity, which is more precisely done with larger scale soil data and high resolution simulation units (e.g., 1:50,000 soil database).

Distribution of the average annual-, total SOC change modeled with the three resolution databases in different counties

The differentiation of the average annual-, total SOC change in P5, C5 and C14 was also shown at the county scale in the Tai-Lake region (**Table 5 and Fig. 7**). In the modeled domain, there were 26 counties that gained SOC and 11 counties that lost SOC from 1982 to 2000 in P5. The highest SOC sequestration rate of P5 were in Dantu, Jurong, Jiading and Baoshan counties which was higher than 200 kg C ha^{-1} yr^{-1}, due to the low initial SOC content (7.1 g kg^{-1}, 9.5 g kg^{-1}, 13 g kg^{-1} and 11 g kg^{-1}, respectively). In addition, the clay content of P5 in Dantu and

Jurong counties were 36% and 30%, respectively. High clay content is associated with high SOC sequestration [22,57,58]. By contrast, the greatest SOC loss rate of P5 in the Huzhou, Songjiang, Linan and Wuxian county was more than 170 kg C ha^{-1} yr^{-1}, due to the high initial SOC content (23 g kg^{-1}, 23 g kg^{-1}, 22 g kg^{-1} and 24 g kg^{-1}, respectively). Moreover, the clay content of P5 in Linan and Songjiang counties were only 22% and 26%, respectively. Low clay content is linked to high CO_2 emissions [22].

However, under the same agricultural practice, there were only 14 counties that gained SOC and 23 counties that lost SOC from 1982 to 2000 in C5. The highest SOC sequestration rate of C5 were in Dantu, Jurong, Jintan, Chongming, and Baoshan counties which was higher than 150 kg C ha^{-1} yr^{-1}. The main reason was that the initial SOC content of C5 in Dantu, Jurong, Jintan, Chongming, and Baoshan counties were 11 g kg^{-1}, 10 g kg^{-1}, 11 g kg^{-1}, 11 g kg^{-1} and 14 g kg^{-1}, respectively; the other was that the average clay content of C5 in Dantu, Jurong, and Baoshan counties ranged from 26% to 31%. Some studies showed that low initial SOC value and high clay content were linked to low CO_2 emissions [22,57,58]. In contrast, the greatest SOC loss rate of C5 in Jinshan, Changshu, Huzhou, Anji, and Kunshan county were more than 400 kg C ha^{-1} yr^{-1}, which possessed high initial SOC and low bulk density [22,35]. Compared with the P5 resolution database, the average annual-, total SOC change modeled with C5 for 28 counties was lower than that from P5. With the total SOC change as modeled with P5 inputs as the baseline, the relative deviations of counties in Jiangyin, Zhangjiagang and Kunshan were relatively high (>1000%). The relative deviations ranged from 50% to 250% in most counties. Only fifteen counties (Wuxian, Wujin, Jintan, Liyang, Dantu, Jurong, Huzhou, Yuhang, Linan, Minhang, Jiading, Chuangsha, Songjiang, Baoshan, and Chongming) had relatively low value of relative deviation (<100%). The SOC changes for the two resolution databases are almost in agreement with the soil feature across the 37 simulated counties (**Table 4 and Table 5**). The average initial SOC content sourced from C5 for 24 counties was higher than

Table 5. Distribution of the average annual SOC change (kg C ha^{-1} yr^{-1}) and the total SOC change (Gg C) in different counties of the Tai-Lake region, China modeled with P5, C5 and C14 from 1982 to 2000.

County	Area 10⁴ha	P5 ASC WA	P5 TSC	C5 ASC Max	C5 ASC Min	C5 ASC AVE	C5 TSC Max	C5 TSC Min	C5 TSC AVE	C14 ASC Max	C14 ASC Min	C14 ASC AVE	C14 TSC Max	C14 TSC Min	C14 TSC AVE
Zhangjiagang	2.54	2	1	138	−50	44	66	−24	21	57	−418	−181	27	−202	−87
Changshu	7.55	−64	−92	246	−1438	−596	353	−2063	−855	92	−404	−156	132	−580	−224
Taicnang	6.14	43	50	241	−360	−60	281	−420	−70	192	−993	−400	224	−1158	−467
Kunshan	7.57	38	55	252	−1057	−402	362	−1520	−579	−75	−838	−456	−108	−1205	−656
Wuxian	14.78	−172	−483	356	−876	−260	998	−2458	−730	72	−403	−166	201	−1132	−466
Wujiang	9.79	53	99	586	−717	−65	1091	−1333	−121	−59	−792	−425	−110	−1472	−791
Wuxi	9.77	49	91	478	−196	141	888	−364	262	−256	−958	−607	−476	−1778	−1127
Jiangyin	8.69	2	4	313	−101	106	517	−167	175	−262	−951	−606	−432	−1569	−1000
Wujin	14.85	91	256	248	−57	96	701	−160	270	−274	−969	−621	−774	−2734	−1754
Jintan	7.10	146	197	245	81	163	331	110	220	136	−364	−114	184	−492	−154
Liyang	10.84	177	365	302	−115	94	623	−237	193	127	−386	−129	263	−796	−267
Yixing	10.34	147	289	514	−1011	−248	1011	−1987	−488	−262	−950	−606	−514	−1867	−1191
Dantu	5.07	371	357	615	−276	170	592	−266	163	141	−367	−113	136	−353	−109
Jurong	8.03	238	363	403	−8	198	614	−12	301	125	−374	−124	191	−570	−190
Danyang	9.58	47	85	416	−152	132	758	−276	241	130	−364	−117	237	−663	−213
Jiaxing	6.57	−124	−155	412	−523	−55	514	−652	−69	−11	−752	−381	−13	−938	−476
Jiashan	4.13	−103	−81	110	−631	−260	87	−494	−204	−48	−786	−417	−38	−616	−327
Pinghu	4.81	127	116	349	−466	−58	319	−425	−53	274	−923	−324	250	−843	−296
Haiyan	2.74	41	21	489	−549	−30	254	−285	−16	279	−943	−332	145	−490	−173
Haining	3.92	151	113	483	−556	−36	359	−413	−27	274	−951	−338	204	−708	−252
Tongxiang	4.42	128	107	504	−634	−65	424	−533	−55	122	−846	−362	102	−711	−304
Huzhou	6.02	−309	−353	163	−1184	−511	186	−1354	−584	92	−915	−411	106	−1046	−470
Changxing	5.62	42	45	422	−863	−221	450	−922	−236	71	−901	−415	76	−962	−443
Anji	4.18	−66	−52	178	−1025	−423	141	−814	−336	58	−873	−408	46	−694	−324
Deqing	3.11	17	10	379	−684	−152	224	−404	−90	77	−930	−426	46	−550	−252
Yuhang	5.27	−91	−92	344	−371	−13	345	−372	−13	120	−1061	−456	120	−1034	−457
Linan	3.06	−220	−128	7	−381	−187	4	−222	−109	218	−754	−268	127	−439	−156
Minhang	3.49	62	41	294	−239	28	195	−158	18	257	−974	−359	171	−645	−238
Jiading	4.29	238	194	387	−197	95	315	−161	77	281	−940	−329	229	−766	−268
Chuangsha	3.71	188	133	339	−295	22	239	−208	15	286	−925	−319	202	−653	−225
Nanhui	4.11	152	119	189	−279	−45	148	−208	−35	287	−939	−326	224	−734	−255
Qingpu	5.68	−52	−56	432	−1042	−305	467	−1125	−329	2	−756	−377	3	−816	−407
Songjiang	5.90	−287	−322	249	−1010	−381	278	−1131	−426	−60	−816	−438	−67	−914	−491
Jinshan	5.63	−77	−83	213	−1481	−634	228	−1585	−679	275	−945	−335	294	−1011	−359
Fengxian	5.87	9	10	171	−270	−49	191	−301	−55	280	−944	−332	321	−1053	−370
Baoshan	3.13	200	119	393	−92	151	234	−55	90	311	−876	−283	185	−522	−168
Chongming	3.73	195	138	283	47	165	201	34	117	165	−108	28	117	−77	20
Tai-Lake region	**232**	**34**	**1483**	**340**	**−521**	**−91**	**14987**	**−22977**	**−3994**	**46**	**−744**	**−349**	**2022**	**−32790**	**−15380**

WA = Weighted average of annual mean SOC change (kg C ha^{-1} yr^{-1}) by the area of each polygon; ASC = Average annual SOC change (kg C ha^{-1} yr^{-1}); TSC = Total SOC change (Gg C); Max = Maximum value of ASC (or TSC); Min = Minimum value of ASC (or TSC); Ave = Average of maximum and minimum ASC (or TSC).

that from P5, and the average clay content sourced from C5 for 24 counties was also lower than that from P5. Some research showed that high initial SOC content and low clay content is favorable for C losses [22,57,58].

As can be seen from the Table 5, a big number of counties where the average annual-, total SOC change modeled with the C14 and P5 differed greatly. There was only one county that gained SOC from 1982 to 2000, while other 36 counties lost SOC in C14. The SOC losses of C14 ranged from 360 kg C ha^{-1} yr^{-1}

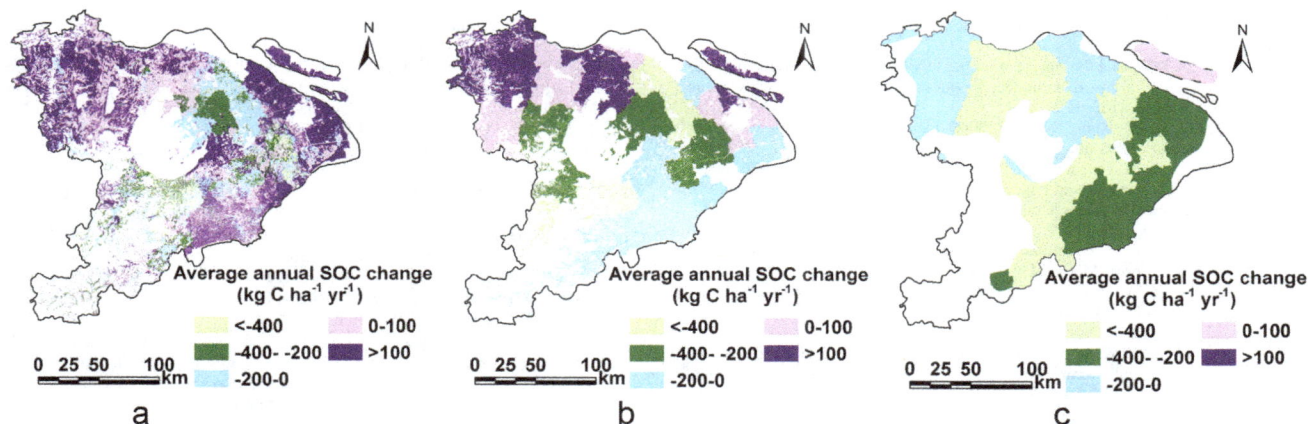

Figure 7. Spatial distribution of average annual SOC change modeled with P5, C5 and C14 in the Tai-Lake region, China (a: P5, b: c5, and c:C14).

to 620 kg C ha^{-1} yr^{-1} in most counties. With the total SOC change as modeled with P5 inputs as the baseline, the relative deviations of counties in Zhangjiagang, Taicang, Kunshan, Wuxi, Jiangyin, Changxing, Deqing, and Fengxian were more than 1000%. Only five counties (Wuxian, Huzhou, Linan, Songjiang, and Chongming) in C14 had relatively low deviation (<100%). The main reasons were that the average pH value of C14 in most counties ranged from 6.5 to 7.5, which were closer to neutral than that from C5 and C14. Moreover, the average initial SOC contents of C14 in most counties were higher than 20 g kg^{-1}, which was also much higher than that from P5 or C5. Therefore, high SOC losses occurred in C14.

The modeled data at county scale in three simulation methods indicated the underestimation with the county-based database was related to its soil data source and simulation unit resolution, especially the coarse soil maps (1:14,000,000) that missed relatively small soil patches containing low or high soil properties (i.e., initial SOC content, pH, and clay content) which were sensitive to SOC change. This would also explain why the precision of soil database plays an important role in elevating the accuracy of modeled SOC change at regional scale.

Conclusions

Using different spatial information, process-based models integrated with GIS databases can play an important role in describing C biogeochemical cycles, such as targeting mitigation efforts to the most beneficial regions. However, SOC models have often been applied to regions with high heterogeneity but limited spatially differentiated soil information and simulated unit resolution.

Simulation results indicate that total SOC change from 1982 to 2000 in the top layer (0–30 cm) of the 2.3 M ha of paddy rice

fields in the Tai-Lake region was +1.48 Tg C for P5. However, discrepancies in the results existed among the three databases, because different soil data and basic simulation units were used. The total SOC changes in the Tai-Lake region were -3.99 Tg C and -15.38 Tg C for C5 (or C14), respectively, from 1982 to 2000. With the total SOC change as modeled with P5 inputs as the baseline, the relative deviation of C5 was lower than C14 due to the more precise soil data. In contrast, the relative deviation of C14 was higher than other databases due to using coarser soil data and low-resolution simulation units. In addition, with the same basic simulation unit, average annual-, total SOC change between C5 and C14 for the Tai-Lake region also had a large discrepancy due to the use of different soil data. The comparison demonstrated that the most sensitive factors (e.g., initial SOC content and pH) for modeling SOC dynamics should be given a high priority during the input data acquisition as they contribute disproportionately to the uncertainties produced during the upscaling process [20]. The results also indicate that improving the performance of the biogeochemical DNDC model is essential in creating accurate models of the soil carbon cycle.

Acknowledgments

We gratefully acknowledge support for this research from the National Natural Science Foundation of China (No. 41001126), and the National Basic Research Program of China (973 Program) (2010CB950702). Sincere thank is also given to Professor Changsheng Li (University of New Hampshire, USA) for his useful advice on DNDC model.

Author Contributions

Conceived and designed the experiments: DSY XZS SHX. Performed the experiments: LMZ SXX. Analyzed the data: LMZ SXX. Contributed reagents/materials/analysis tools: SXX YCZ. Wrote the paper: LMZ.

References

1. Eswaran H, Berg EVD, Reich P (1993) Organic carbon in soil of the world. Soil Science Society of America Journal 57:192–194.
2. Lal R (2006) World soils and greenhouse effect: An overview, in soils and global change. Encyclopedia of Soil Science. doi:10.1081/E-ESS-120042696.
3. Shi XZ, Yang RW, Weindorf DC, Wang HJ, Yu DS, et al. (2010) Simulation of organic carbon dynamics at regional scale for paddy soils in China. Climatic Change 102:579–593.
4. Li QK (1992) Paddy soil of China. Beijing: Science Press. 514 p.
5. Gong ZT (1999) Chinese soil taxonomic classification. Beijing: Science Press. 5–215 p.

6. Liu QH, Shi XZ, Weindorf DC, Yu DS, Zhao YC, et al. (2006) Soil organic carbon storage of paddy soils in China using the 1:1,000,000 soil database and their implications for C sequestration. Global Biogeochemical Cycles 20:GB3024. doi:10.1029/2006GB002731.
7. Jenkinson DS, Rayner JH (1977) The turnover of soil organic matter in some of Rothamsted classical experiments. Soil Science 125:298–305.
8. Smith P, Smith JU, Powlson DS, Arah JRM, Chertov OG, et al. (1997) A comparison of the performance of nine soil organic matter models using datasets from seven long term experiments. Geoderma 81:153–225.

9. Ardö J, Olsson L (2003) Assessment of soil organic carbon in semi-arid Sudan using GIS and the CENTURY model. Journal of Arid Environments 54:633–651.

10. Shirato Y (2005) Testing the suitability of the DNDC model for simulating long-term soil organic carbon dynamics in Japanese paddy soils. Soil Science and Plant Nutrition 51(2):183–192.

11. Tang HJ, Qiu JJ, van Ranst E, Li CS (2006) Estimations of soil organic carbon storage in cropland of China based on DNDC model. Geoderma 134:200–206.

12. Cerri CEP, Easter M, Paustian K, Killian K, Coleman K, et al. (2007) Predicted soil organic carbon stocks and changed in the Brazilian Amazon between 2000 and 2030. Agriculture, Ecosystems and Environment 122:58–72.

13. Huang Y, Yu YQ, Zhang W, Sun WJ, Liu SL, et al. (2009) Agro-C: A biogeophysical model for simulating the carbon budget of agroecosystems. Agricultural and Forest Meteorology 149:106–129.

14. Tang HJ, Qiu JJ, Wang LG, Li H, Li CS, et al. (2010) Modeling soil organic carbon storage and its dynamics in croplands of China. Agricultural Sciences in China 9(5):704–712.

15. Li CS, Frolking S, Frolking TA (1992) A model of nitrous oxide evolution from soil driven by rainfall events: I. Model structure and sensitivity. Journal of Geophysical Research 97:9759–9776.

16. Li CS, Frolking S, Frolking TA (1992) A model of nitrous oxide evolution from soil driven by rainfall events:II. Model applications. Journal of Geophysical Research 97:9777–9783.

17. Tonitto C, David MB, Li CS, Drinkwater LE (2007) Application of the DNDC model to tile-drained Illinois agroecosystems: Model comparison of conventional and diversified rotations. Nutrient Cycling in Agroecosystems 78 (1):65–81.

18. Pathak H, Li CS, Wassmann H (2005) Greenhouse gas emissions from Indian rice fields: calibration and upscaling using the DNDC model. Biogeoscience 2:113–123.

19. Neufeldt H, Schäfe M, Angenendt E, Li CS, Kaltschmitt M, et al. (2006) Disaggregated greenhouse gas emission inventories from agriculture via a coupled economic-ecosystem model. Agriculture, Ecosystems and Environment 112:233–240.

20. Zhang F, Li CS, Wang Z, Wu HB (2006) Modeling impacts of management alternatives on soil carbon storage of farmland in Northwest China. Biogeosciences 3:451–466.

21. Wang LG, Qiu JJ, Tang HJ, Li CS, van Ranst E (2008) Modelling soil organic carbon dynamics in the major agricultural regions of China. Geoderma 147:47–55.

22. Li CS, Mosier A, Wassmann R, Cai ZC, Zheng XH, et al. (2004) Modeling greenhouse gas emissions from rice-based production systems: Sensitivity and upscaling. Global Biogeochemical Cycles 18:GB1043. doi:10.1029/2003 GB002045.

23. Cai ZC, Sawamoto T, Li CS, Kang GD, Boonjawat J, et al. (2003) Field validation of the DNDC model for greenhouse gas emissions in East Asian cropping systems. Global Biogeochemical Cycles 17 (4): GB1107, doi:10.1029/2003 GB002046.

24. Xu Q, Lu YC, Liu YC, Zhu HG (1980) Paddy soil of Tai-Lake region in China. Shanghai: Science Press.

25. Shi XZ, Yu DS, Warner ED, Sun WX, Petersen GW, et al. (2006) Cross-reference system for translating between genetic soil classification of China and Soil Taxonomy. Soil Science Society of America Journal 70:78–83.

26. Soil Survey Staff in USDA (2010) Keys to Soil Taxonomy (11th Edition). Washington: USDA-Natural Resources Conservation Service.

27. Li CS, Frolking S, Harriss R (1994) Modeling carbon biogeochemistry in agricultural soils. Global Biogeochemical Cycles 8 (3):237–254.

28. Li CS, Narayanan V, Harriss R (1996) Model estimates of nitrous oxide emissions from agricultural lands in the United States. Global Biogeochemical Cycles 10 (2):297–306.

29. Li CS, Qiu JJ, Frolking S, Xiao XM, Salas W, et al. (2002) Reduced methane emissions from large-scale changes in water management in China's rice paddies during 1980–2000. Geophysical Research Letters 29 (20):1972, doi:10.1029/2002GL01 5370.

30. Li CS (2007) Quantifying greenhouse gas emissions from soils: Scientific basis and modeling approach. Soil Science and Plant Nutrition 53 (4):344-352.

31. Qiu JJ, Wang LG, Tang HJ, Li H, Li CS (2005) Studies on the situation of soil organic carbon storage in croplands in northeast of China. Agricultural Sciences in China 37 (8):1166–1171.

32. Zhao YC, Shi XZ, Weindorf DC, Yu DS, Sun WX, et al. (2006) Map scale effects on soil organic carbon stock estimation in north China. Soil Science Society of America Journal 70:1377–1386.

33. Institute of Soil Science (1986) The soil atlas of China. Beijing: Institute of Soil Science, Academia Sinica, Cartographic Publishing House.

34. National Soil Survey Office of China (1993–1997) Soils in China (Vol. 1–6). Beijing: Agricultural Publishing House.

35. Zhang LM, Yu DS, Shi XZ, Xu SX, Wang SH, et al. (2012) Simulation soil organic carbon change in China's Tai-Lake paddy soils. Soil and Tillage Research 121:1–9.

36. Zhang LM, Yu DS, Shi XZ, Xu SX, Weindorf DC, et al. (2009) Quantifying methane emissions from rice fields in the Taihu region, China by coupling a detailed soil database with biogeochemical model. Biogeosciences 6:739–749.

37. Xu SX, Zhao YC, Shi XZ, Yu DS, Li CS, et al. (2013) Map scale effects of soil databases on modeling organic carbon dynamics for paddy soils of China. Catena 104:67–76.

38. Yu DS, Yang H, Shi XZ, Warner ED, Zhang LM, et al. (2011) Effects of soil spatial resolution on quantifying CH_4 and N_2O emissions from rice fields in the Tai Lake region of China by DNDC model. Global Biogeochemical Cycles 25:GB2004. doi:10.1029/2010GB003825.

39. Yu DS, Zhang LM, Shi XZ, Warner ED, Zhang ZQ, et al. (2013) Soil assessment unit scale affects quantifying CH_4 emissions from rice fields. Soil Science Society of America Journal 77:664–672.

40. Li CS (2007) Quantifying soil organic carbon sequestration potential with modeling approach. In: Tang HJ, Van Ranst E, Qiu JJ (Eds.) Simulation of soil organic carbon storage and changes in agricultural cropland in China and its impact on food security. Beijing: China Meteorological Press. 1–14 p.

41. Gou J, Zheng XH, Wang MX, Li CS (1999) Modeling N_2O emissions from agriculture fields in Southeast China. Advances in Atmospheric Sciences 16 (4):581–592.

42. China Meteorological Administration (2011) China meteorological data daily value. China Meteorological Data Sharing Service System, Beijing, China. http://cdc.cma.gov.cn/index.jsp.

43. Lu RK, Shi TJ (1982) Agricultural chemical manual. Beijing:China Science Press. 142 p.

44. Whitmore AP, Klein-Gunnewiek H, Crocker GJ, Klir J, Körschens M, et al. (1997) Simulating trends in soil organic carbon in long-term experiments using the Verberne/MOTOR model. Geoderma 81:137–151.

45. Admassu Y, Shakoor A, Wells N (2012) Evaluating selected factors affecting the depth of undercutting in rocks subject to differential weathering. Engineering Geology 124:1–11.

46. Leech NL, Barret KKC, Morgan G (2008) SPSS for intermediate statistics. New York: Lawrence Erlbaum Associates. 270 p.

47. Dielman TE (2001) Applied regression analysis for business and economics. California: Duxbury Thomson Learning. 647 p.

48. Arnold RW (1995) Role of soil survey in obtaining a global carbon budget. In: Lal R, Kimble J, Levine E, Stewart BA (Eds.) Advances in Soil Science: Soils and Global Change. Boca Raton, FL: CRC Press. 57–263 p.

49. Zhong B, Xu YJ (2011) Scale effects of geographical soil datasets on soil carbon estimation in Louisiana, USA: a comparison of STATSGO and SSURGO. Pedosphere 21 (4):491–501.

50. Liu YH, Yu ZR, Chen J, Zhang FR, Reiner D, et al. (2006) Changes of soil organic carbon in an intensively cultivated agricultural region: A denitrification-decomposition (DNDC) modelling approach. Science of the Total Environment 372:203–214.

51. Liu Q, Sun B, Jie XL, Li ZP (2009) The spatial-temporal dynamic change and simulation of county-scale paddy soil organic carbon red soil hilly region. Acta pedologica sinica 46 (6):1059–1067.

52. Xu SX, Shi XZ, Zhao YC, Yu DS, Wang SH, et al. (2012) Spatially explicit simulation of soil organic carbon dynamics in China's paddy soils. Catena 92:113–121.

53. Studdert GA, Monterubbianesi GM, Domínguez GF (2011) Use of RothC to simulate changes of organic carbon stock in the arable layer of a Mollisol of the southeastern Pampas under continuous cropping. Soil and Tillage Research 117:191–200.

54. Yu YQ, Huang Y, Zhang W (2012) Modeling soil organic carbon change in croplands of China, 1980-2009. Global and Planetary Change 82–83:115–128.

55. Wu TY, Schoenau JJ, Li FM, Qian PY, Malhi SS (2004) Influence of cultivation and fertilization on total organic carbon and carbon fractions in soils from the Loess Plateau of China. Soil and Tillage Research 77:59–68.

56. Pacey JG, DeGier JP (1986) The factors influencing landfill gas production. In: Energy from landfill gas. Proceeding of a conference jointly sponsored by the United Kingdom Department of Energy and the United States Department of Energy (October 1986). 51–59 p.

57. Burke IC, Lauenroth WK, Conflin DP (1995) Soil organic matter recovery in semiarid grassland: implications for the conservation reserve program. Ecological Monographs 5:793–801.

58. Kay BD (1998) Soil structure and organic carbon: a review. In: Lal R, Kimble JM, Follett RF. Soil Processes and the carbon cycle. Boca Raton, FL: CRC Press.169–198 p.

11

Maize Yield Response to Water Supply and Fertilizer Input in a Semi-Arid Environment of Northeast China

Guanghua Yin[1], Jian Gu[1], Fasheng Zhang[1]*, Liang Hao[1,2], Peifei Cong[1,2], Zuoxin Liu[1]

1 Institute of Applied Ecology, Chinese Academy of Sciences, Shenyang, China, **2** University of Chinese Academy of Sciences, Beijing, China

Abstract

Maize grain yield varies highly with water availability as well as with fertilization and relevant agricultural management practices. With a 311-A optimized saturation design, field experiments were conducted between 2006 and 2009 to examine the yield response of spring maize (Zhengdan 958, *Zea mays* L) to irrigation (*I*), nitrogen fertilization (total nitrogen, urea-46% nitrogen,) and phosphorus fertilization (P_2O_5, calcium superphosphate-13% P_2O_5) in a semi-arid area environment of Northeast China. According to our estimated yield function, the results showed that *N* is the dominant factor in determining maize grain yield followed by *I*, while *P* plays a relatively minor role. The strength of interaction effects among *I*, *N* and *P* on maize grain yield follows the sequence *N+I* >*P+I*>*N+P*. Individually, the interaction effects of *N+I* and *N+P* on maize grain yield are positive, whereas that of *P+I* is negative. To achieve maximum grain yield (10506.0 kg·ha^{-1}) for spring maize in the study area, the optimum application rates of *I*, *N* and *P* are 930.4 m^3·ha^{-1}, 304.9 kg·ha^{-1} and 133.2 kg·ha^{-1} respectively that leads to a possible economic profit (*EP*) of 10548.4 CNY·ha^{-1} (CNY, Chinese Yuan). Alternately, to obtain the best *EP* (10827.3 CNY·ha^{-1}), the optimum application rates of *I*, *N* and *P* are 682.4 m^3·ha^{-1}, 241.0 kg·ha^{-1} and 111.7 kg·ha^{-1} respectively that produces a potential grain yield of 10289.5 kg·ha^{-1}.

Editor: Hany A. El-Shemy, Cairo University, Egypt

Funding: This work was funded by the Special Fund for Agro-scientific Research in the Public Interest (201303125-9) and National Major Technology Program (2011BAD16B12, 2012BAD09B02, 2013BAD05B07). The funders had no role in study design, data collection and analysis, decision to publish, or preparation of the manuscript.

Competing Interests: The authors have declared that no competing interests exist.

* E-mail: fasheng.zhang@yahoo.com

Introduction

Maize is the third most important grain crop after rice and wheat grown in China [1,2]. In order to ensure food security for its vast population, the Chinese government and its research institutions have made extensive efforts to improve maize grain production in North China since the 1950s [3−5].

Water scarcity and soil infertility are two critical factors limiting maize grain yield over most regions of North China [6−8]. Although irrigation and fertilization are widely applied to improve maize productivity [9,10], maize production in China has not been able to keep pace with grain demand [11,12]. At the same time, low water use efficiency aggravates water stress in North China [13−15] while excessive inputs of chemical fertilizer result in surplus nitrogen and phosphorus in soils that cause eutrophication of surface water as well as greenhouse gas emissions [16−20]. In modern agriculture, such consequences arise mainly from a limited understanding of how irrigation and fertilization affect maize production and a biased estimation of the yield function for identifying maize yield variation. In this context, there is a need to investigate the combined effect of water supply and fertilizer input on maize productivity in North China.

Many field studies have been conducted since the 1990s to examine main and interaction effects of irrigation and fertilization on maize productivity around the world, including North China [21−26]. The optimum coupling or combination of water supply and fertilizer inputs has been derived to seek maximum maize grain yield or to achieve maximum water and fertilizer use

efficiency [27−30]. However, these studies mostly focused on the individual influences of irrigation (*I*), nitrogen application (*N*), phosphorus application (*P*) and/or their binary combination effects on maize productivity. A holistic understanding of the ternary combination effect of *I*, *N* and *P* on maize productivity is still developing. The economic efficiency of growing maize is another important factor influencing maize grain production [31−34]. Farmers will grow more maize if the economic profits of growing maize are higher than for other crops. Profits associated with maize production, however, decrease with improper management practices as well as with increasing energy, material and human labor costs in the context of global climate change [35−39]. The declining profit rate dampens farmers' enthusiasm for growing maize and consequently impacts maize grain production [40,41]. Thus, it is important to improve maize productivity while taking into account the economic evaluation of growing maize.

The relationship between maize grain yield and management practices varies over time and space depending on the maize cultivars, climatic conditions and cropping systems. Knowledge obtained from studies in other regions may not be valid in any specific area of North China. Therefore, the objectives of this study were (1) to construct a yield function to examine the combination effect of *I*, *N* and *P* on maize productivity using field experimental data collected from 2006 to 2009 in a semi-arid environment of Northeast China and (2) to use the estimated yield function for further deriving optimum application rates of *I*, *N* and

P based on the criteria of maximum grain yield and best economic profit.

Materials and Methods

Site and Soil

The field study was conducted from 2006 to 2009 at the field experimental station of Liaoning Key Laboratory of Water-Saving Agriculture in Fuxin County of Northeast China (42°08′14″ N, 121°44′21″ E). This region is a warm temperate zone with a temperate continental monsoon climate. According to the Fuxin Weather Station, the average annual temperature is 7.2°C with an average of 2865.5 hrs of annual sunshine. It is a typical semi-arid area with average annual precipitation of 480 mm, over 60% of which occurs from June to August. The compensation of water resources depends mainly on precipitation of atmosphere. Annual precipitation and precipitation during the maize growing season of Fuxin County are shown in Figure 1.

The main agricultural soil in the region is cinnamon soil which develops through a combination of calcium carbonate leaching, illuviation and humification. It is characterized by a thin humus layer and a medium or thick solum. Its bulk density is 1.51 $g \cdot cm^{-3}$, pH (H_2O) is 7.5−8.5 and the average soil organic matter content is 10.2 $g \cdot kg^{-1}$. The average soil total nitrogen and available phosphorus concentrations are 6.1 $g \cdot kg^{-1}$ and 4.0 $g \cdot kg^{-1}$, respectively.

Experimental Design and Treatments

To reduce cost and size, the experiment in this study was implemented according to a 311-A optimized saturation design [42]. This system consisted of three factors at five levels. There were 11 treatments with 3 replicates each for a total of 33 experimental plots. Each experimental plot was 40 m^2 in size (10 m×4 m). The water supply and fertilizer inputs were standardized for comparability by applying a non-dimensional linear code substitution (Table 1). Rates of *N* and *P* in Table 1 were expressed in format of pure nitrogen and P_2O_5 that were supplemented by urea (46% total nitrogen) and calcium superphosphate (13% P_2O_5). One unit of *I*, *N* and *P* represented 225 $m^3 \cdot ha^{-1}$ of *I*, 112.5 $kg \cdot ha^{-1}$ of total nitrogen and 67.5 $kg \cdot ha^{-1}$ of P_2O_5, which indicated 225 $m^3 \cdot ha^{-1}$ of water supply, 244.6 $kg \cdot ha^{-1}$ of urea and 519.2 $kg \cdot ha^{-1}$ of calcium superphosphate, respectively. One third of the urea used in each treatment was applied at the sowing stage and the remaining amount at the early jointing stage. All calcium superphosphate was applied at the sowing stage in each treatment. Experimental plots were variously irrigated at the jointing stage. All necessary permits were obtained for the described field experiments. The land user and owner approved the field-work activities at each experiment plot. The field employed in this study is not protected in any way, and the study did not involve any endangered or protected species.

Data Collection and Analysis

The maize cultivar in this experiment was Zhengdan 958 (*Zea mays* L). In China, the planting area of Zhengdan 958 was 4.54 million ha in 2009, and the planting area of this variety is still the largest in 2012 [43]. Zhengdan 958 has outstanding yield performance. It can generate relatively stable yield under various environmental conditions and has good disease resistance. This variety can be planted in high density and the ideal planting density is 60,000 to 75,000 plants per ha. In this study, maize plant density was 60,000 plants per ha with 50 cm between rows. The maize was planted in late April and harvested in late September. At maize maturity, the outer two rows in each experimental plot were considered as edge effects and not harvested, while the remaining middle rows were hand-harvested for analysis of maize grain yield. The effective area of each experimental plot was approximately 20 m^2. The average fresh ear weight (G_1, kg) of each treatment was estimated and the average grain yield, Y ($kg \cdot ha^{-1}$), was computed as:

$$Y = k \times G_1/20 \times 10000$$

where *k* is the ratio of grain dry weight to fresh ear weight for each treatment. To estimate the values of *k*, ten medium-sized ears were sampled from each experimental plot and their average fresh ear weight (G_2, kg) and average fresh grain weight (G_3, kg) were measured for each treatment The grain of Zhengdan 958 dries slowly before harvest. Extra moisture in maize grain should be removed before estimating the average maize grain yield. The

Table 1. Experimental treatments of maize using a 311-A optimized saturation design during 2006–2009 in Fuxin County.

Treatments	Code level			Application rate		
	I[†]	*N*	*P*	*I*	*N*	*P*
	X_1[‡]	X_2	X_3	$m^3 \cdot ha^{-1}$	$kg \cdot ha^{-1}$	$kg \cdot ha^{-1}$
1	2	0	0	900	225	135
2	−2	0	0	0	225	135
3	1	−1.414	−1.414	675	66	39.54
4	1	1.414	−1.414	675	384	39.54
5	1	−1.414	1.414	675	66	230.46
6	1	1.414	1.414	675	384	230.46
7	−1	2	0	225	450	135
8	−1	−2	0	225	0	135
9	−1	0	2	225	225	270
10	−1	0	−2	225	225	0
11	0	0	0	450	225	135

[†]*I*, *N* and *P* are abbreviations for irrigation, nitrogen fertilization and phosphorus fertilization, respectively.
[‡]X_1, X_2 and X_3 are non-dimensional linear code for irrigation, nitrogen fertilization and phosphorus fertilization, respectively.

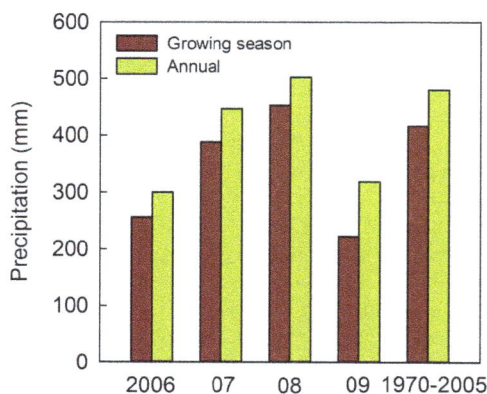

Figure 1. Annual precipitation and precipitation during maize growing season during 2006–2009 in Fuxin County.

average moisture content of fresh grain for each treatment ($A\%$) was therefore determined using a PM-8188 Grain Moisture Tester (Japan). Then k was calculated as:

$$k = G_3/G_2 \times (100 - A)/(100 - 18)$$

A quadratic regression orthogonal design was used to construct the yield function in this study. Regression analysis was conducted by R software and the results were presented in graphs by SigmaPlot 10.0.

Results and Discussion

Yield Function

The average maize grain yield of each treatment from 2006 to 2009 varied between 8468.2 kg·ha^{-1} and 10478.2 kg·ha^{-1}. The best-fitted yield function, which quantified the relationship between maize grain yield (Y_{INP}) and the code levels of I, N and P (i.e., X_1, X_2, X_3), was expressed as:

$$
\begin{aligned}
Y_{INP} = {}& 10130.63^{**} + 232.55X_1^{**} + 363.44X_2^{**} + 151.46X_3^{*} \\
& - 75.19X_1^2 - 436.54X_2^{2**} - 349.33X_3^{2**} \\
& + 120.94X_1X_2^{*} + 55.11X_2X_3 - 98.17X_1X_3
\end{aligned}
\tag{1}
$$

where the coefficient of determination (R^2) for the regression is 0.87 at the significance level of $P<0.01$ (F-test). The reliability of coefficients for X_1, X_2 and X_3 are tested by a t-test. The symbol "**" stands for the significance at the level of $P<0.01$ and the symbol "*" for the significance at the level of $P<0.05$. In addition, the coefficients of X_1^2, X_2X_3 and X_1X_3 were significant at $P<0.20$, $P<0.40$ and $P<0.10$. Considering the importance of I, N and P on maize growth, the parameters in equation (1) were all taken into account when estimating maize grain yield in this study.

The code levels of I, N and P were proportional to their actual rates in each treatment. Thus, equation (1) appropriately described the combination effects of water spply and fertilizer input on maize grain yield in the semi-arid area examined in this study. It accounted for 87% of the variation in maize grain yield. Because of its accuracy and explicitness, equation (1) was used to analyze the main and individual effects as well as the interaction effects of I, N and P on maize grain yield (Text S1).

The main effects of I, N and P on maize grain yield were evaluated by comparing their corresponding coefficients in equation (1). The positive coefficients for X_1, X_2 and X_3 suggested that I, N and P all had positive effects on maize grain yield. The largest value was observed for the coefficient of X_2, indicating that N was the dominant factor influencing maximum maize grain yield. Similarly, I was recognized as a secondary factor determining maize grain yield, whereas P was a relatively minor factor. These results were partially consistent with findings from other studies around the world. Nitrogen input has a large effect on maize grain yield because maize production is an extractive process, with removal of maize equating to removal of nitrogen from the soil [44,45]. The significant effect of water supply on maize grain yield in arid and semi-arid regions is ubiquitous and easily understood. It should be noted that the removal of phosphorus from agricultural soils by maize harvesting was also apparent in this study. It is well-established that supplemental phosphorus can significantly improve maize grain yield as well [7,9,17,22]. In our study, the decreased response of maize grain

yield to P compared to N and I may be due to the dry soil conditions which are characteristics of semi-arid areas in Northeast China.

To examine the individual effects of I, N and P on maize grain yield, each factor was selected as an independent variable with the other two factors fixed at 0 in equation (1) (Text S1). Then a subset of equations of yield function was derived respectively as:

$$Y_I = 10130.63 + 232.55X_1 - 75.19X_1^2 \tag{2}$$

$$Y_N = 10130.63 + 363.44X_2 - 436.54X_2^2 \tag{3}$$

$$Y_P = 10130.63 + 151.46X_3 - 349.33X_3^2 \tag{4}$$

The individual effects of I, N and P on maize grain yield derived from equations (2) ~ (4) are presented in Figure 2. The relationships between maize grain yield and I, N and P can be modeled using second-order parabolic equations. Apexes were observed when examining the trend of maize grain yield as the rates of I, N and P increased. Before the apex, maize grain yield increased as the rates of I, N and P increased. After the apex, maize grain yield decreased as the rates of I, N and P increased. These findings indicate that there must be optimum application rates of I, N and P when implementing agricultural management practices to improve maize grain yield.

The optimum application rates of I, N and P as individual influencing factors on maize grain yield were determined using marginal yield curves (Figure 3), which were the first-order differential analysis of equations (2) ~ (4). The marginal yield showed a monotonic descending trend with the increasing rates of I, N and P and had intersections with the x-axis. The intersection points revealed the optimum application rates of I, N and P. The code values at the intersecting points were +1.546, +0.416 and +0.217 for I, N and P, respectively. According to Table 1, they represented 797.9 m^3·ha^{-1} of I, 271.8 kg·ha^{-1} of N and 149.6 kg·ha^{-1} of P. In addition, levels of the majority of marginal yields before the intersections with the x-axis followed the sequence $N > P > I$. This suggests that the maize grain yield increased most with an increasing application rate of N compared to the increases of I and P. Maize grain yield was more sensitive to P than I.

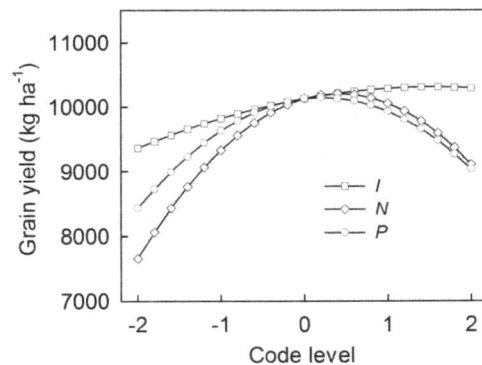

Figure 2. The individual effect of I, N and P on maize grain yield by fixing two factors at 0 level. I, N and P represent irrigation, nitrogen application and phosphorus application, respectively.

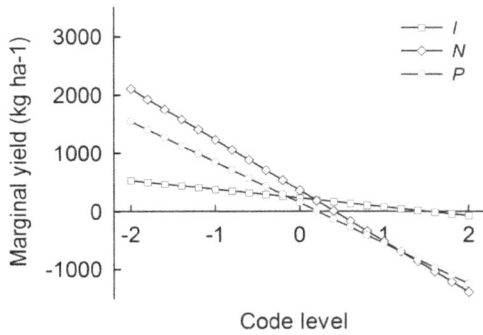

Figure 3. Marginal grain yield with respect to *I*, *N* and *P* through the first-order differential analysis of equations (2)~(4). *I*, *N* and *P* represent irrigation, nitrogen application and phosphorus application, respectively.

The interaction effects of *I*, *N* and *P* on maize grain yield were all less than their main effects in equation (1). By comparing the absolute values of the coefficients of X_1X_2, X_1X_3 and X_2X_3, the strength of interaction effects of *I*, *N* and *P* on maize grain yield followed the sequence *N+I* >*P+I*>*N+P*. The coefficient for X_1X_3 was positive, indicating that the interaction of *I* and *N* had a positive effect on maize grain yield. This finding is presented in Figure 4, which was constructed by fixing *P* at 0 level. In general, the maize grain yield increased as the application rates of *I* and *N* increased: an increase of any one input stimulates maize growth and creates a need for the other input. The optimum application rates of *I* and *N* for the highest grain yield (10,505.7 kg·ha^{-1}) were at code levels of +2.117 and +0.710, which corresponded to 926.3 m^3·ha^{-1} of *I* and 304.9 kg·ha^{-1} of *N*. The negative coefficient for X_1X_3 suggested that the interaction of *I* and *P* was antagonistic and had an inhibitory effect on maize grain yield. This relationship is presented in Figure 5, which was constructed by fixing *N* at 0 level. At code levels of +1.547 and −0.001, equal to 798.1 m^3·ha^{-1} of *I* and 134.9 kg·ha^{-1} of *P*, respectively, maize grain yield reached its highest value of 10,310.4 kg·ha^{-1}. Although the coefficient of X_2X_3 was positive, the interaction of *N* and *P* on maize grain yield was similar to that of *I* and *N*, but weaker and not significant at a relatively high level. This partly reveals the importance of water supply in maize growth in semi-arid areas. No contour plot of this interaction is presented here, although the interaction between *N* and *P* on maize grain yield was used to estimate the optimum schemes in next part.

Optimum Schemes

The findings presented above indicate the complex and sometimes antagonistic interactions of water supply and fertilizer input on maize productivity in semi-arid areas of Northeast China. The maize grain yield does not always increase as the additions of *I*, *N* and *P* improve. The optimum application rates of *I*, *N* and *P* were estimated to obtain the maximum grain yield and best economic profit.

To maximize maize grain yield, we set the first-order partial derivatives of equation (1) to zero (Text S1):

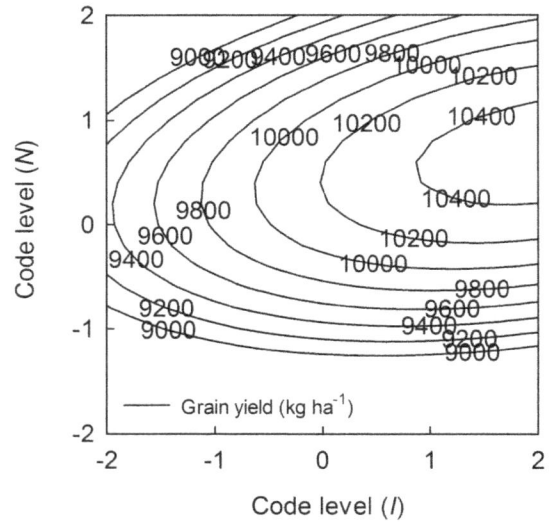

Figure 4. Interaction effects of *I* and *N* on maize grain yield by fixing *P* at 0 level. *I* and *N* represent irrigation and nitrogen application, respectively.

$$\begin{cases} \dfrac{\partial Y}{\partial X_1} = 232.55 - 150.38X_1 + 120.94X_2 - 98.17X_3 = 0 \\ \dfrac{\partial Y}{\partial X_2} = 363.44 + 120.94X_1 - 873.08X_2 + 55.11X_3 = 0 \\ \dfrac{\partial Y}{\partial X_3} = 151.46 - 98.17X_1 + 55.11X_2 - 698.66X_3 = 0 \end{cases}$$

By solving the above equations, the maximum maize grain yield, 10506.0 kg·ha^{-1}, can be obtained when the code levels of *I*, *N* and *P* were at +2.135, +0.710 and −0.027, respectively. According to Table 1, the optimum rates of *I*, *N* and *P* for the

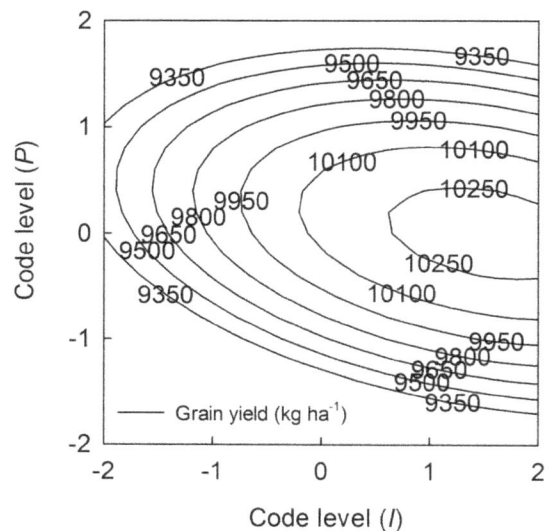

Figure 5. Interaction effects of *I* and *P* on maize grain yield by fixing *N* at 0 level. *I* and *P* represent irrigation and phosphorus application, respectively.

maximum maize grain yield were 930.4 $m^3 \cdot ha^{-1}$, 304.9 $kg \cdot ha^{-1}$ and 133.2 $kg \cdot ha^{-1}$, respectively.

Compared to the optimum application rates of I, N and P based on single-factor and binary combination effects, the optimum application rates based on the ternary combination effect have higher demand for I and N and less demand for P. This should be attributed to the effects of complex interactions of I, N and P on maize grain yield. In addition, maize grain yield showed different sensitivities to I, N and P in Figure 3. In the case of the ternary combination, these factors were considered holistically when estimating the maximum of the yield function. The optimum scheme indicated by the ternary combination enabled farmers to maximize maize grain yield by adjusting management practices without considering the cost of maize growing.

The increase of water supply and fertilizer input may increase maize grain yield as well as the economic profit. The economic profit of growing maize, EP, is decided by the relationship between outputs (i.e., the maize grain yield times the price of maize grain) and inputs (i.e., cost of irrigation, fertilization and seed). This relationship can be expressed as:

$$EP = p \cdot Y_{INP} - p_1 \cdot (225X_1 + 450) - p_2 \cdot (112.5X_2 + 225) \\ - p_3 \cdot (67.5X_3 + 135) - C_{Seed} \tag{5}$$

where p stands for the price of maize grain (1.3 $CNY \cdot kg^{-1}$, CNY, Chinese Yuan), and p_1, p_2 and p_3 for the prices of water (0.750 $CNY \cdot m^{-3}$, i.e., electricity and labor fees for irrigation), urea (3.913 $CNY \cdot kg^{-1}$) and calcium superphosphate (5.769 $CNY \cdot kg^{-1}$), respectively. Note that the prices of urea and calcium superphosphate per kilogram were calculated from the cost of them to provide a kilogram of pure total nitrogen and P_2O_5. C_{Seed} represents the cost of maize seeds for each treatment in this study (450 $CNY \cdot ha^{-1}$). To obtain the maximum economic profit, the highest value of equation (5) was solved using the method of first-order partial derivatives as well. The results show that the highest economic profit (10827.3 $CNY \cdot ha^{-1}$) was obtained when I, N and P were at code levels of 1.033, 0.142 and −0.345, respectively. This optimum scheme consisted of a ternary combination of I, N and P at rates of 682.4 $m^3 \cdot ha^{-1}$, 241.0 $kg \cdot ha^{-1}$ and 111.7 $kg \cdot ha^{-1}$, respectively.

To further discern the results, we compared the two optimum schemes based on the maximum grain yield and the best economic profit. The economic profit of the optimum scheme for maximum grain yield (10506.0 $kg \cdot ha^{-1}$) was 10548.4 $CNY \cdot ha^{-1}$ while the grain yield of the optimum scheme for the best economic profit (10827.3 $CNY \cdot ha^{-1}$) was 10289.5 $kg \cdot ha^{-1}$. Notwithstanding a little higher yield, the optimum scheme for maximum grain yield produced less economic profit than that of the optimum scheme for the best economic profit. In addition, the optimum scheme for maximum grain yield consumed more water and fertilizer than the optimum scheme for the best economic profit. This apparently will increase demand on water and mineral resources and result in leaching of surplus nitrogen and phosphorus into soil and water environments. In contrast, the optimum scheme for the best economic profit produced both relatively high grain yield and economic profit while consumed relatively less water and fertilizer. Considering the water resource and soil conditions, the optimum scheme for the best economic profit was therefore more acceptable and should be recommended in the study area.

Conclusions

Yield response of spring maize (Zhengdan 958, Zea $mays$ L) to water supply and fertilizer input in a semi-arid area of Northeast China was studied. In this field experiment, I, N and P as well as their interaction effects had significant influences on maize grain yield. The yield function derived by the ternary combination was able to describe these influences holistically. To obtain maximum maize grain yield (10506.0 $kg \cdot ha^{-1}$) in the semi-arid areas examined in this study, the optimum application rates of I, N and P based on the present findings were 930.4 $m^3 \cdot ha^{-1}$, 304.9 $kg \cdot ha^{-1}$ and 133.2 $kg \cdot ha^{-1}$, respectively. Alternately, to obtain the best economic profit (10827.3 $CNY \cdot ha^{-1}$), the optimum application rates of I, N and P were 682.4 $m^3 \cdot ha^{-1}$, 241.0 $kg \cdot ha^{-1}$ and 111.7 $kg \cdot ha^{-1}$, respectively. The latter scheme is recommended in the study area because of its relatively high grain yield and economic profit performance.

Author Contributions

Conceived and designed the experiments: GHY FSZ ZXL. Performed the experiments: GHY FSZ JG LH PFC. Analyzed the data: GHY FSZ JG. Contributed reagents/materials/analysis tools: GHY FSZ. Wrote the paper: GHY FSZ.

References

1. Li J (2009) Production, breeding and process of maize in China. In: Bennetzen JL, Hake SC, editors. Handbook of maize: Its biology. New York: Springer. pp. 563–576.
2. Ray DK, Mueller ND, West PC, Foley JA (2013) Yield trends are insufficient to double global crop production by 2050. PLoS ONE 8: e66428.
3. Wang H, Zhang M, Cai Y (2009) Problems, challenges, and strategic options of grain security in China. Advances in Agronomy 103: 101–147.
4. Wang T, Ma X, Li Y, Bai D, Liu C, et al. (2011) Changes in yield and yield components of single-cross maize hybrids released in China between 1964 and 2001. Crop Science 51: 512–525.
5. Ci X, Li M, Xu J, Lu Z, Bai P, et al. (2012) Trends of grain yield and plant traits in Chinese maize cultivars from the 1950s to the 2000s. Euphytica 183: 1–12.
6. Zou C, Gao X, Shi R, Fan X, Zhang F (2008) Micronutrient deficiencies in crop production in China. In: Alloway BJ, editor. Micronutrient deficiencies in global crop production. The Netherlands: Springer. pp. 127–148.
7. Wang CT, Li SK (2010) Assessment of limiting factors and techniques prioritization for maize production in China. Scientia Agricultura Sinica 43: 1136–1146. (in Chinese).
8. Zhang F, Yin G, Wang Z, McLaughlin N, Geng X, et al. (2013) Quantifying spatial variability of selected soil trace elements and their scaling relationships using multifractal techniques. PLoS ONE 8: e69326.9.
9. Fan T, Stewart BA, Wang Y, Luo J, Zhou G (2005) Long–term fertilization effects on grain yield, water–use efficiency and soil fertility in the dryland of Loess Plateau in China. Agriculture, Ecosystems & Environment 106: 313–329.
10. Khan S, Hanjra MA, Mu J (2009) Water management and crop production for food security in China: A review. Agricultural Water Management 96: 349–360.
11. Neumann K, Verburg PH, Stehfest E, Müller C (2010) The yield gap of global grain production: A spatial analysis. Agricultural Systems 103: 316–326.
12. Meng Q, Hou P, Wu L, Chen X, Cui Z, et al. (2013) Understanding production potentials and yield gaps in intensive maize production in China. Field Crops Research 143: 91–97.
13. Liu Z, Gao P (2009) Application basis and technology of water-saving agriculture in semi-arid areas of Northeast China. Beijing: Science Press. p. 314.
14. Fang Q, Ma L, Yu Q, Ahuja LR, Malone RW, et al (2010) Irrigation strategies to improve the water use efficiency of wheat–maize double cropping systems in North China Plain. Agricultural Water Management 97: 1165–1174.
15. Bernacchi CJ, Hollinger SE, Meyers T (2005) The conversion of the corn/soybean ecosystem to no-till agriculture may result in a carbon sink. Global Change Biology 11: 1867–1872.
16. Wang X, Willms WD, Hao X, Zhao M, Han G (2010) Cultivation and reseeding effects on soil organic matter in the mixed prairie. Soil Science Society of America Journal 74: 1348–1355.

17. Ma W, Ma L, Li J, Wang F, Sisák I, et al. (2011) Phosphorus flows and use efficiencies in production and consumption of wheat, rice, and maize in China. Chemosphere 84: 814–821.

18. Wang X, Wang J, Zhang J. (2012) Comparisons of three methods for organic and inorganic carbon in calcareous soils of northwestern China. PLoS ONE 7: e44334.

19. Wang X, Gan YT, Hamel C, Lemke RL, McDonald CL (2012) Water use profiles across the rooting zones of various pulse crops. Field Crops Research 134: 130–137.

20. Cui Z, Yue S, Wang G, Meng Q, Wu L, et al. (2013) Closing the yield gap could reduce projected greenhouse gas emissions: a case study of maize production in China. Global Change Biology 19: 2467–2477.

21. Jokela WE (1992) Nitrogen fertilizer and dairy manure effects on corn yield and soil nitrate. Soil Science Society of America Journal 56: 148–154.

22. Wortmann CS, Dobermann AR, Ferguson RB, Hergert GW, Shapiro CA, et al. (2009) High–yielding corn response to applied phosphorus, potassium, and sulfur in Nebraska. Agronomy Journal 101: 546–555.

23. Ju X, Christie P (2011) Calculation of theoretical nitrogen rate for simple nitrogen recommendations in intensive cropping systems: A case study on the North China Plain. Field Crops Research 124: 450–458.

24. Hou P, Gao Q, Xie R, Li S, Meng Q, et al. (2012) Grain yields in relation to N requirement: Optimizing nitrogen management for spring maize grown in China. Field Crops Research 129: 1–6.

25. Jin L, Cui H, Li B, Zhang J, Dong S, et al. (2012) Effects of integrated agronomic management practices on yield and nitrogen efficiency of summer maize in North China. Field Crops Research 134: 30–35.

26. Hu H, Ning T, Li Z, Han H, Zhang Z, et al. (2013) Coupling effects of urea types and subsoiling on nitrogen–water use and yield of different varieties of maize in northern China. Field Crops Research 142: 85–94.

27. Dang TH, Cai GX, Guo SL, Hao MD, Heng LK (2006) Effect of nitrogen management on yield and water use efficiency of rainfed wheat and maize in Northwest China. Pedosphere 16: 495–504.

28. Liu WZ, Zhang XC (2007) Optimizing water and fertilizer input using an elasticity index: a case study with maize in the loess plateau of china. Field Crops Research 100: 302–310.

29. El-Hendawy SE, Schmidhalter U (2010) Optimal coupling combinations between irrigation frequency and rate for drip–irrigated maize grown on sandy soil. Agricultural Water Management 97: 439–448.

30. Wang X, Dai K, Zhang D, Zhang X, Wang Y, et al. (2011) Dryland maize yields and water use efficiency in response to tillage/crop stubble and nutrient management practices in China. Field Crops Research 120: 47–57.

31. Grassini P, Cassman KG (2012) High–yield maize with large net energy yield and small global warming intensity. Proceedings of the National Academy of Sciences 109: 1074–1079.

32. Mulwa R, Emrouznejad A, Muhammad L (2009) Economic efficiency of smallholder maize producers in Western Kenya: a DEA meta–frontier analysis. International Journal of Operational Research 4: 250–267.

33. Rahman S, Rahman MS, Rahman MH (2012) Joint determination of the choice of growing season and economic efficiency of maize in Bangladesh. Journal of the Asia Pacific Economy 17: 138–150.

34. Sharma AR, Singh R, Dhyani SK, Dube RK (2011) Agronomic and economic evaluation of mulching in rainfed maize–wheat cropping system in the Western Himalayan region of India. Journal of Crop Improvement 25: 392–408.

35. Guo R, Lin Z, Mo X, Yang C (2010) Responses of crop yield and water use efficiency to climate change in the North China Plain. Agricultural Water Management 97: 1185–1194.

36. Xiong W, Holman I, Lin E, Conway D, Jiang J, et al. (2010) Climate change, water availability and future cereal production in China. Agriculture, Ecosystems & Environment 135: 58–69.

37. Piao S, Ciais P, Huang Y, Shen Z, Peng S, et al. (2010) The impacts of climate change on water resources and agriculture in China. Nature 467: 43–51.

38. Li X, Takahashi T, Suzuki N, Kaiser HM (2011) The impact of climate change on maize yields in the United States and China. Agricultural Systems 104: 348–353.

39. Tao F, Zhang Z (2011) Impacts of climate change as a function of global mean temperature: maize productivity and water use in China. Climatic Change 105: 409–432.

40. Tong C, Hall CAS, Wang H (2003) Land use change in rice, wheat and maize production in China (1961–1998). Agriculture, Ecosystems & Environment 95: 523–536.

41. Owombo PT, Akinola AA, Ayodele OO, Koledoye GF (2012) Economic impact of agricultural mechanization adoption: Evidence from maize farmers in Ondo State, Nigeria. Journal of Agriculture and Biodiversity Research 1: 25–32.

42. Ding XQ (1986) The applied regression design of the field experiment. Jilin Science&Technology Press. 199–203. (in Chinese).

43. Jin X, Fu Z, Ding D, Li W, Liu Z, et al. (2013) Proteomic identification of genes associated with maize grain-filling rate. PLoS ONE 8: e59353.

44. Lawlor DW, Lemaire G, Gastal F (2001) Nitrogen, plant growth and crop yield. In: Lea PJ, Morot-Gaudry JF, editors. Plant nitrogen. Berlin Heidelberg: Springer. pp. 343–367.

45. Peng Y, Li X, Li C (2012) Temporal and spatial profiling of root growth revealed novel response of maize roots under various nitrogen supplies in the field. PLoS ONE 7: e37726.

12

Soil Carbon and Nitrogen Fractions and Crop Yields Affected by Residue Placement and Crop Types

Jun Wang[1], Upendra M. Sainju[2]*

1 College of Urban and Environmental Sciences, Northwest University, Xian, Shaanxi Province, China, 2 U.S. Department of Agriculture, Agricultural Research Service, Northern Plains Agricultural Research Laboratory, Sidney, Montana, United States of America

Abstract

Soil labile C and N fractions can change rapidly in response to management practices compared to non-labile fractions. High variability in soil properties in the field, however, results in nonresponse to management practices on these parameters. We evaluated the effects of residue placement (surface application [or simulated no-tillage] and incorporation into the soil [or simulated conventional tillage]) and crop types (spring wheat [*Triticum aestivum* L.], pea [*Pisum sativum* L.], and fallow) on crop yields and soil C and N fractions at the 0–20 cm depth within a crop growing season in the greenhouse and the field. Soil C and N fractions were soil organic C (SOC), total N (STN), particulate organic C and N (POC and PON), microbial biomass C and N (MBC and MBN), potential C and N mineralization (PCM and PNM), NH_4-N, and NO_3-N concentrations. Yields of both wheat and pea varied with residue placement in the greenhouse as well as in the field. In the greenhouse, SOC, PCM, STN, MBN, and NH_4-N concentrations were greater in surface placement than incorporation of residue and greater under wheat than pea or fallow. In the field, MBN and NH_4-N concentrations were greater in no-tillage than conventional tillage, but the trend reversed for NO_3-N. The PNM was greater under pea or fallow than wheat in the greenhouse and the field. Average SOC, POC, MBC, PON, PNM, MBN, and NO_3-N concentrations across treatments were higher, but STN, PCM and NH_4-N concentrations were lower in the greenhouse than the field. The coefficient of variation for soil parameters ranged from 2.6 to 15.9% in the greenhouse and 8.0 to 36.7% in the field. Although crop yields varied, most soil C and N fractions were greater in surface placement than incorporation of residue and greater under wheat than pea or fallow in the greenhouse than the field within a crop growing season. Short-term management effect on soil C and N fractions were readily obtained with reduced variability under controlled soil and environmental conditions in the greenhouse compared to the field. Changes occurred more in soil labile than non-labile C and N fractions in the greenhouse than the field.

Editor: Raffaella Balestrini, Institute for Sustainable Plant Protection, C.N.R., Italy

Funding: Funding came from the U.S. Department of Agriculture-Agricultural Research Service, Sidney, MT, USA and the National Natural Science Foundation of China (No. 31270484). The funders had no role in study design, data collection and analysis, decision to publish, or preparation of the manuscript.

* Email: upendra.sainju@ars.usda.gov

Introduction

Soil organic matter, as indicated by C and N levels, is an important component of soil quality and productivity. Increasing soil organic matter through enhanced C and N sequestration can also reduce the potentials for global warming by mitigating greenhouse gas emissions and N leaching by increasing N storage in the soil [1,2]. Carbon and N sequestration usually occur when non-harvested crop residues, such as stems, leaves, and roots, are placed at the soil surface due to no-tillage [3,4,5]. Carbon and N sequestration rates, however, depend on the balance between the amounts of plant residue C and N inputs and rates of C and N mineralized in the nonmanured soil [6,7]. Other benefits of increasing C and N storage include enhancement of soil structure and soil water-nutrient-crop productivity relationships [8].

Soil and crop management practices can alter the quantity, quality, and placement of crop residues in the soil, thereby influencing soil C and N storage, microbial biomass and activity, and N mineralization–immobilization [9,10]. Residue placement in the soil under different tillage systems can influence C and N levels by affecting soil aggregation, aeration, and C and N

mineralization [9,11]. Crop types can affect the quantity and quality (C/N ratio) of crop residue returned to the soil and therefore on soil C and N levels [9,12]. Legumes, such as pea, because of its higher N concentration and lower C/N ratio, decompose more rapidly in the soil and supply greater amount of N to succeeding crops than nonlegumes [12,13]. As a result, N fertilization rates to crops following pea can be reduced to sustain yields [14,15].

Because of large pool sizes and inherent spatial variability, soil organic C (SOC) and total N (STN) (slow or non-labile fractions) change slowly with management practices [16]. Therefore, measurements of SOC and STN alone may not adequately reflect changes in soil quality and nutrient status [16,17]. Active (or labile) C and N fractions, such as potential C and N mineralization (PCM and PNM) that indicate microbial activity and N mineralization, and microbial biomass C and N (MBC and MBN) that refer to microbial biomass and N immobilization, change seasonally [16,18]. Similarly, particulate organic C and N (POC and PON) that represent coarse organic matter and considered as intermediate C and N levels between slow and active fractions, provide substrates for microbes and influence soil aggregation [19,20].

Available N fractions that influence plant growth and N losses due to leaching, denitrification, or volatilization are NH_4-N and NO_3-N [10,12].

Although active C and N fractions in the soil can change more rapidly than the other fractions, these fractions sometime may not be readily changed within a crop growing season due to high variability in soil properties within a short distance in the field or in regions with limited precipitation, cold weather, and a short growing season [10,12,15]. Under controlled soil and environmental conditions, such as in the greenhouse, it may be possible to detect changes in these fractions more rapidly as affected by management practices than in the field. We hypothesized that surface placement of crop residue (a simulation of no-tillage in the field) under spring wheat can increase soil labile and non-labile C and N fractions and sustain crop yields compared to residue incorporation into the soil (a simulation of conventional tillage) under pea or fallow more in the greenhouse than in the field. Our objectives were to: (1) evaluate the effects of residue placement and crop types on crop yields, residue C and N losses, and soil labile and non-labile C and N fractions within a growing season in the greenhouse and the field and (2) determine if soil C and N fractions change more readily in the greenhouse than the field within a growing season.

Materials and Methods

Greenhouse experiment

The experiment was conducted under controlled soil and environmental conditions in the greenhouse with air temperatures of $25°C$ in the day and $15°C$ in the night. Soil samples were collected manually from an area of 5 m^2 using a shovel to a depth of 20 cm under a mixture of crested wheatgrass [*Agropyron cristatum* (L.) Gaertn] and western wheatgrass [*Pascopyrum smithii* (Rydb.) A. Love] from a dryland farm site, 11 km east of Sidney, Montana, USA. The research farm site where soil samples were collected is under the management of USDA, Agricultural Research Service, Sidney, Montana and no endangered or protected species were involved or was negatively impacted by this research. The soil was a Williams loam (fine-loamy, mixed, frigid, Typic Argiborolls [International classification: Luvisols]) with 350 g kg^{-1} sand, 325 g kg^{-1} silt, 325 g kg^{-1} clay, 1.42 Mg m^{-3} bulk density, and 7.2 pH at the 0–20 cm depth. Soil C and N fractions in the sample before the initiation of the experiment are shown in Table 1. Soil was air-dried and sieved to 4.75 mm after discarding coarse organic materials and rock fragments. Eight kilograms of soil was placed in a plastic pot, 25 cm high by 25 cm diameter, above 3 cm of gravel at the bottom.

Treatments consisted of two residue placements (surface placement vs. incorporation into the soil) and three crop types (spring wheat, pea, and fallow [or no crop]) arranged in a completely randomized design with three replications. In order to match the residue and crop type, spring wheat residue was placed under spring wheat and fallow and pea residue under pea. Residues included nine-week old spring wheat and pea plants collected from the field without grains, chopped to 2 cm, and oven-dried at $60°C$ for 3 d. Fifteen grams of residues per pot (corresponding to 2.6 Mg ha^{-1} of residue found in the field) were either placed uniformly at the soil surface or incorporated into the soil by mixing the residue with the soil by hand. The surface placement of residue corresponded to the simulated no-tillage system in the field, although the soil was disturbed during collection, and incorporated residue to the simulated conventional tillage system. Spring wheat received 0.96 g N pot^{-1} as urea, similar to the recommended N fertilization rate (80 kg N ha^{-1}) in

the field, while pea received 0.11 g N pot^{-1} (or 9 kg N ha^{-1}) while applying monoammonium phosphate as the P fertilizer. Half of 0.96 g N pot^{-1} was applied at planting and other half at four weeks later. Both spring wheat and pea also received P fertilizer (monoammonium phosphate) at 0.25 g P pot^{-1} (or 27 kg P ha^{-1}) and K fertilizer (muriate of potash) at 0.50 g K pot^{-1} (or 29 kg K ha^{-1}). No fertilizers were applied to the fallow treatment.

In July 2012, five spring wheat (cultivar Reeder) and pea (cultivar Majoret) seeds were planted per pot, except in the fallow treatment. At a height of 3 cm, seedlings were thinned to two plants per pot. In order to compensate for the water received as rainfall in the field, water was applied to all treatments in the greenhouse experiment to field capacity (0.25 m^3 m^{-3}) [21] at 300 to 500 mL pot^{-1}. Water was applied at planting and at 3 to 7 d intervals thereafter, depending on soil water content (as determined by a soil water probe [TDR 300, Spectrum Technologies Inc., Aurora, IL] installed to a depth of 15 cm). Since measured amount of water was applied according to soil water content and crop demand, only a negligible amount of water was leached below the pot that was not determined. Herbicides and pesticides were applied to plants as needed. At 105 d after planting, shoot biomass including grains was harvested from the pot, washed with water, oven-dried at $60°C$ for 3 to7 d, and dry matter yield was determined. Because of the small amount of grain production, grains were also included in the shoot biomass. After crop harvest, soil from the entire pot was sieved to 2 mm to separate coarse residue and root fragments, which were picked by hand, washed with water, and oven-dried at $60°C$ for 3 to7 d to determine dry matter yields. A portion (100 g) of residue and root-free soil sample visible to the naked eye was collected from each pot, air-dried, and used for determinations of C and N fractions. The remaining soil samples were further washed in a nest of 1.0 and 0.5 mm sieves under a continuous stream of water to separate fine roots. Roots left in the sieves were picked using a tweezers, oven-dried at $60°C$ for 3 to7 d, and dry matter yield was determined. Total root biomass was determined by adding biomass of coarse and fine roots.

Shoot and root biomass and crop residues added to the soil at the initiation of the experiment and those (>2.00 mm) recovered from the soil at the end were ground to 1 mm and C and N concentrations (g kg^{-1}) were determined with a high induction furnace C and N analyzer (LECO, St. Joseph, MI). Amounts of C and N in the residue added and recovered from the soil were determined by multiplying C and N concentrations by the weight of the soil in the pot. Carbon and N losses from the residue were determined as: Residue C and N losses (g kg^{-1}) = (Residue C and N added – Residue C and N recovered) $\times 1000$/Residue C and N added. While determining the amount of C and N recovered in the residue, it was assumed that fine residue (<2.00 mm) was a part of soil organic matter.

Field experiment

The field experiment was conducted using identical treatments, design, and replications as in the greenhouse from April to August 2012 near the place where soil samples were collected for the greenhouse experiment. As a result, soils were similar in both field and greenhouse experiments. The field site has mean monthly air temperature ranging from $-8°C$ in January to $23°C$ in July and August. The mean annual precipitation (105-yr average) is 340 mm, 80% of which occurs during the crop growing season (April-October). Equivalent amounts of crop residues and fertilizers using the same treatments as in the greenhouse were applied to spring wheat, pea, and fallow in the field. Because the amount of residue applied was similar, the amounts of C and N

Table 1. Average soil organic C (SOC), total N (STN), particulate organic C and N (POC and PON), potential C and N mineralization (PCM and PNM), microbial biomass C and N (MBC and MBN), and NH_4-N and NO_3-N concentrations at the start of the experiment (n = 4).

Parameter	Concentration
SOC (g C kg^{-1})	11.80
POC (g C kg^{-1})	3.18
PCM (mg C kg^{-1})	9.25
MBC (mg C kg^{-1})	117.6
STN (g N kg^{-1})	1.29
PON (g N kg^{-1})	0.34
PNM (mg N kg^{-1})	8.95
MBN (mg N kg^{-1})	69.0
NH_4-N (mg N kg^{-1})	2.86
NO_3-N (mg N kg^{-1})	5.04

added in residue to the soil were also identical in the greenhouse and field. Residues and fertilizers were placed at the soil surface in the no-till system and incorporated to a depth of 10 cm using tillage with a field cultivator in the conventional tillage system. Plot size was 12.2×6.1 m.

Spring wheat and pea were planted in April with a no-till drill at a spacing of 20.3 cm. Growing season weeds were controlled with selective post emergence herbicides appropriate for each crop. Contact herbicides were applied at postharvest and preplanting. Crops were grown under dryland condition receiving only precipitation without irrigation. In August, biomass yield of spring wheat and pea was determined from two 0.5 m^2 areas outside yield rows within each plot and grain yield was determined by harvesting grains from a swath of 1.5 m×12.0 m using a combine harvester. Carbon and nitrogen concentrations in the grain and biomass were determined after oven drying subsamples at 55°C and using the C and N analyzer as above. Carbon and N contents (Mg C or N ha^{-1}) in grain and biomass were determined by multiplying C and N concentrations by grain and biomass yields, respectively. Total aboveground biomass and C and N contents were determined by adding yields and C and N contents of grain and biomass.

Soil samples were collected from five random locations in central rows of the plot to a depth of 20 cm using a truck-mounted hydraulic probe (3.5 cm inside diameter). Samples were composited within a plot, air-dried, ground, and sieved to 2 mm for determining C and N concentrations. No attempts were made to collect the surface residue at soil sampling because of residue loss and contamination with soil and residue from one plot to another due to actions of wind and water. Therefore, residue C and N losses were not determined in the field.

Soil carbon and nitrogen fractions measurements

The SOC concentration in the greenhouse and field soils were determined with a high induction furnace C and N analyzer as above after pretreating the soil with 5% H_2SO_3 to remove inorganic C [22]. The STN concentration was determined by using the analyzer without pretreating the soil with the acid. For determining POC and PON concentrations, 10 g soil sample was dispersed with 30 mL of 5 g L^{-1} sodium hexametaphosphate by shaking for 16 h and the solution was poured through a 0.053 mm sieve [19]. The solution and particles that passed through the sieve and contained mineral-associated and water-soluble C and N were

dried at 50°C for 3 to 4 d and SOC and STN concentrations were determined by using the analyzer as above. The POC and PON concentrations were determined by the difference between SOC and STN in the whole-soil and that in the particles that passed through the sieve after correcting for the sand content.

The PCM and PNM concentrations in air-dried soils were determined by the modified method of Haney et al. [23]. Two 10 g soil subsamples were moistened with water at 50% field capacity [21] and placed in a 1 L jar containing beakers with 4 mL of 0.5 mol L^{-1} NaOH to trap evolved CO_2 and 20 mL of water to maintain high humidity. Soils were incubated in the jar at 21°C for 10 d. At 10 d, the beaker containing NaOH was removed from the jar and PCM was determined by measuring CO_2 absorbed in NaOH, which was back-titrated with 1.5 mol L^{-1} $BaCl_2$ and 0.1 mol L^{-1} HCl. One beaker containing soil was removed from the jar and extracted with 100 mL of 2 mol L^{-1} KCl for 1 h. The NH_4-N and NO_3-N concentrations in the extract were determined by using the autoanalyzer (Lachat Instrument, Loveland, CO). The PNM was calculated as the difference between the sum of NH_4-N and NO_3-N concentrations in the soil before and after incubation.

The other beaker containing moist soil and incubated for 10 d (used for PCM determination above) was used for determining MBC and MBN concentrations by the modified fumigation–incubation method for air-dried soils [24]. The moist soil was fumigated with ethanol-free chloroform for 24 h and placed in a 1 L jar containing beakers with 2 mL of 0.5 mol L^{-1} NaOH and 20 mL water. As with PCM, fumigated moist soil was incubated for 10 d and CO_2 absorbed in NaOH was back-titrated with $BaCl_2$ and HCl. The MBC was calculated by dividing the amount of CO_2–C absorbed in NaOH by a factor of 0.41 [25] without subtracting the values from the nonfumigated control [24]. For MBN, the fumigated–incubated sample at 10 d was extracted with 100 mL of 2 mol L^{-1} KCl for 1 h and NH_4-N and NO_3-N concentrations were determined by using the autoanalyzer as above. The MBN was calculated by the difference between the sum of NH_4-N and NO_3-N concentrations in the sample before and after fumigation–incubation and divided by a factor of 0.41 [25,26]. The NH_4-N and NO_3-N concentrations determined in the nonfumigated–nonincubated samples were used as available fractions of N.

Data analysis

Data for C and N contents in crop biomass and residue and soil C and N fractions were analyzed by using the MIXED model of SAS [27]. Treatment was considered as the fixed effect and replication as the random effect. Means were separated by using the least square means test when treatments and interactions were significant [27]. Statistical significance was evaluated at $P \leq 0.05$, unless otherwise stated.

Results

Greenhouse experiment

Shoot and root biomass yields and carbon and nitrogen contents. Shoot and root biomass yields and C and N contents varied among residue placements and crop types (Table 2). Interaction between residue placement and crop types on these parameters was not significant. Shoot and root biomass yields and C and N contents were greater in surface placement than incorporation of residue into the soil. Shoot biomass yield and C and N contents were also greater in wheat than in pea. Because of the negligible amount of roots, root biomass yield and C and N contents in pea were not determined. Absence of plants in the fallow also resulted in non-existence of crop data in this treatment. The coefficient of variation (CV) for crop parameters ranged from 38.2 to 62.5%.

Residue carbon and nitrogen losses. Total amounts of C and N added through residue application and leaf fall and those recovered in coarse fractions (>2 mm) after crop harvest varied with residue placements and crop types, with the significant residue placement × crop type interaction for C and N recovered in the residue (Table 3). Although the amount of residue applied was similar in all treatments (15 g of wheat or pea residue pot^{-1}), differences in C and N concentrations between residues and those added through leaf fall during crop growth varied residue C and N additions among treatments. Averaged across crop types, residue C addition was greater in surface placement than incorporation of residue into the soil. Averaged across residue placements, residue C addition was greater under wheat than pea or fallow, but residue N addition was greater under pea than wheat or fallow. Residue C recovery was greater in surface placement under wheat and fallow than surface placement under pea and incorporation under fallow. Residue N recovery was also greater in surface placement under wheat and fallow than surface placement under pea and incorporation under fallow and wheat. Averaged across crop types, residue N recovery was greater in surface placement than incorporation of residue into the soil. Averaged across residue placements, residue C recovery was greater under wheat than pea. The coefficient of variation for residue C and N addition and recovery varied from 7.2 to 17.8%.

Residue C and N losses also varied with residue placements and crop species, with the significant residue placement × crop species interaction (Table 3). Residue C loss was greater in surface placement under pea and incorporation under fallow than surface placement under fallow. Residue N loss was in the order: surface placement and incorporation under pea > incorporation under wheat and fallow > surface placement under wheat > surface placement under fallow. Averaged across crop types, residue N loss was greater in residue incorporation than surface placement. Averaged across residue placements, residue N loss was greater under pea than under fallow and wheat. The coefficient of variation for residue C and N losses varied from 14.3 to 31.6%.

Soil carbon and nitrogen fractions. The SOC, POC, and PCM concentrations varied among residue placements and crop types (Table 4). Averaged across crop types, SOC and PCM were greater in surface placement than incorporation of the residue into the soil. Averaged across residue placements, SOC was greater under wheat than pea and fallow and POC was greater under wheat than pea. The MBC was not influenced by treatments. The coefficient of variation for soil C fractions ranged from 2.6 to 14.3%.

The STN, PNM, MBN, NH_4-N, and NO_3-N concentrations also varied among residue placements and crop types (Table 4). Averaged across crop types, PNM and NH_4-N were greater in surface placement than incorporation of residue into the soil. Averaged across residue placements, STN was greater under wheat than pea and MBN was greater under wheat than fallow. In contrast, PNM was greater under pea and fallow than wheat and NO_3-N was greater under fallow than wheat. The PON was not influenced by treatments. The coefficient of variation for soil N fractions ranged from 4.6 to 15.9%.

Field experiment

Aboveground total crop biomass yield and C and N contents varied with crop types (Table 5). Averaged across tillage practices, crop biomass yield and C content were greater in wheat than pea, but the trend reversed for N content. Tillage and its interaction with crop type were not significant for crop biomass yield and C and N contents. The coefficient of variation for crop biomass yield and C and N contents ranged from 28.1 to 41.9%.

Soil MBN, NH_4-N, and NO_3-N concentrations varied with tillage practices and MBC and PNM varied with crop types (Table 5). Averaged across crop types, MBN and NH_4-N were greater in no-tillage than conventional tillage, but NO_3-N was greater in conventional tillage than no-tillage. Averaged across tillage practices, MBC was greater under wheat than fallow and PNM was greater under pea than wheat and fallow. Tillage, crop type, and their interaction were not significant for SOC, POC, PCM, STN, and PON. The coefficient of variations for soil C and N fractions ranged from 8.0 to 36.7%.

Discussion

Enhanced soil water conservation due to mulch action of the residue at the soil surface [28] may have increased shoot and root biomass yields and C and N contents in surface placement compared to incorporation of residue into the soil in the greenhouse (Table 2). It has been reported that surface placement of residue in the no-till system increased spring wheat yield compared to residue incorporation in the conventional till system [3,15]. In our field experiment, crop biomass yield and C and N contents, however, were not influenced by tillage (Table 5). It may be possible that wheat and pea residues applied by hand at the soil surface were more uniformly distributed in the greenhouse than in the field where residues were distributed by a machine sprayer. As a result, soil water was probably conserved more, resulting in increased crop yield and C and N contents with the surface placement than incorporation of residue in the greenhouse compared to the field.

Differences in the amount of N fertilizer applied and N fixation capacity may have resulted in variation in crop biomass yields and C contents among crop species in the greenhouse and the field (Tables 2 and 5). Higher amount of N fertilizer application may have increased biomass yield and C and N contents in wheat than pea in the greenhouse. Higher amount of N fertilizer application also may have increased biomass yield and C content in wheat and pea, but greater N fixation may have increased N content in pea than wheat in the field [14,28]. Grain and biomass yields are usually greater in wheat which receives N fertilizer than pea which

Table 2. Effects of residue placement and crop type on crop shoot (grains+leaves+stems) and root biomass C and N contents in the greenhouse.

Residue placement	Crop type	Shoot biomass	Root biomass	Shoot biomass C	Root biomass C	Shoot biomass N	Root biomass N	Total biomass C	Total biomass N
		——— g pot⁻¹ ———		——— g C pot⁻¹ ———		——— g Npot⁻¹ ———		g C pot⁻¹	g N pot⁻¹
Incorporated		4.51b[a]	2.37b	1.83b	0.74b	0.14b	0.05b	2.91b	0.25b
Surface		7.41a	6.14a	3.07a	1.74a	0.24a	0.11a	5.55a	0.44a
	Pea	4.55b	——[b]	1.91b	——	0.12b	——	1.91b	0.12b
	Wheat	7.36a	4.25	3.00a	1.24	0.26a	0.08	4.23a	0.34a
CV (%)[c]		42.4	61.9	42.4	56.4	55.0	62.5	40.4	38.2
Significance									
Residue placement (R)		*	——	*	*	**	*	*	*
Crop species (C)		*	*	*	——	**	——	*	——
R×C		NS[d]	NS	NS	——	NS	——	——	——

*Significant at $P = 0.05$.
**Significant at $P = 0.01$.
[a]Numbers followed by different letters within a column in a set are significantly different at $P \leq 0.05$ by the least square means test.
[b]Non-measurable values due to negligible amount of root biomass.
[c]Coefficient of variation.
[d]Not significant.

Table 3. Effects of residue placement and crop type on residue C and N addition, recovered in coarse fragments (>2 mm), and losses during the crop growing period in the greenhouse.

Residue placement	Crop type	Crop residue		C recovered	N recovered	C loss	N loss
		C added[a]	N added[a]				
		g C pot^{-1}	g N pot^{-1}	g C pot^{-1}	g N pot^{-1}	g kg^{-1}	g kg^{-1}
Incorporated	Fallow	7.80a[b]	0.40a	3.84b	0.23b	508a	438b
	Pea	7.80a	0.64a	4.24ab	0.26ab	457ab	588a
	Wheat	8.76a	0.44a	4.89ab	0.24b	430ab	456b
Surface	Fallow	7.80a	0.40a	5.16a	0.31a	339b	213d
	Pea	7.80a	0.64a	3.82b	0.23b	511a	640a
	Wheat	8.92a	0.46a	4.95a	0.31a	446ab	325c
CV (%)[c]		7.2	17.8	15.4	16.5	143	316
Means							
Incorporated		8.07b	0.49a	4.32a	0.24b	465a	494a
Surface		8.17a	0.50a	4.64a	0.28a	432a	393b
	Fallow	7.80b	0.40b	4.50ab	0.27a	423a	326b
	Pea	7.80b	0.64a	4.02b	0.25a	484a	614a
	Wheat	8.76a	0.45b	4.92a	0.28a	438a	390b
Significance							
Residue placement (R)		*	NS[d]	NS	*	NS	**
Crop species (C)		***	***	*	NS	*	***
R×C		NS	NS	*	*	*	**

*Significant at $P=0.05$.
**Significant at $P=0.01$.
*** Significant at $P=0.001$.
[a]Includes C and N added from the residue application and leaf fall.
[b]Numbers followed by different letters within a column in a set are significantly different at $P \leq 0.05$ by the least square means test.
[c]Coefficient of variation.
[d]Not significant.

Table 4. Effects of residue placement and crop type on soil organic C (SOC), total N (STN), particulate organic C and N (POC and PON), potential C and N mineralization (PCM and PNM), microbial biomass C and N (MBC and MBN), and NH_4-N and NO_3-N concentrations in the greenhouse.

Residue placement	Crop type	SOC	POC	PCM	MBC	STN	PON	PNM	MBN	NH_4-N	NO_3-N
		— g C kg^{-1} —		— mg C kg^{-1} —		— g N kg^{-1} —		— mg N kg^{-1} —			
Incorporated		12.0b[a]	3.28a	11.6b	128.0a	1.29a	0.33a	5.80b	50.8a	1.14b	11.8a
Surface		12.3a	3.28a	14.4a	143.8a	1.31a	0.31a	9.56a	57.5a	1.65a	16.5a
	Fallow	12.0b	3.26ab	11.4a	118.9a	1.30ab	0.32a	9.36a	46.1b	1.44a	23.7a
	Pea	12.0b	3.18b	13.1a	144.0a	1.26b	0.29a	9.68a	56.3ab	1.38a	12.8ab
	Wheat	12.4a	3.40a	14.5a	145.1a	1.34a	0.34a	5.98b	97.6a	1.36a	6.1b
CV (%)[b]		2.6	11.8	14.3	13.3	4.6	15.9	14.2	14.3	14.8	13.6
Significance											
Residue placement (R)		**	NS[c]	*	NS	NS	NS	*	NS	*	NS
Crop species (C)		**	*	NS	NS	*	NS	*	*	NS	*
R×C		NS	NS	NS	NS	NS	NS	NS	NS	NS	NS

Soil samples were collected at the 0–20 cm depth in the field and used for the greenhouse experiment.

*Significant at $P = 0.05$.
**Significant at $P = 0.01$.
[a]Numbers followed by different letters within a column in a set are significantly different at $P \leq 0.05$ by the least square means test.
[b]Coefficient of variation.
[c]Not significant.

Table 5. Effects of residue placement and crop type on crop aboveground biomass (grains+stems+leaves) yield, C and N contents, and soil organic C (SOC), total N (STN), particulate organic C and N (POC and PON), potential C and N mineralization (PCM and PNM), microbial biomass C and N (MBC and MBN), and NH_4-N and NO_3-N concentrations at the 0–20 cm depth in the field.

Tillage[a]	Crop type	Crop biomass yield	Crop C content	Crop N content	SOC	POC	PCM	MBC	STN	PON	PNM	MBN	NH_4-N	NO_3-N
		Mg ha^{-1}	Mg C ha^{-1}	kg N ha^{-1}	— g C kg^{-1} —		— mg C kg^{-1} —		— g N kg^{-1} —		— mg N kg^{-1} —			
CT		4.91a[b]	2.06a	59.6a	11.0a	2.56a	45.1a	114.4a	1.33a	0.24a	3.21a	12.6b	3.05b	4.54a
NT		5.08a	2.11a	65.3a	11.0a	2.56a	56.8a	122.9a	1.37a	0.28a	4.28a	19.6a	3.82a	2.08b
	Fallow	——[c]	——	——	10.6a	2.55a	45.61a	111.4b	1.30a	0.25a	2.85b	14.2a	2.93a	3.36a
	Pea	4.68b	1.87b	72.3a	11.4a	2.56a	51.0a	118.6ab	1.42a	0.27a	5.56a	15.5a	3.43a	2.87a
	Wheat	5.31a	2.18a	52.6b	11.0a	2.57a	56.0a	126.0a	1.34a	0.26a	2.83b	18.6a	3.93a	3.71a
CV (%)		28.1	30.0	41.9	8.0	18.7	27.1	28.9	9.3	18.4	36.4	36.7	27.9	25.0
Significance														
Tillage (T)		NS[d]	NS	NS	NS	NS	NS	NS	NS	NS	NS	*	*	*
Crop species (C)		*	*	**	NS	NS	NS	*	NS	NS	*	NS	NS	NS
T×C		NS	NS	NS	NS	NS	NS	NS	NS	NS	NS	NS	NS	NS

*Significant at P = 0.05.
**Significant at P = 0.01.
[a]Tillage are CT, conventional tillage; and NT, no-tillage.
[b]Numbers followed by different letters within a column in a set are significantly different at P ≤ 0.05 by the least square means test.
[c]Crop absent in the fallow.
[d]Not significant.

receives no N fertilizer due to increased water-use efficiency, but higher N concentration due to increased atmospheric N fixation can increase N content in pea than wheat [10,15,28]. The fact that different trends in N content in pea vs. wheat occurred in the field and the greenhouse was probably related to root growing soil volume. It may be possible that roots exploited greater soil volume that resulted in increased N fixation by pea and therefore increased its N content in the field compared to the greenhouse where plants were grown in a limited soil volume in the pot.

Greater residue input due to higher biomass yield may have increased residue C addition in surface placement than incorporation of residue into the soil or increased under wheat than pea or fallow in the greenhouse (Tables 2 and 3). In contrast, higher N concentration may have increased residue N addition under pea than wheat or fallow. Greater C and N recovered in the residue placed at the soil surface under wheat and fallow were probably due to reduced mineralization of wheat residue as a result of its higher C/N ratio than pea residue. While surface placement of residue reduces its contact with soil microorganisms that result in reduced mineralization [29,30], increased mineralization of pea residue due to its lower C/N ratio may have resulted in reduced C and N recovery in the residue placed at the soil surface under pea. Residues of legumes, such as pea with lower C/N ratio, decompose more rapidly than those of nonlegumes, such as wheat with higher C/N ratio [12]. When incorporated into the soil, residue C and N recovery were lower under fallow and wheat. As a result, C and N losses were higher in surface placement of residue under pea or residue incorporation under pea and fallow than the other treatments. It may be possible that some of C and N lost from the residue converted into soil C and N fractions, as discussed below.

Reduced mineralization of residue may have increased SOC, PCM, PNM, and NH_4-N concentrations in surface placement than incorporation of residue into the soil in the greenhouse (Table 4). Similar increases in MBN and NH_4-N concentrations in no-tillage compared to conventional tillage were found in the field (Table 5). Several researchers [5,16,31,32,33] have reported greater SOC, POC, MBC, PCM, PNM, and MBN in surface residue placement in the no-tillage system than residue incorporation into the soil in the conventional tillage system. Increased N mineralization due to residue incorporation, however, may have increased NO_3-N concentration in conventional tillage than no-tillage in the field.

Higher C and N substrate availability due to increased yield probably increased SOC, POC, STN, and MBN under wheat than under pea or fallow in the greenhouse (Table 4) or increased MBC under wheat than fallow in the field (Table 5). Root biomass C, residue C addition (Tables 2 and 3), and amount of applied N fertilizer were greater in wheat than pea or fallow. Similar results probably occurred in the field, since treatments were identical in the greenhouse and the field and crop biomass C was higher in wheat than pea in the field (Table 5). Rhizodeposit C released by roots can increase microbial biomass and activity and soil C storage [34]. Liebig et al. [35] also found higher MBC under spring wheat than under fallow. In contrast, greater PNM and NO_3-N under pea and fallow than wheat in the greenhouse were probably either due to increased mineralization of pea residue as a result of its lower C/N ratio than wheat residue [12] or to greater mineralization of soil and wheat residue as a result of enhanced microbial activity from higher soil temperature and water content and absence of plants to uptake N under fallow [11,13,36]. Since residue N loss was greater under pea than wheat and fallow (Table 3), part of N from pea residue may have contributed to increased PNM and NO_3-N concentrations under pea. Similar

result of increased PNM under pea than wheat and fallow was also found in the field, since crop biomass N was greater in pea than wheat (Table 5).

Comparison of soil C and N fractions at the beginning and end of the experiment due to residue placement (Tables 1 and 4) showed that SOC increased by 4.2%, PCM by 55.7%, and PNM by 6.1% with surface residue placement in the greenhouse. Corresponding values in SOC, PCM, and PNM with residue incorporation were 1.7, 25.4, and −35.1%, respectively. In the field, MBN reduced by 71.5% in no-tillage and 81.7% in conventional tillage from the beginning to the end of the experiment. This shows that residue placement at the surface either increased soil C and N fractions in the greenhouse or reduced their losses in the field within a crop growing season compared to residue incorporation. Since soil NH_4-N and NO_3-N concentrations vary seasonally due to N mineralization from crop residue and soil, N fertilization, crop N uptake, and N losses due to leaching, volatilization, and denitrification [10,15], variations in their levels from the beginning to the end of the experiment were not taken into account.

Among crop types, SOC increased by 5.1%, POC by 6.9%, STN by 3.9%, and MBN by 41.4%, but PNM decreased by 33.1% under wheat from the beginning to the end of the experiment in the greenhouse. In the field, MBC increased by 7.1%, but PNM decreased by 68.3% under wheat during this period. The corresponding increases in SOC, POC, STN, and MBN or decrease in PNM during this period were lower under pea and fallow. This suggests that wheat increased more soil C and N fractions, except PNM, than pea or fallow due to increased substrate availability from root and rhizodeposition and/or to slow decomposition of wheat than pea residue due to differences in residue quality (e.g. C/N ratio). The greater PNM under pea than wheat or fallow was due to increased N contribution from its residue (Table 5).

When the greenhouse and field experiments were compared, trends in changes in soil C and N fractions due to treatments within a crop growing season were similar. However, greater changes in labile than nonlabile C and N fractions occurred more in the greenhouse than in the field. Furthermore, the coefficient of variations in soil C and N fractions were lower in the greenhouse (2.6 to 15.9%) than in the field (8.0 to 36.7%) (Tables 4 and 5). This indicates that soil C and N fractions changed more readily but with lower variability with management practices within a crop growing season when soil and environmental conditions are controlled in the greenhouse than in field where soil heterogeneity often results in non-significant differences among treatments in these fractions [16,18,30]. Use of disturbed soil in the greenhouse vs. undisturbed (especially in the no-till system) in the field also may have an influence on differences in changes in soil C and N fractions between the two experiments. The greater changes in labile than nonlabile C and N fractions as influenced by management practices within a short period in the greenhouse and the field suggests that labile C and N fractions are better indicators of changes in soil organic matter quality than nonlabile fractions, a case similar to that reported by various researchers [10,11,13,16,30]. The fact that more changes in labile than nonlabile C and N fractions occurred in the greenhouse than in the field suggests that better measurements of changes in soil organic matter due to management practices within a short period can be observed when soil and environmental conditions are controlled. Greater levels of most soil C and N fractions in the greenhouse than in field was probably a result of increased turnover rate plant C and N into soil C and N, because disturbed soil was used in the greenhouse and environmental condition for microbial transformation was more favorable in the greenhouse than the field.

Greenhouse study provided more information on plant and residue parameters, such as measurement of root biomass and C and N contents and residue C and N losses, which cannot be measured easily in the field. This resulted in the measurement of turnover rate of plant C and N into soil C and N in the greenhouse, a fact that was absent in the field. Because of greater changes in soil C and N fractions, greenhouse study provided a more robust method of evaluating C and N cycling and soil quality within a short period of time as affected by management practices than the field experiment. Such changes can also be measured in the field but it may take longer time. While all results from the greenhouse study may not be readily applied in the field, some information, such as root biomass and residue C and N losses, measured in the greenhouse can be extrapolated to the field condition. The effects of short-term study in the greenhouse can be useful to predict the long-term impact of management practices on soil C and N fractions in the field.

Conclusions

Crop yields, residue C and N losses, and soil C and N fractions varied with residue placement and crop types in the greenhouse and the field. Surface placement of residue increased crop yields, residue C and N losses, and enhanced SOC, PCM, MBN, and NH_4-N concentrations, but residue incorporation increased PNM and NO_3-N concentrations. Similarly, spring wheat had higher yield and increased SOC, POC, MBC, STN, and MBN than pea or fallow, but pea had higher N content and increased PNM than wheat or fallow. Placing nonlegume residue at the soil surface

using no-tillage can increase soil C and N sequestration and microbial biomass and activity that can improve soil health and quality. Using this practice, producers can claim for C credit. Incorporation of legume and nonlegume residues into the soil using conventional tillage can increase N mineralization and availability which can reduce N fertilization rate to succeeding crops, but can degrade soil quality due to reduced organic matter and increased erosion. Although soil labile C and N fractions changed more readily than nonlabile fractions within a crop growing season both in the greenhouse and field, greater changes in labile than nonlabile fractions occurred with reduced variability more in the greenhouse than in the field. Results suggest that greenhouse study provided a more robust measurement of crop growth and changes in soil C and N fractions within a short period as influenced by management practices than the field experiment. Longer time will be probably needed in the field to obtain results similar to those in the greenhouse. Additional information, such as root growth, residue C and N losses, turnover of plant C and N to soil C and N, and results of short-term study on soil C and N fractions as influenced by management practices in the greenhouse can be used to predict the long-term impact in the field.

Acknowledgments

We appreciate the excellent support of Joy Barsotti and Thecan Caesar-TonThat for analyzing soil and plant samples in the laboratory.

Author Contributions

Conceived and designed the experiments: UMS. Performed the experiments: JW. Analyzed the data: UMS. Contributed reagents/materials/analysis tools: UMS. Wrote the paper: UMS.

References

1. Lal R, Kimble JM, Stewart BA (1995) World soils as a source or sink for radiatively-active gases. In: Lal R, editor, Soil management and greenhouse effect. Advances in soil science. CRC Press, Boca Raton, FL, pp. 1–8
2. Paustian K, Robertson GP, Elliott ET (1995) Management impacts on carbon storage and gas fluxes in mid-latitudes cropland. In: Lal R, editor, Soils and global climate change. Advances in soil science. CRC Press, Boca Raton, FL, USA, pp. 69–83.
3. Halvorson AD, Peterson GA, Reule CA (2002a) Tillage system and crop rotation effects on dryland crop yields and soil carbon in the central Great Plains. Agron J 94:1429–1436.
4. Sherrod LA, Peterson GA, Westfall DG, Ahuja LR (2003) Cropping intensity enhances soil organic carbon and nitrogen in a no-till agroecosystem. Soil Sci Soc Am J 67:1533–1543.
5. Sainju UM, Caesar-TonThat T, Lenssen AW, Evans RG, Kolberg R (2007) Long-term tillage and cropping sequence effects on dryland residue and soil carbon fractions. Soil Sci Soc Am J 71:1730–1739.
6. Rasmussen PE, Allmaras RR, Rhoade CR, Roager NC Jr (1980) Crop residue influences on soil carbon and nitrogen in a wheat-fallow system. Soil Sci Soc Am J 44:596–600.
7. Peterson GA, Halvorson AD, Havlin JL, Jones OR, Lyon DG, Tanaka DL (1998) Reduced tillage and increasing cropping intensity in the Great Plains conserve soil carbon. Soil Tillage Res 47:207–218.
8. Bauer A, Black AL (1994) Quantification of the effect of soil organic matter content on soil productivity. Soil Sci Soc Am J 58:185–193.
9. Ghidey F, Alberts EE (1993) Residue type and placement effects on decomposition: Field study and model evaluation. Trans ASAE 36:1611–1617.
10. Sainju UM, Lenssen AW, Caesar-Tonthat T, Waddell J (2006b) Tillage and crop rotation effects on dryland soil and residue carbon and nitrogen. Soil Sci Soc Am J 70:668–678.
11. Halvorson AD, Wienhold BJ, Black AL (2002b) Tillage, nitrogen, and cropping system effects on soil carbon sequestration. Soil Sci Soc Am J 66:906–912.
12. Kuo S, Sainju UM, Jellum EJ (1997) Winter cover cropping influence on nitrogen in soil. Soil Sci Soc Am J 61:1392–1399.
13. Sainju UM, Lenssen AW, Caesar-TonThat T, Waddell J (2006a) Carbon sequestration in dryland soils and plant residue as influenced by tillage and crop rotation. J Environ Qual 35:1341–1349.
14. Miller PR, McConkey B, Clayton GW, Brandt SA, Staricka JA, Johnston AM, Lafond GP, Schatz BG, Baltensperger DD, Neill KE (2002) Pulse crop adaptation in the northern Great Plains. Agron J 94:261–272.
15. Sainju UM, Lenssen AW, Caesar-TonThat T, Evans RG (2009) Dryland crop yields and soil organic matter as influenced by long-term tillage and cropping sequence. Agron J 101:243–251.
16. Franzluebbers AJ, Hons FM, Zuberer DA (1995) Soil organic carbon, microbial biomass, and mineralizable carbon and nitrogen in sorghum. Soil Sci Soc Am J 59:460–466.
17. Bezdicek DF, Papendick DF, Lal R (1996) Introduction: Importance of soil quality to health and sustainable land management. In: Doran JW, Jones AJ, editors, Methods of assessing soil quality, Spec. Publ. 49, Soil Science Society of America, Madison, USA, pp. 1–18.
18. Franzluebbers AJ, Arshad MA (1997) Soil microbial biomass and mineralizable carbon of water-stable aggregates. Soil Sci Soc Am J 67:1090–1097.
19. Cambardella CA, Elliott ET (1992) Particulate soil organic matter changes across a grassland cultivation sequence. Soil Sci Soc Am J 56:777–783.
20. Six J, Elliott ET, Paustian K (1999) Aggregate and soil organic matter dynamics under conventional and no-tillage systems. Soil Sci Soc Am J 63:1350–1358.
21. Pikul JL Jr, Aase JK (2003) Water infiltration and storage affected by subsoiling and subsequent tillage. Soil Sci Soc Am J 67:859–866.
22. Nelson DW, Sommers LE (1996) Total carbon, organic carbon, and organic matter. In: Sparks DL, editor, Methods of soil analysis. Part 3. Chemical method. SSSA Book Ser. 5. Soil Science Society of America, Madison, pp. 961–1010.
23. Haney RL, Franzluebbers AJ, Porter EB, Hons FM, Zuberer DA (2004) Soil carbon and nitrogen mineralization: Influence of drying temperature. Soil Sci Soc Am J 68:489–492.
24. Franzluebbers AJ, Haney RL, Hons FM, Zuberer DA (1996) Determination of microbial biomass and nitrogen mineralization following rewetting of dried soil. Soil Sci Soc Am J 60:1133–1139.
25. Voroney RP, Paul EA (1984) Determination of k_C and k_N in situ for calibration of the chloroform fumigation-incubation method. Soil Biol Biochem 16:9–14.
26. Brookes PC, Landman A, Pruden G, Jenkinson DJ (1985) Chloroform fumigation and the release of soil nitrogen: A rapid direct-extraction method to measure microbial biomass nitrogen in soil. Soil Biol Biochem 17:937–942.
27. Littell RC, Milliken GA, Stroup WW, Wolfinger RR (1996) SAS system for mixed models. SAS Institute Inc., Cary, NC, USA.
28. Lenssen AW, Johnson GD, Carlson GR (2007) Cropping sequence and tillage system influences annual crop production and water use in semiarid Montana. Field Crops Res 100:32–43.

29. Coppens F, Garnier P, de Gryze P, Merckx R, Recous S (2006) Soil moisture, carbon and nitrogen dynamics following incorporation and surface application of labelled crop residues in soil columns. Europ J Soil Sci 57:894–905.

30. Giacomini SJ, Recous S, Mary B, Aita C (2007) Simulating the effects of nitrogen availability, straw particle size and location in soil on carbon and nitrogen mineralization. Plant Soil 301: 289–301.

31. Malhi SS, Lemke R (2007) Tillage, crop residue and nitrogen fertilizer effects on crop yield, nutrient uptake, soil quality and greenhouse gas emissions in the second 4-yr rotation cycle. Soil Tillage Res 96:269–283.

32. Wright AL, Hons FM, Lemon RG, MacFarland ML, Nichols RL (2008) Microbial activity and soil carbon sequestration for reduced and conventional tillage cotton. Appl Soil Ecol 38:168–173.

33. Lupwayi NZ, Lafond GP, Ziadi N, Grant CA (2012) Soil microbial response to nitrogen fertilizer and tillage in barley and corn. Soil Tillage Res 118:139–146.

34. Lu Y, Watanabe A, Kimura M (2002) Contribution of plant-derived carbon to soil microbial biomass dynamics in a paddy rice microcosm. Biol Fertil Soils 36:136–142.

35. Liebig MA, Tanaka DL, Wienhold BJ (2004) Tillage and cropping effects on soil quality indicators in the northern Great Plains. Soil Tillage Res 78:131–141.

36. Kuzyakov Y, Domanski G (2000) Carbon input by plants into the soil: Review. J. Plant Nutri Soil Sci 163:421–431.

13

Conservation of Sandy Calcareous Grassland: What Can Be Learned from the Land Use History?

Anja Madelen Ödman*, Pål Axel Olsson

Biodiversity, Department of Biology, Lund University, Lund, Sweden

Abstract

Understanding the land use history has proven crucial for the conservation of biodiversity in the agricultural landscape. In southern Sweden, very small and fragmented areas of the disturbance-dependent habitat xeric sand calcareous grassland support a large number of threatened and rare plants and animals. In order to find out if historical land use could explain variation in present-day habitat quality, the land use on eight such sites was traced back to the 18th century and compared with key factors such as the amount of bare sand, lime content and P availability. There was no support for the common explanation of the decline in xeric sand calcareous grassland being caused by abandonment of agricultural fields during the last century. Instead, fertilization history was the main explanation for the difference in depletion depth of $CaCO_3$ seen between the sites. The decline in xeric sand calcareous grassland since the 18th century is most probably the result of the drastic changes in land use during the 19th century, which put an end to the extensive sand drift. Since cultivation was shown to have played an important role in the historical land use of xeric sand calcareous grassland, grazing alone may not be the optimal management option for these grasslands. Instead more drastic measures are needed to restore the high calcium content and maintain proper disturbance levels.

Editor: Peter Shaw, Roehampton university, United Kingdom

Funding: Financial support from The County Administrative Board, The Swedish Research Council, and the Gyllenstiernska Krapperup Foundation made this study possible. The funders had no role in study design, data collection and analysis, decision to publish, or preparation of the manuscript.

Competing Interests: The authors have declared that no competing interests exist.

* E-mail: Anja.Odman@biol.lu.se

Introduction

There is a growing awareness among conservation biologists regarding the importance of land use history and local management practices when implementing conservation and restoration measures for biodiversity and threatened species [1,2,3]. It is also recognised that land use changes long ago may have caused the vegetation patterns observed today [4,5]. For example, Eberhardt et al. [2] found that large areas of the sand forests at Cape Cod had been agricultural fields in the middle of the 19th century. They concluded that it was important to "mimic past agricultural practices in order to maintain and restore important sand-plain habitatsÓ. In particular for sandy areas, erosion and vegetation burial have been an important part of habitats history, and plants are adapted to this [6,7].

Throughout Europe, sandy grassland habitats support a large number of threatened species. Perhaps the best example of such sandy habitats is the threatened xeric sand calcareous grassland (Natura 2000 code 6120, 2002/83/EC Habitat Directive), which is home to several endangered species of vascular plants [8,9,10], bryophytes [11], fungi [12] and invertebrates [13]. These habitats are dry, open grasslands on calcareous, more or less humus-free, nutrient-poor and well-drained sandy soils with a discontinuous vegetation cover. The xeric sand calcareous grassland specialists are adapted to high pH [14] and low nutrient content [8,15], and many of the species depend on the open sand for regeneration [8,9]. In eastern Skåne (southern Sweden), these grasslands developed on lime-rich glaciofluvial sand [8,16], which was deposited by glacial meltwater when the ice retreated around

14,000 years ago, and later re-deposited by wind and waves [16]. Sites with natural erosion such as river banks, steep slopes, sand dunes and dune slacks have been proposed as sites where the type of species found in dry grasslands could have occurred before deforestation [17].

Before 3000 BC, eastern Skåne (Fig. 1) was predominantly covered by forest [18]. After 3000 BC, coppice agriculture led to an opening of the landscape, and there was a concentration of settlements to the flat coastal areas, and around 800 BC large areas of the coastal plains were deforested [18]. From 800 BC, permanent settlements replaced the mobile settlements and fields previously practiced [18]. At this time, intensified land use also opened up the landscape on other places in the Baltic region [19]. An infield/outland system with fertilisation was introduced in the area and there was an expansion of cereal cultivation, which also increased erosion. The traditional method of farming was extensive and with long (up to 30 years) periods of fallow when the arable fields were used for grazing [20,21,22]. During the 17th and 18th centuries the human population in southern Sweden grew, which caused further deforestation and intensified agricultural practices [20,22], and in the 18th century, Linné [23] noted an extensive sand drift and the occurrence of sand fields all over eastern Skåne. The problem with sand drift was for the most part solved at the end of the 18th century and during the 19th century by the introduction of ley plants and more efficient use of fertilisers, which increased the production per surface area and thereby stopped over-exploitation. At the same time, the afforestation of Skåne started [22] and pine was planted to

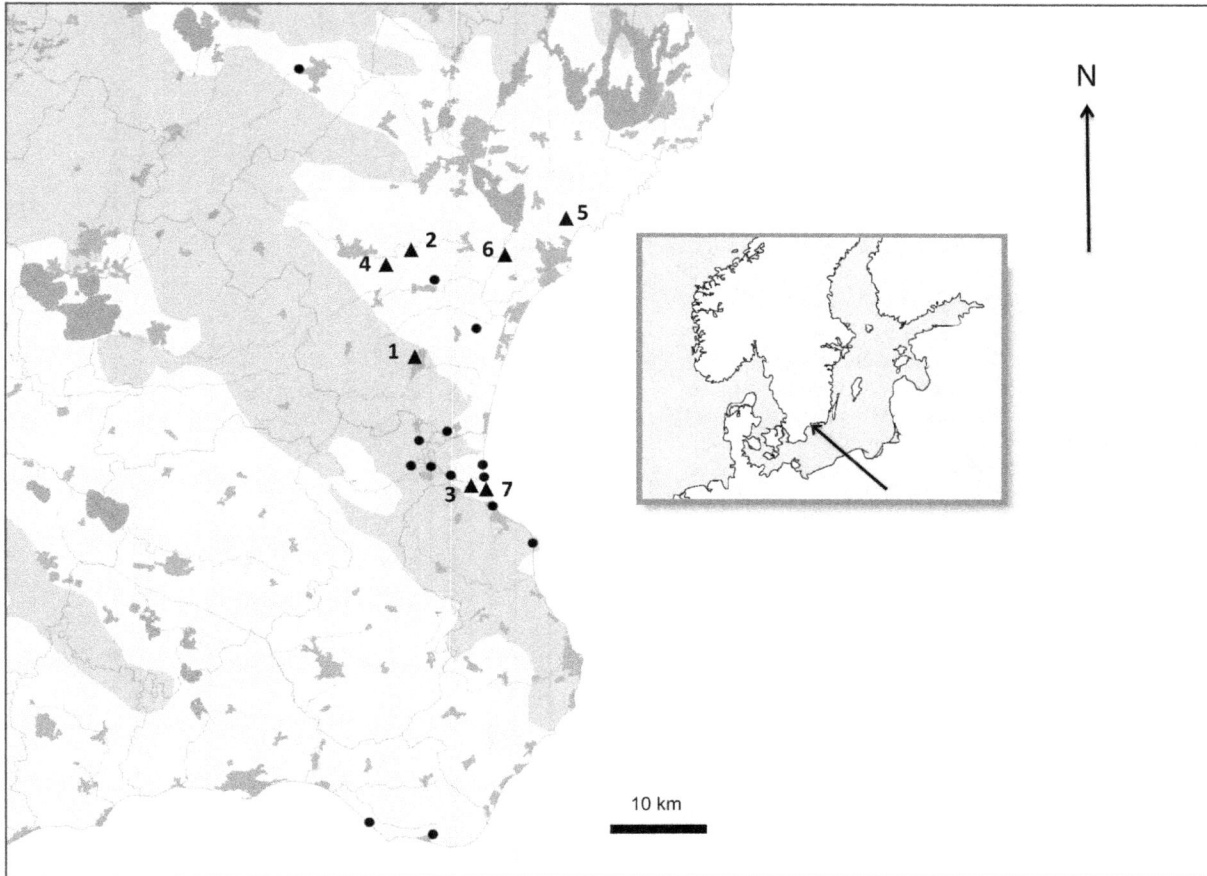

Figure 1. Map showing the existing xeric sand calcareous grassland sites in Skåne. Sites included in the study are represented by triangles and circles represent sites not included. Grey shading indicates forested land and unshaded represents open land. Scale bar represent 10 km. Enclosed is an overview map for southern Scandinavia, where an arrow points out the study area. The sites chosen for the study were: (1) Degeberga, (2) Everöd, (3) Klammersbäck, (4) Lyngsjö, (5) Rinkaby, (6) Ripa and (7) Vitemölla.

prevent wind erosion. Similar trends with deforestation until the middle of the 19th century, and then afforestation again has also been found in other places in Sweden [24] as well as in New England, USA [25].

Today, the old agricultural practices in sandy grasslands have been abandoned, and the arable fields have been turned into pine plantations or pastures [26,27], leaving the vegetation cover to close and thereby ending sand drift [28]. A natural accumulation of nutrients and organic matter follows [10], which can be further accelerated by atmospheric nitrogen deposition [29]. In addition, there is a threat of acidification of the topsoil [8]. The remaining areas of xeric sand calcareous grassland are small and fragmented, which poses a further threat to the rarest species (Fig 1).

Linné [23] described a number of xeric sand calcareous grassland species growing in eastern Skåne. In particular *Dianthus arenarius* ssp *arenarius*, which is a key species in these grasslands and protected by the Natura 2000 network (EU Habitat Directive, Annex II), was mentioned several times and was said to grow on all the sand fields, almost like a weed. In fact, all xeric sand calcareous grassland specialists but two, *Silene conica* and *Alyssum alyssoides*, which came to Skåne in the 19th century, were present in Skåne in 1749 [23,30,31]. Observations by Linné [23] and Campbell [20] suggest that xeric sand calcareous grassland was widespread in eastern Skåne during the 18th century, but there are no records of

the actual area before the 1970s. Since the 1970s, reported areas range between 30 and 50 ha, with no consistent trend between the different studies [26, 27 32]. However, there is a strong indication of decline in xeric sand calcareous grassland species since the 18th century, which is reflected in the high density of nationally red-listed (and thereby declining) plant species [8].

Changes in land use during the last century, i.e. that the agricultural practices on these grasslands were abandoned, has been the common explanation for the decline of xeric sand calcareous grassland in Skåne [26 27, 33]. However, this hypothesis had never been tested prior to this study. The aim of this study was to improve the conservation of dry calcareous grassland by increasing the understanding of how xeric sand calcareous grasslands came to be and why they have disappeared. Land use change and decalcification were compared at eight sites in eastern Skåne in areas with a range of different quantities and qualities of xeric sand calcareous grassland. The amount of bare sand and the $CaCO_3$ content of the soil were used as proxies for favourable conditions and, in the case of open sand, as a proxy for erosion. Historical maps were used to determine the former land use and aerial photographs were used to map the extent of bare sand during the last 70 years. By analysing soil profiles, possible links between land use history and decalcification in the studied grasslands were investigated at specific sites within each studied landscape. Extractable P concentrations were also measured as a proxy for fertilisation history, since P remains in the soil for a longer period than N and causes problems during restoration [34,35].

Materials and Methods

Site descriptions

The sites studied were situated in the coastal area of eastern Skåne in southernmost Sweden. The area is 80 km long and 35 km wide, situated between 6172130 and 6204470 N, and 441330 and 458270 E (SWEREF99 TM). Sites were selected on the basis that they contained at least some remnants of xeric sand calcareous grassland vegetation. They should also be, or be part of, a larger open area. Since the focus of the study was the effects of agriculture, sites known to have been created by sand pits or other digging activities were excluded. Still, the majority of landscapes with xeric sand calcareous grassland in the area were included. The sites chosen for the study were: 1. Degeberga (6188081, 442627), 2. Everöd (6198995, 443126), 3. Klammersbäck (61749171, 448414), 4. Lyngsjö (6198615, 442108), 5. Rinkaby (6202616, 455831), 6. Ripa (6198003, 452133) and 7. Vitemölla (6173465, 449873) (Fig 1). The sites are listed in Table 1, where information about their topography, present day management and the amount and quality of the remaining xeric sand calcareous grassland vegetation can be found. These sites are all under nature conservation management, and the conducted study was made in collaboration with the county administrative board in Skåne and followed site-specific regulations. The Everöd, Klammersbäck, Lyngsjö, Rinkaby and Ripa sites are all more or less flat areas. The vegetation at these sites predominantly consists of Fennoscandian lowland species-rich, dry to mesic grassland (N6270, EU Habitat Directive) as well as small areas of heath-like vegetation resembling continental dunes with open Corynephorus and Agrostis grasslands (N2330, EU Habitat Directive) [8,28]. All of the flat sites contain patches of xeric sand calcareous grassland of varying sizes, depending on the level of degradation. Degeberga and Vitemölla are hilly sites with large presence of xeric sand calcareous grassland vegetation on the south facing slopes, and with Fennoscandian lowland species-rich, dry to mesic grassland on flatter areas.

Soil sampling and analyses

Soil was sampled at the following locations: Everöd (6199099, 443104), Klammersbäck (6174279, 448413), Lyngsjö (6198665, 442138), Rinkaby (6202600, 455808), Ripa pasture (6198191, 452028), and Ripa airfield (6197740, 452277) (SWEREF99 TM). At each of these sites a smaller area was selected that was assumed to consist of degraded xeric sand calcareous grassland. The vegetation at most sampling sites was either Fennoscandian dry to mesic grassland or dry Corynephorus grassland. The soil was sampled on 8-18th June 2009 and soil cores of 20 cm length (Ø 15 cm) were taken down to 1.6 m using a soil auger. Three profiles were sampled at each site, located approximately 10 m from each other in a triangle. At the Ripa airfield site, the soil could only be sampled down to 1.2 m for two of the three profiles because bedrock or boulders were encountered at this depth.

For pH measurements, 10 g of soil was mixed with 50 ml distilled water, and the pH was measured electrometrically in supernatants obtained by 2-h extraction in a rotator. The total amounts of Ca in the samples were measured by digesting the soil in concentrated HNO_3, followed by ICP-AES analysis. One g of soil at field moisture and 50 ml HNO_3 were placed in a 300 ml flask, covered with a watch glass and heated on a hot plate for

Table 1. A summary of the characteristics of the eight sites analyzed in this study, including the total area analysed and the area of xeric sand calcareous grassland (Natura 2000 habitat 6120) according to Olsson [32].

Site	Topography	Present day management/Disturbance	Area (ha)		Vegetation (vascular plants)				
			Total	Habitat type 6120	Total cover (%)	Specialists cover (%)	Red-listed cover (%)	Red-listed (Nr/m²)	N
1. Degeberga	Hilly	Ungulate grazing	13.5	2.6	47±20	23±13	17±14	2.7±0.6	3
2. Everöd	Flat	Ungulate grazing	15	4.9	57±23	14±16	9.2±15	1.0±0.7	10
3. Klammersbäck	Flat	Ungulate grazing	38	0.09	34±16	4±1.4	2.5±3.5	0.5±0.7	2
4. Lyngsjö	Flat	Horse grazing	3	0.1	58±23	12±16	10±16	0.6±0.7	17
5. Rinkaby	Flat	Ungulate grazing/military training	155	4.5	64±20	21±15	14±12	1.4±0.8	25
6a. Ripa airfield	Flat	Fallow/wild grazers (rabbits)	14	0.2	71±23	20±26	17±29	0.3±0.6	3
6b. Ripa pasture	Flat	Ungulate grazing	22	3.0	52±6.4	7.0±5.7	5.5±6.4	1.0±0.0	2
7. Vitemölla	Hilly	Fallow/recreation, wild grazers (rabbits)	38	4.8	59±22	23±21	16±18	1.7±1.3	12

Vegetation data comes from a survey in 2004-2005 [9] and includes the total vegetation cover, as well as the cover of specialist species and nationally red-listed species as defined in Olsson and Ödman [55]. In addition, the number of red-listed species per square meter (Nr/m²) is given. Means are displayed (± SD), and N denotes the number of analysed squares (1 m²) at each site. The dominant red-listed and specialist species were *Koeleria glauca*, *Dianthus arenarius* ssp *arenarius* and *Alyssum alyssoides*.

72 h. The solution was evaporated to 2 ml and diluted to 25 ml with distilled water. Plant-available (extractable) ortho-phosphate (P_{ext}) in the top 20 cm was determined using flow injection analysis, with stannous chloride and ammonium molybdate as reactant (measured at 720 nm, method ISO/FDIS 15681-1). Ten g fresh soil was extracted for 1 min with Bray-1 solution [36].

Soil in Degeberga and Vitemölla was not sampled during this study and samples collected for earlier studies by Olsson et al. [8] and Bahr et al. [37] respectively were used instead. At Degeberga, two samples were used: one from Degeberga south slope (xeric sand calcareous grassland) and one from Degeberga north slope (Fennoscandian dry to mesic grassland), which had been sampled down to 1.5 m. At Vitemölla, four samples were used from a mosaic of xeric sand calcareous grassland and dry Corynephorus grassland. The samples were taken down to 0.6 m at three sites and 0.5 m for one of the sites. Data for soil pH and Ca content comes from data collected for earlier studies by Olsson et al. [8] and Bahr et al. [37]. At Vitemölla, Ca concentration values were only available for the top 0.1 m. Additional measurements of P_{ext} were performed on the soil samples for this study, using the methods described above. The samples from the top 0.1 m for all samples at Vitemölla had unfortunately been used up in the previous study, and the 0.1–0.2 m layer was used for the P_{ext} analysis.

Analysis of land use history

Information on land use in the past was obtained through visual analysis of historical maps and aerial photographs. The area at each site was delimited so that it enclosed the soil sampling plots, had clearly defined borders (arable field margins, roads, water etc.) and only included open land without trees in the 2007 aerial photograph. The sizes of the delimited areas are presented in Table 1. Historical maps were obtained from Lantmäteriet (the Swedish land surveying agency), thru a thorough search in their archives [38]. The historical maps used dated from the middle of the 18th century (oldest maps available from Lantmäteriet) until 1974 and consisted of economic and property redistribution maps (Table 2). The analyses used arable field margins and other objects to locate the sites, and explanatory text or legends were used to determine land use. The land use was either printed on the map or found in a supplementary document accompanying the map. The land uses recorded were: arable field, pasture, meadow, forest and other land uses of interest such as military training or recreation. Areas named "utmark" (outland), "fålad" (old word for common pasture in southern Sweden) and "allmänning" (common land) were classified as pastures, and areas named "wång" (old word for common arable fields in southern Sweden) were classified as arable fields. In the economic map from 1974, only arable fields were mapped and all other land uses were treated as one class (other). The aerial photographs used dated from the 1940s until 2007 (Table 2) and were obtained from Lantmäteriet [39]. The resolution varied between 0.2 m^2 and 1 m^2. The analysis, which was done visually, was based on the following criteria: forest (>50% tree cover), arable field (open, with furrows) or pasture (open, without furrows). Furrows were interpreted as signs of ploughing. The map and aerial photograph analyses were performed at each site, both for the whole area (landscape level) and at the exact sites (plot level) where the soil was sampled. The results from the analysis were used to calculate the number of years since the area was last used for cultivation. The number of years was counted from the last year the area was recorded as having been cultivated. In the case of Degeberga and Vitemölla, which had never been cultivated, the number of years from the first available record of land use was used.

Analysis of the amount of bare sand and forest

The aerial photographs described above were used to analyse how the amount of bare sand changed between the 1940s and 2007. The aerial photographs were rectified and classified using ArcGIS Desktop 9.3 (ESRI 2008). The delimited areas were the same as those described above for the historical land use analysis. The amount of bare sand was estimated by dividing the surface features shown on the aerial photographs into two classes: bare sand and other (anything which was not bare sand). The classes were identified separately for each photograph using supervised classification where areas of bare sand were identified based on colour (and previous knowledge of the areas for the 2007 photographs), and then used as training areas for the software. The delimited areas were then classified using maximum likelihood classification. The classification for 2007 was evaluated in the field by comparing classified and real patches, and the older ones by comparing the classified patches of bare sand with the visual ones in the aerial photographs. The analysis identified patches of bare sand, but did not consider the overall density of the vegetation. Consequently, only patches with close to 100% bare sand could be identified, and areas with partly bare sand of varying amounts were classified as "other" since they could not be distinguished.

The amount of forest (>50% tree cover) in the surrounding landscape was also analysed from the aerial photographs. The size of the aerial photographs varied considerably between sites and years, and four different areas around the sites, 1×1, 1.5×1.5, 2×2 and 3×3 km, were used. The Everöd and Lyngsjö sites are situated within the same area and were analysed together. Analysis showed that, in those areas where squares larger than 1×1 km could be fitted, the amount of forest for all years increased slightly with increasing size of areas (although not significantly so). However, differences between the years were constant and so only results from the analyses using the 1×1 km squares are presented. Polygons were drawn manually around all forest and the percentage of forest in the total area was calculated using ArcGIS.

Statistical analyses

Correlations between land use history, amount of forest, amount of bare sand, depletion depth for $CaCO_3$, maximum $CaCO_3$, pH and P_{ext} content were performed using Pearson correlation or Spearman's correlation when parameters were not normally distributed. For the amount of forest and amount of bare sand, correlations were also performed with year, using the same method as described above. Differences between sites, land use history and depths in terms of the amount of $CaCO_3$ and P were tested using a general linear model followed by Tukey's post hoc test. All statistical analyses were performed using IBM SPSS Statistics (IBM Corporation, New York, US). An Arcsine square root transformation of the parameter amount of bare sand was performed to obtain normal distribution. For the correlation between pH and $CaCO_3$, the $CaCO_3$ values were log transformed. Results presented are mean values and standard deviations.

Results

Historical land use changes

The results from the analysis of land use change at the soil sampling plots showed that the last record of cultivation at the six flat sites was between 1940 and 1957, except for the Ripa pasture, where indications of cultivation could be seen also in aerial photographs from 1985 (Table 2). At the sampling plots at the two

Table 2. Results from visual analysis of land use history through historical maps (economic and property redistribution maps, with original names in Swedish) and aerial photographs for all eight soil-sampling sites.

	1. Degeberga	2. Everöd	3. Klammersbäck	4. Lyngsjö	5. Rinkaby	6a. Ripa airfield	6b. Ripa pasture	7. Vitemölla
Aerial photographs								
2007	Pasture	Pasture	Pasture	Pasture	Military training/Pasture	Pasture	Pasture	Pasture/recreation
2001					Military training/Pasture			
1999	Pasture	Pasture	Pasture	Pasture			Pasture	Pasture	Pasture/recreation
1985	Pasture	Pasture	Pasture	Pasture	Military training/Pasture	Airfield/Pasture	Arable field?	Pasture/recreation
1970					Military training/Pasture	Airfield/Pasture	Arable field?	
1969	Pasture	Pasture	Pasture	Arable field				Pasture/recreation
1957	Pasture	Arable field		Arable field		Pasture?	Arable field	
1956			Pasture		Arable field			Pasture/recreation
1940	Pasture	Arable field	Arable field	Arable field	Arable field	Arable field	Arable field	Pasture/recreation
Historical maps								
1974, Ekonomiska kartan	Other	Other	Other	Other	Other	Other	Other	Other
1926-34, Härads ekonomiska	Pasture	Arable field	Arable field	Arable field	Arable field	Arable field	Arable field	Pasture
1845, Laga skifte					Plantation			Meadow, drift sand
1826, Enskifte		Arable field		Arable field				
1822, Enskifte			Arable field					
1818, Storskifte						Arable field	Arable field	
1811, Enskifte	Pasture							
1803, Avmätning		Arable field		Arable field				
1751, Ägodelning					Arable field			Pasture
Years since cultivation	196	50	67	38	51	67	22	256

"Other" means an area not classified as arable field or real estate in the economic maps (Ekonomiska kartan). Maps and aerial photographs were obtained from Lantmäteriet [38,39].

hilly sites, none of the maps or aerial photographs showed signs of agricultural cultivation (Table 2).

At the landscape level, the six flat areas, Everöd, Lyngsjö, Klammersbäck, Ripa pasture, Ripa airfield and Rinkaby, have all been cultivated but, according to the maps, cultivation ceased between 1940 and the 1990s. They have gone through a historical land use change from a mixture of cultivation and grazing and mowing, to a mixture of forest and grazing. During the 19th century arable fields were common, but at Everöd, Rinkaby and Klammersbäck, pastures occurred as well, and these pastures were managed as commons. At the two coastal sites Rinkaby and Klammersbäck, the common pastures occurred closest to the sea, while fields occurred inland. During the 20th century, the agricultural fields were gradually abandoned and turned into pastures. During this time, forests started to emerge at the Everöd, Rinkaby and Ripa sites.

The two hilly sites, Vitemölla and Degeberga, have a rather similar history. In 1756, the northern part of the site at Vitemölla was classified as very poor arable fields spoiled by sand drift and therefore unusable. The southern part was classified as useless and left as a common in 1778. In 1828 and 1845, the southern part of the area was described as sandy soil and drift sand, and the lower parts close to the sea as useless. According to the maps from 1926-34 and onward, the area was used as pasture and for recreation. The Degeberga site has been used as pasture continually from the first map in 1811 until today. Consequently, according to the maps used in this study, the two hilly sites were not cultivated during the

analysed period. However, the note from 1756 about parts of Vitemölla being arable field spoiled by sand drift indicates that some parts had previously been cultivated.

Forest cover and amount of bare sand from 1940 to 2007 at landscape level

There was no significant temporal trend in the mean forest cover between 1940 and 2007, when analysed in 1×1 km squares within each landscape. In three areas (Ripa, Everöd/Lyngsjö and Rinkaby) there was an increase in forest cover, while the forest cover decreased in Klammersbäck, and almost no changes could be seen in the other two areas (Fig 2a). The area with the highest mean forest cover over the years (n = 6) was Klammersbäck (50±6%), followed by Rinkaby (44±4%), Degeberga (28±1%), Ripa (20±21%), Everöd/Lyngsjö (18±6%) and Vitemölla (11±2%).

The proportion of bare sand was analysed for eight sites, as Ripa airfield and Ripa pasture were treated separately, ranging in size between 3 and 155 ha. At Vitemölla, Rinkaby and Degeberga, the analysis showed a decrease in the amount of bare sand between 1940 and 2007 (Fig 2b). However, the general trend for all sites was that changes since the 1940s were not drastic, and that the mean amount of bare sand at the studied sites was never more than about 3% of the total study area. There was no significant change in the mean amount of bare sand between 1940 and 2007. The largest change was found at Vitemölla, where the amount of bare sand decreased from 11.2% to 4.4%. There was no

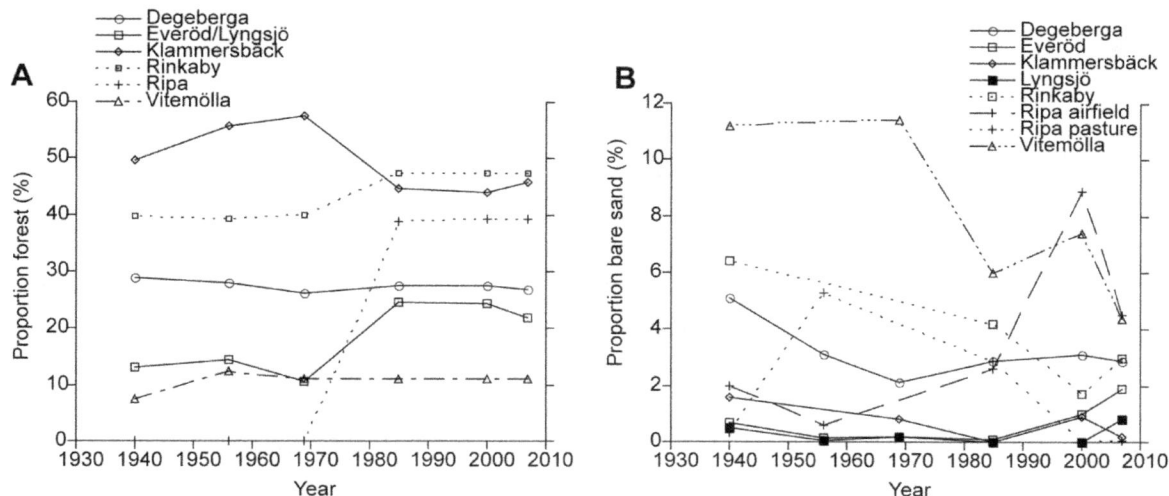

Figure 2. The proportion of land area occupied by forest (a) and bare sand (b) between 1940 and 2009 as analysed from aerial photographs (for simplification we did not differ between 1956/1957 and 1969/1970 in the graph). The proportion of forest was analysed for a 1×1 km square around the sites (Everöd/Lyngsjö and Ripa airfield/Ripa pasture were each contained within the same square), and the proportion of bare sand for the whole area of each of the eight sites.

correlation between the amount of bare sand and the forest cover or the number of years since cultivation.

Historical land use and soil properties

Extractable P (P_{ext}) showed large variation between sites. Lyngsjö had the highest mean (n = 3) concentration of P_{ext} (79±18 µg g^{-1} soil), followed by Ripa pasture (51±4.4 µg g^{-1} soil), Ripa airfield (48±4.4 g^{-1} soil) and Everöd (47±6.8 µg g^{-1} soil). Degeberga north (26 µg g^{-1} soil) and the four sampling points at Vitemölla (13±8.2 µg g^{-1} soil) had intermediate concentrations, while the sites at Rinkaby (6.5±4.4 µg g^{-1} soil), Degeberga south (6.3 µg g^{-1} soil) and Klammersbäck (5.1±1.6 µg g^{-1} soil) showed considerably lower P_{ext} values than the other sites. Lyngsjö had significantly higher P_{ext} values than all the other sites (p<0.05) and the Everöd, Ripa pasture and Ripa airfield sites had significantly higher P_{ext} values than Degeberga, Klammersbäck, Rinkaby and Vitemölla (p<0.05). No significant differences were found between the other sites. There was a positive correlation between the P_{ext} and the depletion depth in the sites that had been cultivated (N = 18, R^2 = 0.53, p<0.001, Fig. 3). There was no significant correlation between the P_{ext} and the number of years since cultivation or the amount of bare sand.

None of the sites had an enrichment of $CaCO_3$ at any particular depth. Most of the topsoils were depleted but, when the lime horizon was reached, the $CaCO_3$ content remained rather constant further down (Fig. 4). The lowest $CaCO_3$ values were found at Lyngsjö (0.2±0.03%), which was also the only site where the $CaCO_3$ content never exceeded 10%, even in the deepest layer. The only sites with a $CaCO_3$ content in the topsoil above 2% were Rinkaby (13.1±3.4%) and Vitemölla (5.4% in sampling point 2), and at 0.2–0.4 m only Everöd, Klammersbäck and Rinkaby had $CaCO_3$ concentrations above 2%. The pH values were strongly correlated with the $CaCO_3$ values (N = 156, R^2 = 0.93, p<0.001).

The maximum $CaCO_3$ concentration for each profile was identified as well as the depletion depth (the first depth where the $CaCO_3$ concentrations exceeded 2%). The maximum $CaCO_3$ value was significantly higher at Rinkaby and Degeberga than at Everöd, Klammersbäck, Lyngsjö and Ripa pasture (p<0.05). The

maximum $CaCO_3$ value at Ripa airfield was significantly higher than that of Lyngsjö (p<0.001), but did not differ from that of Rinkaby and Degeberga. Lyngsjö had a significantly lower maximum $CaCO_3$ value than all other sites except Klammersbäck (p<0.001).

There was a positive correlation between the maximum $CaCO_3$ and the mean amount of open sand (N = 7, R^2 = 0.73, p<0.01). The maximum $CaCO_3$ value and the depletion depth were not significantly correlated but there was a weak negative correlation between maximum $CaCO_3$ and the depth at which the maximum $CaCO_3$ was found (N = 20, R^2 = 0.30, p<0.01). The maximum $CaCO_3$ and the pH in the top 20 cm were also positively correlated (N = 20, R^2 = 0.28, p<0.01). There was no correlation between the number of years since cultivation and the depletion depth, or between the mean amount of bare sand and the depletion depth.

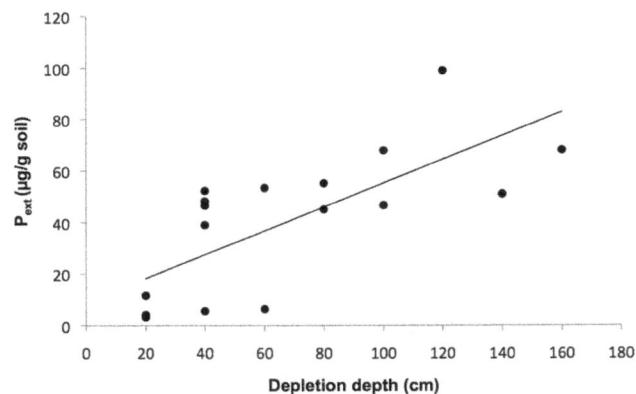

Figure 3. The correlation between P_{ext} (µg g-1 soil) and depletion depth (cm) at the previously cultivated sites (N = 18, R^2 = 0.53, p<0.001). The number of years since cultivation was counted from the last year the area was recorded as having been cultivated. In the case of Degeberga and Vitemölla the number of years from the first available record of land use was used.

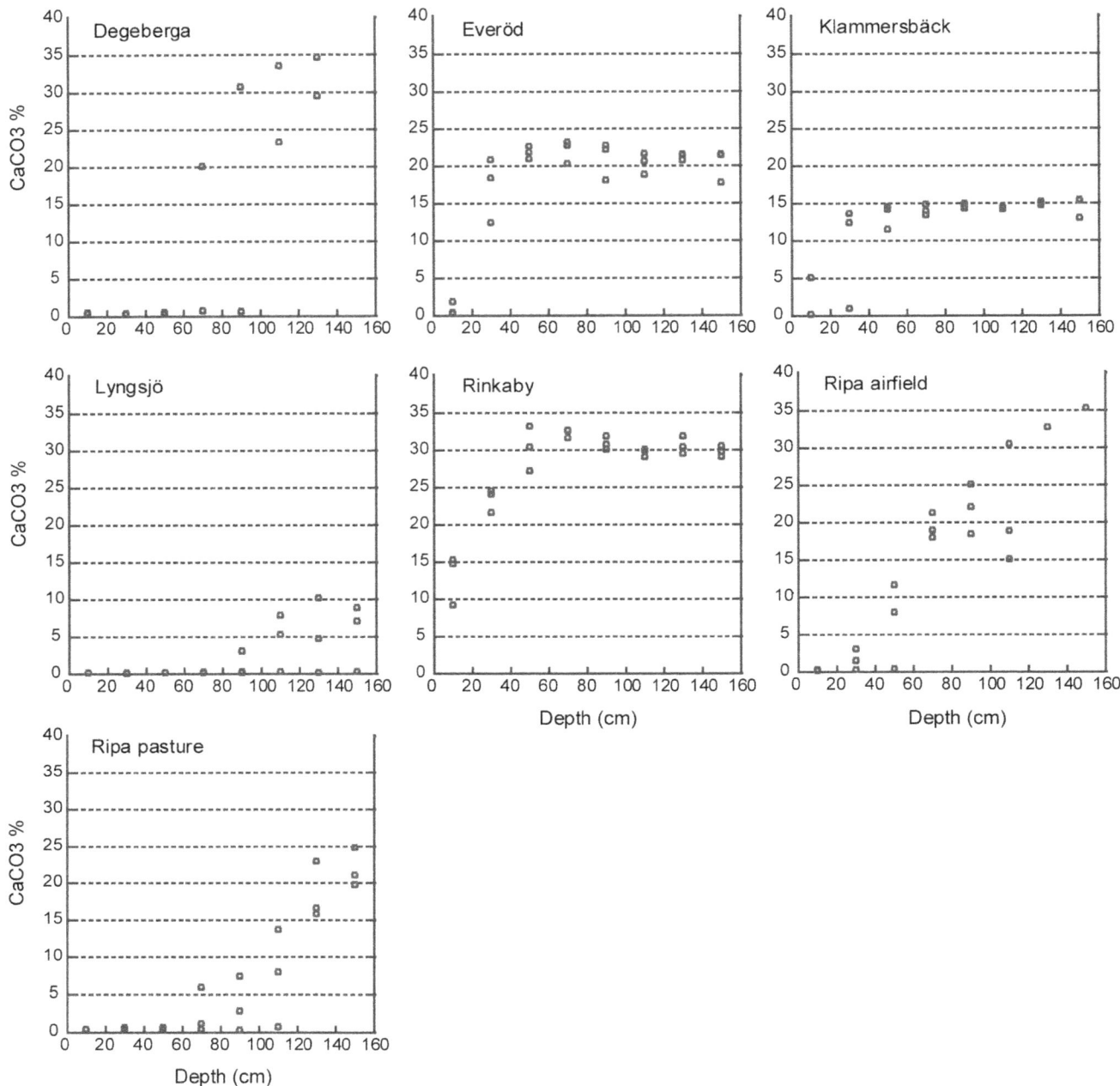

Figure 4. The CaCO₃ content as percentage of the soil weight for 20 cm at a time down to 160 cm. Vitemölla was not included since samples were only available for the top 20 cm.

Discussion

Most of the remaining xeric sand calcareous grasslands in Skåne are found on eroded sites [27], where the underlying lime-rich soil is exposed. This gives an indication about the great importance of erosion and other soil disturbances in the restoration and preservation of these grasslands. The importance of sand pits and military training areas for the conservation of biodiversity and threatened species has also been demonstrated in other parts of Europe [40,41,42,43]. An important finding of this study was that all the flat sites with xeric sand calcareous grassland included in the study had been cultivated in the past, which suggests that former

agricultural practices were important in creating the habitat where natural erosion does not occur. Similarly, the chalk grasslands of Salisbury Plain in the UK, which were previously only thought to have a history as sheepwalks, were found to have a history of cultivation [44]. In other types of semi-natural grasslands, grazing continuity is the most important factor for diversity of grazing-dependent species [45,46]. In the case of sandy grasslands it may instead be that continuous traditional low-intensity cultivation is a very important factor in creating the proper conditions [47], particularly on flat areas where natural erosion is low when covered with vegetation and trampling is not enough to create bare soil. Therefore, the result that disturbance-dependent species were found in areas with long-term continuous soil disturbance created by cultivation should not be surprising, nor should the decline of dry sandy grassland species as soil disturbance decreased.

The cultivation of the soil is thought to have prevented the vegetation cover in the xeric sand calcareous grassland from closing and the $CaCO_3$ from being depleted by mixing the soil [26]. The highest average amount of bare sand during the period 1940 to 2007 was just over 3%, and there was no drastic change in the amount of bare sand during this period. The descriptions of Linné [23] and Campbell [20] indicate that there was much more bare sand in the area in the 18th century, and the most important decline probably occurred during the 19th century. Although there was a difference between the sites in amount of bare sand, there was no negative correlation between the amount of bare sand and the number of years since cultivation, which would be expected if lack of cultivation is an important factor in explaining recent declines in xeric sand calcareous grassland species. In fact, one of the sites not cultivated during the last 260 years had the highest amount of bare sand even in the 1940s. It must be acknowledged that there is a problem when comparing flat and hilly sites, as sloping ground will naturally be more prone to soil erosion. However, even when the two hilly sites were removed from the analysis, no correlation between bare sand and number of years since cultivation could be seen. There was a possible constraint to the method used in this study to detect the amount of bare sand. Assuming that cultivation was the important factor and it only occurred every 8–20 years, the analysis made using aerial photographs from just one date every fifteen years in this study could have failed to spot the bare sand produced. Completely bare sand might have been visible only in the first years, although the vegetation cover could have been sparse for many more. This might still have been enough of disturbance to sustain xeric sand calcareous grassland, and the failure to detect semi-bare sand was a weakness in the method used.

Another explanation for the decline in xeric sand calcareous grassland species could be nutrient enrichment. In the present study, some of the more recently cultivated sites had much higher P_{ext} value than sites that were cultivated a long time ago or not at all. However, Rinkaby and Klammersbäck stood out with their low P_{ext} values, suggesting that they have not been fertilized despite their recent cultivation history. The higher P_{ext} values found at some of the more recently cultivated plots could be a reason for the lack of correlation between the amount of bare sand and the number of years since cultivation, since high P values could speed up succession [48], thereby counteracting the effects of soil disturbance. In addition, the fertilisers would also have contained N, which is known to accelerate the decalcification process [49,50,51]. This was supported by the positive correlation between P_{ext} and depletion depth of $CaCO_3$, which suggests that fertilization could be the main explanation for the difference in depletion depth seen between the sites.

The depletion depth of $CaCO_3$ was not correlated with the time since cultivation, which would be expected if the important factor for retention of $CaCO_3$ in the topsoil was indeed cultivation. No correlation could be found between the mean amount of bare sand and the depletion depth, indicating a lack of connection between depletion and the amount of soil disturbance during the last century. The positive correlation between P and depletion depth of $CaCO_3$ would in fact suggest the opposite, i.e. that cultivation in the last 70 years, if including fertilization, might have had a negative effect. The maximum $CaCO_3$ value and the depth at which it was found were negatively correlated, although the correlation was weak. Still, this suggests that the depletion depth of

$CaCO_3$ partly depended on the soil type. Different glaciofluvial deposits have different $CaCO_3$ content depending on the lithology of the local bedrock and the previously deposited soil types [16], which would explain the observed variation in $CaCO_3$ content between sites. None of the above-mentioned explanations seems entirely satisfactory in explaining differences in depletion depth between sites and they could not explain the presence of $CaCO_3$ in the topsoil of some sites. It is important to keep in mind that depletion has been going on throughout the 14,000 years since the soil was deposited by glacial melt-water. Consequently, there must have been a process counteracting the depletion.

The results discussed above indicate that ploughing was not directly preventing the $CaCO_3$ from being depleted. However, it is probable that the wind erosion resulting from the cultivation could have counteracted depletion by exposing the $CaCO_3$ rich sand. It seems to be a likely scenario, since wind erosion has already been found to remove fertile topsoil and cause organic matter content to decline [52,53]. Some sand drift still occurs in southern Sweden [22], but it is limited compared to the extensive sand drift of the 18th century described by Linné [23] and Campbell [20]. The most important factor determining the amount of sand drift is how much bare sand is available for erosion. During the time period from the 1940s until today, the amount of bare sand has been very low, but in the 18th century extensive areas of bare sand were observed due to the intensified cultivation [20,22,23]. One other factor that would affect wind erosion is the forest cover in the surroundings. Although the amount of forest at some sites increased during the last 70 years, no correlation was found with the amount of bare sand or the $CaCO_3$ content. However, a great change in forest cover took place in the 19th century, when forest was planted to bind the sand [22].

The results strongly suggest that the causes of present-day decline in threatened species may have origins far back in time and that conservation should take this into account. Human activity, such as expansion of arable fields and intensive livestock grazing, also caused sand drift in Germany [9,10], the UK [54] and the Netherlands [52]. In Germany and the Netherlands, dry sandy grasslands, including xeric sand calcareous grassland and Corynephorus grassland, are believed to have developed in areas where arable farming or livestock grazing created bare sand, which was then exposed to wind erosion and resulted in drifting sand [9,52]. Today, these sandy grasslands, in common with similar areas in Sweden, are suffering from changes in land use including afforestation, intensification and abandonment [9,52].

Conclusions

There was no support for the common explanation of the decline in xeric sand calcareous grassland being caused by abandonment of agricultural fields during the last century, i.e. that the old agricultural practices in these grasslands were abandoned. Instead, fertilization history (as indicated by soil P levels) was the main explanation for the difference in depletion depth of $CaCO_3$ seen between the sites. The decline in xeric sand calcareous grassland since the 18th century is most probably the result of the drastic changes in land use during the 19th century, which put an end to the extensive sand drift. The importance of sand movement is shown by the high density of red-listed vascular plant species at the hilly sites (Table 1). Since cultivation played a very important role in the creation of xeric sand calcareous grassland, grazing alone is not a good conservation option for these grasslands [28,55]. Instead more drastic measures are needed to restore the high calcium content, remove the accumulated nutrients and create bare sand for regeneration

[56]. An example of such a restoration measure is topsoil removal, where the decalcified and nutrient rich topsoil is removed much like it would be during sand drift [57]. Not surprisingly, topsoil removal has been shown to be a successful method for restoring xeric sand calcareous grasslands [55].

Acknowledgments

The authors would like to thank Merit Kindström and Micael Runnström for input on the methods used in the map and aerial photograph analysis, and for helpful comments on the manuscript.

References

1. Poschlod P, WallisDeVries MF (2002) The historical and socioeconomic perspective of calcareous grasslands – lessons from the distant and recent past. Biol Conserv 104: 361–376.
2. Eberhardt RW, Foster DR, Motzkin G, Hall B (2003) Conservation of changing landscapes: vegetation and land-use history of Cape Cod National Seashore. Ecol Appl 13: 68–84.
3. Antrop M (2005) Why landscapes of the past are important for the future. Landscape Urban Plan 70: 21–34.
4. Motzkin G, Foster D, Allen A, Harrod J, Boone R (1996) Controlling site to evaluate history: vegetation patterns of New England sand plain. Ecol Monogr 66: 345–365.
5. Lunt ID, Spooner PG (2005) Using historical ecology to understand patterns of biodiversity in fragmented agricultural landscapes. J Biogeogr 32: 1859–1873.
6. Kent M, Owen NW, Dale P, Newnham RM, Giles TM (2001) Studies of vegetation burial: a focus for biogeography and biogeomorphology? Progress in Physical Geography 25: 455–482.
7. Owen NW, Kent M, Dale MP (2004) Plant species and community responses to sand burial on the machair of the Outer Hebrides, Scotland. J Veg Sci 15: 669–678.
8. Olsson PA, Mårtensson L-M, Bruun HH (2009) Acidification of sandy grasslands – consequences for plant diversity. Appl Veg Sci 12: 350–361.
9. Poschlod P, Baumann A, Biedermann H, Bugla B, Neugebauer K (2009) Dry sandy grasslands in Southern Germany – a case study of how to re-develop remnants to achieve the former status of a high nature value landscape. In Veen P, Jefferson R, de Smidt J, van der Straaten J, editors. Grasslands in Europe of High Nature Value. KNNV Publishing, the Netherlands. pp. 113–121.
10. Eichberg C, Storm C, Stroh M, Schwabe A (2010) Is the combination of topsoil replacement and inoculation with plant material an effective tool for the restoration of threatened sandy grassland? Appl Veg Sci 13: 425–438.
11. Tyler T (2005) The Bryophyte flora of Scanian sand-steppe vegetation and its relation to soil pH and phosphate availability. Lindbergia 30: 11–20.
12. Olsson PA, Schnoor TK, Hanson S-Å (2010) pH preferences of red-listed gasteromycetes in calcareous sandy grasslands: Implications for conservation and restoration. Fungal Ecol 3: 357–365.
13. Ljungberg H (1999) Skalbaggar och andra insekter på sandstäppslokaler i östra Skåne (Beetles and other insects on xeric sand calcareous grassland sites in eastern Scania). Länsstyrelsen i Skåne län, Malmö, 1–56.
14. Ellenberg H (1991) Indicator values of vascular plants in Europe 1. Indicator values of plants not including Rubus. Scripta Geobot 18: 9–166
15. Mårtensson L-M, Olsson PA (2010) Soil chemistry of local vegetation gradients in sandy calcareous grasslands. Plant Ecol 206: 127–138.
16. Persson KM (2000) Description to the quaternary map 2D Tomelilla NO. The Swedish Geological Survey, Uppsala, Sweden.
17. Pigott CD, Walters SM (1954) On the interpretation of the discontinuous distributions shown by certain British species of open habitats. J Ecol 42: 95–116.
18. Berglund BE, Larsson L, Lewan N, Olsson EGA, Skansjö S (1991) Ecological and social factors behind the landscape changes, in: Berglund BE (Ed.), The cultural landscape during 6000 years in southern Sweden. Ecol Bull 41: 425–444.
19. Poska A, Saarse L (1999) Holocene vegetation and land-use history in the environs of Lake Kahala, northern Estonia. Veget Hist Archaeobot 8: 185–197.
20. Campbell Å (1929) Skånska bygder under förra hälften av 1700-talet. Etnografisk studie över den skånska allmogens äldre odlingar, hägnader och byggnader (The Scanian countryside during the first half of the 18th century). Thesis, Uppsala University, Sweden.
21. Dahl S (1989) Studier i äldre skånska odlingssystem (Studies of the old Scanian agricultural systems). Meddelande serie B 69, Kulturgeografiska institutionen, Stockholms Universitet. Stockholm, Sweden.
22. Emanuelsson U (2002) Det skånska kulturlandskapet (The Scanian Cultural Landscape). Naturskyddsföreningen i Skåne, Lund.
23. von Linné C (1751) Carl Linnaei skånska resa på höga öfwerhetens befallning förrättad år 1749 (Carl Linnaeus's Journey to Scania as ordered by the high authorities in 1749). Stockholm, Sweden.
24. Björkman L, Sjögren P (2003) Long-term history of land-use and vegetation at Ire, an agriculturally marginal area in Blekinge, south Sweden. Veget Hist Archaeobot 12: 61–74.
25. Foster DR (1992) Land-use history (1730-1990) and vegetation dynamics in central New England, USA. J Ecol 753–772.
26. Mattiasson G (1974) Sandstäpp - Vegetation, dynamik och skötsel (Xeric sand calcareous grassland – vegetation, dynamics and management). Meddelanden från Avdelningen för Ekologisk Botanik, Lunds universitet 2: 1–40.
27. Tyler T (2003) The state of the sand-steppe vegetation of eastern Scania (Skåne) in early spring 2003. Bot Notiser 136: 1–22.
28. Schnoor TK, Olsson PA (2010) Effects of soil disturbance on plant diversity of calcareous grasslands. Agr Ecosyst Environ 139: 714–719.
29. Roem WJ, Berendse F (2000) Soil acidity and nutrient supply ratio as possible factors determining changes in plant species diversity in grassland and heathland communities. Biol Conserv 92: 151–161.
30. Snogerup S, Jörgensen M (2000) Från skånska resa till Skånes flora (From the journey to Scania to the Scanian flora). Lunds Botaniska Förening, BTJ Tryck AB, Lund, Sweden.
31. Tyler T, Olsson K-A, Johansson H, Sonesson M (2007) Floran i Skåne: arterna och deras utbredning (The Scanian flora: the species and their distribution). Lunds Botaniska Förening, Grahns Tryckeri AB, Lund, Sweden.
32. Olsson K-A (1994) Sandstäpp i Skåne – ett upprop (Xeric sand calcareous grassland in Scania – a proclamation). Lunds Botaniska Förenings Medlemsblad 1994: 4–7.
33. Ljungberg KO, Löfroth M, Nitare J (1994) Åtgärdsprogram för Sandstäpp 1994-1996 (Action Plan for Xeric Sand Calcareous Grassland 1994-1996). Report no. 390-2540-94. Swedish Environmental Protection Agency, Stockholm, Sweden.
34. Wassen MJ, Venterink HO, Lapshina ED, Tanneberger F (2005) Endangered plants persist under phosphorus limitation. Nature 437: 547–550
35. Smits NAC, Willems JH, Bobbink R (2008) Long-term after-effects of fertilization on the restoration of calcareous grasslands. Appl Veg Sci 11: 279–286.
36. Bray RH, Kurtz LT (1945) Determination of total, organic, and available forms of phosphorus in soils. Soil Sci 59: 39–45.
37. Bahr A, Ellström M, Schnoor TK, Påhlsson L, Olsson PA (2012) Long-term changes in vegetation and soil chemistry in a calcareous and sandy semi-natural grassland. Flora 207: 379–387.
38. Lantmäteriet (2014a) (Swedish land surveying agency), http://historiskakartor. lantmateriet.se/arken/s/search.html?locale = en_US. Accessed 2014 Feb 15.
39. Lantmäteriet (2014b) (Swedish land surveying agency), http://www. lantmateriet.se/Kartor-och-geografisk-information/Flyg--och-satellitbilder/ Flygbilder/ Accessed 2014 Feb 15.
40. Gazenbeek A (2005) LIFE, Natura 2000 and the Military. LIFE Focus. European Commission, Luxembourg.
41. Warren SD, Holbrook SW, Dale DA, Whelan NL, Elyn M, et al. (2007) Biodiversity and the heterogeneous disturbance regime on military training lands. Restor Ecol 15: 606–612.
42. Řehounková K, Prach K (2008) Spontaneous vegetation succession in gravel-sand pits: a potential for restoration. Restor Ecol 16: 305–312.
43. Jentsch A, Friedrich S, Steinlein T, Beyschlag W, Nezadal W (2009) Assessing conservation action for substitution of missing dynamics on former military training areas in Central Europe. Restor Ecol 17: 107–116.
44. Wells TCE, Sheail J, Ball DF, Ward LK (1976) Ecological studies on the Porton Ranges: relationships between vegetation, soils and land-use history. J Ecol 64: 589–626.
45. Dahlström A, Cousins SAO, Eriksson O (2006) The history (1620–2003) of land use, people and livestock, and the relationship to present plant species diversity in a rural landscape in Sweden. Environment and History 12: 191–212.
46. Johansson LJ, Hall K, Prentice H, Ihse M, Reitalu T, et al. (2008) Semi-natural grassland continuity, long-term land-use changes and plant species richness in an agricultural landscape on Öland, Sweden. Landscape Urban Plan 84: 200–211.
47. Kent M, Dargie T, Reid C (2003) The management and conservation of machair vegetation. Bot J Scot 55: 161–176.
48. Carroll JA, Caporn SJM, Johnson D, Morecroft MD, Lee JA (2003) The interactions between plant growth, vegetation structure and soil processes in semi-natural acidic and calcareous grasslands receiving long-term inputs of simulated pollutant nitrogen deposition. Environ Pollut 121: 363–376.
49. Oenema O (1990) Calculated rates of soil acidification of intensively used grassland in the Netherlands. Fert Res 26: 217–228.

Author Contributions

Conceived and designed the experiments: AMÖ PAO. Performed the experiments: AMÖ. Analyzed the data: AMÖ PAO. Contributed reagents/materials/analysis tools: AMÖ PAO. Wrote the paper: AMÖ PAO.

50. Bolan NS, Adriano DC, Curtin D (2003) Soil acidification and liming interactions with nutrient and heavy metal transformation and bioavailability. Adv Agron 78: 216–272.
51. Stroia C, Morel C, Jouany C (2010) Nitrogen fertilization effects on grassland soil acidification: consequences on diffusive phosphorus ions. Soil Sci Soc Am J 75: 112–120.
52. Riksen M, Ketner-Oostra R, van Turnhout C, Nijssen M, Goossens D, et al. (2006) Will we lose the last inland drift sands of Western Europe? The origin and development of the inland drift-sand ecotype in the Netherlands. Landscape Ecol 21: 431–447.
53. Poortinga A, Visser SM, Riksen MJPM, Stroosnijder L (2011) Beneficial effects of wind erosion: Concepts, measurements and modelling. Aeolian Research 3: 81–86.
54. Bateman MD, Godby SP (2004) Late-Holocene inland dune activity in the UK: a case study from Breckland, East Anglia. The Holocene 14: 579–588.
55. Olsson PA, Ödman AM (2014) Natural establishment of specialist plant species after topsoil removal and soil perturbation in degraded calcareous sandy grassland. Restor Ecol 22: 49-56.
56. Ödman AM, Schnoor TK, Ripa J, Olsson PA (2012) Soil disturbance as a restoration measure in dry sandy grasslands. Biodiv Cons 21: 1921–1935.
57. Riksen M, Spaan W, Stroosnijder L (2008) How to use wind erosion to restore and maintain the inland drift-sand ecotype in the Netherlands? J Nat Conserv 16: 26–43.

14

Dynamics of Potassium Release and Adsorption on Rice Straw Residue

Jifu Li[1,2], Jianwei Lu[1,2]*, Xiaokun Li[1,2], Tao Ren[1,2], Rihuan Cong[1,2], Li Zhou[1,2]

1 College of Resources and Environment, Huazhong Agricultural University, Wuhan, China, **2** Key Laboratory of Arable Land Conservation (Middle and Lower Reaches of Yangtse River), Ministry of Agriculture, Wuhan, China

Abstract

Straw application can not only increase crop yields, improve soil structure and enrich soil fertility, but can also enhance water and nutrient retention. The aim of this study was to ascertain the relationships between straw decomposition and the release-adsorption processes of K^+. This study increases the understanding of the roles played by agricultural crop residues in the soil environment, informs more effective straw recycling and provides a method for reducing potassium loss. The influence of straw decomposition on the K^+ release rate in paddy soil under flooded condition was studied using incubation experiments, which indicated the decomposition process of rice straw could be divided into two main stages: (a) a rapid decomposition stage from 0 to 60 d and (b) a slow decomposition stage from 60 to 110 d. However, the characteristics of the straw potassium release were different from those of the overall straw decomposition, as 90% of total K was released by the third day of the study. The batches of the K sorption experiments showed that crop residues could adsorb K^+ from the ambient environment, which was subject to decomposition periods and extra K^+ concentration. In addition, a number of materials or binding sites were observed on straw residues using IR analysis, indicating possible coupling sites for K^+ ions. The aqueous solution experiments indicated that raw straw could absorb water at 3.88 g g^{-1}, and this rate rose to its maximum 15 d after incubation. All of the experiments demonstrated that crop residues could absorb large amount of aqueous solution to preserve K^+ indirectly during the initial decomposition period. These crop residues could also directly adsorb K^+ via physical and chemical adsorption in the later period, allowing part of this K^+ to be absorbed by plants for the next growing season.

Editor: Jörg Langowski, German Cancer Research Center, Germany

Funding: This work was supported by Non-profit Research Foundation for Agriculture (201103039), National Natural Science Foundation of China (41301319), Fundamental Research Funds for the Central Universities (2012BQ059) and International Potash Institute Co-operation Program. The funders had no role in study design, data collection and analysis, decision to publish, or preparation of the manuscript.

Competing Interests: The authors have declared that no competing interests exist.

* E-mail: lunm@mail.hzau.edu.cn

Introduction

The current Asian population of 4.3 billion is projected to increase by nearly 0.9 billion people, reaching roughly 5.2 billion, by 2050 [1], which will result in significantly increased regional food demand. Of this, population, 80% will be distributed in China, India and the southeast regions of Asia, posing a challenge to the economic development and social stability of these countries [2]. Rice-based cropping systems are the most productive agroecosystems in these areas and produce the most food for the most people [3]. To meet the food demand of the region, intensification and diversification have been applied as the two main strategies for rice-based cropping systems. In addition to the rise of multiple cropping indexes, fertilization consumption has played a very important role in production increases [4]. Compared with nitrogen and phosphorus fertilizer, potash fertilizer is often ignored by farmers, particularly in Asia [5–7]. Potash resources are comparatively limited [8,9], and in recent years, the higher price of potash on the international market has reduced the demand of potassium, as farmers in the area are unwilling to put more potash into the soil [10]. Soil K deficiency has become a major limiting factor in the modern agricultural process [11]. Therefore, it is of great importance to increase

potash supplementation in these regions. K-bearing organic resources such as compost, green manure, farmyard manure and crop straws, particularly abundant crop residues, are again receiving attention from farmers [12].

Annually, the world production of straw is approximately 3.8 billion tons, 74% of which are cereal straws [13]; for rice-based land in Asia, 80% of straw production consists of rice residues [3]. Cereal straws usually have a higher potassium content than other straws (1.2%–1.7%). The results of Kaur and Benipal [14] have shown that returning straw to the field could improve soil available potassium to a significantly great extent than manure. A 30-years field trial conducted by Liao et al. indicated that straw management could increase exchangeable K by 26.4%, nonexchangeable K by 1.8% and SOC 21.0% in comparison to a CK treatment in reddish paddy soil [15]. As straw potassium is primarily present in the form of K^+ ions in the cell fluid [16], the release of K from stubble in field is influenced by rainfall [17]. Duong et al. [18] found that the distance of potassium migration from organic fertilizer is 10 mm. Excepting the K^+ adsorbed or fixed by soil clay particles, 50% of K^+ was retained in the soil solution [14]. Furthermore, farmers prefer to input potash fertilizers one time before sowing. The loss of K in the soil solution from such applications was 1.1- and 14.5- fold that of N and P, respectively

[19]; N and P losses were gradual, while leaching phenomena were observed for K [17]. However, Kozak et al. found that crop residues could intercept a maximum of 29% of the water loss for a given rainfall [20]. Soil water retention is also affected by the organic carbon content, as reported by Rawls et al [21]. These results indicate that crop residues have a positive effect on water absorption. Meanwhile, as a high-quality biological adsorbent, the biochars generated from crop straws can adsorb 0.48–1.40 mol kg^{-1} Cu(II) [22], and even unmodified rice straw can absorb 13.9 mg g^{-1} Cd(II) [23]. However, relatively few data exist regarding the adsorption of potassium by crop residues. Because straw decomposition is a slow and long-term process, a significant quality of plant residues can usually be found on farmland after one season of crop growth.

Since 1980, global warming has received increasing attention [24], bolstered by occasional extreme weather events such as the sustained hot temperature in Europe, North America and Asia in July 2013. Scorching weather causes a water shortage in rice farmland, thereby affecting the absorption of nutrients. The incorporation of residues into the soil may reserve water to slow down seasonal drought and could also adsorb cations. However, the capability of soil residues to preserve water and nutrients during different decomposition periods is unclear, especially, the fixation mechanism of residues for potassium.

Therefore, the aims of this research were as follow: (a) to investigate the characteristics of straw decomposition and K$^+$ release under flooded conditions; (b) to assess the retention capacity of straw for water and potassium during different decomposition periods; and (c) to ascertain the mechanism of K$^+$ adsorption on straw residues.

Materials and Methods

Ethics Statement

The authors of this study hereby confirm that no specific permissions were required for our experimental location and activities, as the experimental field belonged to our institute and is employed for scientific research only. The studies had negligible effects on the functioning of the broader ecosystem. The research did not involve measurements on humans or animals, and no endangered or protected species were involved.

Materials

Soil material. Paddy soil was collected from the plow layer (0–20 cm) after the rapeseed harvest in 2012. The sample was mixed, air-dried at room temperature and ground to pass through a 2 mm sieve. Soil physical and chemical properties were measured using conventional methods [25–27]. Soil pH was measured with a glass electrode in a 1:2.5 soil/water solution. Soil organic matter was measured using the dichromate oxidation method, and total nitrogen was measured using the Kjeldahl acid digestion method. Available phosphorus was determined using the Olsen method, and available potassium was measured by flame photometry after NH$_4$OAc neutral extraction. All parameters were measured three times and are presented as the mean±SD: pH 5.73±0.23, SOM 26.7±1.1 g kg^{-1}, TN 1.09±0.04 g kg^{-1}, Olsen-P 19.4±0.7 mg kg^{-1}, and NH$_4$OAc-K 107.8±4.3 mg kg^{-1}.

Rice straw material. Rice straw was cut into segments of approximately 2–3 cm and then conserved in a dryer until further use. The initial potassium (K) content of the rice straw was 21.87±0.12 mg g^{-1}.

Methods

Straw decomposition trial. The trial was performed at the experimental base of the College of Resources and Environment beginning on July 7, 2012. Three boxes, constructed of PVC (size 50×30×25 cm), were used for the incubation experiment. A total of 15 kg of dry bulk soil was packed in each box. Before the experiment, the rice straw was dried at 40°C for 3 h, and 10.0 g samples were then accurately weighed into 200 mesh (pore diameter 0.075 mm) nylon bag (size 25×20 cm), and sealed [28]. Each box contained 5 bags of straw, totaling 15 bags in the three boxes. According to the growth period of late rice, we removed nylon bags at 5, 15, 30, 60 and 110 d. During the incubation period, deionized water was added to the boxes to maintain a flooding layer of 1 cm. A schematic diagram of the trial is shown in Figure S1.

On each sampling date, one nylon bag was randomly removed from each box and rinsed with distilled water three times to remove any mud that had adhered to the bag. The residue of the rice straw was then dried at 40°C for 48 h until reaching a constant weight, weighed and ground to pass through a 1 mm sieve. Some of the residue powder was ground again using a mortar, until it passed through a 0.149 mm sieve, to obtain micron particles. The particles were dispersed in an aqueous solution of pH 6.0 to test the zeta potential using zeta potential and nanoparticle size analyzer (ZS90, Melvin British Company UK). Zeta potential is the potential difference between the dispersion medium and the stationary layer of fluid attached to the dispersed particle [29], which indicates the electric potential variation of the residue surface during the decomposition period [22]. All of the residue samples were digested with H$_2$SO$_4$-H$_2$O$_2$ to determine their potassium content using a flame photometer (M-410, Cole-Parmer USA) and to calculate their potassium release rates. The formulas employed in this study were as follow:

$$\text{Decomposition amount (g)} = \text{Dry matter at 0 d} - \text{Remaining dry matter at } n \text{ d} \qquad (1)$$

$$\text{Decomposition rate (\%)} = (\text{Decomposition amount/Dry matter at 0 d}) \times 100\% \qquad (2)$$

$$\text{K release amount (K,mg)} = \text{K amount at 0 d} - \text{K remaining at } n \text{ d} \qquad (3)$$

$$\text{K release rate (\%)} = \text{K release amount/K amount at 0 d} \times 100\% \qquad (4)$$

Where n is the day of incubation.

Batches of K sorption experiments. Precise 0.30 g samples of the rice straw residue from different time points were added into 50 mL polythene bottles along with various concentrations of KCl solution (0, 10, 50 and 100 mg K L^{-1}). The bottles were shaken for 4 h at 160 r min^{-1} and filtered to test the K$^+$ concentration of

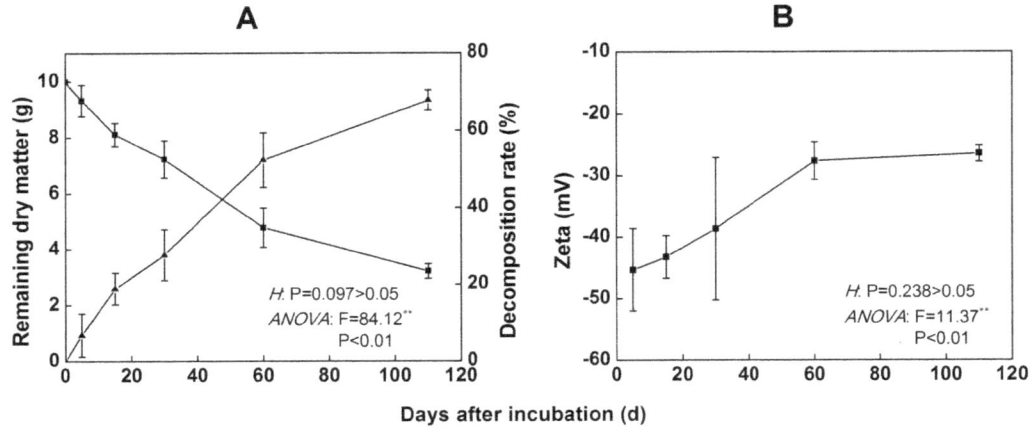

Figure 1. The characteristics of the decomposition and zeta potential of rice straw. The annotations in the panels indicate the homogeneity of variances (H) and ANOVA for the remaining dry matter and zeta potential. The H-test was performed using the Levene test. **indicates significant differences at P<0.01. The values are the means of 3 replicates (±standard deviation).

the equilibrium solution. The experiments were repeated 3 times, and the average values were used for analysis. The K adsorption on straw Q (mg g^{-1}) and the K removing rate R (%) were determined in the following manner [30]:

$$Q = (C_0 - C_t) \times V/m \qquad (5)$$

$$R = (C_0 - C_t)/C_0 \times 100\% \qquad (6)$$

Where C_0 and C_t are the initial and equilibrium K concentration, respectively, mg L^{-1}; V is the volume of the solution, mL; and m is the mass of the straw residue, g.

According to the K sorption experiment, the maximal Q residue was selected to test the changes of residue structure on K$^+$ fixation with the help of Fourier transform infrared spectroscopy. (Nexus, Thermo Nicolet USA).

Kinetics of water absorption on rice straw. The water absorption capacity of untreated dried straw [31] was determined by suspending 0.50 g of natural dried straw through a 1.0 mm sieve in a 50 mL beaker with 40 mL deionized water. three replicates were taken at each time interval and cleaned of gravitational water from the straw surface. The material was then weighed on an electronic balance, and the water absorption capacity was calculated.

The water absorption capacity of rice straw residues from different decomposition time points was also determined. A straw sample of 0.30 g was put into a 50 mL beaker with 40 mL of deionized water. After 300 min of immersion, the residues were removed and cleaned of gravitational water before the determination of water absorption capacity. The water absorption capacity of the straw was calculated as follows:

$$M = (M_t - M_0)/M_0 \qquad (7)$$

Where M is the water absorption capacity of straw, g g−1; and M_t and M_0 are the initial and final mass of straw, respectively, g.

Furthermore, to observe the influence of straw surface pore on the water absorption of rice straw, the surface morphology of

residue from different time points was examined using a scanning electron microscope (JSM-6390LV, NTC Japan).

Data Analysis

All analyses were conducted on three replicates. Statistical analyses were performed using OrignPro8.0 software. The means were the average of three replicates, and analysis of variance (ANOVA) was performed according to standard procedures for factorial randomized block designs. Differences at p<0.05 level were considered statistically significant, as determined using the least significant difference (LSD) test.

Results

Characteristics of Straw Decomposition and Zeta Potential of Rice Straw Surface

The decomposition dynamics of rice straw during different stages of incubation under flooded conditions are shown in Figure 1A. The average rate of rice straw decomposition was 0.09 g d^{-1} from 0 to 60 d. The decomposition amount accounted for 52.3% (remaining mass 4.77 g) of the total mass after 60 d of incubation. The average rate of decomposition was relatively slow, 0.03 g d^{-1}, from 60 to 110 d. During this period, the decomposition amount accounted for 15.5% of the total mass. The remaining amounts of straw and the accumulation rate were 3.22 g and 67.8%, respectively, after the trial was ended at 110 d.

The zeta potential of the untreated rice straw surface was −56.36 mV in the pH 6.0 aqueous phase (Figure 1B). After 5 d of incubation, the zeta potential increased to −43.2 mV, and it continued to increase to −27.65 mV after 60 d of incubation. On the following days, the zeta potential was not significantly different, ranging from −30 to −20 mV.

Characteristics of Straw K Release

The potassium (K) content of untreated rice straw was 21.87 mg kg^{-1}. After 3 d of immersion, 90% of the total K had been released, and the K content of straw was only 1.47 mg g^{-1} after 5 d of immersion (Figure 2A); this value did not change significantly on the following days. These results indicated that 93.7% of the total potassium could be released into the ambient environment during the early stage of immersion (Figure 2B). After 30 d, the K content of the residues increased slightly and remained

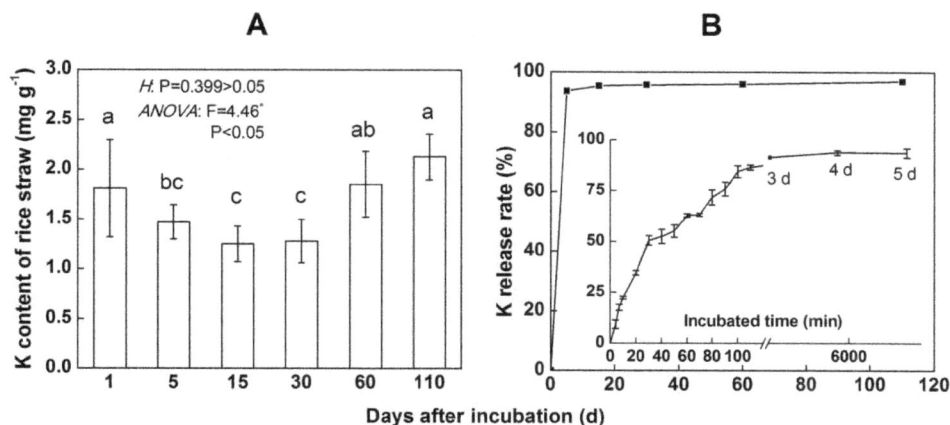

Figure 2. The characteristics of the potassium release of rice straw. The annotation in panel A indicates the homogeneity of variances (H) and ANOVA. The H-test was performed using the Levene test. *indicates significant differences at P<0.05. The values are the means of 3 replicates (±standard deviation). Means with the same letter are not significantly different. The insertion in panel B is the K release rate within 5 d.

at 1–2 mg g^{-1}. At the end of the trial, the K content of the residue and the K release rate were 2.13 mg g^{-1} and 96.9%, respectively.

K Adsorption in Different Decomposition Periods of Straw Residue

Figure 3 shows the adsorption-desorption equilibrium of exogenous potassium over different decomposition periods of straw residue. The results showed that over the entire decomposition period, K$^+$ ions were, on the whole, released in pure water. When the concentration of K$^+$ ions was 10 mg L^{-1} in solution, a positive effect of K$^+$ adsorption was observed after 60 d of decomposition, and the K$^+$ adsorption capacity of the residue reached 0.13 mg g^{-1} on the 110th d of the experiment. The residue at 15 d showed a positive adsorption effect when the concentration of external potassium increased to 50 mg L^{-1}. Meanwhile, the capacity also increased as decomposition continued. The amount of K$^+$ adsorption reached a maximum of 0.76 mg g^{-1} after 110 d of incubation, representing a considerable increase of 76.3% compared to its value on the 15th day. When additional increments of K were applied, up to 100 mg L^{-1}, the straw primarily released potassium into solution within 5 d. After

15 d, the adsorption capacity of the residue was higher than that in the trial with 50 mg L^{-1} K supplied over the same period.

Similarly, results of Figure 3B indicated that the K removing rate (R) increased with the decomposition of rice straw. When the concentration of extra K was 10 mg L^{-1}, for example, the value of R was −51.0% at 5 d after incubation, but reached 15.6% after 110 d. At the same time, when the concentration of external K was increased to 50 mg L^{-1}, the removing rate for each period of decomposition except for the first 5 d reached the maximum value. The values of R tended to decrease with increase external K concentration.

FT-IR Analysis for the K Adsorption on Rice Straw Residue

Figure 4 displays the infrared spectra of untreated dried straw both before and after the K adsorption of the rice straw residue on the 110th day after incubation. The results showed that the chemical structure of the straw had undergone a significant change after 110 d of decomposition (Figure 4A–B). New peaks emerged at 3698, 3619 and 779 cm^{-1}, and the existing peaks at 2851, 1086, 798, 695, 528 and 469 cm^{-1} were strengthened to various degrees.

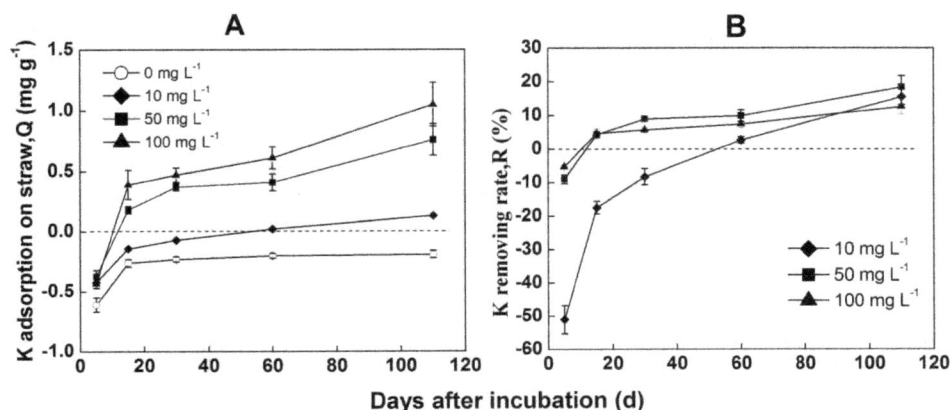

Figure 3. The adsorption of potassium by rice straw residues. The dashed line shows the zero point in the panels. The values are the means of 3 replicates (±standard deviation).

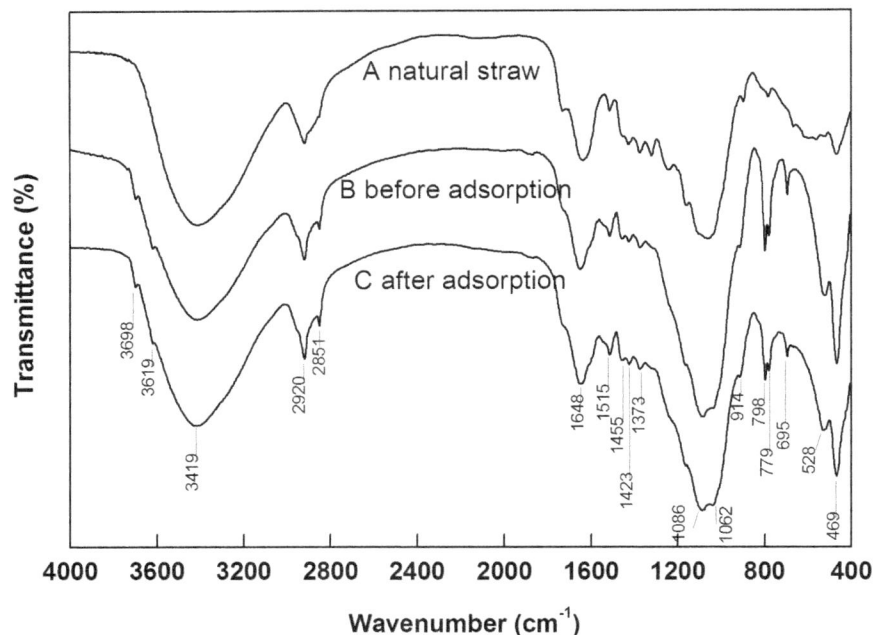

Figure 4. FT-IR spectra of natural dried rice straw and the rice straw residue at 110 d before and after potassium (K) adsorption.
Solid line A represents the infrared spectrum of natural dried rice straw. Solid lines B and C represent the infrared spectra of the rice straw at 110 d before and after K adsorption in 50 mg L^{-1} KCl solution, respectively. The data in the panel are the peak wavenumbers.

Figure 4C illustrates the spectra after K adsorption for rice straw after 110 d. The main peaks of the functional groups at $4000-1300$ cm^{-1} did not show significant changes. However, in the fingerprint region, the peaks at 798, 695, and 469 cm^{-1} decreased significantly compared to those before K adsorption on the residue.

Kinetics of Water Absorption on Rice Straw

The water absorption kinetic of untreated dried straw are shown in Figure 5A. The results indicated that the water storage capacity of straw reached 2.51 g g^{-1} after 5 min of immersions. From 5 to 90 min, the absorption rate continued to increase at an average rate of 0.01 g (g min) $^{-1}$. After 150 min, the rice straw could not absorb more water at the ambient temperature ($23°C$), approaching a saturated value of 3.88 g g^{-1}. The water retention also changed significantly during different periods of straw decomposition (Figure 5B). The results showed that the water absorption capacity of the residue continued to rise from 0 to 15 d, and the maximum water content was 5.17 g g^{-1}. Moreover, the water absorption capacity decreased gradually with the extension of straw decomposition. The water content of rice residue was the lowest at 110 d, with a value of 3.00 g g^{-1}.

Discussions

The Differences between Straw Decomposition and K Release

For field straw management, the rate of straw decomposition is influenced by many factors. These variables include internal factors, such as litter quality, material composition and structure [32,33], and external factors, such as temperature, moisture [34], methods of straw utilization (mulching or incorporation), time, length of straw [31] and the concentration of CO_2 in the soil. Litter quality and climate condition are considered to be key

factors in the regulation of straw decomposition [35]. The present study was conducted from the beginning of July to the end of October. From 0 to 60 d, the soil temperature reached above $35°C$, and this environment did significantly improve the decomposition rate of the straw. However, from 60 to 110 d, the soil temperature decreased to $20–24°C$, and the decomposition rate consequently also decreased. After 110 d of degradation, the straw decomposition rate reached 60%. The results presented in Figure 1 illustrate that the decomposition of straw has two main periods: (1) a rapid decomposition stage from 0 to 60 d, during which the structure of the straw surface changed (Figure 6), and the zeta potential increased significantly; and (2) a slow decomposition stage from 60 to 110 d, during which the decomposition rate is significantly lower than before and the zeta potential is basically stable. The characteristic infrared peaks of cellulose were observed at 3412, 2900, 1425, 1370 and 895 cm^{-1} [30]. The characteristic peak of hemicellulose was observed at 1732 cm^{-1} [36], and the absorption peaks characteristic of lignin were seen at 1595 and 1516 cm^{-1}. All of these peaks were not changed dramatically over the study period, as seen in figure 4. Conversely, easily decomposable plant materials such as lipids, pectin, starch and carbohydrates were degraded over one season of crop growth. In addition, negligible amounts of cellulose and hemicellulose were also degraded. Lignin, along with the majority of the cellulose and hemicellulose, may require more time to be degraded by organisms [35].

The characteristics of K release are dramatically different from those of straw decomposition. As K mainly exists in ionic form in plants, it is able to move easily. After 5 d of immersion, more than 90% of potassium had been released from the straw. This result coincides with that of Rodríguez-Lizana et al. [17], who reported that the K of sunflower residue decreased by 98% over the study period while only 43% of the residue itself degraded over the same

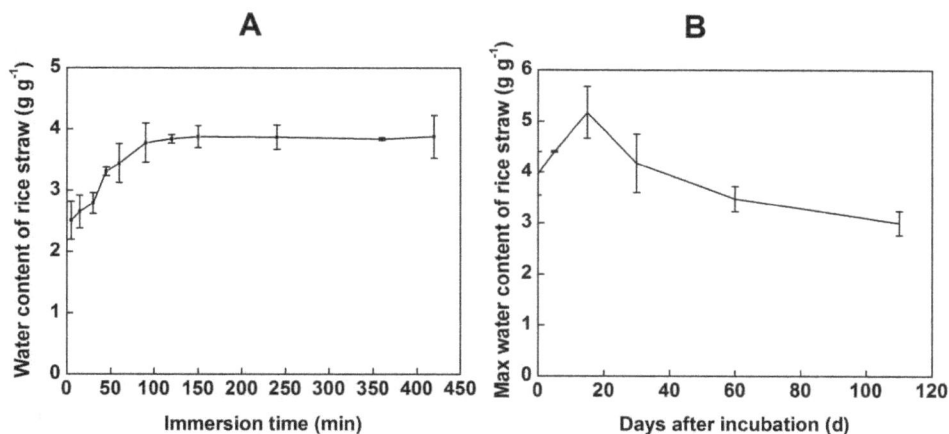

Figure 5. The kinetics of water absorption for rice straw. Panel A shows the changes of water absorption in dried rice straw. Panel B shows the changes of water absorption of straw residue for different decomposition periods. The values are the means of 3 replicates (±standard deviation).

time. Thus, the maintenance of residues under conservation tillage requires the application of large amount of K to the soil.

Mechanism of K Adsorption on Straw Residues

Currently, the ability of crop straw to serve as a bioadsorbent material has been widely studied and has received increasing attention, especially regarding the possibility of reducing the pollution of heavy metals such as Hg(II), Cd(II), and U(VI) [23,37]. The adsorption of metal ions mainly relies on physical adsorption and chemical adsorption. Usually, the latter process is dominated, as reported by Ke et al., who observed that modified rice straw coupled with metal ions to form organic compounds [38]. However, few report exist regarding the coupling interaction

between heavy metals and residues during different decomposition period. In particular, alkali metal atoms (Li, Na, K) rarely form complexes with organic ligands because of their unique structures (which lack an empty orbital in the valence shell). When the active functional groups in rice straw bound Cd, nearly all the infrared bands from 3423 to 1000 cm^{-1} decreased in intensity [23]. In the residues after K adsorption, the functional group signatures at 4000 to 1300 cm^{-1} did not change (Figure 4). These results indicate that potassium chelated with C = C, C-O or carboxylic acids on straw is impossible. Moreover, greater focus has been placed in previous studies on the changes in the physical or chemical properties of soil after straw management [39], or on the conversion among diverse forms of nutrients [11]; the straw

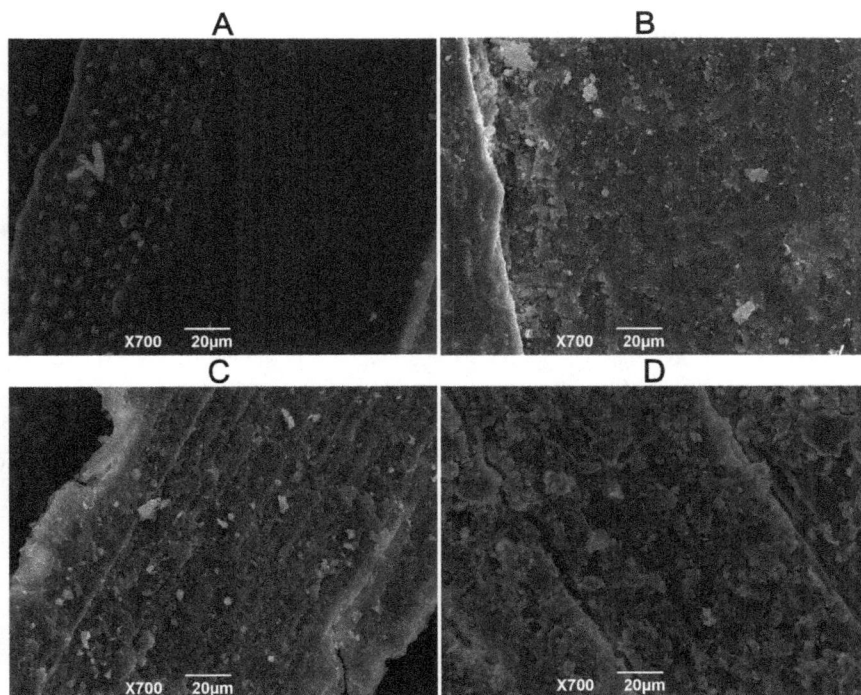

Figure 6. SEM micrographs of the rice straw residue surface on different days after incubation: A 0 d, B 15 d, C 60 d and D 110 d.

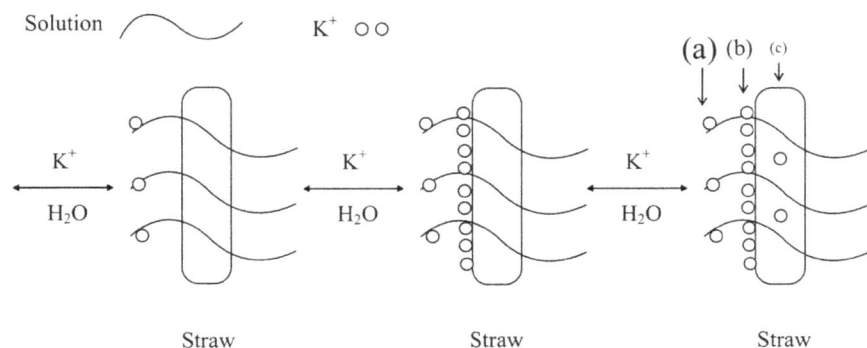

Figure 7. Schematic of the adsorption process of potassium (K⁺) on the rice straw residue surface. (a) K^+ retained on the straw in the form of an aqueous solution absorbed by residue; (b) K^+ adsorbed via electrostatic interaction; and (c) K^+ fixed on the residue surface by chemically interfacial adsorption. The size of the three letters indicates the amount of K^+ reserved on the residue by the corresponding process.

residue itself has therefore often been ignored. The adequate prediction of K dynamics and bioavailability is further complicated by the limited knowledge of the reversibility of K adsorption in soil. However, it was noted in the present study that rice straw residue could adsorb potassium despite releasing most of its potassium at the beginning of the experiment. At the same time, rape (canola) and wheat straw at different points in decomposition could also adsorb a portion of the K^+ ions, indicating that the role of crop residues in potassium adsorption is widespread (Figure S2). The present study provides both direct and indirect observations demonstrating that the adsorption mechanism can be described by the following three aspects:

(1) Physically interfacial adsorption: electrostatic adsorption. Due to higher content of potassium in natural straw, the rate of K release is significantly greater than the rate of K adsorption capacity during the earlier stage of decomposition. As decomposition continues, the zeta potential of the straw surface increases. Straw residues could adsorb K^+ ions from the surrounding environment (Figure 3). The straw K adsorption capacity is a function of the decomposition period and the extra K concentration, indicating an equilibrium between the K pool adsorbed on the residue and that in the bulk solution. Residues had an especially strong adsorption capacity after long-term decomposition. Therefore, as vast majority of K fertilizer inputs to the soil are one-time events, a portion of these inputs could be replaced by accumulated debris (or organic matter). Considering the K uptake by plants when the K concentration in the soil solution is reduced, crop residues could release K to replenish the soil solution [18].

(2) Chemically interfacial adsorption. The infrared spectra of straw residues after K adsorption on the 110th day of decomposition showed decreased intensity of the peaks at 798, 695 and 468 cm^{-1} in the fingerprint region. This results suggested that binding sites existed on the residue surface that could couple with K^+ ions. For peak at 798 cm^{-1} indicates the presence of C-H stretching or Si-O stretching, and Lu et al has demonstrated that the peaks at 795 and 466 cm^{-1} represent symmetric Si-O stretching vibration and Si-O deformation vibration, respectively [40]. The peak at 695 cm^{-1} likely reflects the vibration of nitrogenous or boric compounds [41,42]. These results confirmed that the compounds containing Si, B and N on the residue surface were involved in coupling with K^+.

(3) Solution fixed on residue. A study of the water retention of the 11 main varieties of crop straw, as conducted by Iqbal et al., showed that the water retention of straw was not necessarily linked with content of its components (cellulose, hemicellulose, lignin) but was significant related with its porosity and the variation of this porosity during decomposition. These authors found that the water content of rice straw reached its maximum after 6 h of immersion [31]. According to the present research, the capacity of water absorbed on the straw reached saturation after 150 min of immersion, which was consistent with the pervious study. The water content of the decomposed residue reached its maximum of 5.17 g g^{-1} after 15 d of incubation. Due to the differences of porosity, the change of the water content of the decomposed rice residue differed from that of maize residue, which showed higher water retention as incubation proceeded [31]. On the fifth day, the rice straw displayed a regular and compact surface structure, showing no difference with the raw material [43], which likely relied on its inner surface to absorb water (Figure 6). After 15 d of incubation, pectin, starch and other biodegradable materials on the surface of straw were depredated and fell off; the resulting surface was rougher and retained more water. After 60 d, the easily decomposable materials on the straw surface were completely decomposed, and the lignin and cellulose were exposed to the surface. Due to the reduction of porosity, the water retention decreased.

Briefly, the K adsorption process can be described with the following steps (Figure 7): (1) K^+ is transferred from the solution to the rice straw surface; (2) fractional K^+ ions are transferred from the rice straw surface to active sites; and (3) the water content on the surface is maximized, and the K^+ on the residue surface reaches equilibrium with the bulk solution. Therefore, under the same conditions, the amount of K retention via water absorption in straw is greater than that from physical adsorption and chemical adsorption. The latter two methods may, however, play an additional roles in the potassium retention.

The loss of soil nutrients (especially N and K) occurs primarily through runoff and leakage [44,45]. In China and South Asia, rice-based cropping systems such as two- or three-cropping rotation modes are nearly universally deployed [3]. Because of the large population, food demand and soil nutrient depletion in these regions, potassium deficit has become a serious problem [46]. For these regions in particular, the combination of rain, hot and intensive farming increases the risk of nutrient losses and subsequent loss of soil fertility. The present results indicate that straw residues at different points in the decomposition process have different capacities for adsorbing potassium and retaining water. The accumulation of residues in the soil not only reduces water loss, mitigates soil erosion and provides nutrients [47,48], but can also maintain soil nutrients, enhance soil fertility and promote the development of modern agriculture.

Supporting Information

Figure S1 Schematic diagram of the experimental setup of the straw decomposition trial. The depth of the bulk soil in boxes was approximately 22 cm, and that of the flooding layer was 1 cm. The nylon bags were buried into the soil according to the orientation in the diagram.

Figure S2 The adsorption of potassium (K) on wheat and rape straws for different decomposition periods. The added K concentration is 50 mg L^{-1}. The annotations in the

panels are the homogeneity of variances (H) and ANOVA. The H-test was performed using the Levene test. **indicates significant differences at $P<0.01$. The values are the means of 3 replicates ($\pm$standard deviation). The means with the same letter are not significantly different.

Acknowledgments

We gratefully acknowledge the reviewer and the editor for their constructive comments and suggestions. We also thank Wenjun Zhang for her comments on revising the manuscript.

Author Contributions

Conceived and designed the experiments: JWL JFL TR. Performed the experiments: JFL RHC. Analyzed the data: JFL JWL XKL. Contributed reagents/materials/analysis tools: XKL RHC LZ. Wrote the paper: JFL JWL LZ.

References

1. United Nations DoEaSA, Population Division (2013) World Population Prospects: The 2012 Revision, Highlights and Advance Tables. Working Paper No. ESA/P/WP.228.
2. Alexandratos N, Bruinsma J (2012) World agriculture towards 2030/2050: the 2012 revision. Rome: FAO. pp. ESA Working paper No.12–03.
3. Singh B, Shan YH, Johnson-Beebout SE, Singh Y, Buresh RJ (2008) Crop Residue Management for Lowland Rice-Based Cropping Systems in Asia. Advances in Agronomy 98: 117–199.
4. Stewart WM, Dibb DW, Johnston AE, Smyth TJ (2005) The contribution of commercia l fertilizer nutrients to food production. Agronomy Journal 97: 1–6.
5. Regmi AP, Ladha JK, Pasuquin E, Pathak H, Hobbs PR, et al. (2002) The role of potassium in sustaining yields in a long-term rice-wheat experiment in the Indo-Gangetic Plains of Nepal. Biology and Fertility of Soils 36: 240–247.
6. Panaullah GM, Timsina J, Saleque MA, Ishaque M, Pathan ABMBU, et al. (2006) Nutrient Uptake and Apparent Balances for Rice-Wheat Sequences. III. Potassium. Journal of Plant Nutrition 29: 173–187.
7. Wang XB, Hoogmoed WB, Cai DX, Perdok UD, Oenema O (2007) Crop residue, manure and fertilizer in dryland maize under reduced tillage in northern China: II nutrient balances and soil fertility. Nutrient Cycling in Agroecosystems 79: 17–34.
8. Sheldrick WF, Syers JK, Lingard J (2003) Soil nutrient audits for China to estimate nutrient balances and output input relationships. Agriculture Ecosystem & Environment 94: 341–354.
9. Pathak H, Mohanty S, Jain N, Bhatia A (2009) Nitrogen, phosphorus, and potassium budgets in Indian agriculture. Nutrient Cycling in Agroecosystems 86: 287–299.
10. Evenett SJ, Frédéric J (2012) Trade, competition, and the pricing of commodities. London: Centre for Economic Policy Research.
11. Li XK, Lu JW, Wu LS, Chen F (2009) The difference of potassium dynamics between yellowish red soil and yellow cinnamon soil under rapeseed (Brassica napus L.)–rice (Oryza sativa L.) rotation. Plant and Soil 320: 141–151.
12. Tejada M, Hernandez M, Garcia C (2009) Soil restoration using composted plant residues: Effects on soil properties. Soil and Tillage Research 102: 109–117.
13. Lal R (2005) World crop residues production and implications of its use as a biofuel. Environ Int 31: 575–584.
14. Kaur N, Benipal DS (2006) Effect of crop residue and farmyard manure on K forms on soils of long term fertility experiment. Indian Journal Crop Science 1: 161–164.
15. Liao YL, Zheng SX, Nie J, Xie J, Lu YH, et al. (2013) Long-Term Effect of Fertilizer and Rice Straw on Mineral Composition and Potassium Adsorption in a Reddish Paddy Soil. Journal of Integrative Agriculture 12: 694–710.
16. Jordan CF (1985) Nutrient cycling in tropical forest ecosystems. Principles and their application in management and conservation. New York: Wiley.
17. Rodríguez-Lizana A, Carbonell R, González P, Ordóñez R (2010) N, P and K released by the field decomposition of residues of a pea-wheat-sunflower rotation. Nutr Cycl Agroecosyst 87: 199–208.
18. Duong TTT, Verma SL, Penfold C, Marschner P (2013) Nutrient release from composts into the surrounding soil. Geoderma 195–196: 42–47.
19. Lin CW, Luo CY, Pang LY, Huang JJ, Tu SH (2011) Effect of Different Fertilization Methods and Rain Intensities on Soil Nutrient Loss from a Purple Soil. Scientia Agricultura Sinica 44: 1847–1854.
20. Kozak JA, Ahuja LR, Green TR, Ma LW (2007) Modelling crop canopy and residue rainfall interception effects on soil hydrological components for semi-arid agriculture. Hydrological Processes 21: 229–241.
21. Rawls WJ, Pachepsky YA, Ritchie JC, Sobecki TM, Bloodworth H (2003) Effect of soil organic carbon on soil water retention. Geoderma 116: 61–76.

22. Tong XJ, Li JY, Yuan JH, Xu RK (2011) Adsorption of Cu(II) by biochars generated from three crop straws. Chemical Engineering Journal 172: 828–834.
23. Ding Y, Jing DB, Gong HL, Zhou LB, Yang XS (2012) Biosorption of aquatic cadmium(II) by unmodified rice straw. Bioresource Technology 114: 20–25.
24. Vitousek PM (1994) Beyond Global Warming: Ecology and Global Change. Ecology 75: 1861–1876.
25. Walkley A (1947) A critical examination of a rapid method for determining organic carbon in soils: Effect of variations in digestion conditions and of inorganic soil constituents. Soil Science 63: 251–264.
26. Olsen SR, Cole CV, Watanabe FS, Dean LA (1954) Estimation of available phosphorus in soils by extraction with sodium bicarbonate. Washington, DC: USDA Circ. 939. U.S. Gov. Print. Office. 1–19.
27. van Reeuwijk LP (1992) Procedures for soil analysis 3rd edn. Wageningen, the Netherlands: ISRIC.
28. Daudu CK, Muchaonyerwa P, Mnkeni PNS (2009) Litterbag decomposition of genetically modified maize residues and their constituent Bacillus thuringiensis protein (Cry1Ab) under field conditions in the central region of the Eastern Cape, South Africa. Agriculture, Ecosystems & Environment 134: 153–158.
29. Honary S, Zahir F (2013) Effect of Zeta Potential on the Properties of Nano-Drug Delivery Systems -A Review (Part 1). Tropical Journal of Pharmaceutical Research 12: 255–264.
30. Cao W, Dang Z, Yi XY, Yang C, Lu GN, et al. (2013) Removal of chromium (VI) from electroplating wastewater using an anion exchanger derived from rice straw. Environmental Technology 34: 7–14.
31. Iqbal A, Beaugrand J, Garnier P, Recous S (2013) Tissue density determines the water storage characteristics of crop residues. Plant and Soil 367: 285–299.
32. Thomas MB, Spurway MI, Stewart DPC (1998) A review of factors influencing organic matter decomposition and nitrogen immobilisation in container media. In The International Plant Propagators' Society Combined Proceedings 48: 66–71.
33. Zhu JX, Yang WQ, He XH (2013) Temporal dynamics of abiotic and biotic factors on leaf litter of three plant species in relation to decomposition rate along a subalpine elevation gradient. PLoS ONE 8(4): e62073.
34. Pal D, Broadbent FE (1975) Influence of moisture on rice straw decomposition in soils. Soil Science Society of America Journal 39: 59–63.
35. Wang X, Sun B, Mao J, Sui Y, Cao X (2012) Structural Convergence of Maize and Wheat Straw during Two-Year Decomposition under Different Climate Conditions. Environmental Science & Technology 46: 7159–7165.
36. Chen XL, Yu J, Zhang ZB, Lu CH (2011) Study on structure and thermal stability properties of cellulose fibers from rice straw. Carbohydrate Polymers 85: 245–250.
37. Rocha CG, Zaia DAM, Alfaya RVS, Alfaya AAS (2009) Use of rice straw as biosorbent for removal of Cu(II), Zn(II), Cd(II) and Hg(II) ions in industrial effluents. J Hazard Mater 166: 383–388.
38. Ke X, Zhang Y, Li PJ, Li RD (2009) Study on mechanism of chestnut inner shells removal of heavy metals from acidic solutions. Journal of Shenzhen University (Science & Engineering) 26: 72–76 (In Chinese).
39. Karami A, Homaee M, Afzalinia S, Ruhipour H, Basirat S (2012) Organic resource management: Impacts on soil aggregate stability and other soil physico-chemical properties. Agriculture, Ecosystems & Environment 148: 22–28.
40. Lu P, Hsieh YL (2012) Highly pure amorphous silica nano-disks from rice straw. Powder Technology 225: 149–155.
41. Liao LB, Wang LJ, J.W Y, Fang QF (2010) Mineral material of modern testing technology. Beijing: Chemical industry press (In Chinese).
42. Peter L (2011) Infrared and raman spectroscopy: principles and spectral interpretation. Elsevier.

43. Yu G, Yano S, Inoue H, Inoue S, Endo T, et al. (2010) Pretreatment of Rice Straw by a Hot-Compressed Water Process for Enzymatic Hydrolysis. Applied Biochemistry and Biotechnology 160: 539–551.

44. Sims JT, Goggin N, McDermott J (1999) Nutrient management for water quality protection: Integrating research into environmental policy. Water Science and Technology 39: 291–298.

45. Lai FY, Yu G, Gui F (2006) Preliminary study on assessment of nutrient transport in the Taihu basin based on SWAT modeling. Science in China: Series D Earth Sciences 49: 135–145.

46. Cakmak I (2002) Plant nutrition research: priorities to meet human needs for food in sustainable ways. Plant Soil 247: 3–24.

47. Yang SM, Malhi SS, Li FM, Suo DR, Xu MG, et al. (2007) Long-term effects of manure and fertilization on soil organic matter and quality parameters of a calcareous soil in NW China. Journal of Plant Nutrition and Soil Science 170: 234–243.

48. Liu Y, Gao MS, Wu W, Tanveer SK, Wen XX, et al. (2013) The effects of conservation tillage practices on the soil water-holding capacity of a non-irrigated apple orchard in the Loess Plateau, China. Soil and Tillage Research 130: 7–12.

15

Variation in Chlorophyll Content per Unit Leaf Area in Spring Wheat and Implications for Selection in Segregating Material

John Hamblin[1]*, Katia Stefanova[1], Tefera Tolera Angessa[2]

1 Institute of Agriculture, University of Western Australia, Crawley, Western Australia, Australia, **2** School of Plant Biology, University of Western Australia, Crawley, Western Australia, Australia

Abstract

Reduced levels of leaf chlorophyll content per unit leaf area in crops may be of advantage in the search for higher yields. Possible reasons include better light distribution in the crop canopy and less photochemical damage to leaves absorbing more light energy than required for maximum photosynthesis. Reduced chlorophyll may also reduce the heat load at the top of canopy, reducing water requirements to cool leaves. Chloroplasts are nutrient rich and reducing their number may increase available nutrients for growth and development. To determine whether this hypothesis has any validity in spring wheat requires an understanding of genotypic differences in leaf chlorophyll content per unit area in diverse germplasm. This was measured with a SPAD 502 as SPAD units. The study was conducted in series of environments involving up to 28 genotypes, mainly spring wheat. In general, substantial and repeatable genotypic variation was observed. Consistent SPAD readings were recorded for different sampling positions on leaves, between different leaves on single plant, between different plants of the same genotype, and between different genotypes grown in the same or different environments. Plant nutrition affected SPAD units in nutrient poor environments. Wheat genotypes DBW 10 and Transfer were identified as having consistent and contrasting high and low average SPAD readings of 52 and 32 units, respectively, and a methodology to allow selection in segregating populations has been developed.

Editor: Sara Amancio, ISA, Portugal

Funding: The University of Western Australia funded the salaries of Drs. Stefanove and Angessa. Dr. Hamblin is retired and self funded. The operating and publishing costs were supplied from SuperSeed Technologies Pty Ltd (SST). The funders had no role in study design, data collection and analysis, decision to publish, or preparation of the manuscript.

Competing Interests: John Hamblin is the first author of this paper, and the operatiing and publishing funding comes from SuperSeed Technologies Pty Ltd (SST). This company is not for profit and is wholly owned by John Hamblin. The total support is about \$A2500. SunPalm Australia supplied a 1 kg sample of its slow-release fertilizer, Macrocote Purple, used in some of these experiments.

* E-mail: hamway2@gmail.com

Introduction

A recent review [1] considered that the rapid yield improvements of the last few decades were due primarily to increased Harvest Index of the economically important fraction of the crop with a more or less fixed biomass and/or from improved light harvesting through modified canopy structure. Further they suggested [1] that future increases in yield would have to come from improved photosynthetic efficiency. Several components of photosynthesis were examined for their potential in increasing crop yields. Amongst these was a reduction in leaf chlorophyll per unit area. There is little data on this issue partly because one of the most easily observed effects of N deficiency are pale green leaves, leading to the suggestion that using a SPAD 502 to measure leaf chlorophyll content per unit area could be a useful tool for determining N deficiency in crops [2,3].

Despite the paucity of data, available evidence suggests that the relationship between yield and chlorophyll level per unit leaf area is worth further examination.

Working with a reduced chlorophyll mutant in Soybeans, it was found in some circumstances that the mutant biomass exceeded the wild type by 30% [4]. Crop yields of mutant rice lines were not constrained by reduced levels of chlorophyll [5;6]. Also grain yield increases in wheat over past decades were not due to an increase in the rate of photosynthesis [7], and no significant relationship was observed between photosynthetic rate per unit leaf area and chlorophyll content at high light intensities in a range of C4 plants including maize [8].

Hamblin [unpublished] using a variegated mutant of Arabidopsis, its wild type and the F1 found that chlorophyll per unit leaf weight was additive, but the effect on biomass showed overdominance. The F1 was twice the weight of the wild type and 3 times the variegated mutant parent.

Two reasons were proposed as to why reduced chlorophyll levels might be beneficial [1]. First: less chlorophyll per unit leaf area would lead to improved light transmission though the canopy, potentially increasing photosynthesis of lower leaves and second: when exposed to excess light, more transparent leaves would reduce the level of photochemical damage to the chloroplasts in the upper canopy and thus less energy would be required for their repair.

The reason for excess chlorophyll in leaves above the optimum for photosynthesis may be that during evolution in the wild it was more important for survival to stop neighbours capturing light than it was to reduce photo-chemical damage by increased transmission through leaves [1]. However, this strategy is suboptimal for crops where productivity per unit area rather per plant drives improved production [1,9].

Other possible benefits from reduced chlorophyll levels in leaves include (a) that because chloroplasts are nutrient rich, less chlorophyll per unit leaf area may "spare" nutrients that could be used for crop growth particularly in situations of sub-optimal nutrition and (b) reduced light capture above that needed for maximum photosynthesis may reduce the heat stress on leaves in the upper part of the canopy and less water may be needed for cooling, leaving more for grain filling.

The purpose of this study was (a) to determine within and between plant variation in leaf chlorophyll per unit leaf area in a range of environments, (b) identify genotypes that consistently differ in their levels of chlorophyll per unit leaf area and (c) obtain preliminary data on the impact of different environments on the ranking of genotypes.

Should genotypes differing consistently in chlorophyll per unit leaf area be identified, then a breeding programme, using a SPAD 502 can be developed to produce comparable lines that would allow testing of the hypothesis that lower levels of chlorophyll per unit leaf area improve crop yields.

Materials and Methods

Germplasm

Up to 26 wheat and 2 barley genotypes were used in a series of experiments to assess variation in chlorophyll per unit leaf area. These included between different positions on the leaf, between different leaves per plant, between plants, between genotypes and between environments to determine the impact of these factors on chlorophyll level per unit leaf area, measured with a SPAD 502. The details of the meter and underlying information on its use in chlorophyll measurement are available from the manufacturer [10].

The source, passport data and experimental coding of the genotypes used in each experiment are listed in Table 1. Limited seed was available for the 23 wheat and 1 barley genotypes obtained from the Australian Winter Cereals Collection (AWCC), Tamworth, NSW, Australia (now at Horsham, Victoria, Australia). Of the other 4 genotypes, W5; W6; and B2, (varieties grown in Western Australia) were obtained from InterGrain Pty Ltd and 1 genotype (W7) from Dr David Bowran.

Experimental conditions and treatments

Four experiments were conducted in the winter growing season of 2012 at the University of Western Australia's Field Station, Shenton Park, Western Australia. Experiment 3 was also conducted during the 2012/13 summer in a glasshouse at Shenton Park.

Experiment 1. Three seeds per pot of 9 genotypes (Table 1) were planted with 3 replications on 30.4.2012 in 15 cm pots in an evaporatively cooled glasshouse. All pots had 18 grams of Macrocote Purple slow release fertilizer and 6 grams of Hortico soil wetter granules. The main tiller of one plant per pot was tagged on germination and all measurements were made on these tillers.

Chlorophyll levels per unit area were estimated using a SPAD 502 on the youngest fully expanded leaf (YFEL1) and the second youngest fully expended leaf (YFEL2) on the tagged tillers. Measurements were made on five dates (04.06.12; 11.06.12; 22.06.12; 05.07.12; and 31.07.12). As time progressed and plants

grew, the initial YFEL1 became YFEL2 and so on up the plant. All measurements were made between 10.00 and 12.00 a.m. The degree of cloud cover varied between and within dates. After each measurement date the pots were re-randomised.

Five evenly spaced SPAD readings, as judged by eye, were taken per leaf at each sample date, and overall 1350 measurements were made on the 27 tagged plants. These were recorded individually providing estimates of chlorophyll levels for YFEL1 and YFEL2 between different points on a leaf, between different leaves on a tiller, between different plants of the same genotype and between genotypes. Measurements stopped after the 5th date when ear emergence occurred on the two earliest genotypes and these lines changed from vegetative to reproductive growth.

Experiment 2. This experiment was conducted with 13 wheat genotypes (Table 1) across 5 environments. The environments were 2 planting dates (30.04.2012 and 29.05.2012) each with two 2 fertiliser rates under field conditions. The two fertilizer rates were a low fertility environment with nil and a high fertility environment with 20 grams of Macrocote Purple slow release fertilizer applied per row at seeding.

The genotypes were sown in spaced East/West rows 50 cm long (approximately 10 seeds per row). Rows were 100 cm apart and 50 cm between bays. Rows were thinned at 4 weeks to 5 plants per row (per genotype). As seed was severely limited there was only one entry for each genotype in each environment. Although this causes problems with interpretation (see later discussion) it reflects breeding reality where effective selection should be as early as possible to reduce the amount of material requiring field testing in plots at commercial densities.

The fifth environment was planted on 02.07.2012 into 15 cm pots with 20 grams of Bunnings slow release complete fertilizer and 6 grams of Hortico soil wetter granules with 4 seeds per genotype and grown in an evaporatively cooled greenhouse.

Chlorophyll per unit area was estimated for each genotype with a SPAD 502. Measurements were made on 6 dates, 29.06.2012 (planting date 1 only), 23.07.2012, 07.08.2012, 23.08.2012 (all environments); on 29.08.2012 (planting date 2 and glasshouse) and on 14.09.2012 (glasshouse only). The first measuring date for any environment depended on whether the plants in that environment had reached a suitable size for measuring with SPAD 502 and the final date on whether any genotypes in that environment has changed from vegetative to reproductive growth. At each date 10 chlorophyll measurements per plot (genotype) were made on random YFELs in the field environments, while 5 measurements per pot were made on random YFELs in the glasshouse.

Experiment 3. After the first planting date of Experiment 2, a further 13 wheat genotypes were received (Table 1) and were included in the 2nd planting date, high fertilizer treatment of experiment 2, which was planted on 29.05.2012 in the field and on 02.07.2012 in the glasshouse.

As in Experiment 2, chlorophyll per unit area was estimated with a SPAD 502. Measurements were made on 4 dates (26.07.2012; 07.08.2012; 23.08.2012; 29.08.2012) in the field and glasshouse environments. Additional measurements were taken on 13.09.2012 in the glasshouse environment only.

To provide an extreme alternative environment in terms of temperature and day length, the 26 wheat genotypes from Experiment 3 were planted on 30.1.2013 in an evaporatively cooled glasshouse and grown through February, the hottest month of the year in Perth, particularly so in 2013. Apart from the summer growing season, conditions were the same as those used in the glasshouse during the winter. SPAD readings were taken on the 24.02.2013 and 19.03.2013 when the earliest genotypes became reproductive.

Table 1. List of the genotypes used in the four experiments and obtained from the Australian Winter Cereals Collection (AWCC), InterGrain Pty Ltd (IGPL) and Dr. D. Bowran (DBW).

Genotype	Sources (AWCC No)	Exp Code	Exp 1	Exp 2	Exp 3	Exp 4
Wheat						
Ranee	AWCC (Aus1001)	W1	x	x	x	
Ranee (Variegated)	AWCC (Aus90113)	W2	x	x	x	
Alberta Red	AWCC (Aus1761)	W3	x	x	x	
Alberta Red (Variegated)	AWCC (Aus1762)	W4	x	x	x	
Emu Rock	IGPL	W5	x	x	x	x
Magenta	IGPL	W6	x	x	x	x
DBW 10	DBW	W7	x	x	x	
Kharchia	AWCC (Aus20741)	W8		x	x	
Janz	AWCC (Aus24794)	W9		x	x	
Stilletto	AWCC (AUS25923)	W10		x	x	
Pitic 62	AWCC (Aus804)	W11		x	x	
Tobari 66	AWCC (Aus1395)	W12		x	x	
Transfer	AWCC (Aus1406)	W13		x	x	x
Siete Cerros	AWCC (Aus1214)	W14			x	
Super X	AWCC (Aus6623)	W15			x	
Neepawa	AWCC (Aus12120)	W16			x	
Alfa	AWCC (Aus13900)	W17			x	
Uniculm 492	AWCC (Aus20430)	W18			x	
Oligoculm 112-76	AWCC (Aus20431)	W19			x	
CMH77A.917-1B-7Y-1B2Y-7B-0Y	AWCC (Aus21205)	W20			x	
Alfa	AWCC (Aus24324)	W21			x	
81W28-12	AWCC (Aus25186)	W22			x	
81W29-130	AWCC (Aus25192)	W23			x	
81W30-2	AWCC (Aus25194)	W24			x	
81W31-13	AWCC (Aus25209)	W25			x	
Excalibur	AWCC (Aus25292)	W26			x	
Barley						
BGS 306 va/3*Bowman (Variegated)	AWCC (Aus490473)	B1	x			
Bass	IGPL	B2	x			x

Experiment 4. The SPAD 502 has been recommended for use in determining whether wheat crops are deficient in applied N and whether more should be applied [2;3]. Experiment 4 was established with 3 nutrient levels and 4 genotypes (Table 1) in pots in the glasshouse to assess the potential impact of fertility on SPAD units.

The 3 nutritional levels used were (1) potting compost with no added slow release fertilizer but with 100 ml every 10 days post-planting of Thrive nutrient solution, (2) Potting compost + 20 grams Bunnings slow release fertilizer and (3) Potting compost + 20 grams Bunnings slow release fertilizer plus 100 ml of Thrive nutrient solution every 10 days post-planting. All pots had 6 grams of Hortico Soil Wetter Granules.

Three of the genotypes used are currently grown by Western Australia farmers, two wheat genotypes (W5 and W6), one barley genotype (B2). W13 was also grown as it was consistently low in chlorophyll per unit area. Due to insufficient seed, W13 was planted in the high fertilizer treatment only. This was to test if improved fertility reduced the difference between W13 and the other genotypes. Five seeds per genotype per pot were planted on

03.07.2012 in 3 replications. The SPAD measurements were taken on 5 random YFELs on 30.08.2012 and 13.09.2012.

Statistical analysis

Linear mixed models and ANOVA techniques were used to analyse the SPAD unit data generated from the four experiments varying in number of genotypes, replications and environments. All statistical analyses were performed using R (R Core Team, 2011) and GenStat, 15th Edition (VSN International Ltd, UK, 2012).

Experiment 1. Chlorophyll levels were measured on the two youngest fully expanded leaves (YFEL1 and YFEL2) at 5 approximately evenly spaced points per leaf of 3 plants (replicates) for 9 varieties (7 wheats and 2 barley) over time. This is a typical repeated measurement experiment and as such the linear mixed model used to analyse data from this experiment accounted for possible correlations between the plants' measurements over time. The fixed part of the model comprised the main effects and the interaction of *Day* and *Variety*. The random part of the linear mixed model was specified by including terms *Plant+Plant.Day*. In addition, instead of assuming uniform correlations between the measurement dates over time, a more realistic model was employed assuming a decreasing correlation as time between measurements increases. This was done by fitting a first order auto-regressive model.

Experiments 2 and 3. These were conducted to assess variety chlorophyll rankings, and interactions over time. Due to seed shortages, these two experiments were not replicated. The SPAD measurements per genotype were averaged over all readings and the data was subjected only to exploratory data analysis (correlations and graphs).

Experiment 4. The experiment comprised 3 fertilizer treatments (F0, F1 and F2) applied to 4 genotypes Emu Rock (W5), Magenta (W6), Transfer (W13) and Bass (B2) planted in 3 replications. SPAD measurements taken were on 2 dates. Transfer was only planted in F2 fertiliser treatment due to seed limitations. As a result of the unbalanced data, the data was grouped into two fully balanced subsets and analysed separately. The first subset included 3 varieties by 3 fertiliser treatments. This data was analysed using a two-way ANOVA, where the *Variety* and *Treatment* main and interaction effects were fitted in the model along with *Date/Replication* blocking structure. The model was re-fitted after deleting outliers which did not alter the conclusions and will not be considered further. The second subset included four varieties and treatment *F2* only. This data was analysed using one-way ANOVA modelling only *Treatment* effect and included the same blocking structure.

Results

Experiment 1

The SPAD unit measurements were highly consistent between measuring points on leaves and with the leaf mean within a date, between leaves on a tiller and over measuring dates. This is shown by the very good agreement of SPAD units between YFEL1 and YFEL2 for all genotypes and replications, for each of the 5 measurement dates (Fig. 1).

In all cases, the SPAD correlations were highly significant (p<0.001). The SPAD for date 5 YFEL1, (Table 2) illustrates the relationship between sampling positions on leaves over varieties and replicates on a given date. The variation accounted at each of the 5 recording dates/leaves correlations ranged from a low of 72% to a high of 92%.

When the SPAD data from experiment 1 was subjected to linear mixed model analysis, the main effects, of *Variety* and *Date of SPAD unit* by *Variety* interaction were highly significant (p<0.001). The SPAD units for both YFEL1 and YFEL2 for all varieties fell over time (Fig. 2).

Substantial and statistically significant genotypic differences were observed for the SPAD units as an indicator of chlorophyll content per unit area of leaf. The highest recording was from wheat genotype W7 (DBW 10) and the lowest reading was from wheat genotype W4 (Alberta Red Variegated) reflecting phenotypic variability of SPAD units, but attributable to genetic differences (Fig. 2, Table 3).

Experiment 2

The SPAD 502 can automatically calculate the means of a series of readings. Ten measurements were made per YFEL of each genotype in each environment in the field and 5 in the glasshouse. These were averaged and the mean recorded as a genotype's SPAD value. The pattern of genotypic response to different environments changed with the environment (Fig. 3). In environments E1, E2 and E4, only slight changes in SPAD units occurred over time; but there was a marked decrease of SPAD units in environment E3 and a marked increase in environment E5 (Fig. 3). Environment E5 was conducted in the glasshouse and its SPAD pattern over time was the opposite of that found in Experiment 1, which was also grown in the glasshouse but planted some 2 months earlier where for both YFEL1 and YFEL2, the SPAD units predominantly fell with time (Fig. 2). Wheat genotype W7 recorded the highest SPAD units in experiments 1 and 2 (Figs. 2 and 3).

Correlation analysis revealed that SPAD units between genotypes in different environments were highly and significantly correlated p<0.001 (Table 4). In terms of an individual genotype's SPAD units as an indicator of chlorophyll content per unit leaf area, wheat genotype W7 had the highest level (50.3) and genotype W13 had the lowest (28.5) averaged across all five environments.

Experiment 3

Two distinctive patterns were observed in environments 1 and 2 over time. There was a steady increase in SPAD units over time in environment 1 for all genotypes whereas in Environment 2 there was a decrease (Fig. 4).

Despite different patterns of response over time for the genotypes in the different environments (Winter field, winter glasshouse and summer glasshouse), the correlations were remarkably stable not only between the 2 winter environments (Fig. 4) but also with the summer ranking of varieties for chlorophyll content. The correlation between winter field and winter glasshouse environments was (r = 0.84*** significant at p≤ 0.001), between winter glasshouse and summer glasshouse was (r = 0.60** significant at p≤0.01), and between winter field and summer glasshouse was (r = 0.51** significant at p≤0.01).

Experiment 4

Due to the unbalanced nature of the design of this experiment, two separate analyses were performed. The first analysis involved 3 genotypes, Emu Rock (W5) Magenta (W6) and Bass (B2), and 3 fertiliser treatments (F0, F1 and F2). The treatment effect was highly significant (p<0.001), whilst the genotype effect was not significant (p = 0.131) Table 5. The genotype by treatment interaction was significant (p = 0.002). Genotypic means were very similar under fertiliser treatments F1 and F2 but significantly different in treatment F0. Within treatment F0, barley (Bass, B2) had a significantly higher SPAD reading than wheat at the first SPAD reading date (Table 5). This was because barley was initially much more vigorous than wheat in treatment F0.

The second analysis involved treatment F2 only, where the wheat genotype Transfer was tested alongside three other genotypes (Table 5). The result revealed that the genotypic effect was highly significant (p<0.001). This was due to Transfer's SPAD reading being substantially less than the other 3 genotypes in fertiliser treatment F2 (Table 5).

Discussion

SPAD units as estimates of leaf chlorophyll content per unit area from different points on a leaf and from different leaves of the same plant were highly correlated to SPAD units of other leaves on the same plant in Experiment 1 (Fig. 1; Table 2). Differences between genotypes for SPAD readings were large. From a plant selection perspective, a series of measurements using a SPAD 502 on a single leaf and averaging the measurements provides an adequate ranking assessment of genotypes for their level of leaf chlorophyll per unit area. Averaging several measurements to

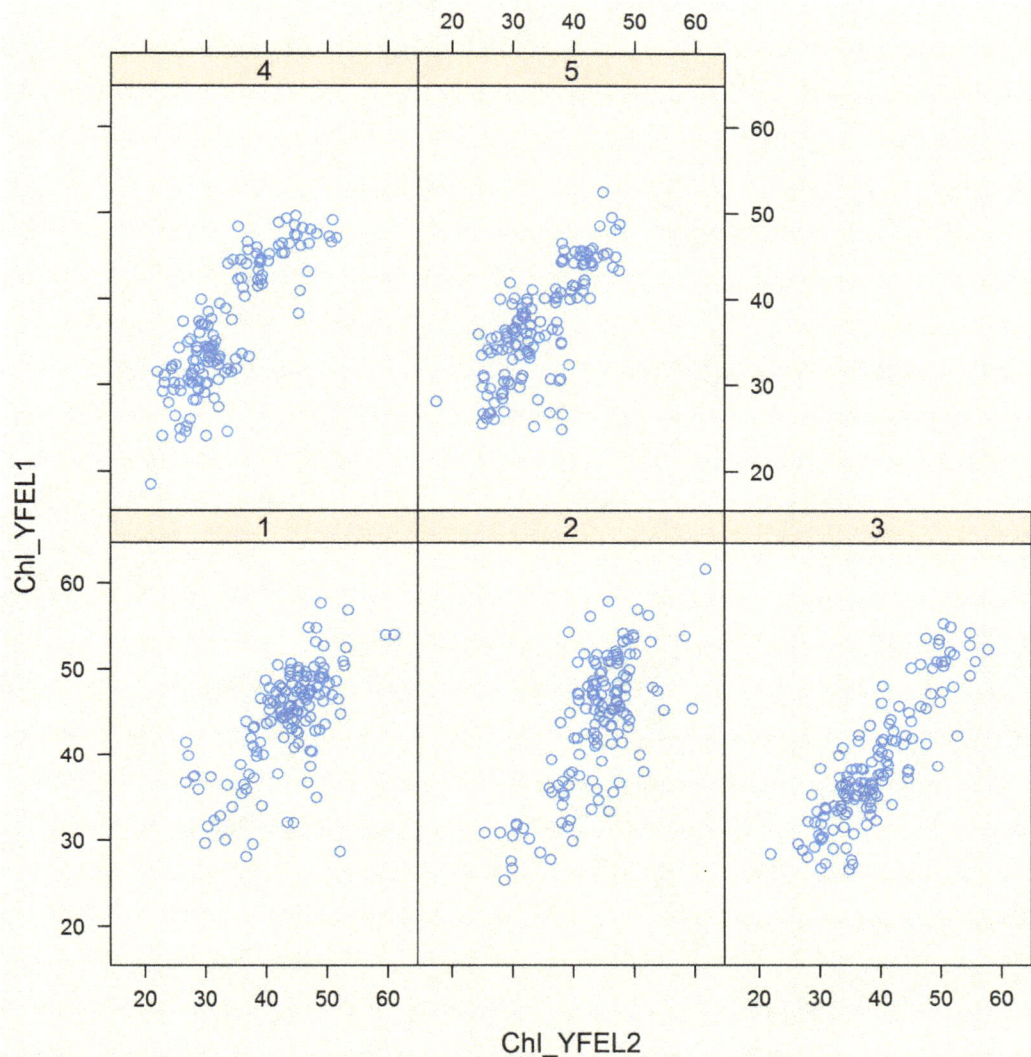

Figure 1. SPAD units (Chlorophyll) of first young fully expanded leaf (YFEL1) versus second young fully expanded leaf (YFEL2) for all five SPAD reading dates (1, 2, 3, 4 and 5) of experiment 1.

obtain a mean improves selection efficiency and reduces measurement and recording times and costs [11].

Despite the fact that there was only one entry per genotype in the different environments of Experiments 2 and 3, and there were easily observable interactions between genotypes and environ-

ments (Figs. 3, 4), nonetheless the ranking of varieties between environments were remarkably similar in both experiments (Table 4 and results section Experiment 3). This consistency was independent of the time of measurement and leaf position when the measurements were taken (Tables 2, 4). Such high consistency

Table 2. Correlations of date 5 YEFL (Young Fully Expanded Leaf) SPAD readings of chlorophyll content per unit area between 5 measurement positions and leaf mean in Experiment 1.

Position on leaf	2	3	4	5	leaf mean
1	0.87***	0.88***	0.83***	0.81***	0.92***
2		0.91***	0.85***	0.82***	0.94***
3			0.93***	0.92***	0.98***
4				0.97***	0.96***
5					0.95***

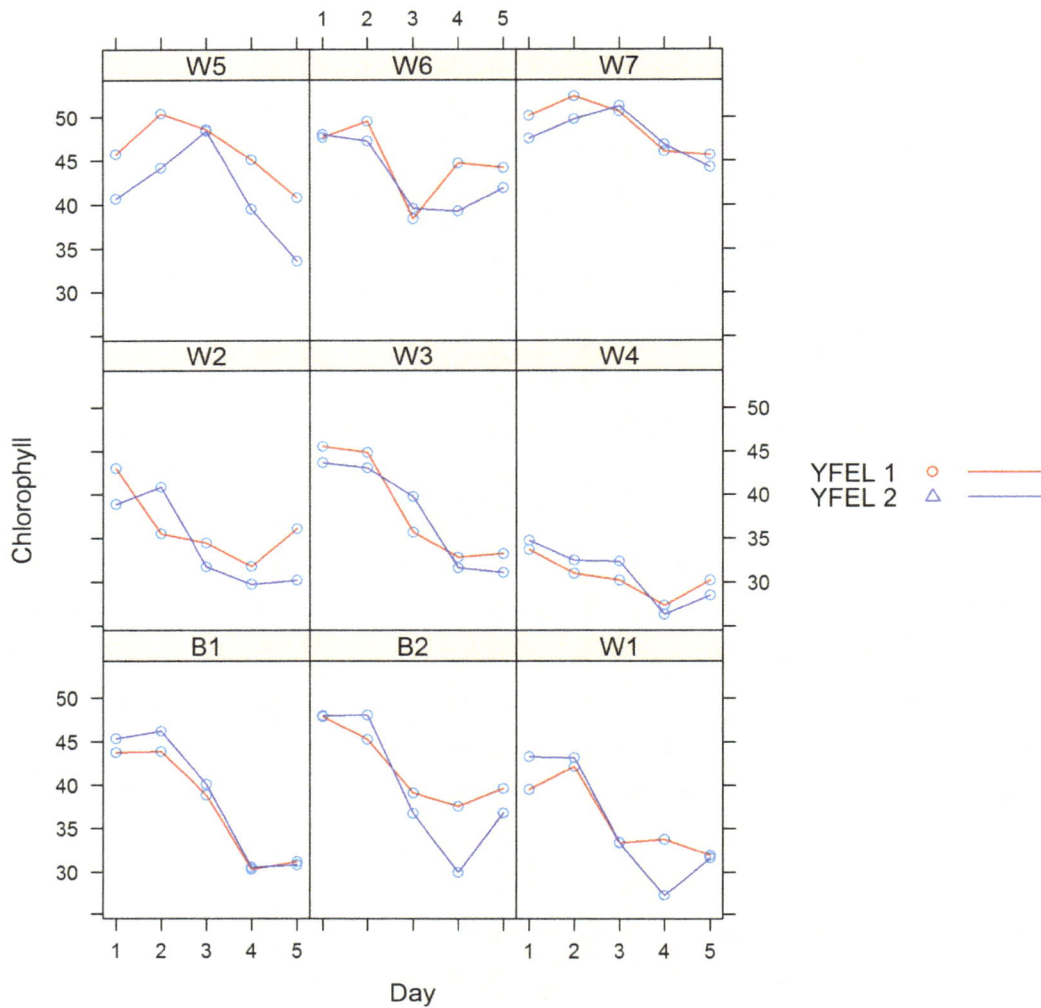

Figure 2. Declining pattern of SPAD units (chlorophyll) for the first young fully expanded leaf (YEFL1) and the second young fully expanded leaf (YEFL2) from the first through to the fifth (1, 2, 3, 4, 5) day of SPAD measurements for 7 wheat (W1–W7) and 2 barley (B1 and B2) genotypes in experiment 1.

Table 3. Predicted means and SE for the chlorophyll content of YEFL1 and YEFL2.

Variety	Chlorophyll YFEL1	Chlorophyll YFEL2
B1	37.6	38.6
B2	41.9	40.0
W1	36.2	35.8
W2	36.2	34.3
W3	38.5	37.9
W4	30.5	30.9
W5	46.2	41.3
W6	44.9	43.2
W7	49.0	48.0
SE	0.46	0.60

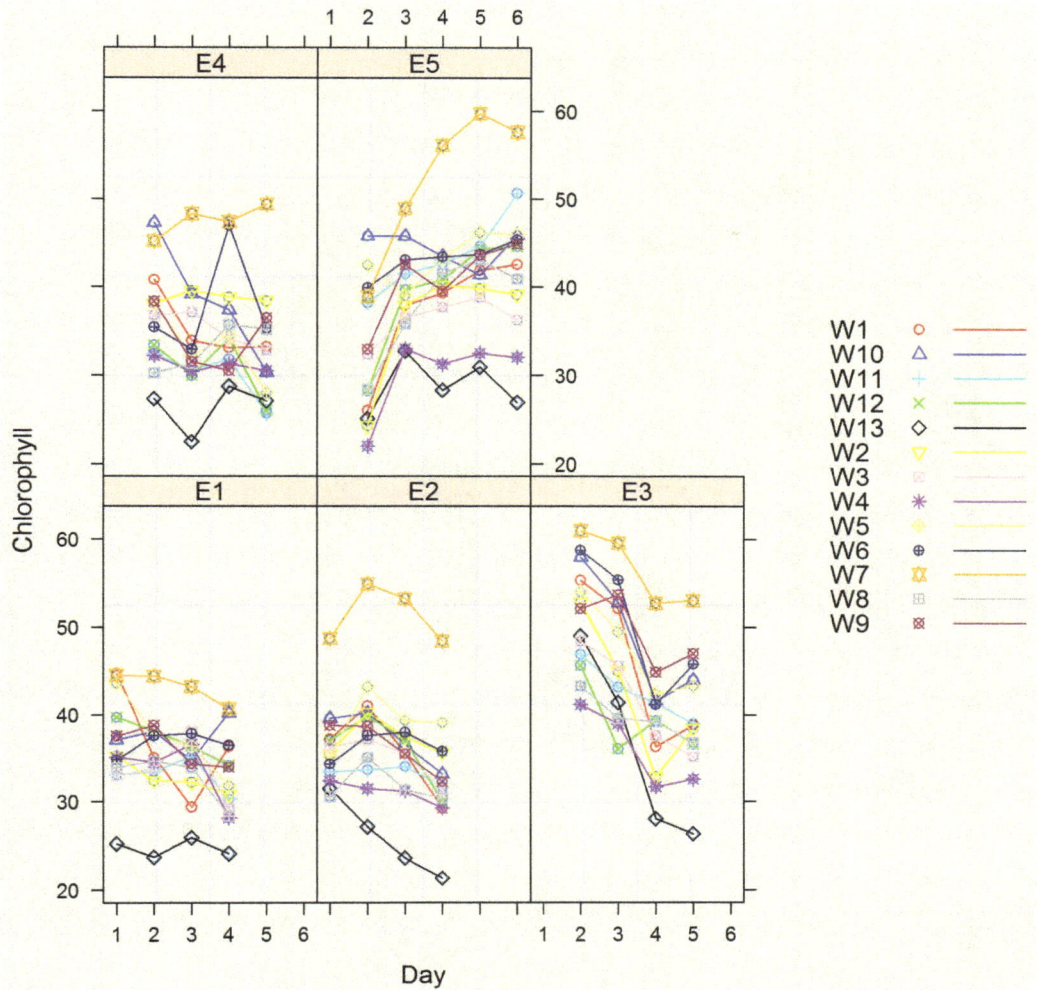

Figure 3. Wheat genotypes (W1–W13) SPAD units (Chlorophyll) in five environments (E1 Early sown High nutrition, E2 Early sown, Low nutrition, E3 Late sown, High nutrition, E4 Late sown low nutrition & E5 Glasshouse) taken on up to 5 days (1, 2, 3, 4, 5) in experiment 2.

in genotype ranking was unexpected as there are reported to be at least 17 additive and 9 epistatic QTLs for chlorophyll content in wheat and that 10 of the additive QTLs are expressed at different growth stages [12]. In this circumstance a large G x E interaction in genotype rankings and no significant correlations might be expected but were not observed. The results show that it should be possible to select for genotypic differences in levels of chlorophyll content per unit leaf area in spring wheat either on single plants or

Table 4. Correlation matrix of rankings of mean SPAD readings over dates in Experiment 2 for all genotypes and environments (E1–E4) in the field arising from combination of 2 planting dates (DOP) and 2 fertility levels (Low and High) and in the glasshouse (GH) as the fifth environment (E5).

DOP	DOP	01.05.12	29.05.12	29.05.12	02.07.12	
	Fertility	Low	High	Low	GH	Mean
01.05.12 (E1)	High	0.89***	0.80***	0.76***	0.83***	0.91***
01.05.12 (E2)	Low		0.85***	0.89***	0.85***	0.96***
29.05.12 (E3)	High			0.82***	0.88***	0.94***
29.05.12 (E4)	Low				0.71**	0.90***
GH (E5)						0.93***

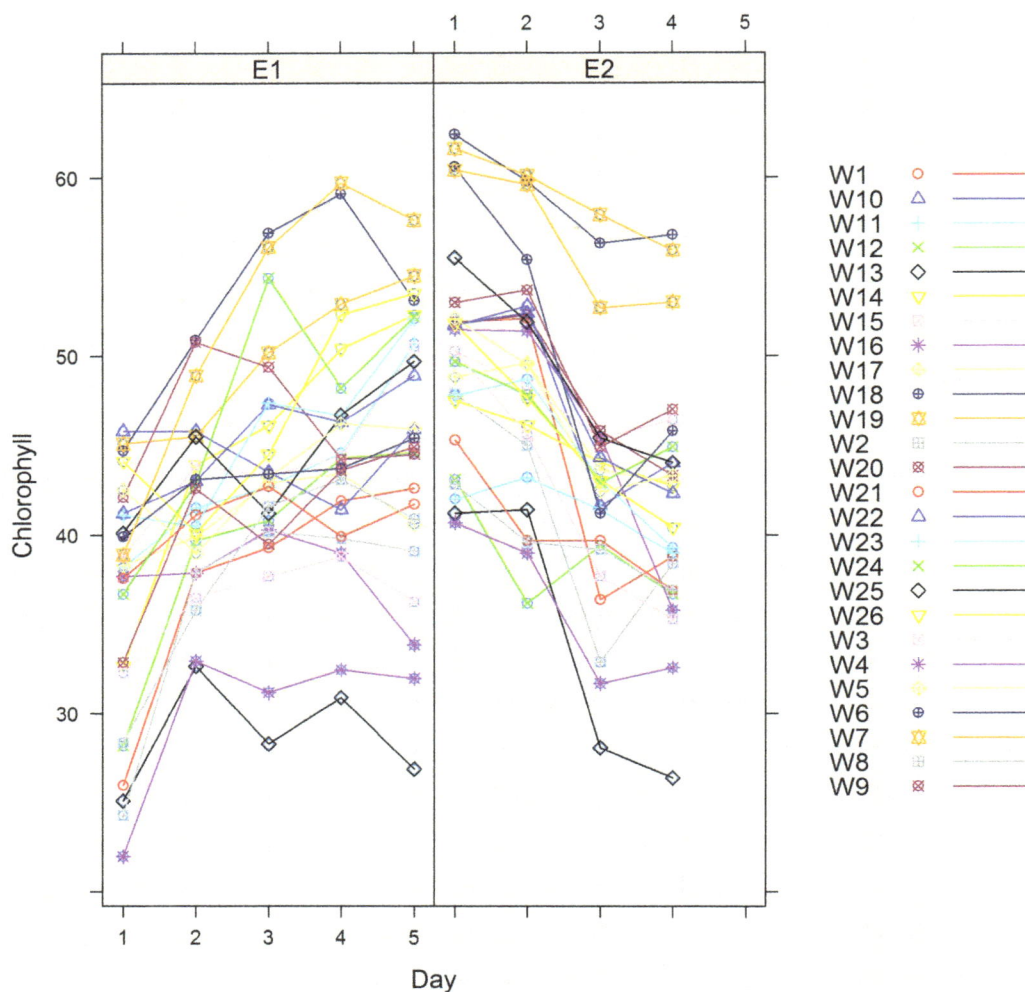

Figure 4. Wheat genotypes (W1–W26) SPAD units (Chlorophyll) in environments E1 (field based high fertility) and E2 (winter glasshouse) taken on 5 days in experiment 3.

Table 5. Mean SPAD units of chlorophyll content per unit leaf area of wheat genotypes Emu Rock, Magenta and Transfer and barley genotype Bass assessed under fertiliser treatments F0, F1 and F2 in Experiment 4.

Treatment/Genotype	Bass	Emu Rock	Magenta	Transfer
F0	32.4	19.8	24.1	N.T.
F1	47.1	49.1	46.2	N.T.
F2	48.1	49.8	50.0	33.9
[a]SED	2.6			N.A.
[a]LSD (0.05)	5.3			N.A.
[b]SED	1.9			
[b]LSD (0.05)	4.0			

[a]indicates SED or LSD when 3 genotypes tested in all 3 fertiliser treatments;
[b]indicates SED or LSD when all 4 genotypes tested in only F2 fertiliser treatment);
N.T. - Not Tested;
N.A. - Not Applicable.

in short rows at one time using a SPAD 502. This is in agreement with the observation that SPAD chlorophyll measurements have intermediate heritability and are of use in selection [7].

Unlike all the other genotypes used with stable SPAD units across environments, wheat genotype W4 (Variegated Alberta Red) produced seedlings that were variegated, albino (up to 40%) and green. Although this genotype had consistently low mean SPAD readings of 32 units across all environments, its unstable leaf variegation indicates that W4 is likely to be a chloroplast mutant [13] and not suitable for a breeding program aimed examining the hypothesis that less chlorophyll may be a path to higher yields.

From a practical point of view, SPAD measurements should be taken from fully expanded young upper leaves. When genotypes differing in days to flowering are studied, measurements should be taken before flowering of the earliest genotypes or genotypes should be grouped according to maturity before SPAD measurements are taken.

Results for the 3 modern genotypes, Emu rock, Magenta and Bass, grown at three fertiliser levels showed significant genotypes by fertiliser interactions in the F0 treatment as compared to the F1 and F2 treatments. In the no fertiliser (F0) environment, the barley genotype, Bass, had more vigorous growth and a higher SPAD reading at the first measurement date, where all genotypes had low SPAD readings (Table 5). Under environments with fertilisers (F1 and F2), wheat genotypes improved and their SPAD units were the same as those of barley. Despite the high nutrition of the F2 environment, W13 (Transfer) was significantly lower in SPAD units than the other wheat genotypes.

The SPAD 502 has been recommended for use in determining whether wheat crops are deficient in N [2,3]. Results from experiment 4, which was conducted to establish the impact of soil fertility on the level of SPAD units showed that the wheat genotype W13 maintained its significantly lower SPAD readings at high levels of plant nutrition. This result was in agreement with observations in experiments 2 and 3 where W13 always had low SPAD readings.

Two wheat genotypes, W7 (DBW 10) with high mean SPAD units of 52 and W13 (Transfer) with low mean SPAD units of 32 have been identified as having consistent and contrasting levels of leaf chlorophyll content per unit area.

Future Prospects

The substantial and consistent differences in SPAD units between W7 and W13 will be the basis of future agronomic, breeding, genetic, and mapping studies of chlorophyll content in spring wheat. Together with two commercial wheat cultivars (Emu Rock and Magenta with intermediate SPAD units) they have been crossed to develop pure lines to be used to determine the effect of different chlorophyll levels on yield in spring wheat.

Acknowledgments

For stimulating our research in this area: David Bowran, Willey Erskine, Don Ort, Tom Payne and Richard Richards. For part time technical assistance: Freda Blakeway. For provision of germplasm: The Australian Winter Cereals Collection formerly at Tamworth, now at Horsham, InterGrain Pty Ltd and Dr. D. Bowran. Peter Hiscock of Sunpalm Australia for supplying 1 kg of Macrocote purple fertilizer used in some of these experiments. We would also like to acknowledge our reviewers and editors whose helpful suggestions have significantly improved the manuscript.

Author Contributions

Conceived and designed the experiments: JH KS TTA. Performed the experiments: JH. Analyzed the data: KS TTA JH. Contributed reagents/materials/analysis tools: JH KS. Wrote the paper: JH KS TTA. Sourced seed: JH.

References

1. Zhu Xin-G, Long SP, Ort DR (2010) Improving photosynthetic efficiency for greater yield. Annual Rev Plant Biol 61: 235–261.
2. Chunjiang Z, Aning J, Wenjiang H, Keli L, Liangyun L, et al. (2007) Evaluation of variable-rate nitrogen recommendation of winter wheat based on SPAD chlorophyll meter measurement. New Zealand Journal of Agricultural Research 50(5): 735–741. Available: http://dx.doi.org/10.1080/00288230709510345.
3. Murdock L, Call D, Ames J (2004) Comparison and use of chlorophyll meters on wheat. University of Kentucky College of Agriculture Cooperation extension service AGR 181. Available: www.ca.uky.edu/agc/pubs/agr/agr181/agr181. pdf. Accessed 14.5.2013 and 5.3.2014.
4. Pettigrew WT, Hesketh JD, Peters DB, Wooley JT (1989) Characterization of canopy photosynthesis of chlorophyll-deficient soybean isolines. Crop Sci 29: 1025–1029.
5. Chen X, Zhang W, Xie Y, Lu W, Zhang R (2007) Comparative proteomics of thylakoid membrane from a chlorophyll b-less rice mutant and its wild type. Plant Sci 173: 397–407.
6. Li Y, Ren B, Gao L, Ding L, Jiang D, et al. (2013) Less Chlorophyll Does not Necessarily Restrain Light Capture Ability and Photosynthesis in a Chlorophyll-Deficient Rice Mutant. Journal of Agronomy and Crop Science 199: 49–56.
7. Richards RA (2000) Selectable traits to increase crop photosynthesis and yield of grain crops. J Exp Bot 51: 447–458.
8. Usuda H, Ku MSB, Edwards GE (1984) Rates of photosynthesis relative to activity of photosynthetic enzymes, chlorophyll and soluble protein content among ten C4 Species. Aust. J. Plant Physiol. 11: 509–517.
9. Donald CM, Hamblin J (1976) The Biological Yield and Harvest Index of Cereals as Agronomic and Plant Breeding Criteria. Advances in Agronomy 28: 361–405.
10. Konica Minolta products website (2014) Available: http://www.konicaminolta. eu/fileadmin/content/eu/Measuring_Instruments/2_Products/1_Colour_ Measurement/6_Chlorophyll_Meter/PDF/Spad502plus_EN.pdf. Accessed 12.5.2013, 5.3.2014.
11. Giunta F, Motzo R, Deidda M (2002) SPAD readings and associated leaf traits in durum wheat, barley and triticale cultivars. Euphytica 125: 197–205.
12. Zhang K, Fang Z, Liang Y, Tian J (2009) Genetic dissection of chlorophyll content at different growth stages in common wheat. Journal of Genetics 88 (2): 183–189.
13. Robertson DW (1937) Maternal inheritance in barley. Genetics 22: 104–112.

16

Fungi Benefit from Two Decades of Increased Nutrient Availability in Tundra Heath Soil

Riikka Rinnan[1,2,3], **Anders Michelsen**[2,3], **Erland Bååth**[1]*

1 Department of Biology, Lund University, Lund, Sweden, 2 Department of Biology, University of Copenhagen, Copenhagen, Denmark, 3 Center for Permafrost (CENPERM), University of Copenhagen, Copenhagen, Denmark

Abstract

If microbial degradation of carbon substrates in arctic soil is stimulated by climatic warming, this would be a significant positive feedback on global change. With data from a climate change experiment in Northern Sweden we show that warming and enhanced soil nutrient availability, which is a predicted long-term consequence of climatic warming and mimicked by fertilization, both increase soil microbial biomass. However, while fertilization increased the relative abundance of fungi, warming caused only a minimal shift in the microbial community composition based on the phospholipid fatty acid (PLFA) and neutral lipid fatty acid (NLFA) profiles. The function of the microbial community was also differently affected, as indicated by stable isotope probing of PLFA and NLFA. We demonstrate that two decades of fertilization have favored fungi relative to bacteria, and increased the turnover of complex organic compounds such as vanillin, while warming has had no such effects. Furthermore, the NLFA-to-PLFA ratio for ^{13}C-incorporation from acetate increased in warmed plots but not in fertilized ones. Thus, fertilization cannot be used as a proxy for effects on warming in arctic tundra soils. Furthermore, the different functional responses suggest that the biomass increase found in both fertilized and warmed plots was mediated via different mechanisms.

Editor: Kathleen Treseder, University of California Irvine, United States of America

Funding: RR was supported by the Academy of Finland (decision no. 108277) and the European Commission (MEIF-CT-20065-024364). EB was supported by the Swedish Research Council (grant 621-2009-4503). AM was supported by the Danish Council for Independent Research/Nature and Universe (grant 10-084285). This work was part of LUCCI (Lund University Centre for Studies of Carbon Cycle and Climate Interactions) and the Center for Permafrost (CENPERM) funded by the Danish National Research Foundation. The funders had no role in study design, data collection and analysis, decision to publish, or preparation of the manuscript.

Competing Interests: The authors have declared that no competing interests exist.

* E-mail: erland.baath@biol.lu.se

Introduction

Microbial use of carbon substrates in soil is of critical importance when estimating impacts of climate change on the fate of soil carbon. Especially in the Arctic, where the amount of soil carbon exceeds the amount of carbon in aboveground plant biomass by an order of magnitude [1], microbial processes transforming the carbon compounds are crucial. In this region, climate change is predicted to increase the mean annual temperature by 2–9°C within this century [2].

Warming generally stimulates microbial growth [3] and thereby increases nutrient mineralization [4] in soil. As enhanced nutrient availability thus is an expected long-term consequence of warming, fertilization is often used as a long-term proxy for warming in experimental studies [5,6]. Both warming and enhanced nutrient availability affect turnover of soil carbon, directly through effects on microbial activity [3] and indirectly by increasing plant growth [7] and altering vegetation composition [8,9].

Changes in plant growth and vegetation composition influence carbon inputs to soil in several ways. In the short-term, the quantity and the chemical quality of root exudates may change as these differ among plant species [10]. Also, the distribution of these fresh carbon inputs in the soil horizon may change, as different plant groups have different rooting depths [6,11]. In the long-term, vegetation changes will influence litter input. For example,

an increase in deciduous shrubs [12] would lead to higher leaf litter deposition each autumn, while an increase in graminoids [8] would add easily decomposable plant litter to soil [13]. Climate change may also alter the chemical composition of a plant, often increasing the concentration of secondary metabolites in the plant tissue [14].

We aimed to unravel how the composition of microbial communities and their use of carbon substrates in subarctic tundra soil respond to long-term warming and enhanced nutrient availability using SIP-PLFA (stable isotope probing of phospholipid fatty acids). Soil samples were taken from a field experiment which had been running for 18 years prior to the sampling on a mesic/dry heath just above the tree line near Abisko Scientific Research Station in Northern Sweden. The warming treatment, which was accomplished by open-top chambers, initially resulted in an increased biomass of deciduous shrubs [15], decreased abundance of bryophytes [8], and reduced soil microbial biomass [5]. After a decade, warming had thus reduced the frequency of mosses by 50% and that of lichens by 35%, but there was no effect on graminoids [8]. The NPK (nitrogen, phosphorous and potassium) fertilization treatment, which aimed at mimicking increased soil nutrient availability, had also increased vascular plant biomass at the expense of bryophyte biomass [15], but increased soil microbial biomass including fungal biomass [5]. After 14 years, both warming and fertilization had increased the biomass of deciduous shrubs, but the increase in graminoids was still confined to the fertilized plots [15]. After 22 years, both warming and fertilization treatments had 35% higher vascular plant cover than the control [16].

In the laboratory, the soil samples were amended with a range of ^{13}C-labelled substrates, and the uptake of these substrates into various fatty acids indicating different microbial groups and either growth (PLFAs) or storage (neutral lipid fatty acids, NLFAs) was followed. The selected substrates represented carbon sources present in soil. Glucose, acetic acid and glycine are simple compounds common in plant root exudates, and glycine is also a source of nitrogen for subarctic plants and microbes [11]. Starch is a very common polysaccharide in plant residues. Vanillin is a common product of lignin depolymerisation [17], containing a phenol ring, and is often used as a model substance indicating lignin degradation. Starch and vanillin are therefore examples of more complex substrates and supposedly more difficult to decompose.

We expected that, in line with the previous results from the field experiment [5], fertilization would increase and warming decrease microbial biomass. We hypothesized that fertilization would increase the uptake of carbon substrates into PLFA as compared to NLFA due to alleviated nutrient limitation. Furthermore, we expected that vanillin uptake would decrease with increased N availability, since it has earlier been shown that fertilization can suppress ligninolytic enzymes in forest soil [18–20]. Warming was expected to have the same effects as fertilization if this treatment would lead to increased nutrient availability.

Results

Treatment Effects on Fatty Acid Concentrations

Substrate additions did not cause any significant changes in the total PLFA or NLFA concentration, and therefore an average of each substrate is presented (Fig. 1). Fertilization and warming significantly increased the total PLFA concentration in soil by about 31% and 20%, respectively (Fig. 1; P<0.001), which indicates that these treatments increased microbial biomass. The total NLFA concentration was only increased by fertilization (P<0.001), although less when fertilization was combined with warming (Fig. 1; P<0.01 for warming×fertilization interaction). There were no significant differences between the two incubation times.

In the PCA of the PLFA profiles, the first PC (explained variance 32%) described treatment effects on the PLFA patterns (Fig. 2). The control soil had the highest values along this PC axis, followed by partly overlapping warming and warming + fertilization treatments close to the origin, and the fertilized soil had the lowest values (Fig. 2a; P<0.001 for effects of both warming and fertilization). The warming×fertilization interaction term was highly significant (P<0.001) because of the relatively similar PLFA patterns in the warming and warming + fertilization treatments as compared to fertilization alone. The fertilized soil was characterized by relatively higher amounts of the fungal biomarker 18:2ω6,9 than the unfertilized soil (Fig. 2b). The second PC (explained variance 15%) mainly accounted for the difference between the incubation times (Fig. 2a; P<0.001); the relative abundance of the straight-chained PLFAs 14:0, 15:0 and 16:0, and the markers for Gram-positive bacteria i14:0, i15:0, a15:0 and i16:0 increased from one to seven-day samples (Fig. 2b). The fungal biomarker 18:2ω6,9 was not affected by the incubation time.

The PCA of the NLFA mole percentage profiles showed that while the 1-day incubation time produced relatively similar patterns except for the fertilization treatment, there was a large

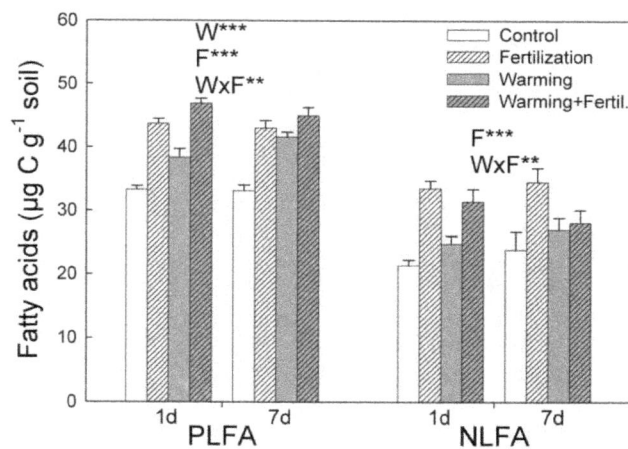

Figure 1. Total concentrations of phospholipid (PLFA) and neutral lipid fatty acids (NLFA) in tundra heath soil after 18 years of fertilization and warming treatments. The bars represent mean ± SE (n = 6) for 1-day and 7-day-long incubations averaged across the substrate additions, which did not affect the fatty acid concentrations. Significant effects of factors warming (W) and fertilizer addition (F), and their interaction are shown at **P<0.01, ***P<0.001 (Linear Mixed Models with Warming, Fertilization and Time as fixed factors and Block as a random factor).

spread in the patterns for the 7-day-long incubation (Fig. 3a). As for PLFA, the first PC (explained variance 38%) accounted for the difference between the fertilized and the unfertilized soils, and showed a warming×fertilization interaction (Fig. 3a; P<0.001). As for PLFAs, the difference was mainly due to a higher abundance of the fungal markers, in this case the NLFAs 18:2ω6,9 and 18:1ω9 (Fig. 3b). The second PC mainly separated the incubation times from each other (Fig. 3a; P<0.001).

Treatment Effects on ^{13}C-incorporation into PLFA

There were no significant effects of treatments or incubation time on total ^{13}C-incorporation from glucose, acetic acid, glycine (Table 1). Fertilization increased total ^{13}C-incorporation from vanillin (P<0.001), but did not significantly affect incorporation from starch. The uptake of starch and vanillin increased from one to seven-day-long incubation (Table 1; P<0.01).

The PCAs for the ^{13}C-incorporation patterns showed that for all substrate additions, the first two PCs separated the fertilized soil from the unfertilized soil (Fig. 4; P<0.001). A common explanation for this difference was an increased incorporation of ^{13}C into the fungal biomarker PLFAs, 18:2ω6,9 and 18:1ω9 and a decreased incorporation into most of the bacterial biomarker PLFAs under fertilization (see Figs S1, S2, S3, S4, S5, S6, S8). As an exception, the Gram-negative bacteria indicated by cy17:0 and cy19:0 took up significantly more glycine, starch and vanillin in the fertilized than in the unfertilized soil (Fig. S3). In addition, utilization of starch was higher in the fertilized than in the unfertilized soil for Gram-positive bacteria containing i16:0 and the actinomycetes indicated by the methylated PLFA 10Me16:0, especially after incubation for seven days (Figs S6 and S7).

The substrate utilization patterns in the warming treatment were rather similar to those in the control samples, except for starch addition with a significant warming effect on the second PC (Fig. 4, P<0.01). The incorporation of ^{13}C from starch into the PLFAs i16:0 and 10Me18:0 was higher in the warming treatment than in the control soil after incubation for seven days (Figs S6 and

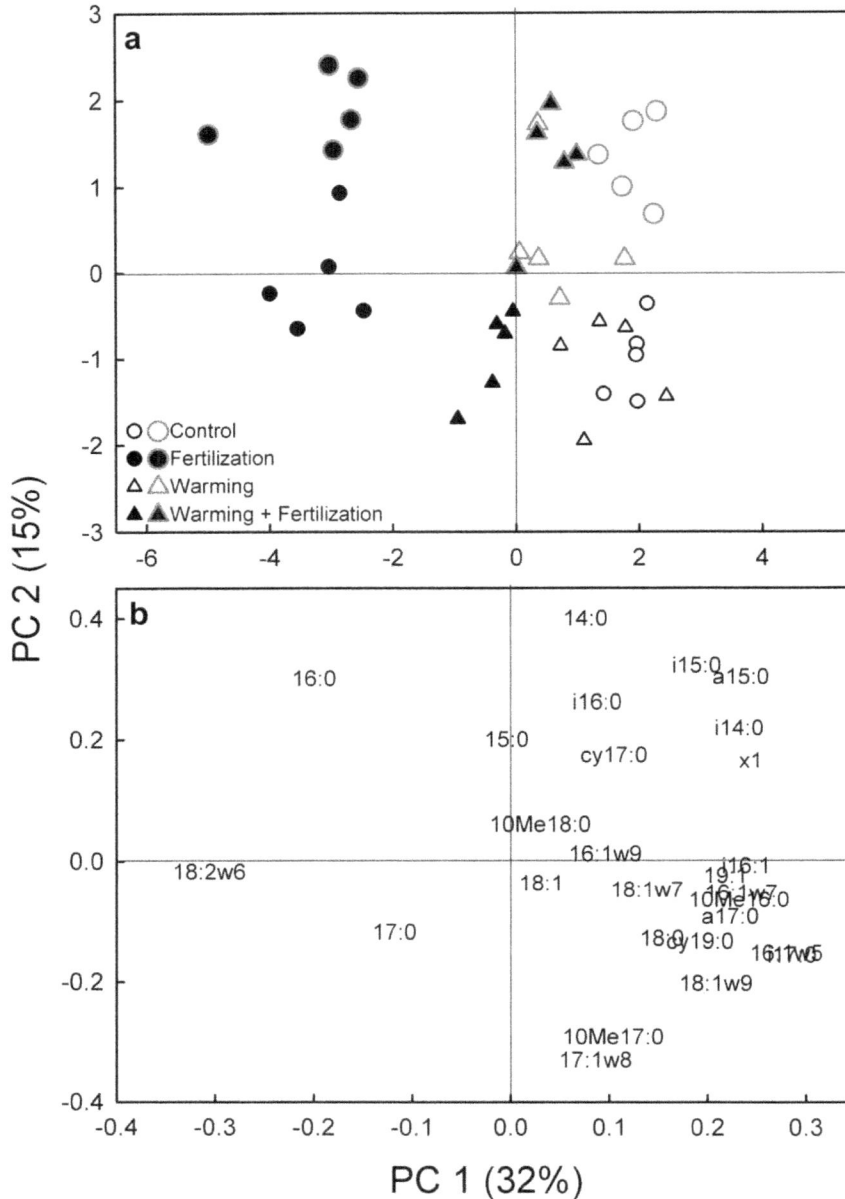

Figure 2. Principal component analysis of the phospholipid fatty acid (PLFA) mole percentage profiles in tundra heath soil from a field experiment with long-term fertilization and warming treatments. (a) A score plot showing mean values for substrate additions after a 1-day-long incubation (small symbols drawn with black line) and after a 7-day-long incubation (large symbols drawn with grey line). The mean values for all substrates within each treatment (glucose, glycine, acetic acid, starch and vanillin) are marked with similar symbols since the substrate additions did not affect the PLFA pattern. $P<0.001$ for effects of Warming, Fertilization, Time and Warming×Fertilization on the PC scores (Linear Mixed Models with Warming, Fertilization and Time as fixed factors and Block as a random factor). (b) A loading plot showing the individual PLFAs. Variance explained by principal component (PC) 1 and 2 in parentheses.

S8). Uptake of glucose by organisms containing the PLFA 18:1ω9 was significantly reduced by warming (Fig. S2). The warming+fertilization treatment had relatively similar substrate utilization patterns as fertilization alone (Fig. 4), which led to warming×fertilization interactions in the incorporation to some individual PLFAs (Figs S1, S2, S3, S4, S5, S6, S7, S8).

A clear difference in the ^{13}C-incorporation patterns between the two incubation times could be observed on the first two PCs for all substrates (Fig. 4; $P<0.001$). This owed mainly to the increased incorporation into the PLFAs cy17:0, cy19:0, 10Me16:0 and 10Me18:0 over time (Figs S3, S7, S8).

Treatment Effects on ^{13}C-incorporation into NLFA

The incorporation of ^{13}C to NLFA decreased over time for glucose, acetic acid, and glycine (Table 1; $P<0.01$). Fertilization increased the total incorporation of ^{13}C from vanillin ($P<0.01$) but did not significantly affect incorporation from other substrates.

Incorporation of ^{13}C to the fungal biomarker NLFAs, 18:2ω6,9 and 18:1ω9, was highest from acetic acid, followed by glucose and

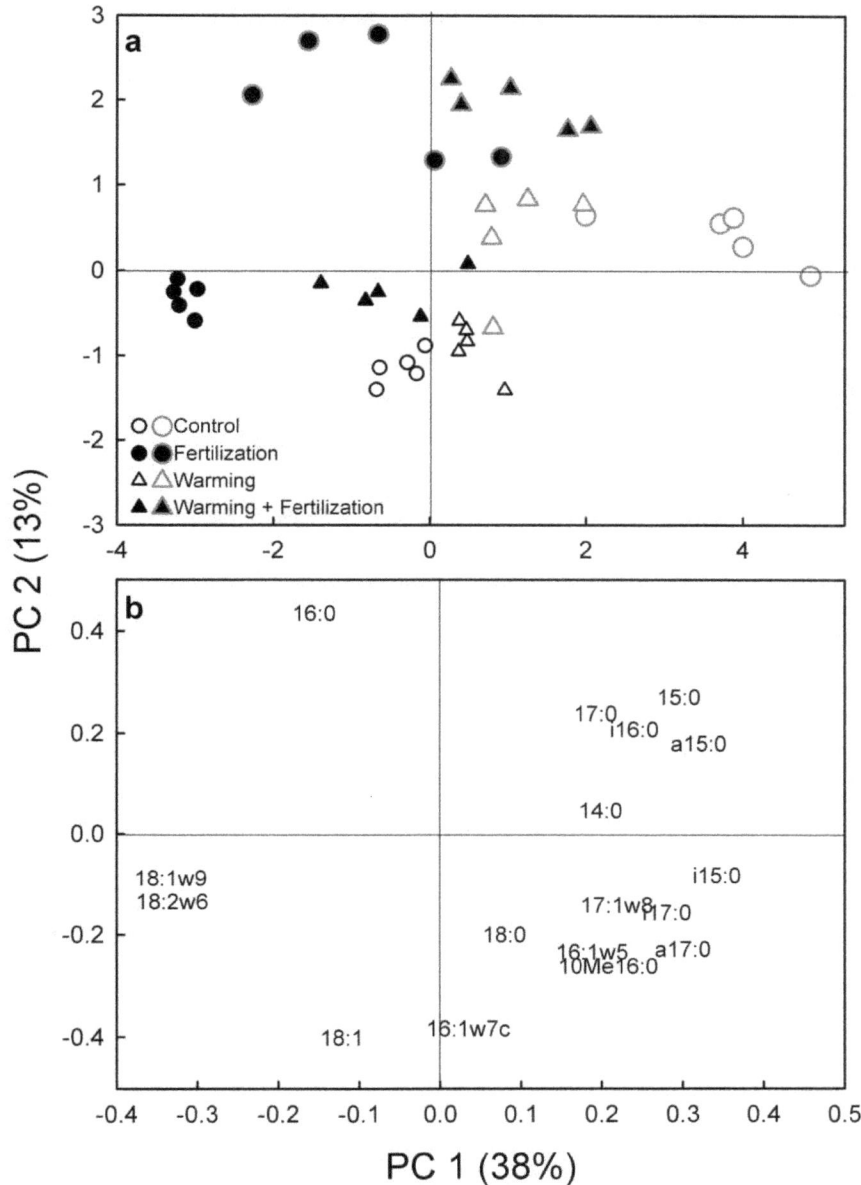

Figure 3. Principal component analysis of the NLFA mole percentage profiles in tundra heath soil from a field experiment with long-term fertilization and warming treatments. (a) A score plot showing mean values for substrate additions after a 1-day-long incubation (small symbols drawn with black line) and after a 7-day-long incubation (large symbols drawn with grey line). The mean values for all substrates within each treatment (glucose, glycine, acetic acid, starch and vanillin) are marked with similar symbols since the substrate additions did not affect the NLFA pattern. (b) A loading plot showing the individual NLFAs. Variance explained by principal component (PC) 1 and 2 in parentheses. $P < 0.001$ for effects of Fertilization, Time and Warming×Fertilization on the PC scores (Linear Mixed Models with Warming, Fertilization and Time as fixed factors and Block as a random factor).

vanillin (Fig. 5a, b). Incorporation of ^{13}C to these NLFAs from glycine and starch was minimal. The response patterns for 18:2ω6,9 and 18:1ω9 were similar, but because of the higher variance, only vanillin-amended samples showed significant differences at $P < 0.01$ for 18:2ω6,9; fertilization increased ($P < 0.001$) and warming decreased ($P < 0.01$) ^{13}C-incorporation from vanillin (Fig. 5a). Similar responses were observed for ^{13}C-incorporation from vanillin into the NLFA 18:1ω9 (Fig. 5b; $P < 0.001$ for both warming and fertilization). Fertilization also significantly increased ^{13}C-incorporation from glucose, when not

combined with warming treatment ($P < 0.01$ for warming×fertilization).

The NLFA-to-PLFA ratios for ^{13}C-incorporation into 18:2ω6,9 and 18:1ω9 were not significantly affected by treatments for other substrates than acetic acid (data for glucose and acetic acid shown in Table 1). For acetic acid, the ratio for incorporation into 18:2ω6,9 was significantly increased by warming after incubation for 7 days (with a similar insignificant trend for 18:1ω9), while it was unaffected by fertilization (Table 1).

Table 1. Total ^{13}C-incorporation into phospholipid (PLFA) and neutral lipid fatty acids (NLFA) and NLFA/PLFA ratios for ^{13}C-incorporation into fungal biomarkers (mean$\pm$SE, $n = 6$) in tundra heath soil following incubation with added glucose, glycine, acetic acid, starch or vanillin for 1 or 7 days.

	1-day incubation				7-day incubation				Statistical
	C[a]	F	W	W+F	C	F	W	W+F	significance
Total ^{13}C-incorporation into PLFA									
(μg ^{13}C per mg added ^{13}C g^{-1} soil)									
Glucose	1.48±0.13	1.35±0.06	1.19±0.12	1.27±0.12	1.45±0.20	1.23±0.06	1.32±0.08	1.40±0.14	n.s.[b]
Glycine	0.92±0.11	0.85±0.06	0.85±0.12	1.05±0.09	1.08±0.08	0.85±0.07	1.03±0.06	1.15±0.12	n.s.
Acetic acid	2.38±0.24	2.50±0.11	2.20±0.08	2.62±0.13	2.32±0.47	2.06±0.17	2.63±0.14	2.11±0.71	n.s.
Starch	0.40±0.06	0.33±0.06	0.28±0.03	0.39±0.08	1.24±0.04	1.20±0.11	1.03±0.07	1.36±0.25	T***
Vanillin	0.32±0.03	0.41±0.01	0.24±0.01	0.73±0.16	0.54±0.08	0.89±0.09	0.48±0.05	0.57±0.11	F***, T**
Total ^{13}C-incorporation into NLFA									
(μg ^{13}C per mg added ^{13}C g^{-1} soil)									
Glucose	0.51±0.06	0.70±0.08	0.51±0.04	0.43±0.07	0.26±0.06	0.31±0.05	0.32±0.07	0.24±0.02	T***
Glycine	0.15±0.02	0.14±0.02	0.23±0.05	0.21±0.03	0.08±0.02	0.07±0.01	0.11±0.03	0.13±0.05	T**
Acetic acid	1.27±0.11	1.45±0.13	1.51±0.12	1.41±0.18	0.51±0.20	0.52±0.12	0.99±0.11	0.65±0.19	T***
Starch	0.06±0.01	0.11±0.02	0.08±0.01	0.12±0.01	0.12±0.05	0.10±0.01	0.15±0.03	0.15±0.01	n.s.
Vanillin	0.05±0.01	0.12±0.01	0.10±0.04	0.10±0.02	0.07±0.04	0.19±0.02	0.05±0.01	0.08±0.02	F**
NLFA/PLFA for ^{13}C-incorporation from									
Glucose to 18:2ω6,9	0.09±0.01	0.15±0.03	0.15±0.03	0.11±0.03	0.08±0.02	0.11±0.02	0.12±0.04	0.10±0.02	n.s.
Glucose to 18:1ω9	1.09±0.16	1.78±0.15	1.65±0.25	1.25±0.09	0.58±0.07	0.57±0.06	0.80±0.15	0.68±0.02	T***
Acetic acid to 18:2ω6,9	0.16±0.01	0.12±0.02	0.15±0.02	0.12±0.02	0.05±0.02	0.08±0.01	0.15±0.03	0.15±0.01	W**, W×T**
Acetic acid to 18:1ω9	1.65±0.23	1.53±0.20	2.21±0.42	1.68±0.26	0.55±0.03	0.60±0.09	0.88±0.05	0.71±0.02	T***

[a]C, control; F, fertilization; W, warming; W+F, warming+fertilization.
[b]n.s., no significant effects at $P<0.01$.
Significant effects of factors warming (W), fertilizer addition (F), time (T) and their interactions are shown at **$P<0.01$, ***$P<0.001$.

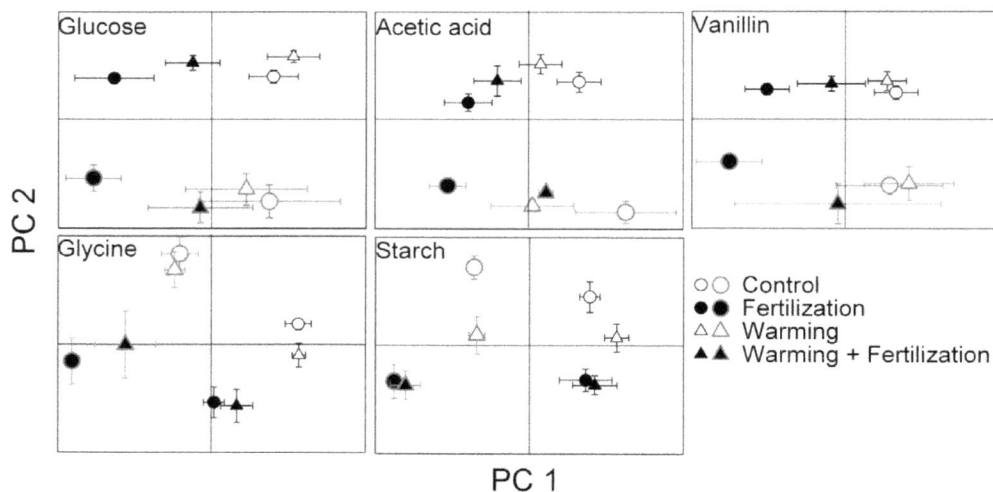

Figure 4. Principal component analysis (PCA) of the relative ^{13}C-incorporation of added ^{13}C-labelled glucose, glycine, acetic acid, starch and vanillin into phospholipid fatty acids (PLFAs) of tundra heath soil. Scores (mean ± SE) of the principal components (PC) 1 and 2 showing effects of warming and fertilization, and differences between the incubation times (1day, small symbols drawn with black line; 7 days, large symbols drawn with grey line). The PCAs were run separately for each substrate. The explained variances were 33–46% for PC1 and 18–27% for PC2. For glucose, glycine, acetic acid and vanillin $P<0.001$ for Fertilization and Time effect on the PC scores (Linear Mixed Models with Warming, Fertilization and Time as fixed factors and Block as a random factor). For starch $P<0.01$ for Warming, $P<0.001$ for Fertilization and Time, and $P<0.01$ for Warming×Fertilization.

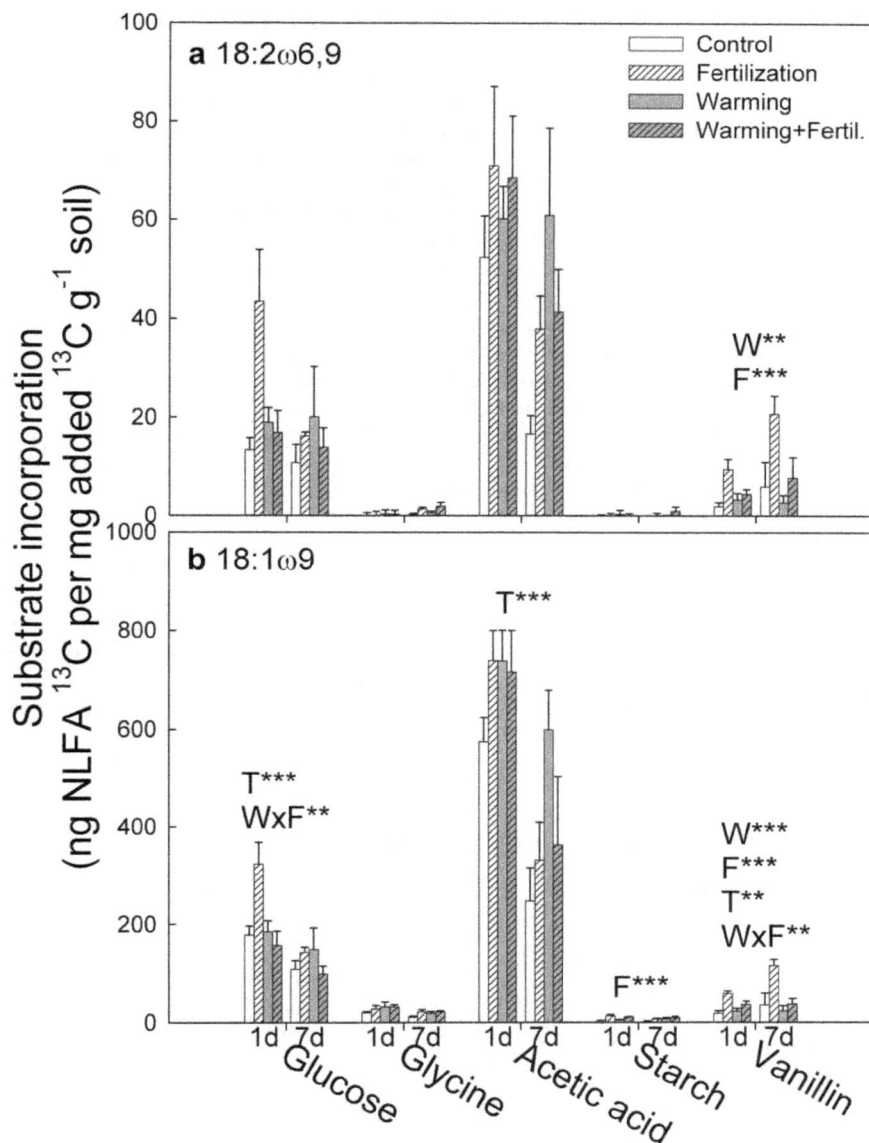

Figure 5. Incorporation of ^{13}C from glucose, glycine, acetic acid, starch and vanillin into the individual NLFAs (a) 18:2ω6,9 and (b) 18:1ω9. Tundra heath soils from a field experiment with long-term warming and fertilization treatments were incubated 1 and 7 days. The bars represent mean+SE. Note different y-axis scales. Significant effects of factors warming (W), fertilizer addition (F), time (T) and their interactions for each substrate are shown at **$P < 0.01$, ***$P < 0.001$ (Linear Mixed Models with Warming, Fertilization and Time as fixed factors and Block as a random factor).

Discussion

Fertilization of the subarctic heath for 18 years had increased the total PLFA concentration in soil by 31%, which indicates increased microbial biomass. This increase is in agreement with the results after 15 treatment years, when fertilization also had increased the total PLFA concentration, however only by 16% [5]. In contrast to the reduction of total PLFA by the warming treatment after 15 years [5], warming for the additional three years up to the sampling for the present study had led to a significantly increased total PLFA concentration. This implies that the total microbial biomass responses in the warming treatment now, for the first time, resemble those in the fertilization treatment [21]. The result probably reflects the long-term increase in the

total cover of vascular vegetation. After 22 years, both warming and fertilization treatments had 35% higher vascular plant cover than the control [16], in contrast to previous measurements after 5 and 10 years, with varying responses in the two treatments [22]. Increased primary production typically leads to an increase in soil microbial biomass [23], thanks to enhanced substrate inputs.

Based on a PCA of the PLFA profiles, the microbial community composition in the fertilized soil was characterized by a significantly higher abundance of the fungal biomarker, 18:2ω6,9, as compared to the unfertilized soil, which was similar to the results after 15 treatment years [5]. The fertilization-induced increase in fungi was further evidenced by the significantly higher total concentration of NLFAs, which are mainly found in eukaryotic organisms and thus are indicative of fungi in soil [24]. More specifically, the fungal marker NLFAs, 18:2ω6,9 and 18:1ω9 [24] had a significantly higher relative abundance in the fertilized than in the unfertilized soil. Rinnan et al. [5] discussed possible effects of adding N on soil fungi, also comparing with other habitats like coniferous forest soil. They suggested that the fertilizer-induced increase in fungi in this peaty subarctic soil resulted from 1) an alleviated N or P limitation of the fungal exoenzyme production, 2) a greater predation pressure on the bacterial than on the fungal community or 3) a higher abundance of mycorrhizal fungal mycelium. An altered community composition of ectomycorrhizal fungi was also found in a fertilization experiment in Arctic soils in North America [25].

Direct effects of a higher temperature on the PLFA pattern is well studied, with increased length and lower degree of unsaturation of fatty acids at higher temperatures [26]. Such changes were not found in the present study, indicating no direct effect of temperature. Instead warming caused only a minimal shift in the fatty acid profiles, and the effect of warming overrode that of fertilization as the combined warming+fertilization treatment had a similar PLFA profile to warming alone. Thus, although warming had increased the soil microbial biomass (total PLFA) indicating similar effects as fertilization, the different responses in the community composition suggest that the mechanisms behind the increase in microbial biomass under these treatments are different. This is in line with the vegetation responses, which show that despite the identical increase in the total cover of vascular plants in warming and fertilization treatments, the responses in the dominant plant species under these treatment are contrasting. While warming enhanced the cover of ectomycorrhizal *Betula nana*, fertilization benefited the ericoid mycorrhizal *Empetrum hermaphroditum* [16]. It is highly unlikely that the air warming by open-top chambers has had any direct effect on soil microbial communities in the long term, because the expansion of the vascular plant cover has over time led to shading and no soil temperature differences between the control and warming treatments [27].

The most obvious difference between the warming and fertilization treatments is that the supplied NPK fertilization provides the microbes with extra N and P, alleviating limitation by these nutrients, whereas soil warming increases mineralization of carbon as well as other nutrients. Earlier results have shown that microbes are N-limited or limited by both C and N in northern soils [6,28], including soils from the Abisko region [3,5]. Findings from another experiment in Abisko suggest, however, that warming shifts the originally N-limited soil microbial community slightly towards carbon limitation, although no significant changes in mineralization are observed [3]. This indicates that the warming-induced increase in net mineralization may be small, as also suggested by *in situ* buried bag studies at the experimental site [29,30].

Utilization of the amended substrates by soil microbial communities was mainly affected by fertilization, while warming only had few effects. The most pronounced response was that the fertilized soil was characterized by a relatively higher ^{13}C-incorporation into fungi than into bacteria as compared with the unfertilized soil, irrespective of the added substrate. Incorporation into the fungal NLFA 18:2ω6,9 followed a similar response, but because of high variance, only vanillin showed a statistically significant fertilization effect. The ^{13}C-recovery in the fatty acid 18:1ω9, which also indicates fungi [24], was increased by fertilization for vanillin (both PLFA and NLFA) and starch (only NLFA).

Fungi store carbon in excess as triacylglycerols, which are detected in the NLFA fraction [24]. Allocation to fungal energy storage can thus be estimated with NLFA-to-PLFA ratios, and a lower ratio after adding C-rich, labile, substrates indicates available N in the soil [24]. Contrary to our hypothesis, there were no significant fertilization effects on the NLFA-to-PLFA ratios for ^{13}C-incorporation. However, the NLFA-to-PLFA ratios for ^{13}C-incorporation from acetic acid into 18:2ω6,9 and 18:1ω9 were higher in the warmed soil but unaffected by fertilization compared to the control. The increased allocation to storage under warming suggests that the increased temperature actually led to limitation of growth by the lack of nutrients when a simple carbon-rich substrate, acetic acid, was easily available. The contrasting responses of warming and fertilization clearly show that fertilization cannot be used as a proxy for effects of warming, as also indicated by the PLFA pattern.

That fertilizer addition increased ^{13}C-incorporation from vanillin (a model substance indicating lignin depolymerisation), while warming had no significant effect, is additional evidence for this conclusion. This is also in agreement with the results of Neff et al. [31], who observed that fertilization of an alpine meadow had triggered increased decomposition of plant-derived lignin compounds. White-rot fungi of coniferous forest soil are known to down-regulate the production of lignolytic extracellular enzymes under high soil N concentrations [32,33], suggesting that N fertilization should result in decreased lignin degradation. This was not the case in the present study. However, the increased vanillin utilization was mainly due to the stimulated uptake of ^{13}C to the cyclic and fungal fatty acids, suggesting that both bacteria and fungi were involved in vanillin degradation in the soil studied.

The pattern of ^{13}C incorporation from starch into PLFAs differed from the other substrates, which was also reported earlier [34]. Thus, although starch degradation is fairly common among microorganisms [35], a different subset of microorganisms appears to degrade starch as compared to the other studied substrates. Also, in contrast to the other substrates used in this study, utilization of starch was significantly affected by both fertilization and warming. More specifically, the incorporation of ^{13}C from starch into indicators of Gram-positive bacteria (i16:0) and actinomycetes (10Me16:0 for fertilization and 10Me18:0 for warming) increased by the treatments. It appears that the starch-degrading microorganisms were favored by the changes that had occurred both in the fertilized and warmed plots, possibly by the increase in deciduous shrubs and the concomitant increase in leaf litter [8,15].

Increased biomass and activity of soil fungi relative to those of bacteria in response to enhanced nutrient availability may lead to altered ecosystem functioning. Most importantly, an increased importance of fungi relative to bacteria has been suggested to result in an increased carbon sequestration in soil ([36], but see also discussion by [37]). Because of the enormous size of the arctic soil carbon pool, even tiny changes in soil carbon processing have a potential to feed back on climate change. Effects of climate change on microbial use of carbon substrates would influence soil carbon storage both directly, as altered efflux of carbon from the soil, and indirectly, via priming effects (for a review, see [38]) that could lead to altered degradation of older carbon.

Several of our results indicate that – at least in the 18-year time horizon – warming does not result in similar effects as experimentally increased soil nutrient availability. Thus, climate warming will not increase carbon sequestration in tundra heaths via beneficial effects on soil fungi to the same degree as fertilization unless soil nutrient availability increases more than the increase in moderately warmer soil. As the N deposition rates in the region north of 60°N are low [39], a pronounced increase in nutrient availability via atmospheric deposition is, however, unlikely.

Materials and Methods

Experimental Design and Soil Sampling

A long-term climate change simulation experiment has been maintained on a tree-line heath at Abisko, northern Sweden (68°21′N, 18°49′E; 450 m a.s.l.), since 1989. The vegetation is dominated by ericoid dwarf shrubs (mainly *Cassiope tetragona* (L.) D. Don) and mosses, with deciduous shrubs and graminoids as subdominants. The experiment hosts various treatments (originally described in [21]) of which 1) warming, 2) fertilization, 3) the combination of warming and fertilization and 4) unmanipulated control plots were selected for the present study. For the warming treatment, 3–4°C air and 1–2°C soil temperature (4 cm depth; [40,41]) increases were achieved with the help of dome-shaped open-top plastic greenhouses, that were erected on the plots for the growing season from early June to late August. The fertilization treatment simulated enhanced nutrient availability and it was maintained by annual additions of 10 g m^{-2} N, 2.6 g m^{-2} P and 9 g m^{-2} K as NH_4NO_3, KH_2PO_4 and KCl (except 1989, when half of this amount was applied, and 1993 and 1998, when no fertilization was performed). Each of the plots covered a 1.2×1.2 m area and was randomly distributed in six blocks, i.e. there were six replicate plots of each treatment. No specific permits were required for the described field studies, as the location is not privately-owned and not protected and the field studies did not involve endangered or protected species.

Three soil cores were taken from each field plot on August 24, 2006, i.e. after 18 years of treatments. The 5-cm-deep cores with a diameter of 4 cm were mixed into a single sample per treatment plot (total volume of 63 cm^3), while removing roots and stones by hand sorting. The soil has a c. 15 cm deep organic layer with a pH of 7.1, which has not been affected by the treatments [41] and a soil organic matter content of 89% in the unfertilized and 92% in the fertilized soil in the top 5 cm [5]. The organic layer rests on rocky mineral soil without permafrost.

Incubation with ^{13}C-labelled Substrates

Uptake of label from ^{13}C-labelled substrates into fatty acids was determined by SIP-PLFA as described by [34]. Five different substrates were used: universally ^{13}C-labelled glucose (99%), glycine (99%) and starch (98%), ring-labelled vanillin (99%) and a 1:1 mixture of acetic acid labelled with ^{13}C in either one or the other of the two C atoms (Cambridge Isotope Laboratories, Andover, MA, USA). Subsamples of 0.4-g wet soil were distributed into ten small minigrip bags (duplicates of each substrate), and amended with 100 μl water containing a substrate. The added amount was 0.5 mg substrate g^{-1} wet soil ($\approx$2 mg substrate g^{-1} d.w. organic matter), which corresponds to 0.21 mg ^{13}C for glucose, 0.17 mg ^{13}C for glycine, 0.11 mg ^{13}C for acetic acid, 0.23 mg ^{13}C for starch, and 0.25 mg ^{13}C for vanillin per gram wet soil.

Half of the subsamples were incubated for 24 h and another half for 7 days, to enable comparison of immediate with longer-term uptake, at 15°C in dark followed by freezing and freeze-drying.

Fatty Acid Analysis

Fatty acids were extracted from the freeze-dried soil following a modified Bligh and Dyer method according to [42] as described in [34]. Shortly, lipids were extracted and fractionated on silicic acid columns (Bond Elut, Varian Inc., Palo Alto, CA, USA) with chloroform (neutral lipids, the NLFA fraction), acetone (glycolipids) and methanol (phospholipids, the PLFA fraction). The NLFA and PLFA fractions were subjected to mild alkaline methanolysis, and the resulting fatty acid methyl esters were analyzed on a Hewlett-Packard 6890 gas chromatograph equipped with a 50-m HP5 capillary column (Hewlett-Packard, Palo Alto, CA, USA) and helium as the carrier gas. The GC was interfaced with a Europa 20/20 isotope ratio mass spectrometer (Sercon Ltd., Cheshire, UK) used for determination of the δ^{13}C values [43]. Methyl nonadecanoate fatty acid (19:0) was used as an internal standard.

Isotope data was calculated following Boschker [44]. The δ^{13}C values of each lipid were corrected for the methyl group added during methanolysis [45]. For each PLFA and NLFA, the absolute amount of ^{13}C incorporated was calculated by relating the increase in the fraction ^{13}C after labeling [F = R/(R+1), where R is a ^{13}C/^{12}C-ratio] to the fatty acid concentration and the amount of ^{13}C in the added substrate.

The fatty acids are presented as the total number of carbon atoms followed by a colon and the number of double bonds. The prefixes a and i signify anteiso- and isobranching, respectively. The prefix cy indicates cyclopropyl fatty acids, while 10Me is a methyl group on the 10th carbon atom from the carboxyl end of the molecule. Terminally and mid-chain branched fatty acids (i15:0, a15:0, i16:0, 10Me16:0, i17:0, a17:0, 10Me17:0, 10Me18:0), cyclopropyl saturated (cy-17:0 and cy-19:0) and some monoun-saturated (16:1ω7 and 18:1ω7) fatty acids were considered indicative of bacteria [46]. The fatty acid 18:2ω6,9 was considered to represent fungi [46]. The PLFA 18:1ω9 is mainly of fungal origin (see discussion in [47]), and the NLFA 18:1ω9 is of fungal origin, and often found to increase under excess of carbon [24].

Statistical Analysis

The fatty acid concentrations and the total ^{13}C-incorporation to fatty acids were analyzed for treatment effects by Linear Mixed Models of SPSS 14.0 for Windows with warming, fertilization, substrate and incubation time as fixed factors. The block variable from the field experiment was used as a random factor. As the differences between the ^{13}C-labelled substrates were not of primary interest in the present study and have been described before [34], the analysis was run separately for each substrate. To reduce the chance for type I errors, the alpha-level of 0.01 was used to indicate statistical significance.

The unit-variance scaled and centered fatty acid mole percentage profiles and the patterns of ^{13}C-incorporation were subjected to principal component analyses (PCA) using Simca-P 11.0 (Umetrics, Umeå, Sweden). The extracted principal components (PC) were analyzed for treatment effects using Linear Mixed Models in a similar manner as described above.

Supporting Information

Figure S1 Incorporation of ^{13}C from glucose, glycine, acetic acid, starch and vanillin into the PLFA 18:2ω6,9. Tundra heath soils from a field experiment with long-term warming and fertilization treatments were incubated 1 and 7 days. The bars represent mean+SE. Note different y-axis scales. Significant effects of factors warming (W), fertilizer addition (F), time (T) and their interactions for each substrate are shown at **$P<0.01$, ***$P<0.001$ (Linear Mixed Models with Warming, Fertilization and Time as fixed factors and Block as a random factor).

Figure S2 Incorporation of ^{13}C from glucose, glycine, acetic acid, starch and vanillin into the PLFA 18:1ω9. Tundra heath soils from a field experiment with long-term warming and fertilization treatments were incubated 1 and 7 days. The bars represent mean+SE. Note different y-axis scales.

Significant effects of factors warming (W), fertilizer addition (F), time (T) and their interactions for each substrate are shown at **$P<0.01$, ***$P<0.001$ (Linear Mixed Models with Warming, Fertilization and Time as fixed factors and Block as a random factor).

Figure S3 Incorporation of ^{13}C from glucose, glycine, acetic acid, starch and vanillin into the PLFA cy19:0. Tundra heath soils from a field experiment with long-term warming and fertilization treatments were incubated 1 and 7 days. The bars represent mean+SE. Note different y-axis scales. Significant effects of factors warming (W), fertilizer addition (F), time (T) and their interactions for each substrate are shown at **$P<0.01$, ***$P<0.001$ (Linear Mixed Models with Warming, Fertilization and Time as fixed factors and Block as a random factor).

Figure S4 Incorporation of ^{13}C from glucose, glycine, acetic acid, starch and vanillin into the PLFA 18:1ω7. Tundra heath soils from a field experiment with long-term warming and fertilization treatments were incubated 1 and 7 days. The bars represent mean+SE. Note different y-axis scales. Significant effects of factors warming (W), fertilizer addition (F), time (T) and their interactions for each substrate are shown at **$P<0.01$, ***$P<0.001$ (Linear Mixed Models with Warming, Fertilization and Time as fixed factors and Block as a random factor).

Figure S5 Incorporation of ^{13}C from glucose, glycine, acetic acid, starch and vanillin into the PLFA i15:0. Tundra heath soils from a field experiment with long-term warming and fertilization treatments were incubated 1 and 7 days. The bars represent mean+SE. Note different y-axis scales. Significant effects of factors warming (W), fertilizer addition (F), time (T) and their interactions for each substrate are shown at **$P<0.01$, ***$P<0.001$ (Linear Mixed Models with Warming, Fertilization and Time as fixed factors and Block as a random factor).

Figure S6 Incorporation of ^{13}C from glucose, glycine, acetic acid, starch and vanillin into the PLFA i16:0.

Tundra heath soils from a field experiment with long-term warming and fertilization treatments were incubated 1 and 7 days. The bars represent mean+SE. Note different y-axis scales. Significant effects of factors warming (W), fertilizer addition (F), time (T) and their interactions for each substrate are shown at **$P<0.01$, ***$P<0.001$ (Linear Mixed Models with Warming, Fertilization and Time as fixed factors and Block as a random factor).

Figure S7 Incorporation of ^{13}C from glucose, glycine, acetic acid, starch and vanillin into the PLFA 10Me16:0. Tundra heath soils from a field experiment with long-term warming and fertilization treatments were incubated 1 and 7 days. The bars represent mean+SE. Note different y-axis scales. Significant effects of factors warming (W), fertilizer addition (F), time (T) and their interactions for each substrate are shown at **$P<0.01$, ***$P<0.001$ (Linear Mixed Models with Warming, Fertilization and Time as fixed factors and Block as a random factor).

Figure S8 Incorporation of ^{13}C from glucose, glycine, acetic acid, starch and vanillin into the PLFA 10Me18:0. Tundra heath soils from a field experiment with long-term warming and fertilization treatments were incubated 1 and 7 days. The bars represent mean+SE. Note different y-axis scales. Significant effects of factors warming (W), fertilizer addition (F), time (T) and their interactions for each substrate are shown at **$P<0.01$, ***$P<0.001$ (Linear Mixed Models with Warming, Fertilization and Time as fixed factors and Block as a random factor).

Acknowledgments

Logistical support and facilities were provided by the Abisko Scientific Research Station.

Author Contributions

Conceived and designed the experiments: RR AM EB. Performed the experiments: RR. Analyzed the data: RR EB. Wrote the paper: RR AM EB.

References

1. Jonasson S, Chapin FS,III, Shaver GR (2001) Biogeochemistry in the Arctic: Patterns, processes and controls. In: Schulze E-D, Heimann M, Harrison S, Holland E, Lloyd J, Prentice IC, Schimel D, editors. Global Biogeochemical Cycles in the Climate System. San Diego: Academic Press. 139–150.

2. IPCC (2007) Climate change 2007: the physical science basis. Contribution of working group I to the fourth assessment report of the intergovernmental panel on climate change. Cambridge: Cambridge University Press.

3. Rinnan R, Michelsen A, Bååth E, Jonasson S (2007) Mineralization and carbon turnover in subarctic heath soil as affected by warming and additional litter. Soil Biol Biochem 39: 3014–3023.

4. Rustad LE, Campbell JL, Marion GM, Norby RJ, Mitchell MJ et al. (2001) A meta-analysis of the response of soil respiration, net nitrogen mineralization, and aboveground plant growth to experimental ecosystem warming. Oecologia 126: 543–562.

5. Rinnan R, Michelsen A, Bååth E, Jonasson S (2007) Fifteen years of climate change manipulations alter soil microbial communities in a subarctic heath ecosystem. Global Change Biol 13: 28–39.

6. Mack MC, Schuur EAG, Bret-Harte MS, Shaver GR, Chapin FS (2004) Ecosystem carbon storage in arctic tundra reduced by long-term nutrient fertilization. Nature 431: 440–443.

7. van Wijk MT, Clemmensen KE, Shaver GR, Williams M, Callaghan TV, et al. (2004) Long-term ecosystem level experiments at Toolik Lake, Alaska, and at Abisko, Northern Sweden: generalizations and differences in ecosystem and plant type responses to global change. Global Change Biol 10: 105–123.

8. Graglia E, Jonasson S, Michelsen A, Schmidt IK, Havström M, et al. (2001) Effects of environmental perturbations on abundance of subarctic plants after three, seven and ten years of treatments. Ecography 24: 5–12.

9. Walker MD, Wahren CH, Hollister RD, Henry GHR, Ahlquist LE, et al. (2006) Plant community responses to experimental warming across the tundra biome. P Natl Acad Sci Biol 103: 1342–1346.

10. Bais HP, Weir TL, Perry LG, Gilroy S, Vivanco JM (2006) The role of root exudates in rhizosphere interactions with plants and other organisms. Annu Rev Plant Biol 57: 233–266.

11. Andresen L, Jonasson S, Ström L, Michelsen A (2008) Uptake of pulse injected nitrogen by soil microbes and mycorrhizal and non-mycorrhizal plants in a species-diverse subarctic heath ecosystem. Plant Soil 313: 283–295.

12. Sturm M, Racine C, Tape K (2001) Climate change: Increasing shrub abundance in the Arctic. Nature 411: 546–547.

13. Cornelissen JH, Van Bodegom PM, Aerts R, Callaghan TV, Van Logtestijn RSP, et al. (2007) Global negative vegetation feedback to climate warming responses of leaf litter decomposition rates in cold biomes. Ecol Lett 10: 619–627.

14. Hansen AH, Jonasson S, Michelsen A, Julkunen-Tiitto R (2006) Long-term experimental warming, shading and nutrient addition affect the concentration of phenolic compounds in arctic-alpine deciduous and evergreen dwarf shrubs. Oecologia 147: 1–11.

15. Sorensen PL, Michelsen A, Jonasson S (2008) Nitrogen uptake during one year in subarctic plant functional groups and in microbes after long-term warming and fertilization. Ecosystems 11: 1223–1233.
16. Campioli M, Leblans N, Michelsen A (2012) Twenty-two years of warming, fertilisation and shading of subarctic heath shrubs promote secondary growth and plasticity but not primary growth. PLoS ONE 7: e34842. doi:10.1371/journal.pone.0034842.
17. Flaig W (1964) Effects of micro-organisms in the transformation of lignin to humic substances. Geochim Cosmochim Ac 28: 1523–1535.
18. Carreiro MM, Sinsabaugh RL, Repert DA, Parkhurst DF (2000) Microbial enzyme shifts explain litter decay responses to simulated nitrogen deposition. Ecology 81: 2359–2365.
19. DeForest JL, Zak DR, Pregitzer KS, Burton AJ (2004) Atmospheric nitrate deposition, microbial community composition, and enzyme activity in northern hardwood forests. Soil Sci Soc Am J 68: 132–138.
20. Edwards IP, Zak DR, Kellner H, Eisenlord SD, Pregitzer KS (2011) Simulated atmospheric N deposition alters fungal community composition and suppresses ligninolytic gene expression in a northern hardwood forest. PLoS ONE 6: e20421. doi:10.1371/journal.pone.0020421.
21. Havström M, Callaghan TV, Jonasson S (1993) Differential growth responses of Cassiope tetragona, an arctic dwarf-shrub, to environmental perturbations among three contrasting high- and sub-arctic sites. Oikos 66: 389–402.
22. Michelsen A, Rinnan R, Jonasson S (2012) Two decades of experimental manipulations of heaths and forest understory in the Subarctic. Ambio 41: 218–230.
23. Wardle DA (2002) Communities and Ecosystems: Linking the Aboveground and Belowground Components. Monographs in Population Biology 34. Princeton University Press, Princeton, New Jersey.
24. Bååth E (2003) The use of neutral lipid fatty acids to indicate the physiological conditions of soil fungi. Microbial Ecol 45: 373–383.
25. Deslippe JR, Hartmann M, Mohn WW, Simard SW (2011) Long-term experimental manipulation of climate alters the ectomycorrhizal community of Betula nana in Arctic tundra. Global Change Biol 17: 1625–1636.
26. Wixon DL, Balser TC (2013) Toward conceptual clarity: PLFA in warmed soils. Soil Biol Biochem 57: 769–774.
27. Sorensen PL, Lett S, Michelsen A (2012) Moss-specific changes in nitrogen fixation following two decades of warming, shading, and fertilizer addition. Plant Ecol 213: 695–706.
28. Sørensen L, Holmstrup M, Maraldo K, Christensen S, Christensen B (2006) Soil fauna communities and microbial respiration in high Arctic tundra soils at Zackenberg, Northeast Greenland. Polar Biol 29: 189–195.
29. Schmidt IK, Jonasson S, Michelsen A (1999) Mineralization and microbial immobilization of N and P in arctic soils in relation to season, temperature and nutrient amendment. Appl Soil Ecol 11: 147–160.
30. Jonasson S, Castro J, Michelsen A (2006) Interactions between plants, litter and microbes in cycling of nitrogen and phosphorus in the Arctic. Soil Biol Biochem 38: 526–532.
31. Neff JC, Townsend AR, Gleixner G, Lehman SJ, Turnbull J, et al. (2002) Variable effects of nitrogen additions on the stability and turnover of soil carbon. Nature 419: 915–917.
32. Waldrop M, Zak D (2006) Response of oxidative enzyme activities to nitrogen deposition affects soil concentrations of dissolved organic carbon. Ecosystems 9: 921–933.
33. Berg B, Matzner E (1997) Effect of N deposition on decomposition of plant litter and soil organic matter in forest systems. Environ Rev 5: 1–25.
34. Rinnan R, Bååth E (2009) Differential utilization of carbon substrates by bacteria and fungi in tundra soil. Appl Environ Microbiol 75: 3611–3620.
35. Alexander M (1977) Introduction to soil microbiology. New York: John Wiley and Sons. 467 p.
36. Jastrow JD, Amonette JE, Bailey VL (2007) Mechanisms controlling soil carbon turnover and their potential application for enhancing carbon sequestration. Climatic Change 80: 5–23.
37. Strickland MS, Rousk J (2010) Considering fungal:bacterial dominance in soils – Methods, controls, and ecosystem implications. Soil Biol Biochem 42: 1385–1395.
38. Kuzyakov Y (2010) Priming effects: Interactions between living and dead organic matter. Soil Biol Biochem 42: 1363–1371.
39. Forsius M, Posch M, Aherne J, Reinds G, Christensen J, et al. (2010) Assessing the impacts of long-range sulfur and nitrogen deposition on arctic and sub-arctic ecosystems. AMBIO 39: 136–147.
40. Michelsen A, Jonasson S, Sleep D, Havström M, Callaghan TV (1996) Shoot biomass, δ13C, nitrogen and chlorophyll responses of two arctic dwarf shrubs to in situ shading, nutrient application and warming simulating climatic change. Oecologia 105: 1–12.
41. Ruess L, Michelsen A, Schmidt IK, Jonasson S (1999) Simulated climate change affecting microorganisms, nematode density and biodiversity in subarctic soils. Plant Soil 212: 63–73.
42. Frostegård Å, Tunlid A, Bååth E (1991) Microbial biomass measured as total lipid phosphate in soils of different organic content. J Microbiol Meth 14: 151–163.
43. Olsson PA, van Aarle IM, Gavito ME, Bengtson P, Bengtsson G (2005) 13C incorporation into signature fatty acids as an assay for carbon allocation in arbuscular mycorrhiza. Appl Environ Microbiol 71: 2592–2599.
44. Boschker HTS (2004) Linking microbial community structure and functioning: stable isotope 13C labeling in combination with PLFA analysis. In: Kowalchuk GA, de Bruijn FJ, Head IM, Akkermans AD, van Elsas JD (2004) editors. Molecular Microbial Ecology Manual II. Dordrecht: Kluwer Academic Publishers. 1673–1688.
45. Abraham W-R, Hesse C, Pelz O (1998) Ratios of carbon isotopes in microbial lipids as an indicator of substrate usage. Appl Environ Microbiol 64: 4202–4209.
46. Frostegård Å, Bååth E (1996) The use of phospholipid fatty acid analysis to estimate bacterial and fungal biomass in soil. Biol Fert Soils 22: 59–65.
47. Frostegård Å, Tunlid A, Bååth E (2011) Use and misuse of PLFA measurements in soils. Soil Biol Biochem 43: 1621–1625.

17

Impact of Fertilizing Pattern on the Biodiversity of a Weed Community and Wheat Growth

Leilei Tang[1,2], Chuanpeng Cheng[1], Kaiyuan Wan[1,3]*, Ruhai Li[4], Daozhong Wang[5], Yong Tao[1], Junfeng Pan[1], Juan Xie[1], Fang Chen[1,6]*

1 Key Laboratory of Aquatic Botany and Watershed Ecology, Wuhan Botanical Garden, Chinese Academy of Sciences, Wuhan, China, **2** Hainan Modern Agriculture Inspection and Testing Precaution & Control Center, Agricultural Department of Hainan Province, Haikou, China, **3** Ecological Restoration (ECORES) Lab, Chengdu Institute of Biology, Chinese Academy of Sciences, Chengdu, China, **4** Institute of Plant Protection and Soil Science, Hubei Academy of Agricultural Sciences, Wuhan, China, **5** Institute of Soil and Fertilizer Science, Anhui Academy of Agricultural Sciences, Hefei, China, **6** China Program of International Plant Nutrition Institute (IPNI), Wuhan, China

Abstract

Weeding and fertilization are important farming practices. Integrated weed management should protect or improve the biodiversity of farmland weed communities for a better ecological environment with not only increased crop yield, but also reduced use of herbicides. This study hypothesized that appropriate fertilization would benefit both crop growth and the biodiversity of farmland weed communities. To study the effects of different fertilizing patterns on the biodiversity of a farmland weed community and their adaptive mechanisms, indices of species diversity and responses of weed species and wheat were investigated in a 17-year field trial with a winter wheat-soybean rotation. This long term field trial includes six fertilizing treatments with different N, P and K application rates. The results indicated that wheat and the four prevalent weed species (*Galium aparine*, *Vicia sativa*, *Veronica persica* and *Geranium carolinianum*) showed different responses to fertilizer treatment in terms of density, plant height, shoot biomass, and nutrient accumulations. Each individual weed population exhibited its own adaptive mechanisms, such as increased internode length for growth advantages and increased light interception. The PK treatment had higher density, shoot biomass, Shannon-Wiener and Pielou Indices of weed community than N plus P fertilizer treatments. The N1/2PK treatment showed the same weed species number as the PK treatment. It also showed higher Shannon-Wiener and Pielou Indices of the weed community, although it had a lower wheat yield than the NPK treatment. The negative effects of the N1/2PK treatment on wheat yield could be balanced by the simultaneous positive effects on weed communities, which are intermediate in terms of the effects on wheat and weeds.

Editor: Shuijin Hu, North Carolina State University, United States of America

Funding: This work was supported by the China Program of International Plant Nutrition Institute (IPNI-HB-34) and the Opening Project of Hubei Key Laboratory of Wetland Evolution & Ecological Restoration(2011-02). The funders had no role in study design, data collection and analysis, decision to publish, or preparation of the manuscript.

Competing Interests: The authors have declared that no competing interests exist.

* E-mail: kaiyuanwan@126.com (KW); fchen@ipni.ac.cn (FC)

Introduction

Weeds are one of the major constraints to crop yields and quality [1,2]. However, as one of the primary producers within farming systems, weeds are of central importance to the arable system's food web. The weed community provides a range of resources for higher trophic groups, supports a high diversity of insect species and birds [3,4], and therefore plays an important role in the biological diversity of agroecosystems. Weeds also have other ecosystem functions, including nutrient cycling and soil preservation [5]. In addition, there are correlations between an impoverished landscape and the appearance of pests and diseases [6]. Moreover, an overreliance on herbicides has imposed a cost upon society and the environment [7]. The biodiversity of weed communities in a cropland can therefore be an important element for the reliable and sustainable provision of agroecosystem services. However, encouraging in-field biodiversity is unpopular among farmers because of the risk of decreased crop production as a result of weed competition. It is important to match crop production with conservation of biological resources to develop more sustainable systems [8].

Crop nutrient management practices may serve as an important component of more robust weed management programs [9–11]. Fertilization alters soil fertility, thus affecting weed density, nutrient uptake, and biomass yield, which in turn affects species composition and biodiversity [9,12–14]. For example, Mahn [15] observed a general decline in the number of weed individuals and an increase in weed biomass with increasing rates of nitrogen (N) fertilizer. In another study, *Digitaria ischaemum* Shreb (smooth crabgrass) was found to be the dominant species under N + potassium (K) and non-fertilized treatments, *Cyperus rotundus* L (purple nutsedge) dominated under phosphorus (P)+K treatment, and more weed species and higher Shannon's diversity (*H'*) values were detected in the balanced fertilization treatment [14]. These effects, however, indirectly depend on the light penetration caused by crop competition variation [16]. Balanced fertilization promotes the growth of crops, resulting in closed crop stands and light limitation for the weed communities growing underneath, thereby affecting weed species diversity [14,17].

Varying physiological responses of weed species to soil amendments are one of the explanations for weed community biodiversity [18]. Murphy and Lemerle [18] reported that the type and rate of fertilizers applied and the physiology of the species involved play an important role in weed population shifts. Haas and Streibig [19] also showed that high N levels will favor weed species that possess either physiological shade tolerance (esp. *Stellaria media*) or are able to climb into more favorable light conditions (esp. *Galium aparine*). Pyšek and Lepš [20] and Bengtsson et al. [21] also found that N fertilization favors shade-tolerant, climbing, and competitive weed species, but suppresses other kinds of species and results in a decrease of weed species diversity. With cessation of mineral fertilizer application during the period of conversion in a cropland, van Elsen [6] observed a gradual decline of weeds like *G. aparine*, which need a high nitrate level, and an increase in leguminous weeds (esp. *Vicia spp.*).

Documentation of the effects of a farming system on vegetation diversity is an important step toward understanding ecosystem function in agricultural landscapes [22], but only a handful of studies have examined the effect of fertilization on weed community biodiversity in agroecosystems. Biodiversity promotes ecosystem productivity, sustainability and stability in grassland [23–24]. Based on these reports, it is hypothesized that appropriate fertilization of agroecosystems would not only provide desirable crop productivity but also maintain the biodiversity of weed communities.

Long-term field experiments would ensure that proper data are accumulated and that confounding experimental effects are reduced, thereby providing better insights into the effects of prolonged fertilization over time [25]. It was also hypothesized that weed species could develop physiological adaptive mechanisms under long-term fertilization conditions. This study was conducted in a winter wheat field under continuous fertilization since 1994 with the objective of examining the effects of different N, P and K fertilization patterns on crop growth and yield. Effects on weeds were not examined at initiation, so baseline data are unavailable. Therefore, the objective of this study was to evaluate the cumulative effects of different fertilizing patterns on weed community biodiversity and their adaptive mechanisms with data from one-year sampling in this long-term experiment.

Materials and Methods

Site and Experimental Design

The long-term fertilization field experiment used in this study has been run by the Anhui Academy of Agricultural Sciences since 1994 and is located in Mengcheng county in the Anhui province of China (33°13′38 N, 116°36′58 E). The field studies did not involve endangered or protected species and no specific permits were required. This region has a warm temperate to sub-humid monsoon climate. The mean temperature, precipitation, evaporation and selected soil properties of the experimental site are given in Table 1. Since 1999, there has been a crop rotation of winter wheat (*Triticum aestivum* L.) and soybean (*Glycine max* [L.] Merr.).

Fertilizer treatments consisted of six combinations with different rates of N, P and K (Table 1) applied as urea, calcium superphosphate, and potassium chloride, respectively. All fertilizers were applied by soil surface broadcasting before the sowing of wheat. No fertilizer was applied during the soybean production seasons. The plot size was 20 m^2 (4×5 m) with three replications in a randomized complete block design. Soybean ('Zhonghuang 13') and winter wheat ('Yannong 19') were planted at a density of 249,800 seeds and 4,685,100 plants per hectare on 15 June and 20 October 2010, respectively.

Table 1. Experimental conditions at the field site.

Item	Value
Weather	
Mean annual temperature (1994–2010)	15°C
Rainfall (annual mean, 1994–2009)	872 mm
Rainfall (2010)	821 mm
Evaporation (annual mean, 1994–2010)†	1026 mm
Soil conditions	
Soil type	Lime concretion black soil
Soil pH	8.0
Soil organic matter	9.9 g kg^{-1}
Total soil N	0.79 g kg^{-1}
Soil available P	7.8 mg kg^{-1}
Soil available K	111 mg kg^{-1}
Crop rotation	
Winter wheat-maize cropping system	1994–1998
Winter wheat- Soybean cropping system	1999–2010
Fertilizer treatments	N, P$_2$O$_5$, K$_2$O kg ha^{-1}
Control	0, 0, 0
PK	0, 90, 135
NP	188, 90, 0
NK	188, 0, 135
NPK	188, 90, 135
N1/2PK*	188, 45, 135

†Evaporation was measured using the Penman formula.
*The N1/2PK treatment was designed to further evaluate the effect of P on winter wheat as it was the most critical soil limiting nutrient at the beginning of the experiment.

Herbicides were used according to the weed spectrum and expert recommendations. All treatments had a broad application of herbicides, and manual weeding was implemented before 2008. To better examine the effects of fertilization on weed community biodiversity, only a broadcast application of herbicides was continued in each growing season, and without manual weeding after 2008. Tribenuron methyl (Anhui Research Institute of Chemical Industry, China) at 12 g ai.ha^{-1} for wheat and Acetochlor (Anhui Huilong Group Rmf Agrochemical Co., Ltd, China) at 1500 g ai.ha^{-1} for soybean were used once every growing season in all the experimental plots.

Sampling Procedures

Weed evaluations were conducted three times during wheat growing season on April 10 (177 days after sowing [d]), May 8 (205 d), and May 23 (221 d), 2010. At each sampling date (growing period), three 0.5×0.5 m quadrats were systematically positioned to avoid edge effects and re-sampling of the previously sampled areas in each plot. Five wheat plants and all weeds present in the quadrat were clipped, collected, sorted by species, counted, and oven dried at 70°C for 48 h before weighing. Wheat tillers were also counted, and plant heights of wheat and weed species were measured.

At 205 d and 221 d, the tallest plant of *Galium aparine* (a prevalent broadleaved weed species) in each quadrat was selected for the measurement of internode length. Light measurements were also made in the quadrats (using Minolta Illuminance Meter T-1H) above the wheat and on the soil surface. At 221 d, each individual weed species and wheat plant was sampled for total N, P and K analysis. The N was measured using the Kjeldahl method, and P and K were measured using an inductively coupled plasma optical emission spectrometer (ICP-OES) after the plant samples were digested [26].

Statistical Procedures

Weed density was calculated as the total plant number of a particular weed species per square meter, total weed density was calculated as the total of all weed species per square meter, and winter wheat density was calculated as the total number of tillers per square meter. The species diversity of weed communities was assessed by calculating different indices. For example, Species richness (S) was measured by the mean number of species in each treatment [27]. Shannon's diversity index (H'), Simpson index (D) and Pielou index (E) were calculated for the weed communities using the following equations:

$$H' = \frac{N \ln N - \sum n \ln n}{N}$$

$$D = \sum (n/N)(n/N)$$

$$E = \frac{H'}{\ln S}$$

where N is the total number of individuals per plot and n is the number of individuals per species per plot. Shannon's diversity index (H') takes not only the number of species (community richness) into account but also the relative distribution of each species in the community (community evenness). The Simpson index was applied to measure the degree of dominance of individual weed species.

Rank-abundance plots were used to display species relative abundance data. Abundance distributions provide a complete description of the community diversity and simultaneously show both components of species diversity, species number, and evenness. The relative abundance of a species indicates its degree of dominance or subordination in the weed community (i.e., the greater the relative abundance of a species in the weed community, the higher its dominance, [28]). Relative growth rate (RGR, g m^{-2} d^{-1}), the increase of plant biomass per unit time, was used to compare crop growth with that of weed species.

Statistical analysis was done using SPSS version 16.0. Weed density and biomass data were log$_e$ (x+1) transformed before analysis to meet homogeneity of variance assumptions. For normally distributed parameters, the General Linear Model Univariate was used, and means were compared based on Tukey's multiple comparison tests P<0.05. Otherwise, nonparametric tests were used, and median values were presented.

Results

Structural Changes and Biological Diversity at the Community Level

The results indicated that total weed density was influenced by fertilization (P<0.0001), but was not affected by growing periods (P=0.59) (Fig. 1). A general decline in the number of weed individuals was observed in the different fertilization treatments except in the NK treatment where the number increased slightly. The PK treatment showed the highest total mean weed density (711 plants m^{-2}) followed by the control and NK treatments. Weed densities in these three treatments were significantly higher than the 201 plants m^{-2} in the treatments with N and P (72% less than that in the PK treatment).

Fertilization increased weed biomass in the winter wheat field (Fig. 2). The PK treatment obtained the highest (P<0.05) total weed biomass of 108.03 and 117.94 g m^{-2} at 205 and 221 d, respectively. Treatments with N and P showed a lower weed shoot biomass.

The species diversity of the weed community was modified by fertilizing patterns. Species number and evenness were affected differently by the fertilization treatments, which were shown in (a) the rank-abundance plots and (b) the vertical bar chart (Fig. 3). Among all the recorded species (13 spp.), the species numbers in the PK (11 spp.) and N1/2PK (11 spp.) treatments reached the highest, followed by the control (9 spp.) and NK (9 spp.) treatments, while the lowest value (7 spp.) was observed in the NPK treatment.

The evenness of the weed community also varied with the fertilization treatments (Fig. 3a, b). Fertilizing patterns greatly shaped the equitability in the partitioning of total biomass among species in the community as shown by the slope of the rank-abundance plots. A higher Shannon-Wiener Index (Fig. 3b) and weed community evenness (Pielou Index, Fig. 3a, b) was observed in the PK treatment, followed by the NK and control treatments. However, in these treatments, the Simpson Index was lower compared with that in the NP and NPK treatments (Fig. 3b). Compared with the NPK treatment, the half P rate treatment (N1/2PK) showed a higher equitability of total biomass partitioning (Fig. 3a) with higher Pielou Index and Shannon-Wiener Index values (Fig. 3b).

Changes at the Population Level

Densities of Weed Population. *Galium aparine* L. (tender catchweed bedstraw), *Vicia sativa* L. (common vetch), *Veronica persica* Poir (iran speedwell), and *Geranium carolinianum* L. (carolina geranium) were the most dominant species in the weed community. These four dominant species comprised >90% of the total weed density. Average densities of the four prevalent broadleaved weeds during the three growing periods in the different fertilization treatments are shown in Fig. 4. Densities of these four weed species varied significantly among the treatments. *Galium aparine* densities were greatly affected by fertilization (P<0.0001) and growing period (P=0.02). No significant difference was detected in the interactions of treatments with growing period (P=0.43). The densities of *G. aparine* (138 plants m^{-2}) were higher in the treatments with N and P fertilizers, while no *G. aparine* was found in the control treatment.

Vicia sativa density was significantly affected by fertilization (P<0.0001) and growing period (P=0.008). However, in contrast to *G. aparine*, *Vicia sativa* grew better in treatments without N and/or P with an average density of ~200 plants m^{-2}, while it had only 10–30 plants m^{-2} in the treatments with N and P. *Geranium carolinianum* density was also affected by fertilization (P<0.0001) and growing period (P<0.001), and the interactions of fertilization with growing period (P<0.001). Only three plants of *G. carolinianum* were found in the NP treatment at 205 d and no plants were found at 221 d. During the early growing period, *G. carolinianum* densities showed no significant differences among the treatments, with an average value of 72 plants m^{-2}. However, its densities responded differently to fertilization treatments in the

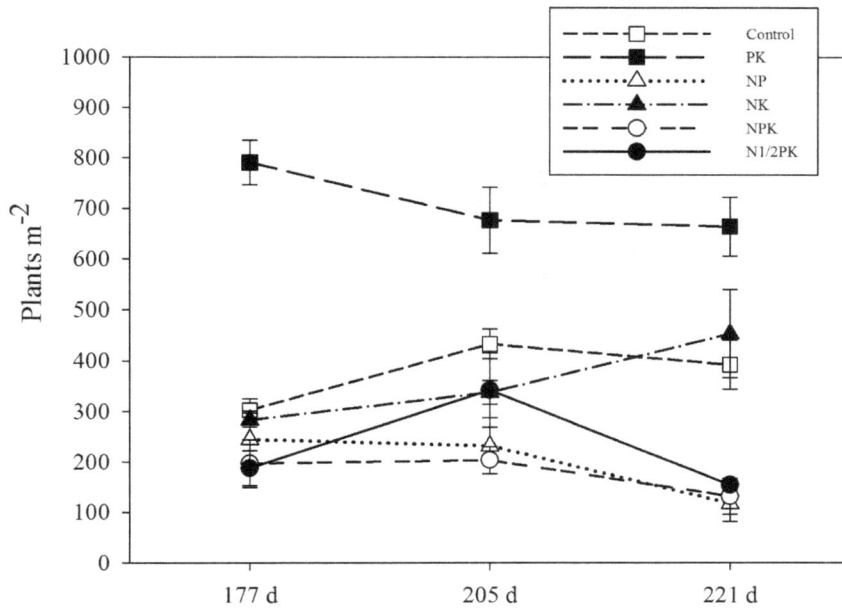

Figure 1. The influence of fertilization on the density of the weed community.

latter growing period, with the same tendency as *V. sativa* in each treatment. *Veronica persica* grew well only in the PK treatment, and few or no plants were detected in the other treatments.

Plant Height of Weed Population. Among the four prevalent broadleaved weed species, *V. persica* grew underneath the crop canopy, and its maximum plant height was <50 cm (Fig. 5D). On the contrary, other species were erect for capturing adequate light. Fertilization significantly affected plant height of *G. carolinianum* (Fig. 5A), *V. sativa* (Fig. 5B) and *G. aparine* (Fig. 5C). *Geranium carolinianum* was taller in the PK and NK treatments, even taller than in the NPK treatment at 205 d. Plant height of *V. sativa* in the PK treatment was at a maximum, reaching 73.8 cm at

maturity stage, followed by the NK treatment where it reached 64.1 cm, while lower values were observed in treatments with N and P. Conversely, *G. aparine* showed higher plant heights of 83.8 cm, 85.3 cm and 73.3 cm in the NP, NPK and N1/2PK treatments, respectively. The study confirmed that fertilization had little effect on the number of *G. aparine* shoot internodes. However, fertilization significantly modified the internodal length of *G. aparine*, especially from the 4th to the 12th internode (Table 2).

Shoot Biomass of the Weed Population. Shoot biomass of the four weed species varied with different fertilization treatments (Table 3). Like the weed density, *G. aparine* shoot biomass in the NP treatment was highest at each growing period, followed by the

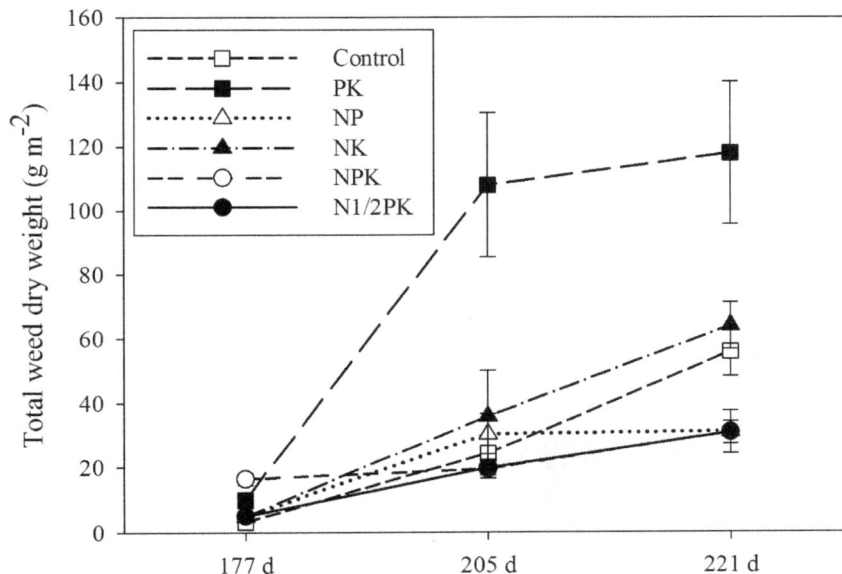

Figure 2. The influence of fertilization on the biomass of the weed community.

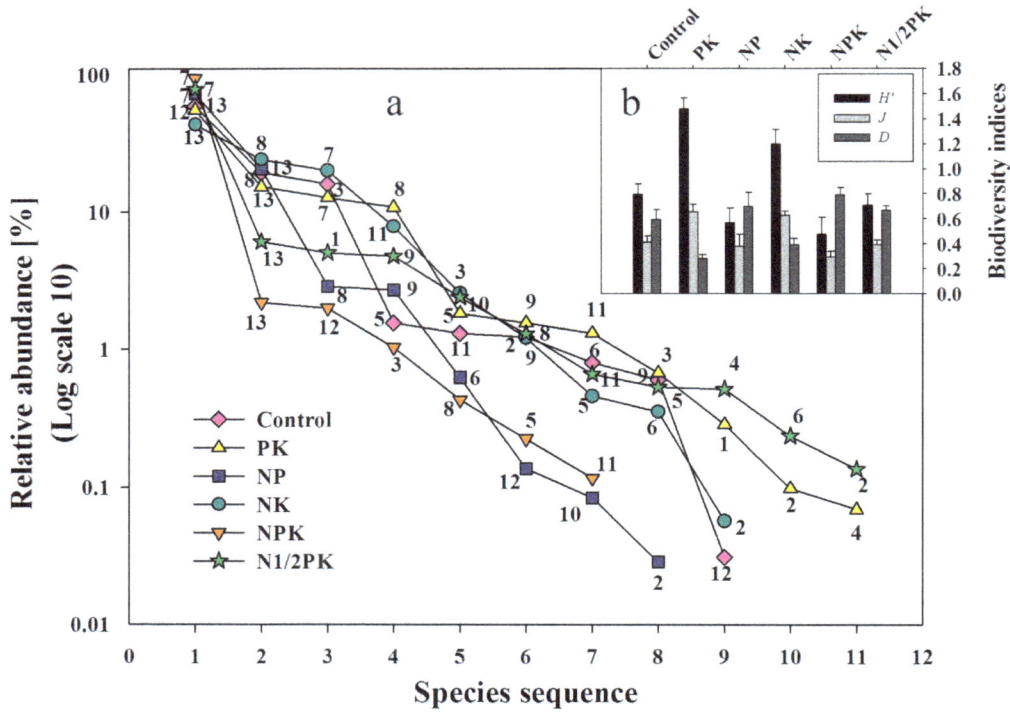

Figure 3. The influence of fertilization on the species diversity of the weed community. (a) Rank-abundance plots corresponding to different fertilization treatments in a winter wheat field. (b) The influence of fertilization on the biodiversity indices of weed communities (H', Shannon-Winner Index; J, Pielou Index; and D, Simpson index). Species reference: (1) *Avena fatua* L., (2) *Calystegia hederacea* Wall., (3) *Cirsium segetum* Bge., (4) *Cyperus rotundus* L., (5) *Erigeron annuus* (L.) Pers., (6) *Euphorbia helioscopia* L., (7) *Galium aparine* L. var. *tenerum* (Gren.et Godr.) Rcbb., (8) *Geranium carolinianum* L., (9) *Lithospermum arvense* L., (10) *Mazus pumilus* (Burm. f) V. Steenis., (11) *Plantago virginica* L., (12) *Veronica persica* Poir., (13) *Vicia sativa* L.

NPK and N1/2PK treatments. The relative growth rates of *G. aparine* in the treatments without N and/or P were lower, especially in the NK treatment (0.11 g m^{-2} d^{-1}), than that in the NP treatment (0.83 g m^{-2} d^{-1}). The highest proportion of *G. aparine* shoot biomass to total weed biomass was obtained in the NPK treatment (91%), followed by the NP and N1/2PK treatments at

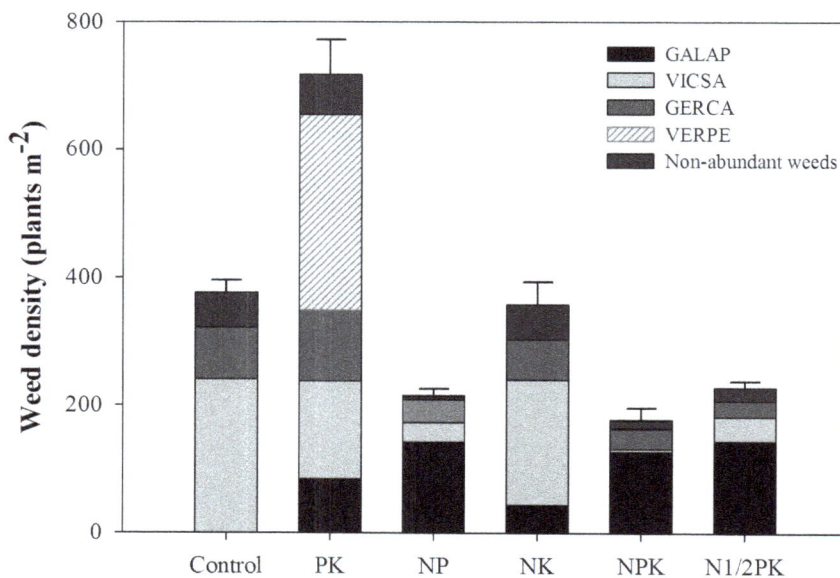

Figure 4. The effects of fertilization on average weed densities. Analyses of variance were performed on $\log_e$-transformed data, but means of untransformed data are shown. Error bars are the SE of the mean.

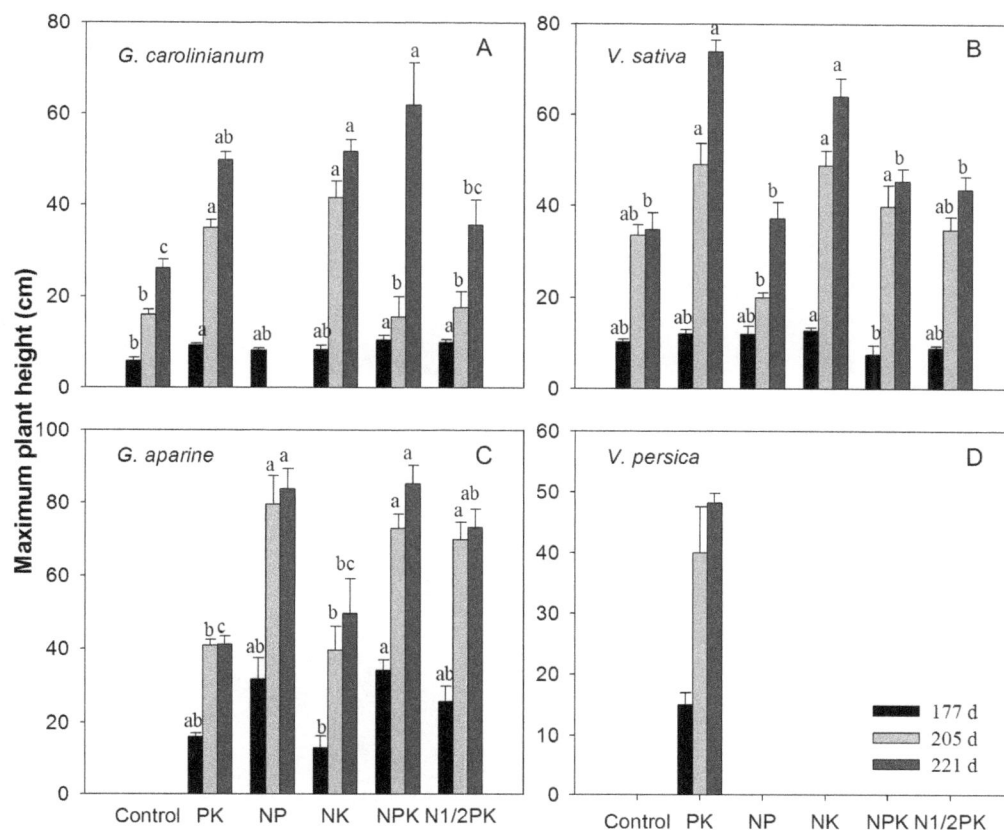

Figure 5. The influence of fertilization on plant height of *G. carolinianum* **(A),** *V. sativa* **(B),** *G. aparine* **(C) and** *V. persica* **(D).**

89% and 77%, respectively, while treatments without N and/or P had the lowest value (8%). These results indicated that *G. aparine* had an absolute growth advantage under higher soil fertility conditions, and its growth could be significantly promoted by the application of N and P fertilizers.

Biomass accumulation of *V. sativa* followed an opposite trend to that of *G. aparine* for the different fertilization treatments. The average shoot biomass of *V. sativa* in the control, NK and PK treatments were 3.32, 14.81, and 25.17 g m^{-2}, respectively in contrast to values under the NP, NPK and N1/2PK treatments where the values were 0.63, 0.69, and 1.33 g m^{-2} at 177, 205 and 221 d, respectively. The relative growth rate showed that *V. sativa* also grew rapidly in the control, PK and NK treatments (0.50 g m^{-2} d^{-1}). The highest proportion of *V. sativa* shoot biomass to the total weed biomass was observed in the control treatment with 68% and 41% at 205 and 221 d, respectively, followed by the NK and PK treatments, while treatments incorporating N and/or P fertilizers showed the lowest values (4%).

The average shoot biomass of *G. carolinianum* at 177 d reached 5.05 g m^{-2}, which was much higher than the corresponding values for *G. aparine*, *V. sativa* and *V. persica* (1.72, 1.97 and 0.55 g m^{-2}, respectively). The *G. carolinianum* shoot biomass under the NPK treatment (9.11 g m^{-2}) was higher than that under control. However, in the middle and latter growing periods, *G. carolinianum* shoot biomass performed similarly to *V. sativa* with the different fertilization treatments. Specifically, biomass accumulation of *V. persica* was only higher in the PK treatment. Its biomass at 205 d accounted for 81% of the total biomass at 221 d, and it also

accounted for >50% of the total shoot biomass of weeds present in the PK treatment.

Nutrient Uptake by Weeds. Nutrient uptake by the four weed species differed among treatments (Table 3). In general, nutrient uptake was higher in *G. aparine* and *V. sativa* than in *G. carolinianum*, suggesting that the two weed species were stronger nutrient competitors to wheat. Nutrient uptake of weed species in different treatments followed a pattern similar to shoot biomass in the corresponding treatments. Average N, P, and K uptakes by *G. aparine* in the NP, NPK, and N1/2PK treatments were 0.353, 0.090, and 0.447 g m^{-2}, respectively, which were 435, 375, and 425% higher than those under the PK treatment, and 45, 158, and 62% higher than those under the NK treatment. The N, P, and K uptakes under the control, PK, and NK treatments were 0.426, 0.076, and 0.361 g m^{-2}, respectively for *V. sativa* and 0.163, 0.052, and 0.187 g m^{-2}, respectively for *G. carolinianum*. For *V. persica*, the highest nutrient uptake was found under the PK treatment, and its P and K uptakes were 127 and 241% higher, respectively than those of *V. sativa*.

Winter Wheat Responses to Fertilization

Winter Wheat Density. Significant differences in wheat density were found among the fertilization treatments (P<0.001). Wheat density increased in the treatments that included N and P. Wheat density in the NP treatment was significantly (P<0.01) higher than that in the control, PK and NK treatments. However, there were no significant differences in winter wheat density within the NP, NPK and N1/2PK treatments. The average density was

Table 2. Effects of different fertilization treatments on internode length (cm) and number of *G. aparine* plant.

Treatment	Internode number	The 4th Internode	The 5th Internode	The 6th Internode	The 7th Internode	The 8th Internode	The 9th Internode	The 10th Internode	The 11th Internode	The 12th Internode
177 d										
CK	–	–	–	–	–	–	–	–	–	–
PK	9 a	1.5 b	2.0 ab	2.1 bc	2.1 b	2.0 bc	1.3 b	–	–	–
NP	11 a	1.8 b	2.5 ab	2.8 ab	4.3 a	3.9 ab	3.2 ab	–	–	–
NK	9 a	1.4 b	1.6 b	1.4 c	1.6 b	1.5 c	1.0 b	–	–	–
NPK	11a	2.8 a	3.1 a	2.8 ab	2.5 ab	4.0 ab	5.5 a	–	–	–
N1/2PK	10 a	2.4 a	2.9 a	3.2 a	3.3 ab	3.3 a	3.4 ab	–	–	–
205 d										
CK	–	–	–	–	–	–	–	–	–	–
PK	16 a	1.4 b	1.9 b	2.5 c	3.2 c	3.8 b	4.7 bc	4.7 ab	4.3 bc	4.4 b
NP	18 a	3.8 ab	4.4 ab	5.8 ab	6.1 ab	6.3 a	6.2 abc	5.9 ab	6.1 ab	5.1 ab
NK	16 a	2.6 ab	3.0 ab	3.6 bc	4.1 bc	4.2 b	4.2 c	3.8 b	3.7 c	3.4 b
NPK	18 a	3.7 a	4.9 a	6.2 a	7.0 a	7.2 a	6.9 a	6.1 a	6.7 a	6.8 a
N1/2PK	18 a	3.3 ab	4.3 ab	5.2 ab	6.1 ab	6.4 a	6.3 ab	6.0 a	5.8 ab	5.3 ab

Different letters within each day after sowing (d) indicate significant differences among treatments (Tukey test; $P < 0.05$).
The internode lengths (heights) are taken between the previous node and the listed node.

Table 3. Shoot biomass of the four prevalent weed species at 177, 205 and 221 d and nutrient uptake at 221 d in the different fertilization treatments.

Treatments	Weed biomass (g m^{-2})			Biomass proportion (%)[e]		Nutrient uptake (g m^{-2})		
	177 d	205 d	221 d	205 d	221 d	N	P	K
G. aparine								
Control	0.00	0.00	0.00	0.00	0.00	—	—	—
PK	1.39 ab	10.38 bc	10.38 bc	12.72 b	8.80 b	0.066 b	0.019 c	0.085 b
NP	2.85 a	27.05 a	27.63 a	89.06 a	89.13 a	0.331 a	0.093 ab	0.337 b
NK	0.85 b	2.68 c	5.21 c	7.48 b	8.14 b	0.243 ab	0.035 c	0.276 b
NPK	2.83 a	18.24 ab	27.33 a	94.03 a	88.41 a	0.455 a	0.114 a	0.611 a
N1/2PK	2.43 a	15.43 ab	23.42 ab	78.41 a	76.02 a	0.273 ab	0.064 ab	0.392 ab
V. sativa								
Control	2.68 ab	16.42 a	22.75 a	67.71 a	40.72 a	0.365 a	0.042 b	0.352 a
PK	3.73 a	12.41 a	28.58 a	15.20 b	24.24 a	0.538 a	0.133 a	0.400 a
NP	0.48 bc	0.47 b	1.53 b	1.56 c	5.26 b	0.028 b	0.006 c	0.013 b
NK	3.54 a	15.60 a	24.17 a	43.55 a	37.76 a	0.442 a	0.061 b	0.412 a
NPK	0.11 c	0.42 b	0.64 b	2.19 c	2.07 b	0.009 b	0.002 c	0.006 b
N1/2PK	1.31 bc	1.19 b	1.83 b	6.06 bc	5.94 b	0.035 b	0.006 c	0.032 b
G. carolinianum								
Control	0.65 b	4.68 abc	8.68 abc	19.30 a	15.53 ab	0.124	0.016	0.093
PK	5.59 ab	8.83 a	16.42 a	10.82 ab	13.93 abc	0.172	0.107	0.253
NP	7.50 ab	1.06 bc	0.00 d	3.50 ab	0.00 c	—	—	—
NK	2.94 ab	8.62 ab	12.94 ab	24.07 a	20.22 a	0.200	0.034	0.214
NPK	9.11 a	0.08 c	1.75 bcd	0.43 b	5.65 bc	0.024	0.008	0.027
N1/2PK	4.53 ab	0.25 d	0.28 cd	1.29 ab	0.26 c	0.001	0.000	0.002
V. persica								
Control	0.00	0.00	0.00	0.00	0.00	—	—	—
PK	4.68	65.94	81.19	59.69	58.52	0.762	0.415	1.903
NP	0.00	0.00	0.68	0.00	2.25	—	—	—
NK	0.00	0.00	0.33	0.00	0.51	—	—	—
NPK	0.01	0.13	0.12	0.81	0.30	—	—	—
N1/2PK	0.00	0.00	0.00	0.00	0.00	—	—	—

[e]indicates the proportion of shoot biomass within a species to the total weeds biomass per treatment.
Different letters within a species indicate significant differences among treatments (Tukey test; $P<0.05$).

524 tillers per m^2 in the latter three treatments, which was 82% higher than that in the control treatment (data not shown).

Plant Height of Winter Wheat. Fertilization significantly increased wheat height at all the growing periods and the final heights ranged from 60.0 to 79.4 cm at maturity stage (Fig. 6). Wheat heights of the NPK (78.2 cm) and N1/2PK treatments (79.4 cm) were 31 and 33% higher, respectively than the control treatment.

Winter Wheat Biomass, Nutrient Uptakes and Grain Yield. Fertilization significantly increased the shoot biomass accumulation of winter wheat (Table 4). Higher wheat shoot biomasses were observed in the treatments with N and P. The highest wheat shoot biomass was found in the N1/2PK treatment, and significant differences were also found in two of the three growing periods within the treatments containing N and P.

Wheat nutrient uptake is usually positively correlated with its shoot biomass accumulation. However, the wheat P uptake had no significant difference between the N and P treatments in this study (Table 4). The highest N and K uptakes were in the N1/2PK

treatment. They were 18 and 78% higher than those in the NP treatment and 35 and 31%, higher respectively than those in the NPK treatment. These results might be attributed to the lower shoot biomass and nutrient accumulations of the weed community in the N1/2PK treatment (Fig. 2, Table 4). It also suggested that reducing P application in the NPK treatment decreased the productivity of weeds, such as *G. aparine* (Table 3), and improved nutrient uptake and biomass accumulation of wheat.

The NPK treatment resulted in the highest wheat yield, followed by the N1/2PK treatment (Fig. 7). In contrast, the control treatment obtained the highest light penetration (55%). The average light transmittance rate in the treatments with N and P was 5% at 205 d, which was 92% lower than the light transmittance value of the control treatment (Fig. 7).

Discussion

Weed community structure changed sharply among the different fertilization treatments and the results concurred with previous research conducted under field conditions by Mahn [15].

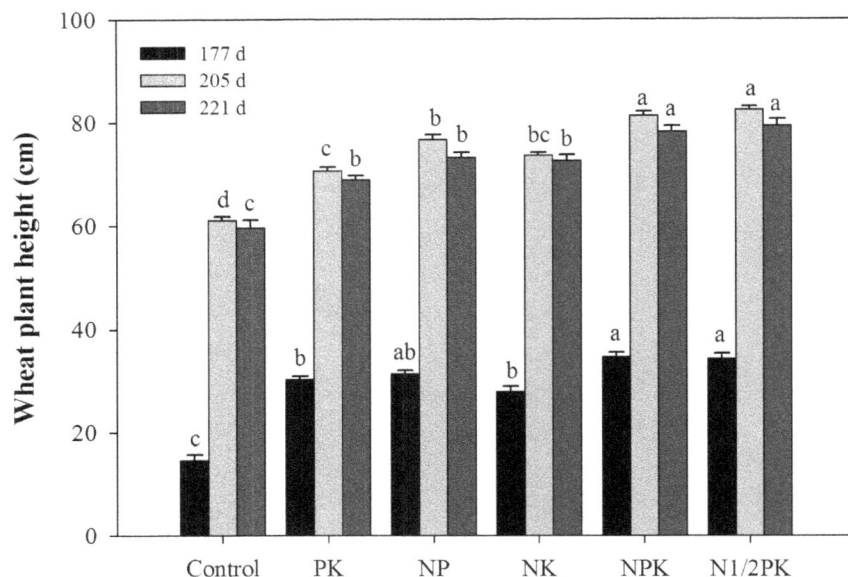

Figure 6. The influence of fertilization on wheat plant height at three growing periods.

The PK treatment showed the highest weed density and shoot biomass, followed by the NK and control treatments, while the treatments with N and P resulted in lower values. Yin et al. [14] and Nie et al. [17] also reported similar results. The structural variability of the weed community across treatments could be directly explained by the different soil fertility conditions. For example, soil available N concentration plays an important role in terminating dormancy by removing germination constraints and promoting the emergence of seedlings [29]. Soil available P was the primary nutrient regulating the species composition and floristic construction of the weed communities, followed by N and K [30]. Moreover, solar radiation reaching weeds was indirectly modulated by different fertilization treatments (Fig. 7). Wheat growth was promoted in the treatments with N and P, especially in the balanced fertilization treatments (Table 4, Fig. 7), increasing its ability to intercept solar radiation (Fig. 7), and thereby affecting the growth of weed species. The structural variability of the weed community may also be significantly affected by rainfall and other environmental factors, such as mean annual temperature. Ochoa-

Hueso and Manrique [31] showed that apart from nutrients, plant germination and growth are also influenced by water availability. Pal et al. [32] and Šilc et al. [33] also reported that environmental factors, such as mean annual precipitation and temperature showed significantly positive and negative correlation coefficients with weed species composition, respectively.

The four prevalent weed populations and wheat differed substantially in their responses to different fertilization treatments. *Galium aparine* grew well in high fertility soil conditions, and was taller than wheat. *Vicia sativa* and *G. carolinianum* grew well under the control, NK, and PK treatments, and they were also taller than wheat in these treatments. *Veronica persica* grew well only in the PK treatment. These results show that the different weed species had formed their own adaptive mechanisms. Morphological plasticity enables plants to cope with a wide variety of ecological conditions, and nutrient availability has been widely shown to induce important anatomical responses [34]. For instance, *G. aparine* grew well under higher soil fertility conditions through climbing into more favorable light conditions by internode elongation (Table 2),

Table 4. The effects of different fertilization treatments on wheat shoot biomass at 177, 205 and 221 d, and nutrient uptake and harvest indices at 221 d.

Treatments	Biomass (g m^{-2})			Nutrient uptake (g m^{-2})			Weed parameters percentages (%)[f]				
	177 d	205 d	221 d	N	P	K	weed biomass	N	P	K	Weed number
Control	149 d	405 c	457 d	4.11 c	0.62 d	3.05 d	47.37	52.59	11.59	22.47	81.8
PK	411 c	1003 b	1032 c	5.12 c	2.37 bc	5.56 cd	100.00	100.00	100.00	100.00	100.0
NP	512 bc	1302 a	1503 b	14.52 a	4.10 a	8.12 bc	26.39	25.19	14.49	12.73	72.7
NK	395 c	843 b	1069 c	6.91 bc	1.08 cd	6.02 cd	54.27	65.93	18.84	32.96	81.8
NPK	601 ab	1337 a	1556 b	12.68 ab	3.75 ab	11.05 ab	26.21	36.30	18.84	23.97	63.6
N1/2PK	700 a	1476 a	2234 a	17.17 a	3.26 ab	14.46 a	26.12	25.19	11.59	17.60	100.0

[f]indicates the percentage of the value of a treatment to that in the PK treatment.
Different letters indicate significant differences among treatments (Tukey test; P<0.05).

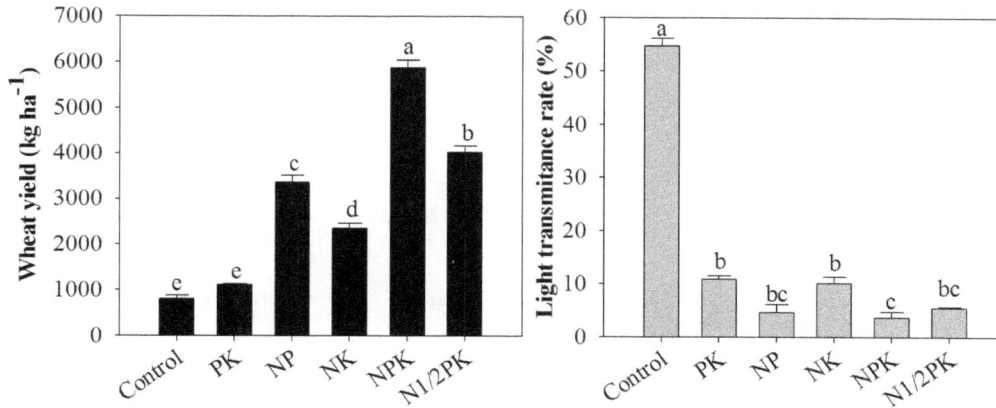

Figure 7. The influence of fertilization on wheat yield and light transmittance at the ground surface.

and responded to soil nutrient supply in the order of P>N>K. In contrast, *V. sativa* and *G. carolinianum* increased their heights for receiving more light under lower soil fertility and limited nutrient conditions. Li et al. [35] reported similar results for *G. carolinianum*, but Yin et al. [13] reported different results for *G. aparine* versus *V. sativa* and *V. persica*. These differences in the effects of fertilization on cropland weeds might be due to herbicide usage. A broadcast application of herbicides was used by Li et al. [35], while no herbicide was used by Yin et al. [13]. Furthermore, environmental conditions, such as plot location, temperature, crop type, precipitation, soil texture, neighboring habitat, and soil pH, which vary in time and space, may also cause variation in weed species responses [36–38].

The PK treatment showed the highest Shannon-Wiener and Pielou Indices, followed by the control and NK treatments. In contrast, these three treatments had lower Simpson Indices than the treatments with N and P. Solar penetration on the ground surface was also one of the reasons for the variation in weed community biodiversity, which is similar to the findings of other studies [16,39,28]. Biomass is often used as an indicator of the amount of resources captured by a crop. Higher biomass and grain yield of wheat in the NP, NPK and N1/2PK treatments indicated that wheat intercepted a larger proportion of light radiation (Fig. 7), thereby reducing the amount reaching the weeds. Wheat yield showed a significant negative correlation with light penetration (Pearson correlations >0.880, P<0.0001). The species richness, Shannon-Wiener and Pielou Indices and the Simpson Index of weed communities were proved to have significantly negative and positive linear function relationships with wheat yield [30]. Shading leads to thinning mortality in individuals of subordinate weed species, and consequently reduces evenness and biodiversity among weed species [39,40]. Only those weed species which possess either physiological shade tolerance or are able to climb into more favorable light conditions could grow well [18]. In contrast to other species, *G. aparine* had higher plant density, shoot biomass and nutrient accumulations in the NP, NPK and N1/2PK treatments with lower light penetration. Mahn [15] also reported that the mean biomass values for *G. aparine* in higher N fertilization treatments within a spring barley field were almost two times more than the values under the no N treatment. Under high soil fertility conditions in this study, *G. aparine* could enhance its plant height through internode elongation, especially from the 4th to the 12th internodes for capturing more sun light. Its plants were 14% and 9% taller than wheat plants in the NP and NPK treatments, respectively (Fig. 5). Therefore, weed vegetation

under higher soil fertility conditions was predominantly composed of a few species, such as *G. aparine*, resulting in a higher Simpson Index and lower Pielou and Shannon-Wiener Indices.

More weed biomass and nutrient accumulations were detected in the PK treatment. Weed biomass under the NP, NPK and N1/2PK treatments was 26.4, 26.2, and 26.1%, respectively, of that under the PK treatment. Similarly, the N, P and K accumulations, as a proportion of that in the PK treatment, were 25, 14, and 13%, respectively in the NP treatment, 36, 19, and 24%, respectively in the NPK treatment and 25, 12, and 18%, respectively in the N1/2PK treatment (Table 4). These results indicated that fewer resources were captured by the weed community in the treatments with N and P, especially in the N1/2PK treatment. If a single weed species has a shoot biomass to total weed biomass ratio > 10%, it is regarded as a dominant weed species. More than two dominant weed species were found under the control, PK and NK treatments, with four species present in the PK treatment. Only one dominant weed species (*G. aparine*) was present in the balanced fertilization treatment. Thus, the results also indicated that balanced fertilization could not only maintain a stable crop yield, but also greatly reduce weed productivity.

Lower solar radiation under high soil fertility conditions resulted in fewer weed species compared with the control and PK treatments. However, the number of weed species was only 73% and 64% in the NP and NPK treatments, respectively and their number in the N1/2PK treatment was the same as in the PK treatment (Table 4). These results demonstrated that *G. aparine* is sensitive to soil P, and reducing the P application rate indirectly improved the growth of other weed species. Blackshaw and Brandt [41] also reported that the competitiveness of the high P-responsive species was progressively improved as the P dose increased. Likewise, with a lower P dose in the N1/2PK treatment, its competitive ability would decrease and more opportunities would be available for other species. The N1/2PK treatment could be considered as a balanced fertilization treatment in adjusting the interactions of wheat and weeds in agroecosystems. Although wheat yield in the N1/2PK treatment was 31% less than that in the NPK treatment, its weed species number, Shannon-Wiener and Pielou Indices were significantly higher than those in the NPK treatment. It is noteworthy that a certain amount of *G. aparine* would have a high biodiversity value for invertebrates and is important for seed-eating birds [22]. Thus, applying half the P rate in the NPK treatment could be considered as a better strategy to provide desirable farming benefits and maintain the biodiversity of weed communities. The negative effects on wheat yield could be balanced by the simultaneous positive effects of reduced herbicide usage. Appropriate combinations and rates of fertilization would not only be helpful for keeping desirable crop productivity, but also maintaining the biodiversity of weed communities in an agroecosystem.

References

1. Vollmann J, Wagentristl H, Hartl W (2010) The effects of simulated weed pressure on early maturity soybeans. Eur J Agron 32: 243–248.
2. Odero DC, Mesbah AO, Miller SD, Kniss AR (2011) Interference of redstem filaree (*Erodium cicutarium*) in sugarbeet. Weed Sci 59: 310–313.
3. Marshall EJP, Brown VK, Boatman ND, Lutman PJW, Squire GR, et al. (2003) The role of weeds in supporting biological diversity within crop fields. Weed Res 43: 77–89.
4. Fried G, Petit S, Dessaint F, Reboud X (2009) Arable weed decline in Northern France: Crop edges as refugia for weed conservation. Biol Conserv 142: 238–243.
5. Altieri MA (1999) The ecological role of biodiversity in agroecosystems. Agric Ecosyst Environ 74: 19–31.
6. Van Elsen T (2000) Species diversity as a task for organic agriculture in Europe. Agric Ecosyst Environ 77: 101–109.
7. Marsh SP, Llewellyn RS, Powles SB (2006) Socail costs of herbicide resistance: the case of resistance to glyphosate. International Association of Agricultural Economists in its series 2006 Annual Meeting, August 12–18, 2006, Queensland, Australia with number 25413.
8. Storkey J, Cussans JW (2007) Reconciling the conservation of in-field biodiversity with crop production using a simulation model of weed growth and competition. Agric Ecosyst Environ 122: 173–182.
9. Di Tomaso JM (1995) Approaches for improving crop competitiveness through the manipulation of fertilization strategies. Weed Sci 43: 491–497.
10. Angonin C, Caussanel JP, Meynard JM (1996) Competition between winter wheat and *Veronica hederifolia*: influence of plant density and the amount and timing of nitrogen application. Weed Res 36: 175–187.
11. Blackshaw RE, Semach G, Janzen HH (2002) Fertilizer application method affects nitrogen uptake in weeds and wheat. Weed Sci 50: 634–641.
12. Blackshaw RE, Molnar IJ, Larney FJ (2005) Fertilizer, manure and compost effects on weed growth and competition with winter wheat in western Canada. Crop Prot 24: 971–980.
13. Yin LC, Cai ZC, Zhong WH (2005) Changes in weed composition of winter wheat crops due to long-term fertilization. Agric Ecosyst Environ 107: 181–186.
14. Yin LC, Cai ZC, Zhong WH (2006) Changes in weed community diversity of maize crops due to long-term fertilization. Crop Prot 25: 910–914.
15. Mahn EG (1988) Changes in the structure of weed communities affected by agro-chemicals: what role does nitrogen play? Ecological Bulletins, No. 39, Ecological Implication of Contemporary Agriculture: Proceeding of the 4th European Ecology Symposium, 7–12 September 1986, Wageningen, 71–73.
16. Kleijn D, van der Voort LAC (1997) Conservation headlands for rare arable weeds: the effects of fertilizer application and light penetration on plant growth. Biol Conserv 81: 57–67.
17. Nie J, Yin LC, Liao YL, Zheng SX, Xie J (2009) Weed community composition after 26 years of fertilization of late rice. Weed Sci 57: 256–260.
18. Murphy C, Lemerle D (2006) Continuous cropping systems and weed selection. Euphytica 148: 61–73.
19. Haas H, Streibig JC (1982) Changing patterns of weed distribution as a result of herbicide use and other agronomic factors, in: LeBaron HM, Gressel J (Eds.), Herbicide Resistance in Plants. John Wiley & Sons, New York, pp. 57–79.
20. Pyšek P, Lepš J (1991) Response of a weed community to nitrogen fertilization: a multivariate analysis. J Veg Sci 2: 237–244.
21. Bengtsson J, Ahnström J, Weibull AC (2005) The effects of organic agriculture on biodiversity and abundance: a meta-analysis. J Appl Ecol 42: 261–269.

22. Rassam G, Latifi N, Soltani A, Kamkar B (2011) Impact of crop management on weed species diversity and community composition of winter wheat fields in Iran. Weed Biol Manag 11: 83–90.
23. Tilman D, Wedin D, Knops J (1996) Productivity and sustainability influenced by biodiversity in grassland ecosystems. Nature 379: 718–720.
24. Tilman D, Reich PB, Knops JMH (2006) Biodiversity and ecosystem stability in a decade-long grassland experiment. Letters 441, doi: 10.1038/nature04742.
25. Derksen DA (1996) Weed community ecology: tedious sampling or relevant science? A Canadian perspective. Phytoprotection 77: 29–39.
26. Yoshida S, Forno DA, Cock JH, Gomez KA (1976) Laboratory Manual for Physiological Studies of Rice. International Rice Research Institute, Philippines.
27. Magurran AE (1988) Ecological diversity and its measurements. Princeton, USA: Princeton University Press.
28. Poggio SL (2005) Structure of weed communities occurring in monoculture and intercropping of field pea and barley. Agric Ecosyst Environ 109: 48–58.
29. Forcella F, Benech Arnold RL, Sanchez R, Ghersa CM (2000) Modeling seedling emergence. Field Crop Res 67: 123–139.
30. Tang LL, Wan KY, Cheng CP, Li RH, Wang DZ, et al. (2013) Effect of fertilization patterns on the assemblage of weed communities in an upland winter wheat field. J Plant Ecol doi: 10.1093/jpe/rtt018.
31. Ochoa-Hueso R, Manrique E (2010) Nitrogen fertilization and water supply affect germination and plant establishment of the soil seed bank present in a semi-arid Mediterranean scrubland. Plant Ecol 210: 263–273.
32. Pal RW, Pinke G, Botta-Dukát Z, Campetella G, Bartha S, et al. (2013) Can management intensity be more important than environmental factors? A case study along an extreme elevation gradient from central Italian cereal fields, Plant Biosystems - An international Journal Dealing with all Aspects of Plant Biology: Official Journal of the Societa Botanica Italiana 147: 2, 343–353.
33. Šilc U, Vrbnicanin S, Bozic D, Carni A, Dajic Stevanovic Z (2009) Weed vegetation in the north-western Balkans: diversity and species composition. Weed Res 49: 602–612.
34. Lamberti-Raverot B, Puijalon S (2012) Nutrient enrichment affects the mechanical resistance of aquatic plants. J Exp Bot 63: 6115–6123.
35. Li RH, Qiang S, Qiu DS, Chu QH, Pan GX (2008) Effects of long-term different fertilization regimes on the diversity of weed communities in oilseed rape fields under rice-oilseed rape cropping system. Biodivers Sci 16: 118–125. (In Chinese with English abstract)
36. Fried G, Norton LR, Reboud X (2008) Environmental and management factors determining weed species composition and diversity in France. Agric Ecosyst Environ 128: 68–76.
37. Hanzlik K, Gerowitt B (2011) The importance of climate, site and management on weed vegetation in oilseed rape in Germany. Agric Ecosyst Environ 141: 323–331.
38. Pinke G, Karácsony P, Czúcz B, Botta-Dukát Z, Lengyel A (2012) The influence of environment, management and site context on species composition of summer arable weed vegetation in Hungary. Appl Veg Sci 15: 136–144.
39. Goldberg DE, Miller TE (1990) Effects of different resource additions of species diversity in an annual plant community. Ecology 71: 213–225.
40. Tilman D, Pacala S (1993) The maintenance of species richness in plant communities, in: Ricklefs RE, Schulter D (Eds.), Species Diversity in Ecological Communities. Chicago, USA: University of Chicago Press, 13–25.
41. Blackshaw RE, Brandt RN (2009) Phosphorus fertilizer effects on the composition between wheat and several weed species. Weed Biol Manag 9: 46–53.

Author Contributions

Conceived and designed the experiments: FC DZW. Performed the experiments: CPC KYW RHL YT JFP LLT. Analyzed the data: LLT. Contributed reagents/materials/analysis tools: JX LLT. Wrote the paper: LLT.

18

Ecoinformatics Reveals Effects of Crop Rotational Histories on Cotton Yield

Matthew H. Meisner[1,2]*, **Jay A. Rosenheim**[3]

1 Department of Evolution and Ecology, University of California Davis, Davis, California, United States of America, 2 Department of Statistics, University of California Davis, Davis, California, United States of America, 3 Department of Entomology and Nematology, University of California Davis, Davis, California, United States of America

Abstract

Crop rotation has been practiced for centuries in an effort to improve agricultural yield. However, the directions, magnitudes, and mechanisms of the yield effects of various crop rotations remain poorly understood in many systems. In order to better understand how crop rotation influences cotton yield, we used hierarchical Bayesian models to analyze a large ecoinformatics database consisting of records of commercial cotton crops grown in California's San Joaquin Valley. We identified several crops that, when grown in a field the year before a cotton crop, were associated with increased or decreased cotton yield. Furthermore, there was a negative association between the effect of the prior year's crop on June densities of the pest *Lygus hesperus* and the effect of the prior year's crop on cotton yield. This suggested that some crops may enhance *L. hesperus* densities in the surrounding agricultural landscape, because residual *L. hesperus* populations from the previous year cannot continuously inhabit a focal field and attack a subsequent cotton crop. In addition, we found that cotton yield declined approximately 2.4% for each additional year in which cotton was grown consecutively in a field prior to the focal cotton crop. Because *L. hesperus* is quite mobile, the effects of crop rotation on *L. hesperus* would likely not be revealed by small plot experimentation. These results provide an example of how ecoinformatics datasets, which capture the true spatial scale of commercial agriculture, can be used to enhance agricultural productivity.

Editor: Raul Narciso Carvalho Guedes, Federal University of Viçosa, Brazil

Funding: Funding Sources: 1. California State Support Committee of Cotton Incorporated. URL: http://www.ccgga.org/cotton_research/cssc.htm. 2. University of California Statewide IPM Program. URL: http://www.ipm.ucdavis.edu. 3. USDA-NRICGP (Grant 2006-01761). URL: http://www.csrees.usda.gov/funding/rfas/nri_rfa.html. 4. California Department of Pesticide Regulation. URL: http://www.cdpr.ca.gov. 5. National Science Foundation GRFP (Grant DGE-1148897). URL: http://www.nsfgrfp.org. The funders had no role in study design, data collection and analysis, decision to publish, or preparation of the manuscript.

Competing Interests: The authors have declared that no competing interests exist.

* E-mail: mhmeisner@ucdavis.edu

Introduction

Maximizing agricultural crop yield is an important goal for several reasons. First, a growing worldwide population will generate increased demand for agricultural resources [1]. Since expanding the land area devoted to agriculture is often unfeasible, or would involve the destruction of sensitive landscapes such as forests and wetlands, the only way to meet this demand will be to increase the crop yield generated from existing farmland. Second, there are substantial economic incentives for profit-seeking farmers to maximize the yield of their crops, especially given the low profit margins typical of commercial agriculture [2].

Farmers make a wide range of decisions regarding the management of their crops, involving pest management, planting/harvest dates, fertilization, irrigation, and, as we focus on in this study, crop rotation. These decisions are, along with external factors that fall outside farmers' control, such as weather, likely to affect crop performance and yield substantially. A rigorous quantitative understanding of the factors, including farmer management decisions, that affect crop yield is an essential prerequisite for developing management strategies that maximize yield.

A critical factor known to affect crop yield in a given field is the crop rotational history of that field [3]. There are several possible mechanisms by which the crops previously grown in a field can affect crop yield. First, different crops have different effects on the nutrient composition of the soil, so the identities of crops previously grown in a field can affect nutrient availability and crop yield [3]. For example, nitrogen-limited crops can benefit from rotation with nitrogen-fixing legumes [4], and phosphorus nutrition in California cotton is shaped by whether or not the previous crop received phosphorus fertilizer [5]. Second, certain crops may increase the local abundance of particular insect pests and pathogens [6–8]. Since different crops are often susceptible and resistant to different pathogens and pests, the identities of the crops recently grown in a field can affect yield. For example, if one crop increases local abundances of an insect pest that also attacks a second crop, planting the second crop immediately following the first may lead to decreased yield resulting from attack from the built up local pest population. In contrast, such a yield depression could potentially be averted if the second crop were planted following a crop that does not lead to local accumulation of the pest. In monocultures of wheat, substantial yield declines have been noted and attributed to the buildup of the soil-borne fungal pathogen *Gaeumannomyces graminis* [9]. Third, many studies have shown that a field's crop rotational history can strongly affect weed densities [10]. Numerous other mechanistic explanations for the yield effects of crop rotation have also been suggested [3].

Crop rotation has been practiced for thousands of years; evidence for its inception dates back to ancient Roman and Greek societies [11,12]. Experimental studies on the effects of crop rotation first appeared in the early 20th century, revealing that growing crops in rotation led to increased crop yields of up to 100% compared to continuous planting of a single crop [13,14]. Interest in the yield effects of crop rotation waned during the middle of the 20th century, due to the increasing availability of cheap fertilizers, insecticides, and herbicides [3,14]. However, crop rotation continues to be a relevant and important practice; low-input farming remains desirable due to the costs of fertilizers and pesticides, and fertilizer and pesticide applications can often not fully compensate for the benefits afforded by crop rotation [3]. In addition, the significant environmental and public health concerns surrounding fertilizer and pesticide use [1,15] highlight the desirability of methods of increasing crop yield through alternative methods such as crop rotation.

The effects of rotational histories on yield are well understood for some crops, such as corn, where rotation is recognized to be crticial in avoiding the buildup of corn rootworms [16]. However, for many crops, the direction, magnitude, and mechanism of the effect of crop rotational histories on crop yield remain poorly understood [3]. Cotton is one such crop. Experimental field studies of the effect of crop rotation on cotton yield have demonstrated increased cotton yield, compared to continuous cultivation of cotton, when cotton is grown in rotation with sorghum [17,18], corn [19], and wheat [20,21]. Despite these useful results, only a small subset of possible rotations has been studied, experiments have been restricted to plots significantly smaller than typical commercial cotton fields, and mechanisms for these effects remain poorly understood. To help address these limitations, we seek to expand upon this work by exploring the effects of crop rotational histories on yield in commercial cotton fields in California, using an "ecoinformatics" approach [22] capitalizing on existing observational data gathered by growers and professional agricultural pest consultants.

In recent years, there has been a surge in research and interest involving the rapidly emerging field of "big data." The big data movement has been fueled by several developments, including a dramatic increase in the magnitude of data generation, an improved ability to cheaply store, manipulate, and explore massive datasets, and the development of new analytic methods [23]. Most importantly, the movement has been driven by a growing realization that existing data, and data generated as a byproduct of our everyday lives, can be leveraged to explore key questions about nature and human behavior, even if the data were not collected for this purpose [24]. Ecoinformatics is a nascent field focused on harnessing the power of big data to address questions in environmental biology. Ecoinformatics approaches typically involve the analysis of large datasets, the synthesis of diverse data sources, and the analysis of pre-existing, observational datasets [22]. In some commercial agricultural settings, farmers, along with hired consultants, collect a great deal of regular data about their fields that are used to guide real-time crop management decisions, such as the timing of pesticide applications. By capitalizing on data that are already generated as a byproduct of commercial agriculture, ecoinformatics provides a low-cost means of obtaining a large dataset that can be used to explore key questions in agricultural biology, some of which might be too difficult or too costly to explore experimentally. Furthermore, the large size of datasets created for ecoinformatics can afford greater statistical power than could possibly be generated through experimental work.

Experimentally studying the yield effects of crop rotational histories is challenging for several reasons. There are a plethora of possible rotational histories, which means that a large number of treatments would be required to explore the space of possible rotational histories thoroughly. Furthermore, experimentally studying effects of crop rotations requires experiments spanning several growing seasons, which may be logistically challenging. Finally, in order to maintain realism and applicability to commercial fields, which are typically quite large, sizeable experimental plots would be required, especially in light of research suggesting that landscape composition as far as 20 km from a focal field can affect the densities of agricultural pests in that field [25]. While yield effects of non-mobile factors such as soil characteristics may be readily detected through small plot experimentation, the effects of highly mobile arthropods may only be detected at much larger spatial scales.

An ecoinformatics approach offers attractive solutions to these challenges. Since we analyze a large preexisting dataset that includes over a thousand records, a diversity of the possible crop rotational histories already exists in the dataset. In addition, our dataset spans 11 years of data, so the data span the temporal scale necessary to ask questions regarding effects of multi-year rotational histories. And, since the data come from the exact setting where we wish to apply our results, the data are realistic and capture the appropriate spatial scale of commercial agriculture.

First, we sought to identify which crop rotational histories are associated with increased and decreased cotton yield, and to quantify these yield effects. We then explored possible explanations for the yield effects identified in the previous step by examining the associations between crop rotational histories and pest abundance.

Materials and Methods

Dataset

The dataset was constructed by collecting existing crop records from commercial cotton fields in California's San Joaquin Valley. The data were shared by both growers and pest control advisors (PCAs), professional consultants hired to monitor field conditions and provide crop management recommendations. The dataset contains records of 1498 unique field-year instances from 566 unique fields, ranging from 1997 to 2008. Growers and PCAs collect and maintain detailed records of the conditions in their fields; numerous variables were recorded for each field-year record, and the following were used in our analyses:

1. Cotton yield. Measured once for each field-year instance, cotton lint yield was measured in bales/acre (converted to kg/ha for our analyses) and recorded for 1240 of the 1498 total records.

2. Crop rotational histories. The identity of the crop grown in the same field in previous growing seasons was recorded. For some fields, records extended back for 10 years. However, the vast majority of fields did not have records extending this far into the past. There were 15 unique crops that appeared in rotational histories: alfalfa, barley, carrots, corn, cotton, garbanzo beans, garlic, lettuce, melons, onions, potatoes, safflower, sugarbeets, tomatoes, and wheat.

3. Surrounding crops. For 1026 of the 1498 crops, we had data on the identity of the crop grown in each of the 8 fields immediately adjoining the focal field (to the North, Northeast, East, Southeast, South, Southwest, West, and Northwest).

4. Cotton variety. The database consisted of records of two different cotton species: *Gossypium barbadense* L. ("Pima cotton") and *Gossypium hirsutum* L. ("upland cotton").

5. *Lygus hesperus* densities. The plant bug *L. hesperus* is one of the most damaging pests of cotton, and a frequent target of

insecticide applications [26,27]. PCAs measured *L. hesperus* densities approximately weekly, primarily during June and July. The PCAs' sampling procedure consisted of 50 swings of a sweep net across the top of the plant canopy. Since not all PCAs sampled on the same days or at exactly the same intervals for all fields, we transformed successive samples into mean *L. hesperus* density estimates by calculating the area under the linear curve of *L. hesperus* density versus time and dividing by the number of days in the sampling interval.

Modeling approach

We employed a hierarchical Bayesian modeling approach, fitting linear mixed models to explore our questions about the effects of crop rotational histories on cotton yield. Mixed models combine the use of random effects and fixed effects, making them ideally suited for analysis of data that are structured, or clustered, in some known way, such that separate observations from within clusters are expected to be similar to one another [28]. When we model a source of clustering using a random effect, we assume that each cluster-specific parameter was drawn from a common distribution, and we estimate the parameters of this distribution from the data. We use this common distribution as the prior when calculating the posterior distribution of each cluster-specific parameter. The parameters (often called hyperparameters) of the distribution of cluster-specific parameters have posteriors that are estimated from the data, typically after assuming uninformative priors for the hyperparameters [28]. Using a common, empirical prior for all cluster-specific parameters allows pooling of information across clusters, so that data from all clusters can help inform estimates of every other per-cluster parameter. Assuming all clusters are the same introduces high bias and tends to underfit the data, whereas estimating fixed effects for each cluster introduces high variance and tends to overfit the data; however, using a random effect provides an optimal compromise between introducing bias and introducing variance [28]. In this dataset, there are several plausible sources of clustering.

1. First, we expect the data to be clustered by field, since there likely exist field-specific factors that affect yield, such as soil characteristics, local climate, and grower agronomic and pest management practices. We controlled for variable yield potential between fields by including field identity as a random effect in our models. Random effects allow pooling of information across clusters, so they are particularly useful when there are few observations from some clusters - a situation in which it is difficult to accurately estimate each per-cluster parameter with only the data from that one cluster [28]. Since there are three or fewer records for 78% of the fields in our database, we feel that including field as a random effect was preferable to trying to estimate field-specific fixed effects with very few observations per field.

Additionally, including field as a random effect provides a straightforward way to make predictions for fields not represented in our database. Since modeling field as a random effect involves estimating a distribution of per-field parameters, we can simply sample a field-specific parameter from this distribution if we wish to make predictions about a previously unobserved field. Uncertainty in this field-specific parameter can be propagated by simulating many samples from this distribution, while simultaneously accounting for uncertainty in the parameters of this distribution. However, if we were to model field as a fixed effect, we would not estimate a distribution of field-specific parameters. We would only estimate parameters for the specific

fields in our database, leaving us with no obvious way to make inferences about new fields.

2. Second, we expect that our data are clustered by year, since there is substantial between-year variability in climate, particularly in the winter and early spring. Climatic variables can affect crop performance, planting date, and insect pest populations, all of which can in turn affect cotton yield. To control for and quantify variation in yield due to year-specific factors, we included year as a random effect in our models. Our reasons for including year as a random effect are the same as those for field: there are few observations from some years, and we may wish to make predictions for future years not covered by the existing database.

All models were fit using a No-U-Turn Sampler variant of Hamiltonian Markov Chain Monte Carlo [29] implemented in Stan version 1.3.0, accessed through the rstan packing in R [30,31]. We ran three chains from random initializations, each with 10,000 samples, and discarded the first 5,000 samples from each as burn-in. Inferences were based upon the remaining 15,000 samples. We checked convergence by making sure that $\hat{R}$, an estimate of the potential scale reduction of the posterior if sampling were to be infinitely continued, was near 1 [32].

Models

Model 1. To explore the yield effects of the crop grown in the same field the previous year, we fit a linear mixed model with yield as the response variable. The predictor variable of primary interest was the identity of the crop grown in that field the previous year, which was included as a fixed effect.

Given that we are working with an observational dataset, a critical step in order to make meaningful inferences about the variable of primary interest - the crop grown the year before - was to control, to the extent possible, for potentially confounding variables that could generate spurious correlations and taint the validity of our inferences about crop rotation. To control for variable yield potential between fields and years, field and year were included in the model as random effects. The field terms control for the possibility that some fields may have higher yield potential due to their location, soil characteristics, or growing practices; the year terms control for the substantial year-to-year variation in cotton yield, which likely results from yearly weather differences. A term indicating cotton species (Pima or upland) was included in the model to account for yield differences between cotton species. Cotton species was modeled as a fixed effect, since there are only two possible categories - not enough to meaningfully estimate a random effects distribution [28]. We also included 15 real-valued fixed effect predictor variables that indicate the number of fields, out of the 8 surrounding fields, planted with each of the 15 crops we analyzed. The goal was to control for effects of the surrounding landscape, and thereby avoid spurious correlations between rotational history (which may be correlated with the crops surrounding the focal crop) and yield.

Our Bayesian modeling approach required the specification of priors for all parameters whose posteriors were estimated using MCMC. Noninformative priors (normal distributions with mean 0 and standard deviation of 100) were used for all fixed effects. The random effects for both field and year were assumed to follow a normal distribution with mean 0 (allowing means of these distributions to be estimated from the data would lead to nonidentifiability with the fixed effects for prior crop identity) and variance hyperparameters estimated from the data. Since the support of variance parameters is constrained to positive real numbers, noninformative inverse gamma distributions with shape and scale parameters set to 0.001 were used as the prior for the variance parameter of the top-level stochastic node, and as the priors for the variance hyperparameters of the field and year random effects distributions.

Model 2. To help us understand whether any effects of the crop grown in the field the previous year on cotton yield could be due to effects on *L. hesperus*, we fit the same model as Model 1, but with average June *L. hesperus* abundance as the response variable.

Model 3. Next, to formally assess whether there was an association between the effects of crop rotation on yield and the effects of crop rotation on *L. hesperus* abundance, we performed a linear regression of the estimated effects on yield (measured as the posterior means from Model 1) against the estimated effects on *L. hesperus* abundance (measured as posterior means from Model 2). Noninformative $\mathcal{N}(0,100^2)$ priors were used for the mean and intercept, and a noninformative inverse gamma distribution with shape and scale parameters set to 0.001 was used as the prior for the variance.

Model 4. A great deal of experimental evidence has demonstrated that crop rotation leads to increased yield compared to successive plantings of a single crop [3]; therefore, we explored whether or not a yield loss was incurred by cotton crops grown in fields where cotton was grown in previous years. For the 782 fields that had complete crop rotational records for the previous 4 years, we calculated the number of consecutive cotton plantings (from 1 to 4) in the 4 years preceding the focal cotton crop. We then fit a model, with yield as the response variable, using the number of consecutive prior cotton plantings as a predictor (again with the same noninformative prior of $\mathcal{N}(0,100^2)$). Field, year, and cotton type were included as they were in Models 1 and 2. Since the number of prior consecutive cotton plantings could be correlated with the number of cotton fields in the surrounding landscape during the focal year, we avoided a possible spurious correlation between consecutive cotton plantings and yield by also including a fixed effect for the number of cotton fields in the 8 fields adjacent to the focal field. We chose not to explore rotational histories of specific crops (and instead just grouped all crops into "cotton" or "not cotton") for longer than one previous year, since the number of possible rotational histories becomes very large and the number of records for each possible history becomes too small to allow for robust statistical analysis.

Model 5. To see if the number of consecutive years of cotton cultivation preceding the focal year was associated with June *L. hesperus* densities, we fit the same model as Model 4, but with June *L. hesperus* as the response variable.

Results

Model 1

Using our samples from the joint posterior of Model 1, we calculated, for each crop other than cotton, the posterior distribution of the difference in mean cotton yield in fields where that crop was grown the year before compared to mean yield in fields where cotton was grown the year before. The posterior means of these comparisons, as well as 95% highest posterior density intervals (HPDIs), are displayed in Figure 1A. Highest posterior density intervals are a Bayesian analogue of frequentist confidence intervals; they denote the narrowest region of parameter space containing 95% of the posterior probability [28]. Three crops had 95% HPDIs that did not overlap 0. Garlic (lower limit = 42.0 kg/ha, upper limit = 213.7 kg/ha), tomatoes (57.9 kg/ha 178.1 kg/ha), and melons (92.9 kg/ha, 793.7 kg/ha) had entirely positive 95% HPDIs, suggesting that previous

cultivation of these crops was associated with increased cotton yield. While no crops had entirely negative 95% HPDIs, the posterior probability of safflower and sugarbeets having negative effects on yield was 96% and 95%, respectively, suggesting that cultivation of these crops the previous year was associated with decreased cotton yield. Yield was 153.0 kg/ha higher, with a 95% HPDI of (115.0 kg/ha, 192.8 kg/ha), for upland cotton than for Pima cotton.

Model 2

Using the joint posterior of Model 2, we calculated the posterior distribution of the difference in mean June *L. hesperus* densities between fields where cotton was the year grown before and where other specific crops were grown the year before. The posterior means of these comparisons, as well as 95% HPDIs, are displayed in Figure 1B. Corn (0.10 insects/sweep, 1.41 insects/sweep), onions (0.09 insects/sweep, 0.76 insects/sweep), and garlic (0.06 insects/sweep, 0.50 insects/sweep) all had 95% HPDIs that were entirely positive, suggesting that previous cultivation of these crops was associated with increased *L. hesperus* abundance. June *L. hesperus* density was 0.35 insects/sweep lower, with a 95% HPDI for this decrease of (0.26 insects/sweep, 0.44 insects/sweep), for upland cotton than for Pima cotton.

Model 3

While there were exceptions, we noticed that there was a trend for crops associated with increased pest abundances to also be associated with decreased yield. To more rigorously quantify this trend, for the 14 crops other than cotton, we regressed the posterior mean of the yield difference from cotton against the posterior mean of the *L. hesperus* difference from cotton. There was a negative slope with posterior mean −0.49 and 95% HPDI of (−1.16, 0.15) that marginally overlapped 0; the posterior probability of there being a negative slope was 93.4%. This provided evidence that crops associated with increased June *L. hesperus* densities were also associated with negative effects on yield (Figure 2).

Model 4

Model 4 suggested that every additional consecutive year of prior cotton cultivation in a field led to reduced cotton yield. Figure 3A displays the posterior distribution of the change in yield for each additional year that cotton was consecutively grown in the field prior to the focal year; the posterior mean for this change in yield was −40.9 kg/ha, with 95% HPDI (−57.5,−23.4 kg/ha). This translates to a mean of the percentage change in yield of −2.4% per year and 95% HPDI of (−1.4%,−3.4%) per year. We refit Model 4 without the term for consecutive cotton plantings; boxplots of the residuals are plotted against consecutive cotton plantings in Figure 3B, where a decreasing trend can be observed. Yield was 169.2 kg/ha higher, with a 95% HPDI of (123.4 kg/ha, 211.5 kg/ha), for upland cotton than for Pima cotton.

Model 5

Model 5 revealed a positive association between the number of preceding consecutive cotton plantings and June *L. hesperus* densities; the posterior mean of the slope regressing June *L. hesperus* on consecutive cotton plantings was 0.037 insects/sweep with a 95% HPDI that slightly overlapped 0 (−0.007,0.079). The posterior probability of there being a positive relationship between consecutive cotton plantings and *L. hesperus* densities was 95.3%. June *L. hesperus* density was 0.32 insects/sweep lower, with a 95% HPDI of (0.20 insects/sweep, 0.43 insects/sweep), for upland cotton than for Pima cotton.

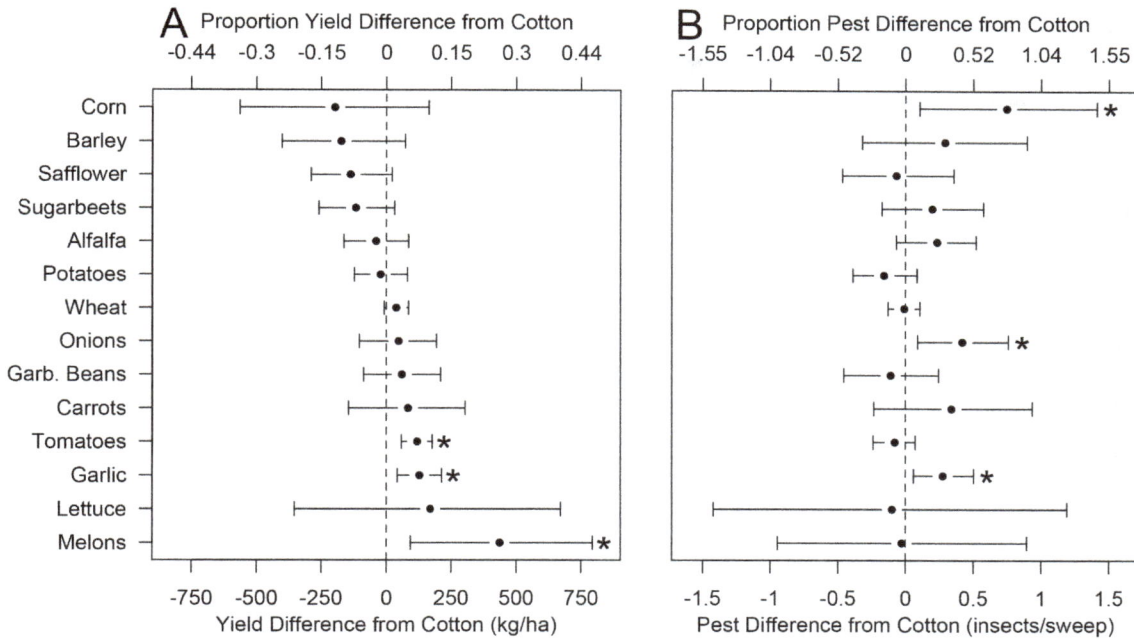

Figure 1. Means and 95% HPDIs of the differences in mean yield (A) and mean June *L. hesperus* density (B) between fields where a certain crop was grown the previous year and where cotton was grown the previous year. 95% HPDIs that do not overlap 0 are marked with a (*).

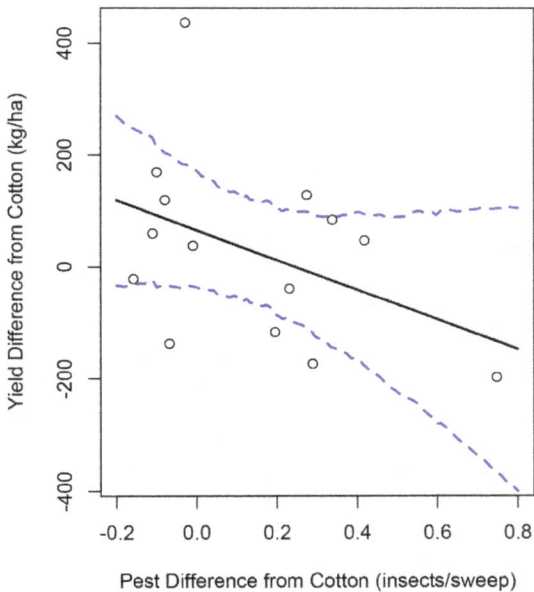

Figure 2. For each of the 14 crops other than cotton, we calculated the posterior mean of the mean difference in yield when that crop was grown in the field the year before compared to when cotton was grown in the field the year before (y-axis; these estimates are also displayed in Figure 1A). We also calculated the posterior mean of the difference in mean June *L. hesperus* densities between fields where a specific crop was grown the year before and where cotton was grown the year before (x-axis; these estimates are also displayed in Figure 1B). These estimates are plotted above (open circles). Then, we fit a linear model by regressing the mean yield differences on the mean *L. hesperus* differences. The posterior mean of the model fit (solid black) and 95% HPDI (dashed blue) are overlaid.

Discussion

Capitalizing on a large existing set of crop records from commercial cotton fields in California, we employed an ecoinformatics approach to explore the effects of crop rotational histories on cotton yield. Our hierarchical Bayesian analyses revealed evidence that several crops, when grown in the same field the year before the focal cotton planting, were associated with either decreased or increased cotton yield (Figure 1A), and either increased or decreased early season densities of the pest *L. hesperus* (Figure 1B). Furthermore, crops associated with decreased yield were generally also associated with increased *L. hesperus* densities, while those associated with increased yield were also associated with decreased *L. hesperus* densities (Figure 2).

These results suggest a possible mechanism for the observed yield effects of these rotational histories. Since *L. hesperus* preferentially attacks certain crops [33], a field cultivated with a crop that is heavily attacked by *L. hesperus* may, if *L. hesperus* disperse from the focal field, increase the abundance of *L. hesperus* in nearby fields. These populations may subsequently attack the crop planted in the focal field the following year, explaining the increase in early-season *L. hesperus* densities that we detected following certain crops. In turn, these increased *L. hesperus* populations may exert strong herbivorous pressure on focal cotton crops, possibly explaining the corresponding decrease in yield.

We believe that the effect of rotational history on early-season *L. hesperus* likely operates at a landscape scale that is larger than the within-field scale. If cotton was grown in a field the previous year, then farmers in the San Joaquin Valley are required to maintain a 90-day plant-free period prior to 10 March of the following year [27]. This prevents *L. hesperus*, which overwinter as adults on live host plants, from overwintering in a focal field where cotton was grown the year before. If a crop other than cotton was grown the previous year, then it could be possible for *L. hesperus* to overwinter in the focal field on residual plant or weed populations; however,

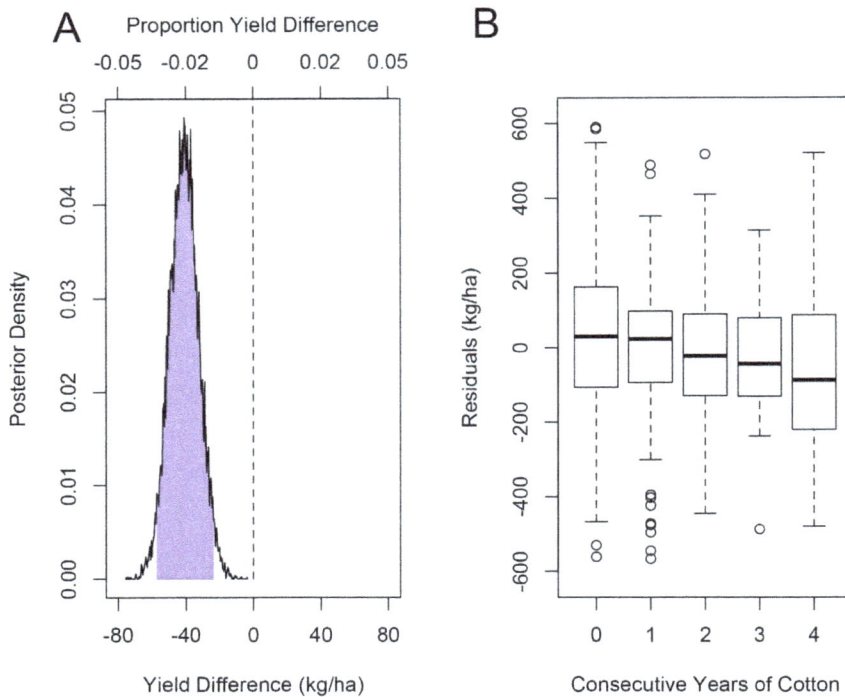

Figure 3. The posterior distribution of the change in yield for every additional year that cotton was grown consecutively in a field prior to the focal year (A). We refit the model without consecutive cotton plantings in the model and display boxplots of the residuals vs. consecutive cotton plantings, where a decreasing trend can be observed (B).

since fields are completely plowed prior to planting cotton in the spring, *L. hesperus* adults would still need to temporarily leave the focal field. Therefore, we believe that the preferred host crops for *L. hesperus* increase *L. hesperus* populations at a landscape scale. Then, when cotton, another target of *L. hesperus*, is planted in the same field the following year, the cotton field is attacked by this regional population. If regional populations are already large due to lingering effects from crops grown the previous year, *L. hesperus* populations may move into cotton early in the growing season; this could be particularly damaging given research suggesting that cotton yield is particularly sensitive to *L. hesperus* densities early in the growing season [34]. Using our data, we were not able to determine at exactly what scale the effects of rotation on *L. hesperus* likely operate. We do not believe a within-field scale is plausible, but determining a more precise spatial scale for these effects could be an interesting topic for future research.

Our findings match expectations of crop yield effects based on previous research on *L. hesperus* host crop preferences, lending support to our hypothesis that yield effects of crop rotational histories are, at least partially, mediated by effects on *L. hesperus*. Alfalfa and sugarbeets, both crops for which we found negative effects on yield and positive effects on *L. hesperus* when grown in a field the previous growing season, are all considered preferred hosts for *L. hesperus* [27], and have been shown to also increase *L. hesperus* populations in nearby cotton fields during an individual growing season [33,35]. Presumably, this effect is due to these crops supporting large *L. hesperus* populations. Large *L. hesperus* populations are known to build up in alfalfa [36], and their dispersal following alfalfa harvesting can threaten nearby cotton crops [27,37]. *L. hesperus* is also known to emigrate to nearby cotton fields when safflower begins to dry in mid-summer [38]. While the potential for nearby alfalfa [27,33] and safflower [33,35]

fields to increase *L. hesperus* populations in cotton fields in a given year has been recognized, our results are the first indication that these landscape effects may extend temporally, affecting *L. hesperus* populations, and yield, in the next growing season. Tomatoes, associated with increased yield and decreased pest abundance in our data, have likewise been shown to decrease *L. hesperus* abundances in nearby cotton fields within a given year [33].

While previous experimental work has examined the effects of crop rotations on cotton yield [14,17,18,20,21], our work expands on these studies in several ways. First, we explore a much wider diversity of possible crop rotational histories, providing quantitative estimates of the cotton yield effects of cultivating 14 different crops the previous year. Second, since we analyze records from commercial cotton fields, our data have the potential to capture yield effects (such as those due to highly mobile arthropods) that could only be detected at this realistic spatial scale. Third, since we have collected data on pest abundances, not only yield, we have also been able to use our data to generate and build evidence for a hypothesized mechanistic explanation of the yield effects we identify.

We also found that farmers incurred a decline in cotton yield of about 2.4% for every additional year cotton was grown consecutively in a field preceding the focal season (Figure 3). This is consistent with previous research suggesting that continuous cultivation of cotton in the same location can reduce yield compared to interspersing cotton with other crops [14,17,18, 20,21]. We also found some evidence that the number of years cotton was grown consecutively in a field was associated with higher June *L. hesperus* densities: the posterior probability of there being a positive association was about 95%. Identifying the actual mechanism underlying this yield effect is beyond the scope of this study, but would be an interesting avenue for future research. It is possible that the yield decline is not caused by changes in *L. hesperus* densities, and instead results from the buildup of soil pathogens, especially in light of previous research showing that continuous cotton cultivation increases the densities of fungal pathogens in the soil [18].

When interpreting our results, it is important to remain cognizant of the challenges of drawing causal inferences from observational data. The key assumption required to make causal inferences from regression coefficients is that all variables that affect both the treatment assignment (crop rotation, in our analyses) and the response variable (yield and *L. hersperus* density, in our analyses) are included in the model; this ensures that the probability of receiving each treatment becomes, conditional on the predictor variables included in the model, conditionally independent of the response variable [28]. In experimental studies, the treatment assignment is typically controlled by the experimenter, so one can be confident that the only difference between treatment and control groups is in fact the treatment. However, in observational studies, it is impossible to prove definitively that there was no other factor that affected both the treatment assignment and the response variable (thus spuriously suggesting a treatment effect).

As such, we want to be very clear that our hypothesis that the effects of rotation on yield are mediated by effects on *L. hesperus* densities is exactly that - a hypothesis. While our data do support a negative *association* between effects on *L. hesperus* and effects on yield, we cannot prove with observational data that the varying effects on yield are *caused* by the varying effects on *L. hesperus*. This could be a fruitful topic for future experimental work.

Although causality is impossible to prove using observational data, ecoinformatics paves the way for implementing data-driven agricultural strategies and allows us to mine large datasets to explore important questions that are difficult to address experimentally. While by no means a replacement for experimentation, ecoinformatics can be a cost-effective and realistic complementary approach. In particular, our result identifying the effects of crop rotation on *L. hesperus* density would have been extremely difficult to reach experimentally. Since *L. hesperus* readily disperse across spatial scales of more than 1000 meters [37], an experimental study would have required massive plots comparable to the size of commercial fields in order to adequately capture their spatial dynamics.

Our results have numerous practical applications for commercial cotton growers. Growers with knowledge of the crop rotations associated with depressed cotton yield could make more informed decisions, selecting the sequence of crop cultivations that lead to maximized yield. When feasible, cotton plantings could be avoided following crops that decrease cotton yield, and instead limited to fields where crops that increase cotton yield were previously planted. In some cases, market conditions may lead a grower to plant cotton following a yield-depressing crop, even given the knowledge of likely yield loss. In those situations, our results may still be helpful, as an early warning sign of a potential pest problem in a particular field could allow the grower and PCA to focus pest detection efforts on that field and provide time to eliminate the problem before severe yield loss was incurred.

Our results suggest that the yield effects of crop rotational histories in cotton are relatively modest in magnitude: the posterior means for effects of any specific crop were mostly under 15%. However, given the tight profit margins of commercial agriculture, a 15% change in yield could translate into a far greater percentage change in profit, and could therefore be of substantial economic significance to a grower. As we seek to feed a growing worldwide population while doing minimal harm to the environment, crop management practices that increase yield while reducing the need for costly and damaging pesticides and fertilizers are of great value. Crop rotation is one such method, and we are optimistic that ecoinformatics approaches may be helpful in elucidating the details of how to optimally implement crop rotation.

Acknowledgments

We would like to sincerely thank the growers and PCAs who generously donated their data and time to help make this research possible, and R.F. Denison for valuable feedback on this manuscript.

Author Contributions

Conceived and designed the experiments: MM JR. Performed the experiments: MM JR. Analyzed the data: MM JR. Contributed reagents/materials/analysis tools: JR. Wrote the paper: MM JR.

References

1. Godfray HCJ, Beddington JR, Crute IR, Haddad L, Lawrence D, et al. (2010) Food security: the challenge of feeding 9 billion people. Science 327: 812–818.

2. Crookston RK (1984) The rotation effect: What causes it to boost yields? Crops Soils 36: 12–14.

3. Karlen DL, Varvel GE, Bullock DG, Cruse RM (1994) Crop rotations for the 21st century. Adv Agron 53: 1–45.

4. Russelle MP, Hesterman OB, Shaeffer CC, Heichel GH (1987) Estimating nitrogen and rotation effects in legume-corn rotations. In: Power JF, editor, The role of legumes in conservation tillage systems, Soil Conserv Soc Am. pp. 41–42.

5. Forbes AA, Rosenheim JA (2011) Plant responses to insect herbivore damage are modulated by phosphorus nutrition. Entomol Exp Appl 139: 242–249.

6. Benson GO (1985) Why the reduced yields when corn follows corn and possible management responses. Proc 40th Corn Sorghum Res Conf: 161–174.

7. Cook RJ (1984) Root health: Importance and relationship to farming practices. In: Bezdicek DF, editor, Organic Farming: Current Technology and its Role in a Sustainable Agriculture, Am Soc Agron. pp. 111–127.

8. Edwards JH, Thurlow DL, Eason JT (1988) Inuence of tillage and crop rotation on yields of corn, soybean, and wheat. Agron J 80: 76–80.

9. Rothamsted Research Center (2006) Guide to the classical and other long-term experiments, datasets and sample archive. Technical report, Rothamsted Research Center.

10. Liebman M, Dyck E (1993) Crop rotation and intercropping strategies for weed management. Ecol Appl 3: 92–122.

11. White KD (1970) Fallowing, crop rotation, and crop yields in Roman times. Agric Hist 44: 281–290.

12. White KD (1970) Roman farming. Ithaca: Cornell University Press.

13. Johnson TC (1927) Crop rotation in relation to soil productivity. J Am Soc Agron 19: 518–527.

14. Mitchell CC, Westerman RL, Brown JR, Peck TR (1991) Overview of long-term agronomic re-search. Agron J 83: 24–29.

15. Rosner D, Markowitz G (2013) Persistent pollutants: a brief history of the discovery of the widespread toxicity of chlorinated hydrocarbons. Environ Res 120: 126–133.

16. Peairs FB, Pilcher SD (2013) Western Corn Rootworm. Colorado State University Extension. URL http://www.ext.colostate.edu/pubs/insect/05570.html. Accessed 6 November 2013.

17. Bordovsky JP, Mustian JT, Cranmer AM, Emerson CL (2011) Cotton-grain sorghum rotation under extreme deficit irrigation conditions. Appl Eng Agric 27: 359–371.

18. Wheeler TA, Bordovsky JP, Keeling JW, Mullinix BG, Woodward JE (2012) Effects of crop ro-tation, cultivar, and irrigation and nitrogen rate on verticiullium wilt in cotton. Plant Dis 96: 985–989.

19. Mitchell CC, Delaney DP, Balkcom KS (2008) A historical summary of Alabama's Old Rotation (circa 1896): the world's oldest, continuous cotton experiment. Agron J 100: 1493–1498.

20. Constable GA, Rochester IJ, Daniells IG (1992) Cotton yield and nitrogen requirement is modi_ed by crop rotation and tillage method. Soil Tillage Res 23: 41–59.

21. Bordovsky JP, Lyle WM, Keeling JW (1994) Crop rotation and tillage effects on soil water and cotton yield. Agron J 86: 1–6.

22. Rosenheim JA, Parsa S, Forbes AA, Krimmel WA, Law YH, et al. (2011) Ecoinformatics for integrated pest management: Expanding the applied insect ecologist's tool-kit. J Econ Entomol 102: 331–342.

23. Streibich K (2013) Using big data to drive commercial advantage: excellent processes, excellent results. Database Netw J 43: 12–13.

24. Gobble MM (2013) Big data: The next big thing in innovation. Res-Technol Manage 56: 64–66.

25. O'Rourke ME, Rienzo-Stack K, Power AG (2011) A multi-scale, landscape approach to predicting insect populations in agroecosystems. Ecol Appl 21: 1782–1791.

26. Rosenheim JA, Steinmann K, Langellotto GA, Zink AG (2006) Estimating the impact of Lygus hesperus on cotton: the insect, plant, and human observer as sources of variability. Environ Entomol 35: 1141–1153.

27. Godfrey LD, Goodell PB, Natwick ET, Haviland DR, Barlow VM (2013) UC IPM pest management guidelines: cotton. University of California Division of Agriculture and Natural Resources. URL http://www.ipm.ucdavis.edu/PMG/r114301611.html. Accessed 17 September 2013.

28. Gelman A, Hill J (2009) Data analysis using regression and multilevel/hierarchical models. Cam- bridge: Cambridge University Press.

29. Hoffman MD, Gelman A (2013) The No-U-Turn sampler: Adaptively setting path lengths in Hamiltonian Monte Carlo. J Mach Learn Res: In press.

30. Stan Development Team (2013) Stan: A C++ library for probability and sampling, Version 1.3. URL http://mc-stan.org/. Accessed 17 September 2013.

31. Stan Development Team (2013) Stan modeling language user's guide and reference manual, Version 1.3. URL http://mc-stan.org/. Accessed 17 September 2013.

32. Gelman A, Rubin DB (1992) Inference from iterative simulation using multiple sequences. Stat Sci 7: 457–511.

33. Carriere Y, Goodell PB, Ellers-Kirk C, Larocque G, Dutilleul P, et al. (2012) Effects of local and landscape factors on population dynamics of a cotton pest. PLOS ONE 7: e39862.

34. Rosenheim JA, Meisner MH (2013) Ecoinformatics can reveal yield gaps associated with crop-pest interactions: a proof-of-concept. PLOS ONE 8: e80518.

35. Carriere Y, Ellsworth P, Dutilleul P, Ellers-Kirk C, Barkley C, et al. (2006) A GIS-based approach for areawide pest management: the scales of Lygus hesperus movements to cotton from alfalfa, weeds, and cotton. Entomol Exp Appl 118: 203–210.

36. Sevacherian B, Stern VM (1975) Movement of Lygus bugs between alfalfa and cotton. Environ Entomol 4: 163–165.

37. Sivakoff FS, Rosenheim JA, Hagler JR (2012) Relative dispersal ability of a key agricultural pest and its predators in an annual agroecosystem. Biol Control 63: 296–303.

38. Mueller AJ, Stern VM (1974) Timing of pesticide treatments on safflower to prevent Lygus from dispersing to cotton. J Econ Entomol 67: 77–80.

19

Rapid and Accurate Evaluation of the Quality of Commercial Organic Fertilizers Using Near Infrared Spectroscopy

Chang Wang[1], Chichao Huang[1], Jian Qian[1], Jian Xiao[1], Huan Li[1], Yongli Wen[1], Xinhua He[2], Wei Ran[1], Qirong Shen[1], Guanghui Yu[1]*

1 National Engineering Research Center for Organic-based Fertilizers, Jiangsu Collaborative Innovation Center for Solid Organic Waste Resource Utilization, Nanjing Agricultural University, Nanjing, PR China, 2 School of Plant Biology, University of Western Australia, Crawley, Australia

Abstract

The composting industry has been growing rapidly in China because of a boom in the animal industry. Therefore, a rapid and accurate assessment of the quality of commercial organic fertilizers is of the utmost importance. In this study, a novel technique that combines near infrared (NIR) spectroscopy with partial least squares (PLS) analysis is developed for rapidly and accurately assessing commercial organic fertilizers quality. A total of 104 commercial organic fertilizers were collected from full-scale compost factories in Jiangsu Province, east China. In general, the NIR-PLS technique showed accurate predictions of the total organic matter, water soluble organic nitrogen, pH, and germination index; less accurate results of the moisture, total nitrogen, and electrical conductivity; and the least accurate results for water soluble organic carbon. Our results suggested the combined NIR-PLS technique could be applied as a valuable tool to rapidly and accurately assess the quality of commercial organic fertilizers.

Editor: Andrea Motta, National Research Council of Italy, Italy

Funding: This work was funded by the National Natural Science Foundation of China (41371248 and 41371299), the National Basic Research Program of China (2011CB100503), the Natural Science Foundation of Jiangsu Province of China (BK20131321), the 111 Project (B12009), Qing Lan Project, Innovative Research Team Development Plan of the Ministry of Education of China (IRT1256), and the Priority Academic Program Development of Jiangsu Higher Education Institutions (PAPD). The funders had no role in study design, data collection and analysis, decision to publish, or preparation of the manuscript.

Competing Interests: The authors have declared that no competing interests exist.

* E-mail: yuguanghui@njau.edu.cn

Introduction

Composting is an inexpensive, efficient, and sustainable treatment for solid wastes [1–4]. The composting industry has been growing rapidly because of a boom in the animal industry in China over the past ten years [5]. Numerous large scale animal farms with more than 10,000 pigs or more than 5,000 dairy cattle have been or are being established in east China, resulting in a large amount of animal manure, which is a major pollutant if not treated and utilized as a fertilizer [5]. Each year, more than 100 factories produce over 5 thousand tons of commercial organic fertilizers in Jiangsu Province, China. Consequently, the total amount of commercial organic fertilizers produced by subsidized composting facilities is more than 2 million tons per year over there (Fig. 1). The difference between commercial organic fertilizer and compost is that the former is referred to compost products entered into the market and having a trademark on the package, while the latter is referred to materials produced during the composting process.

Nevertheless, the application of immature compost can result in inhibited seed germination, root destruction, suppressed plant growth, and a decrease in the oxygen concentration and redox potential [5–7]. The main difference between composts and commercial organic fertilizers is the complexity of the raw materials of the latter, and no studies have investigated if such inhibitions apply to the commercial organic fertilizers that are currently used in China. Therefore, assessing the quality of commercial organic fertilizers is of the utmost importance for achieving high quality marketable fertilizers.

Various parameters, such as the moisture, total organic matter (TOM) content, pH, water soluble organic carbon (WSOC), water soluble organic nitrogen (WSON), pH, electrical conductivity (EC), and germination index (GI), are commonly used to evaluate the compost quality [1,3–5,8]. However, all these approaches are time-consuming or expensive when a large number of samples are involved [3,4,9]. Near infrared (NIR) spectroscopy has many advantages over traditional chemical analyses, such as its ease of sample preparation, rapid spectrum acquisition, non-destructive nature of the analysis, and the portability of the technology [9]. A number of investigators have shown that NIRS (near infrared reflectance spectroscopy) can be applied to rapidly assess compost quality during composting [10–18]. However, it is unclear whether NIRS can also be used to rapid assess the quality indices of commercial organic fertilizers.

The objectives of this study were therefore to search the feasibility of rapidly assessing the essential quality indices of commercial organic fertilizers produced from different raw materials using NIR. For this purpose, a total of 104 commercial organic fertilizers were collected from full-scale compost factories, which are distributed in 13 regions in Jiangsu Province, China (Fig. 1). The measured chemical and biological parameters include

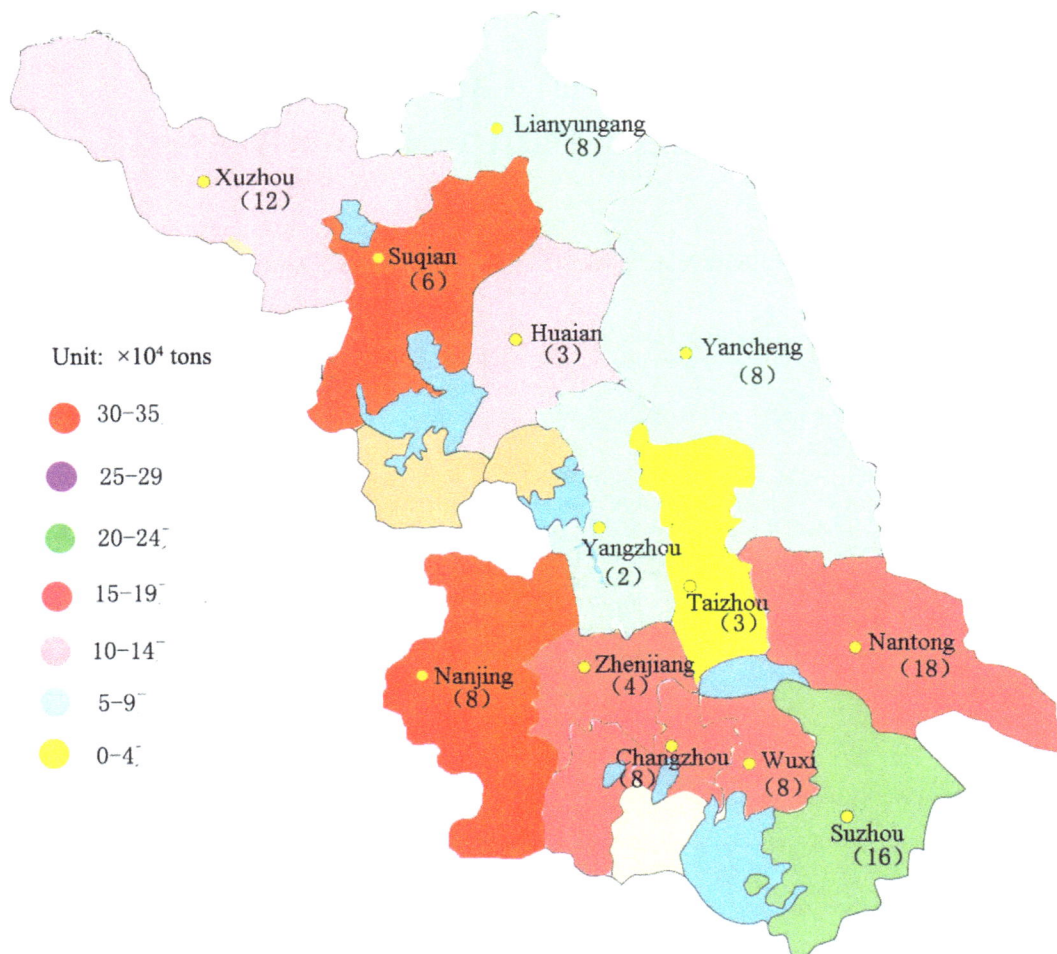

Figure 1. Summary of factory locations to produce commercial organic fertilizers in Jiangsu Province, China.

the moisture content, TOM, TN, WSOC, WSON, pH, EC, and GI. A combination of NIR spectra with partial least squares (PLS) analysis was applied to rapidly evaluate the quality of commercial organic fertilizers.

Materials and Methods

Sample collection and pretreatment

A total of 104 commercial organic fertilizers were collected from full-scale compost factories (Fig. 1). These factories treat organic matter from animal (chicken, cattle, duck, pig, etc.) manure and other agricultural organic (straw from wheat, corn, rice, etc.) residues. No specific field permits were required for this study. The land accessed is not privately owned or protected. These factories produce approximately 0.5 to 1.5 million tons of commercial organic fertilizers per year.

All organic fertilizer products were identified and collected during factory packaging after being screened and crashed. The collected products, as granular and powered fertilizers that were thus more uniform than those during composting (see Figure S1), were then transported to the lab, air-dried, and stored at 4°C for further analysis. Meanwhile, the samples were mixed and divided into four equal subsamples by the quartile method. The subsamples were then ground (0.15 mm sieve) for the determina-

tion of their chemical and biological indices and the near infrared spectra.

Chemical and biological indices analysis

All chemical analyses were conducted in duplicate using analytical grade chemicals. The TOM in these commercial organic fertilizers was determined by the loss on ignition at 550°C for 4 h, according to [3]. The TN was analyzed using a Perkin-Elmer 2400 CHN elemental analyzer [3,4]. The moisture was measured by oven drying at 105°C to a constant weight. The EC and pH were measured in a 1:10 (w/v) water extract [5]. For the WSOC and WSON analysis, 5 g samples were shaken with 50 ml of deionized water (1:10 w:v) for 2 h, the resulting extracts were centrifuged at 3,500 rpm for 30 min, filtered through 0.45 μm membranes and then determined using a TOC/TN analyzer (multi N/C 3000, Analytik Jena AG, Germany) [3,4,5]. The GI measurement was with *Lepidium sativum* L. seeds [3].

Spectroscopic measurement

The spectra of NIR were recorded at 1.4 nm intervals from 350 to 1,000 nm and 2 nm intervals from 1,000 to 2,500 nm using a FieldSpec@ 3 NIR spectrometer (ASD Inc, USA). Approximately 30 g of dried sample were scanned in a 6 cm diameter sample cell with a quartz window at 18–22°C. A dark reference measurement

Table 1. Composition statistics for quality indices in commercial organic fertilizers.

Parameters	Calibration set (n=78)					Validation set (n=26)			
	Outliers	Min	Max	Mean	SD	Min	Max	Mean	SD
Moisture (%)	2	2.43	7.34	3.73	0.99	2.57	6.64	3.84	0.82
TOM (g/kg)	2	280.82	534.54	347.86	51.18	288.30	498.35	353.82	42.30
TN (g/kg)	0	4.30	128.56	19.89	19.07	5.74	46.36	18.41	8.18
WSOC (g/kg)	7	0.11	28.34	3.75	5.66	0.17	17.47	3.80	4.58
WSON (g/kg)	10	0.08	13.97	1.95	2.79	0.18	8.44	1.89	2.35
pH	2	5.09	9.90	6.36	0.97	5.23	9.22	6.48	0.80
EC (mS/cm)	0	3.27	21.91	5.61	2.86	3.49	9.58	5.39	1.23
GI (%)	0	33.98	137.96	61.18	21.18	35.95	117.99	61.66	16.48

Abbreviations: EC, electrical conductivity; GI, germination index; SD, standard deviation, TN, total nitrogen; TOM, total organic matter content; WSOC, water soluble organic carbon; WSON, water soluble organic nitrogen.

was conducted for noise and ambient temperature correction every 30 min. Background correction was performed as a total of 64 scans which were averaged before each sample being scanned. After using the ViewSpecPro software (ASD Inc, USA), each sample was remixed, rescanned five times, and then averaged. The recorded spectral data were processed and stored as the reflectance (R) and then converted to the absorbance A ($A = \log 1/R$).

Partial least squares regression (PLS) analysis

The PLS regressions were used to construct the model between the laboratory parameters and NIR spectral data. A cross-validation was performed to select the optimal number of terms in the equation and to avoid over-fitting. The data pre-processing and model development was conducted using the spectroscopic software Unscrambler Trial 9.7 (CAMO Inc, Norway). A validation set, composed of independent samples, was applied to estimate the prediction accuracy of the calibration models. In this study, all 104 samples were randomly divided into a calibration set (78 samples) and a validation set (26 samples). This method had been applied by several publications [19–21]. To restrain invariable background signals and to improve the visual resolution, the second derivative spectra of each sample were used for further calibration and validation. The selection of the models developed was based on the values of the coefficient of determination for the calibration set (R^2) and the root mean square error in the cross validation (RMSECV), the determination coefficient of the validation (r^2) and the RPD (ratio of standard error of performance to standard deviation). The RPD is the ratio of the standard deviation (SD) in the validation set over the root mean squared error of the prediction (RMSEP). The bias and the slope value were used to evaluate the usefulness of NIRS for determining the selected quality indicators of commercial organic fertilizers.

The formula for the root mean standard error of calibration (RMSEC) is:

$$RMSEC = \sqrt{\left(\frac{\sum_{i=1}^{N} (C_i - C_i')^2}{N - 1 - p} \right)} \qquad (1)$$

where C_i is the known value, C_i' is the value calculated using the calibration equation, N is the number of samples, and p is the number of independent variables in the regression.

The root mean standard error of prediction (RMSEP) estimates the prediction performance during the validation step of the calibration equation:

$$RMSEP = \sqrt{\left(\frac{\sum_{i=1}^{M} (C_i - C_i')^2}{M - 1} \right)} \qquad (2)$$

where M is the number of samples in the prediction set.

Results and Discussion

Chemical and biological indices

The minimum, maximum, mean, and standard deviation (SD) of the quality indices of commercial organic fertilizers in the calibration sets and validation sets are shown in Table 1. This data set represented a wide range of compositions. For all the 104 sampled fertilizers, the mean value and SD of the quality indices were 3.95±1.70% of moisture, 357.63±76.90 g TOM/kg, 19.52±16.99 g TN/kg, 4.87±9.54 g WSOC/kg, 2.48±5.09 g WSN/kg, 6.40±0.92 pH, 5.56±2.55 mS/cm of EC, and 61.30±20.03% of GI.

This wide variability in the quality indices of the commercial organic fertilizers allowed us to successfully build a correlation between the NIR spectra and the compost quality indices [3]. Samples with a difference between the reference and predicted values were considered outliers and thus excluded during the calibration process. The removal of outliers was on the basis of being labeled as compositional outliers based on the criterion that if the predicted versus actual difference for a sample was 3 SD or more from the mean difference [13]. For the quality indices, two outliers were removed for the moisture, TOM, and pH; 0 for TN, EC, and GI; 7 for WSOC; and 10 for WSON (Table 1).

NIR spectra

All the NIR spectra of the collected commercial organic fertilizers could be divided into two groups of signal with different slopes under 1,400 nm, i.e., one group presented an increased curvature and the another one was more flat. Meanwhile, the former had a significant absorbance peak at wavelengths of approximately ~1,420 nm, but while the latter had only a small absorption at this position. The second significant spectral peak

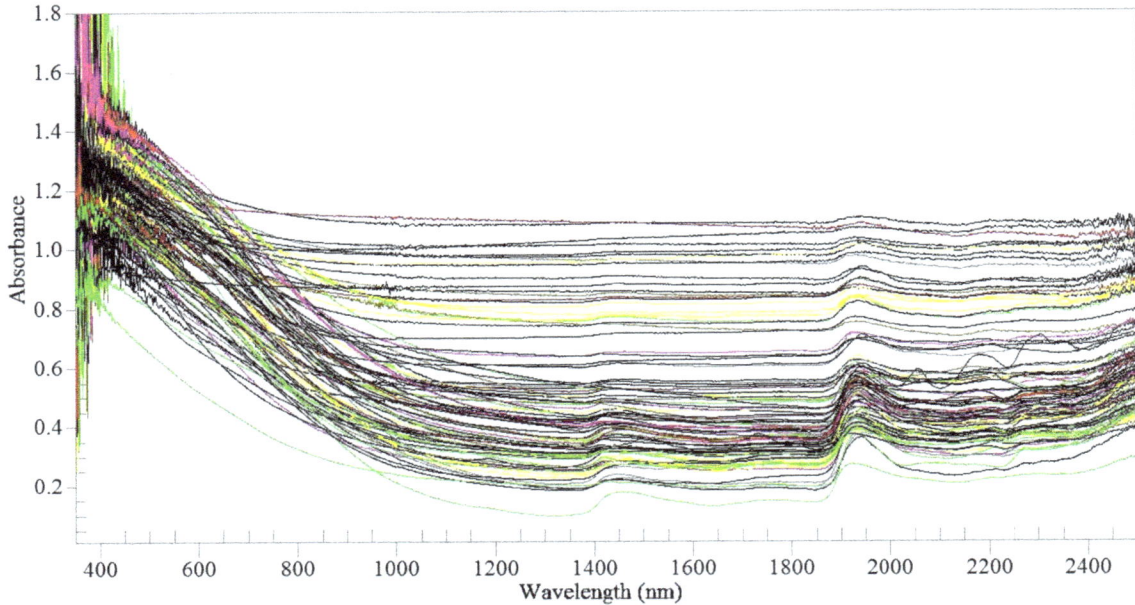

Figure 2. Spectra of NIR of a total 104 commercial organic fertilizers.

was at approximately ~1,950 nm (Fig. 2). The absorbance band at 1,420 nm is usually assigned to the O–H and aliphatic C–H, while the band at 1,950 nm is associated with the amide N–H and O–H [20,22,23]. Note that the absorption peaks were heavily overlapped, mainly because the near-infrared spectrum contains all strength information of the chemical bond, chemical composition, electronegativity, etc. Meanwhile, other interference information, such as scattering, diffusion, special reflection, surface gloss, refractive index, and reflected light polarization, affects the near-infrared spectrum [24,25]. Thus, the quantitative predictions are difficult directly through NIR spectra alone. Multivariate analyses are required to discern the response of properties of commercial organic fertilizers from spectral characteristics with the support of chemometric methods, e.g., PLS analysis.

Partial least square calibration and validation

Based on the guideline proposed by Saeys et al. [26], the accuracy of the predictions for the calibration model is classified as excellent when $r^2 > 0.90$, good when $0.81 < r^2 < 0.90$, moderately successful when $0.66 < r^2 < 0.80$, and unsuccessful when $0.50 < r^2 < 0.65$. Meanwhile, according to Albrecht [9] and Chang et al. [27], the accuracy of the PLS model and prediction was considered good for RPD > 2, acceptable for $1.4 < $ RPD < 2, and unreliable for RPD < 1.4.

In this study, the results of the NIRS calibration and validation for the quality indices of commercial organic fertilizers are listed in Table 2 and Figures 3–4. The NIR calibrations allowed accurate predictions of the TOM, WSON, pH, and GI ($R^2 = 0.73$–0.93 and RPD $= 1.47$–2.96). The results were less accurate for the moisture ($R^2 = 0.91$, $r^2 = 0.79$, RPD $= 2.22$), TN ($R^2 = 0.98$, $r^2 = 0.80$, RPD

Table 2. NIRS calibration and validation results for quality indices of commercial organic fertilizers.

Parameters	Calibration set			Validation set				
	PC	R^2	RMSECV	r^2	RMSEP	RPD	Bias	Slope
Moisture content (%)	8	0.91	0.58	0.79	0.37	2.22	−0.07	0.86
TOM (g/kg)	8	0.93	22.92	0.78	19.38	2.18	−3.60	0.91
TN (g/kg)	17	0.98	3.08	0.80	3.63	2.25	−0.29	0.77
WSOC (g/kg)	10	0.88	3.64	0.76	2.18	2.10	0.24	1.04
WSON (g/kg)	8	0.86	1.81	0.77	1.60	1.47	0.05	0.74
pH	4	0.73	0.96	0.88	0.27	2.96	−0.07	0.84
EC (mS/cm)	17	0.99	0.46	0.74	0.54	2.27	−0.04	0.78
GI (%)	18	0.84	4.81	0.68	9.52	1.73	−2.68	0.63

Note that: TOM, total organic matter; TN, total nitrogen; WSOC, water soluble organic carbon; WSON, water soluble organic nitrogen; EC, electrical conductivity; GI, germination index; PC, number of principal component; R^2, the coefficient of determination for the calibration set; RMSECV, the root mean square error in cross validation; r^2, the coefficient of determination for the validation set; RMSEP, room mean squared error of prediction; RPD, the ratio of the standard deviation in the validation set over the room mean squared error of prediction. The calibration and validation of some indices (i.e., moisture content, TOM, WSOC, pH) was conducted using the second derivative with SNV, but that for the others was conducted using only the second derivative.

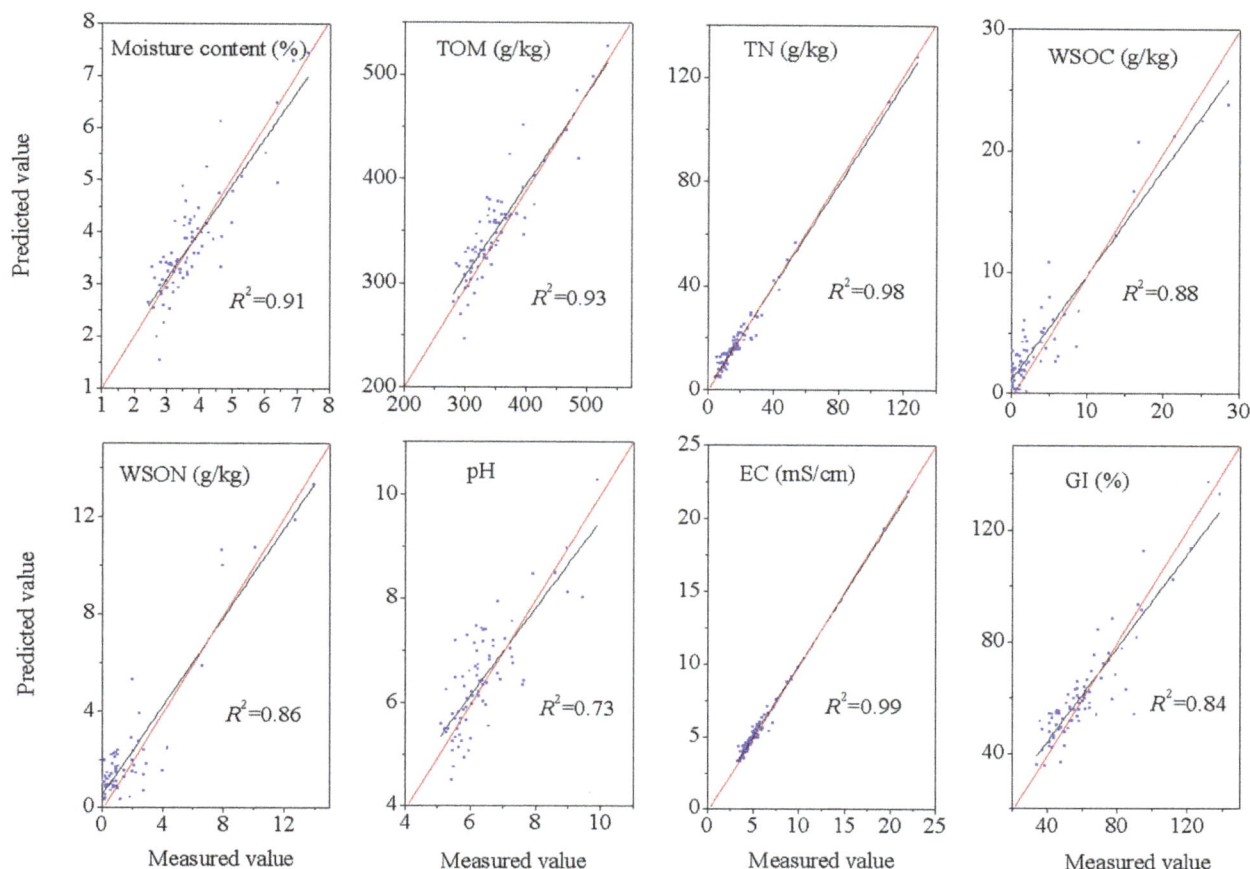

Figure 3. Relationships between the measured and predicted values of the quality indices of commercial organic fertilizers for the calibration data set. The red line represents the best fit. Abbreviations: EC, electrical conductivity; GI, germination index; SD, standard deviation, TN, total nitrogen; TOM, total organic matter content; WSOC, water soluble organic carbon; WSON, water soluble organic nitrogen

= 2.25) and EC ($R^2 = 0.99$, $r^2 = 0.74$, RPD = 2.27). However, the WSOC had the worst prediction, with $R^2 = 0.88$, $r^2 = 0.76$ and RPD = 2.10. Therefore, predictions were moderately successful for the moisture, TOM, TN, WSON, pH, EC, and GI, but unsuccessful for WSOC.

Previous studies have demonstrated that the NIR-PLS was successful in predicting some parameters such as nitrogen (N), carbon (C), C/N, humic acid, pH, respiration, and composting time during the composting process. For example, Saeys et al. [26] developed calibrations using the PCA and PLS regressions for the moisture ($r^2 = 0.91$, RPD = 3.22), TOM ($r^2 = 0.90$, RPD = 3.00) and TN ($r^2 = 0.86$, RPD = 2.63) in pig manure using a mobile spectroscopy instrument. Huang et al. [28] obtained calibrations for the moisture ($r^2 = 0.98$, RPD = 7.48), pH ($r^2 = 0.62$, RPD = 1.63), EC ($r^2 = 0.90$, RPD = 3.10), and TN ($r^2 = 0.97$, RPD = 6.11) in animal manure (cattle, chicken, and pig manures) composts using the NIR-PLS method. Vergnoux et al. [18] obtained excellent calibrations using PCA and PLS regressions in sewage sludge compost for the moisture content ($r^2 = 0.91$), TN ($r^2 = 0.98$), and pH ($r^2 = 0.92$). Albrecht et al. [9] evaluated the biological and chemical changes during the composting process of green waste and sewage sludge using NIR and found that the NIR calibrations successfully allowed accurate predictions of N, C, the C/N ratio, humic acid (HA), pH, and composting time, but were less accurate for the OM, protease, acid, and alkaline phosphatase and unsatisfactory for fulvic acid. Soriano-Disla et al. [29]

obtained moderately successful predictions for WSOC in compost ($r^2 = 0.75$, RPD = 1.70) and in sewage sludge ($r^2 = 0.60$, RPD = 1.60). However, these investigations were conducted in samples during composting, and no report has applied the NIR-PLS to predict the indices of commercial organic fertilizers.

An obvious difference between samples from the whole composting process and those from commercial organic fertilizers is the wide variability of the ingredients used in the elaboration of the composting heaps. Therefore, obtaining good correlations was more difficult for commercial organic fertilizers used in this study. The schematic of rapidly evaluating the quality of commercial organic fertilizers using near infrared spectrometer was given in Figure S2. The results in this study indicated for the first time that the indices of commercial organic fertilizers could be well-evaluated by the NIR with PLS regression method.

Conclusions

In this study the NIR spectroscopy combined with PLS analysis has been developed as an alternative method to traditional chemical analysis for rapidly and accurately predicting the essential quality indices of commercial organic fertilizers. In general, the NIR-PLS technique provided accurate predictions of the TOM, WSON, pH, and GI; less accurate results for the moisture, TN, and EC; and the worst results for WSOC. As a result, we suggest the NIR spectroscopy with PLS analysis may be

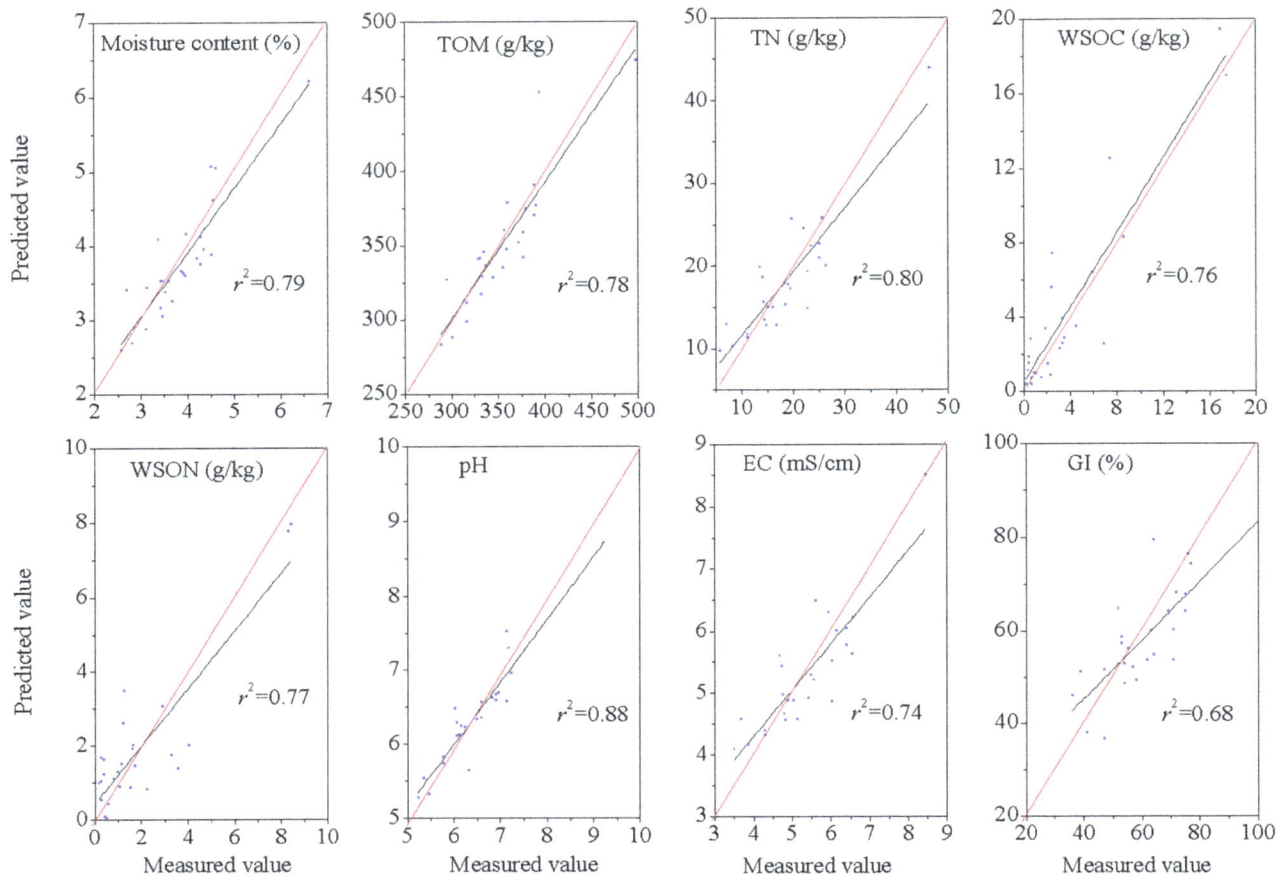

Figure 4. Relationships between the measured and predicted values of the quality indices of commercial organic fertilizers for the prediction data set. The red line represents the best fit. Abbreviations: TOM, total organic matter content; TN, total nitrogen; WSOC, water soluble organic carbon; WSN, water soluble nitrogen; EC, electrical conductivity; GI, germination index.

used as a valuable industrial and research tool to rapidly and accurately assess the quality of commercial organic fertilizers.

Supporting Information

Figure S1 Typical commercial organic fertilizers, including powered (A, C) and granular (B, D) fertilizers. These photos suggest that the commercial organic fertilizers are more evenly than samples from the composting process.

Figure S2 Schematic of rapid evaluating the quality of commercial organic fertilizers using near infrared spectrometer.

Author Contributions

Conceived and designed the experiments: GY. Performed the experiments: CW CH JQ JX HL. Analyzed the data: GY XH QS WR YW. Contributed reagents/materials/analysis tools: GY QS WR. Wrote the paper: GY XH CW.

References

1. Gajalakshmi S, Abbasi SA (2008) Solid waste management by composting: state of the art. Crit Rev Environ Sci Technol 38: 311–400.
2. Moral R, Paredes C, Bustamante MA, Marhuenda-Egea F, Bernal MP (2009) Utilisation of manure composts by high-value crops: safety and environmental challenges. Bioresour Technol 100: 5454–5460.
3. Yu GH, Luo YH, Wu MJ, Tang Z, Liu DY, et al. (2010) PARAFAC modeling of fluorescence excitation-emission spectra for rapid assessment of compost maturity. Bioresour. Technol 101: 8244–8251.
4. Yu GH, Wu MJ, Luo YH, Yang XM, Ran W, et al. (2011) Fluorescence excitation-emission spectroscopy with regional integration analysis for assessment of compost maturity. Waste Manage 31: 1729–1736.
5. Tang Z, Yu GH, Liu DY, Xu DB, Shen QR (2011) Different analysis techniques for fluorescence excitation-emission matrix spectroscopy to assess compost maturity. Chemosphere 82: 1202–1208.
6. Said-Pullicino D, Erriquens FG, Gigliotti G (2007) Changes in the chemical characteristics of water-extractable organic matter during composting and their influence on compost stability and maturity. Bioresour Technol 98: 1822–1831.
7. Smith DC, Hughes JG (2004) Changes in maturity indicators during the degradation of organic wastes subjected to simple composting procedures. Biol Fert Soils 39: 280–286.
8. Bernal MP, Alburquerque JA, Moral R (2009) Composting of animal manures and chemical criteria for compost maturity assessment. A review. Bioresour Technol 100: 5444–5453.
9. Albrecht R, Joffre R, Le Petit J, Terrom G, Périssol C (2009) Calibration of chemical and biological changes in cocomposting of biowastes using near-infrared spectroscopy. Environ Sci Technol 43: 804–811.
10. Saeys W, Darius P, Ramon H (2004) Potential for on site analysis of hog manure using a visual and near-infrared diode array reflectance spectrometer. J Near Infra Spectrosc 12: 299–309.
11. Francou C, Poitrenaud M, Houot S (2005) Stabilization of organic matter during composting: influence of the process and of the composted wastes. Compost Sci Util 13: 72–83.

12. Malley D, McClure C, Martin PD, Buckley K, McCaughey WP (2005) Compositional analysis of cattle manure during composting using a field-portable near infrared spectrometer. Commun Soil Sci Plant Anal 36: 455–475.

13. Yang ZL, Han LJ, Fan X (2006) Rapidly estimating nutrient contents of fattening pig manure from floor scrapings by near infrared reflectance spectroscopy. J Near Infrared Spectrosc 14: 261–268.

14. Fujiwara T, Murakami K (2007) Application of near infrared spectroscopy for estimating available nitrogen in poultry manure compost. Soil Sci Plant Nutr 53: 102–107.

15. Moral R, Galvez-Sola L, Moreno-Caselles J, Perez-Murcia MD, Perez-Espinosa A, et al. (2007) Can Near Infrared Reflectance Spectroscopy (NIRS) Predict Heavy Metals in Sewage Sludge? In: Kungolos A, Aravossis K, Karagiannidis A, Samaras P, editors. First Conference on Environmental Management, Engineering, Planning and Economics, 24–28 June 2007, Skiathos island, Greece; p. 1683–1688.

16. Albrecht R, Joffre R, Gros R, Le Petit J, Terrom G, et al. (2008) Efficiency of near-infrared reflectance spectroscopy to assess and predict the stage of transformation of organic matter in the composting process. Bioresour Technol 99: 448–455.

17. Xing L, Chen LJ, Han LJ (2008) Rapid analysis of layer manure using near-infrared reflectance spectroscopy. Poult Sci 87: 1281–1286.

18. Vergnoux A, Guiliano M, Le Dréau Y, Kister J, Dupuy N, et al. (2009) Monitoring of the evolution of an industrial compost and prediction of some compost properties by NIR spectroscopy. Sci Total Environ 407: 2390–2403.

19. Miriam F, Malcolm RB, Louise RW, Natalie AM (2011) Predicting glycogen concentration in the foot muscle of abalone using near infrared reflectance spectroscopy (NIRS). Food Chem 126: 1817–1820.

20. Peltre C, Thuriès L, Barthès B, Brunet D, Morvan T, et al. (2011) Near infrared reflectance spectroscopy: A tool to characterize the composition of different types of exogenous organic matter and their behaviour in soil. Soil Biol Biochem 43: 197–205.

21. Daniel JMH (2012) Development of near infrared spectroscopy models for the quantitative prediction of the lignocellulosic components of wet *Miscanthus* samples. Bioresour Technol 119: 393–405.

22. Fidencio PH, Poppi RJ, Andrade JC, Cantarella H (2002) Determination of organic matter in soil using near-infrared spectroscopy and partial least squares regression. Commun Soil Sci Plant Anal 33: 1607–1615.

23. Cozzolino D, Moron A (2003) The potential of near-infrared reflectance spectroscopy to analyse soil chemical and physical characteristics. J Agric Sci 140: 65–71.

24. Lu WZ, Yuan HF, Xu GT (2000) Modern Near Infrared Spectroscopy Analytical Technology. China Petrochemical Press, Beijing, China.

25. Meissl K, Smidt E, Schwanninger M, Tintner J (2008) Determination of humic acids content in composts by means of near- and mid- infrared spectroscopy and partial least squares regression models. Appl Spectrosc 62: 873–880.

26. Saeys W, Mouazen AM, Ramon H (2005) Potential for onsite and online analysis of pig manure using visible and near infrared reflectance spectroscopy. Biosyst Eng 91: 393–402.

27. Chang CW, Laird DA, Mausbach MJ, Hurburgh CR (2001) Near-infrared spectroscopy- principal components regression analyses of soil properties. Soil Sci Soc Am J 65: 480–490.

28. Huang G, Han L, Liu X (2007) Rapid estimation of the composition of animal manure compost by near infrared spectroscopy. J Near Infrared Spectrosc 15: 387–394.

29. Soriano-Disla JM, Gómez I, Guerrero C, Navarro-Pedreno J, García-Orenes F (2010) The potential of NIR spectroscopy to predict stability parameters in sewage sludge and derived compost. Geoderma 158: 93–100.

20

Responses of Bacterial Communities in Arable Soils in a Rice-Wheat Cropping System to Different Fertilizer Regimes and Sampling Times

Jun Zhao[1,2], **Tian Ni**[1,2], **Yong Li**[3], **Wu Xiong**[1,2], **Wei Ran**[1,2], **Biao Shen**[1,2], **Qirong Shen**[1,2,4], **Ruifu Zhang**[1,2,4*]

1 Key Laboratory of Plant Nutrition and Fertilization in Low-Middle Reaches of the Yangtze River, Ministry of Agriculture, Nanjing, Jiangsu, China, **2** Jiangsu Key Lab and Engineering Center for Solid Organic Waste Utilization, Nanjing Agricultural University, Nanjing, Jiangsu, China, **3** Soil and Fertilizer Technical Guidance Station of Jintan City, Jintan Agricultural and Forestry Bureau, Jintan, Jiangsu, China, **4** Jiangsu Collaborative Innovation Center for Solid Organic Waste Resource Utilization, Nanjing Agricultural University, Nanjing, Jiangsu, China

Abstract

Soil physicochemical properties, soil microbial biomass and bacterial community structures in a rice-wheat cropping system subjected to different fertilizer regimes were investigated in two seasons (June and October). All fertilizer regimes increased the soil microbial biomass carbon and nitrogen. Both fertilizer regime and time had a significant effect on soil physicochemical properties and bacterial community structure. The combined application of inorganic fertilizer and manure organic-inorganic fertilizer significantly enhanced the bacterial diversity in both seasons. The bacterial communities across all samples were dominated by *Proteobacteria*, *Acidobacteria* and *Chloroflexi* at the phylum level. Permutational multivariate analysis confirmed that both fertilizer treatment and season were significant factors in the variation of the composition of the bacterial community. Hierarchical cluster analysis based on Bray-Curtis distances further revealed that bacterial communities were separated primarily by season. The effect of fertilizer treatment is significant ($P = 0.005$) and accounts for 7.43% of the total variation in bacterial community. Soil nutrients (e.g., available K, total N, total P and organic matter) rather than pH showed significant correlation with the majority of abundant taxa. In conclusion, both fertilizer treatment and seasonal changes affect soil properties, microbial biomass and bacterial community structure. The application of NPK plus manure organic-inorganic fertilizer may be a sound fertilizer practice for sustainable food production.

Editor: Hauke Smidt, Wageningen University, Netherlands

Funding: This research was financially supported by the Agricultural Ministry of China (201103004) and the Chinese Ministry of Science and Technology (2011BAD11B03), RZ and QS were also supported by the 111 Project (B12009) and the Priority Academic Program Development (PAPD) of Jiangsu Higher Education Institutions. The funders had no role in study design, data collection and analysis, decision to publish, or preparation of the manuscript.

Competing Interests: The authors have declared that no competing interests exist.

* E-mail: rfzhang@njau.edu.cn

Introduction

Soils are considered to be the most diverse microbial habitats on Earth with respect to species diversity, community size and microbial biomass [1,2]. Bacteria are the most abundant and diverse group of soil microorganisms [3], playing a vital role in agroecosystems through participation in recycling soil nutrients, maintaining soil structure and promoting plant growth [4,5]. It has long been recognized that the appropriate community structure, abundant diversity and a high activity of microorganisms are significant factors in maintaining the sustainability and productivity of ecosystems [6–9].

Fertilization is an important agricultural practice for improving plant nutrition, increasing soil organic matter and achieving high crop yields, but it can also affect soil microbial abundance, activity and community structure [10–13]. Significant differences in soil microbial biomass and microbial diversity have been observed following fertilization [14–16]. Organic fertilizers usually increase soil microbial biomass [17,18], enzyme activities [13,19] and functional diversity [16], while microbial biomass and enzyme activities are decreased in response to inorganic fertilizers [20–22].

Previous studies have indicated that seasonal changes impact soil microbial communities in agroecosystems [23–27]. Moreover, temporal variation of conditions is a very common feature of agroecosystems, including temperature, rainfall, plant type, nutrient levels, etc. Hence, we reason that study using single time point sampling cannot tease apart the effect of temporal variation on microbial community structure.

Soil microorganisms respond quickly to environmental changes (e.g., application of fertilizer or herbicide, tillage, crop rotation and seasonal variation), resulting in dynamic changes in microbial biomass, activity, diversity, abundance and composition. Microbial biomass, activity and diversity are effective indicators of soil quality and health [28,29]. Therefore, understanding the shifts of microbial biomass, community structure and diversity following different agricultural management practices is important for selecting suitable management strategies to improve ecosystem service [30,31].

The rice-wheat cropping system is one of the main cropping systems for cereal food production in China. Understanding the bacterial community in response to the different fertilizer treatments and seasonal variations will help us disclose the "real" effect of fertilizer regimes and guide us in selecting an appropriate

Table 1. Effects of fertilizer regime, sample time and the interaction between them on soil physicochemical characteristics.

	pH	Organic matter (g/kg)	Available N (mg/kg)	Available P (mg/kg)	Available K (mg/kg)	Total N (g/kg)	Total P (g/kg)	Total K (g/kg)
Fertilizer Regime§(FR)								
CK	7.01±0.15 ab	24.9±3.5 a	117.0±7.6 b	8.1±2.9 b	57.0±15.9 c	1.43±0.11 b	0.83±0.24 a	15.3±1.4 a
NPK	6.97±0.14 b	26.4±4.2 a	134.2±9.7 a	13.8±4.3 ab	64.5±21.2 bc	1.52±0.10 ab	0.87±0.25 a	15.3±1.2 a
NPKM	7.11±0.19 ab	26.6±3.4 a	136.6±7.5 a	14.5±1.8 ab	66.1±21.7 bc	1.59±0.09 a	0.90±0.24 a	16.1±1.5 a
NPKS	7.13±0.07 ab	27.3±4.5 a	137.5±7.0 a	15.2±3.5 a	76.3±28.3 ab	1.62±0.11 a	0.89±0.21 a	15.4±2.2 a
NPKMS	7.27±0.19 a	27.7±4.2 a	135.0±6.5 a	16.6±4.4 a	70.7±23.9 b	1.64±0.17 a	0.92±0.22 a	16.0±1.7 a
NPKMOI	7.07±0.16 ab	27.3±3.6 a	138.5±5.9 a	15.1±3.1 a	86.5±36.0 a	1.59±0.07 a	0.89±0.22 a	15.6±1.7 a
Sample Time (ST)								
June	7.04±0.17 a	24.0±2.2 b	131.2±10.2 a	12.4±3.1 a	91.9±17.1 a	1.50±0.10 b	1.09±0.07 a	15.2±1.0 a
October	7.12±0.18 a	29.3±3.0 a	137.7±8.0 a	13.8±4.1 a	48.5±5.4 b	1.63±0.12 a	0.68±0.05 b	16.0±1.9 a
ANOVA P-values								
FR	0.021	NS	< 0.001	0.009	< 0.001	0.003	NS	NS
ST	NS	< 0.001	NS	NS	< 0.001	< 0.001	< 0.001	NS
FR × ST	NS	NS	NS	NS	0.008	NS	NS	NS

Values are means ± standard deviation (n = 6 or n = 18).
NS: not significant (P>0.05).
§Fertilizer regimes: CK: control without fertilizers; NPK: chemical fertilizers; NPKM: 50% NPK fertilizer plus 400 kg/ha manure; NPKS: 100% NPK fertilizer plus crop straw; NPKMS: 50% NPK fertilizer plus 400 kg/ha manure and crop straw; NPKMOI: 30% NPK fertilizer plus 240 kg/ha manure organic-inorganic compound fertilizer.
Means followed by the same letter for a given factor are not significantly different (P<0.05; Turkey's HSD test where there are more than two treatment levels).

management practice for more stable and sustainable agroecosystem for food production. In the present study, we aimed 1) to study the effect of different fertilizer regimes and seasonal variations on the bacterial community and microbial biomass; 2) to determine whether the fertilizer regime or seasonal change is the major determinant of bacterial community structure; and 3) to determine the contribution of environmental factors on changes in the bacterial community. To address this aim, measurement of biological properties (microbial biomass) and molecular analysis (pyrosequencing) were used to assess the variation in bacterial community.

Materials and Methods

Ethic statement

No specific permits were required for the described field studies. The locations are not protected. The field studied did not involve endangered or protected species.

Field description and experimental design

The field experiment is located in Jintan city, Jiangsu Province, China (31°39′N, 119°28′E, 3 m a.s.l.). This region has a northern subtropical monsoon climate, with an average annual temperature of 15.3°C and a mean annual precipitation of 1063.6 mm. The soil type is classified as typical Clay loamy Fe-leachic-gleyic-stagnic anthrosol, with a long-term annual rotation of winter wheat (*Triticum aestivum* L.) and summer rice (*Oryza sativa* L.), which is the typical cropping system in this region. The fertilization experiment has been in operation since 2010, including four replicates of six treatments in a random block design. The treatments were control without fertilizers (CK), chemical fertilizers (NPK), 50% NPK fertilizer plus 400 kg/ha manure (NPKM), 100% NPK fertilizer plus crop straw (NPKS), 50% NPK fertilizer plus 400 kg/ha manure and crop straw (NPKMS) and 30% NPK fertilizer plus 240 kg/ha manure organic-inorganic compound fertilizer

(NPKMOI). It is noted that the manure organic-inorganic compound fertilizer were comprised of pig manure compost and suitable amount of chemical fertilizer. Each plot was 5 m × 8 m and received the same levels (except for the control plot) of nitrogen, phosphorus and potassium (N: 240 kg/ha, P_2O_5: 120 kg/ha and K_2O: 100 kg/ha) from fertilizers in each cropping season. All P, K and manure fertilizers were applied as basal fertilizers before planting, whereas N fertilizer (urea) was used both as basal fertilizer before planting and supplementary fertilizer at tillering and panicle stage. The wheat was planted in October and harvested in June, followed by rice, which was planted in June and harvested in October.

Soil sampling and analysis

Soil samples were taken in summer (June 2012) after the wheat harvest and in autumn (October 2012) after the rice harvest. Ten cores (2.5 cm in diameter) were randomly collected from the plough layer of soil (0–20 cm) in each replicate plot. The cores from each replicate plot were mixed together, pooled in a sterile plastic bag and transported to the laboratory on ice. All the samples were sieved (2 mm), thoroughly homogenized, and divided into three subsamples: one was air-dried for soil characteristic analysis; another was stored at −4°C for the measurement of biological properties; the rest was stored at −80°C for DNA extraction and subsequent molecular analysis. The soil analysis of each sample was performed by the soil testing lab in Qiyang at the red soil experimental station of the Chinese Academy of Agricultural Sciences.

Microbial biomass

Microbial biomass C (MBC) and biomass N (MBN) were measured using the chloroform fumigation-extraction method [32,33]. After 24 h fumigation, soils were extracted using 0.5 M K_2SO_4 with a 1:5 ratio for 60 min on a rotary shaker, and C and N were determined on a LiquiTOC element analyzer II

Table 2. Effects of fertilizer regime, sample time and the interaction between them on soil microbial biomass carbon and nitrogen.

	Microbial biomass C (mg/kg)	Microbial biomass N(mg/kg)
Fertilizer Regime[§] (FR)		
CK	310.76±89.51 b	20.29±6.45 d
NPK	387.75±88.13 ab	28.36±2.49 c
NPKM	428.73±100.23 a	33.71±7.24 abc
NPKS	398.00±117.64 ab	30.22±3.07 bc
NPKMS	447.68±103.02 a	36.00±5.48 ab
NPKMOI	471.20±87.32 a	38.85±4.93 a
Sample Time (ST)		
June	479.04±96.25 a	32.77±8.18 a
October	325.04±93.96 b	29.71±7.29 b
ANOVA P-values		
FR	< 0.001	< 0.001
ST	< 0.001	0.043
FR × ST	NS	NS

Values are means ± standard deviation (n = 8 or n = 24).
NS: not significant ($P>0.05$).
Means followed by the same letter for a given factor are not significantly different ($P<0.05$; Turkey's HSD test where there are more than two treatment levels).
[§]Fertilizer regimes as described in Table 1.

(Elementar Analysen-systeme GmbH,Hanau, Germany). Finally, the result was calculated using 0.45 (k_{EC}) and 0.54 (k_{EN}) correction factors [34,35].

DNA extraction

Three samples (biological replicates) of each fertilizer treatment for each season were used for DNA extraction. Total soil DNA was extracted from 0.5 g of fresh soil using a PowerSoil DNA Isolation Kit (Mo Bio Laboratories Inc., Carlsbad, CA, USA) according to the manufacturer's instructions. The extracted DNA was evaluated on a 1% agarose gel, and the concentration and quality (A_{260}/A_{280}) of the extracts were determined using a NanoDrop ND-2000 spectrophotometer (NanoDrop, Wilmington, DE, USA). To minimize the DNA extraction bias, three successive DNA extracts of each sample were pooled. The DNA of each sample was diluted 50-fold for further use.

16S rRNA gene amplification and 454 pyrosequencing

An aliquot (10 ng) of purified DNA from each sample (one biological replicate) was used as template for amplification. The V1-V3 hypervariable region of the bacterial 16S rRNA was amplified using the primer set 27F: AGAGTTTGATCCTGGCT-CAG, with the Roche-454 B sequencing adapter, and 533R: TTACCGCGGCTGCTGGCAC, which contained the Roche-454 A sequencing adapter and a unique, error-correcting 10-bp barcode sequence at the 5′-end of each primer. Each sample was amplified in triplicate in a 20 µl reaction using the following program: 2 min of initial denaturation at 95°C followed by 25 cycles of denaturation (95°C for 30 s), annealing (55°C for 30 s), extension (72°C for 30 s), and a final extension at 72°C for 5 min. The PCR products of each sample were pooled together and purified using a PCR Purification Kit (Axygen Bio, USA). The amplicons of each sample were then combined in equimolar concentrations into a single tube prior to 454 pyrosequencing. Pyrosequencing was performed on a 454 GS-FLX Titanium System (Roche, Switzerland) by Majorbio Bio-pharm Technology Co., Ltd (Shanghai, China).

Processing of pyrosequencing data

Raw pyrosequencing data were processed using Mothur (version 1.29.2) [36] following the Schloss standard operating procedure [37]. Specially, sequences having a minimum flow length of 360 flows and a maximum flow length of 450 flows were de-noised using the Mothur based re-implementation of the PyroNoise algorithm [38] with the default parameters. De-noised sequences with more than two mismatches to the forward primer sequence, one mismatch to the barcode sequence, containing a homopolymer longer than eight nucleotides, any ambiguous base calls (Ns) or a sequence length shorter than 200 bases were eliminated. Then, the filtered sequences were assigned to soil samples based on the unique 10-bp barcodes. After removing the barcode and primer sequences, the unique sequences were aligned against the Silva 106 database [39]. Through screening, filtering, pre-clustering processes and chimera removal using UCHIME [40], the retained sequences were used to build a distance matrix with a distance threshold of 0.2. Using the average neighbor algorithm with a cutoff of 97% similarity, bacterial sequences were clustered to operational taxonomic units (OTUs) (hereby defined as identified OTUs) and the most abundant sequence in each OTU was selected as the representative sequence. Representative sequences were taxonomically classified using an RDP naïve Bayesian rRNA Classifier [41] with a confidence threshold of 60%. Relative abundance of a given phylogenetic group was set as the number of sequences affiliated with that group divided by the total number of sequences per sample. To correct for sampling effort, we used a randomly selected subset of 4,000 sequences per sample for α-diversity analysis. To increase the sampling intensity, we pooled the sequences belonging to each fertilizer treatment for each season. In addition, to downweight the effects of rare species on ß-diversity analysis, only OTUs with ≥20 counts summed across all samples were retained (hereby defined as retained OTUs).

Table 3. Effect of fertilizer regime, sample time and the interaction between them on the relative abundance (%) of abundant bacterial classes (subgroups) [£] (relative abundance >1%).

Phylum	Proteobaceria			Acidobacteria				Chloroflexi	Bacteroidetes
Class (subgroups)	α-proteobacteria	ß-proteobacteria	γ-proteobacteria	6	4	7	3	Anaerolineae	Sphingobacteria
Fertilizer Regime [§] (FR)									
CK	4.8±0.9 ab	13.4±0.9 ab	4.3±0.6 c	6.1±0.4 a	3.4±1.2 a	2.2±0.5 a	1.3±0.2 a	10.3±2.6 a	4.6±1.6 a
NPK	4.5±1.1 b	13.6±1.7 a	5.3±0.9 bc	5.1±0.9 ab	3.0±1.2 a	1.8±0.4 ab	1.5±0.5 a	11.3±3.0 a	5.2±1.4 a
NPKM	4.2±1.2 b	14.0±1.2 a	5.8±1.0 b	5.9±0.6 a	3.3±1.0 a	2.1±0.2 a	1.5±0.3 a	9.8±1.8 a	5.4±1.4 a
NPKS	4.0±1.4 b	14.8±0.7 a	5.8±0.9 b	5.9±0.6 a	3.3±0.9 a	2.1±0.2 a	1.2±0.4 a	9.0±1.9 a	5.2±1.6 a
NPKMS	5.9±2.1 ab	13.2±1.0 ab	5.2±0.9 bc	5.4±0.6 a	2.8±1.1 ab	1.8±0.3 ab	1.7±0.4 a	10.8±3.6 a	5.0±1.7 a
NPKMOI	7.1±1.6 a	11.7±1.0 b	7.6±1.4 a	4.0±1.1 b	2.1±0.7 b	1.5±0.2 b	1.8±0.4 a	9.7±3.1 a	5.1±2.1 a
Sample Time (ST)									
June	4.5±1.9 b	13.9±1.2 a	6.3±1.3 a	5.5±0.8 a	3.8±0.7 a	2.0±0.4 a	1.4±0.3 b	8.3±1.3 b	6.6±0.6 a
October	5.7±1.4 a	13.0±1.4 b	5.0±1.1 b	5.2±1.1 a	2.1±0.5 b	1.9±0.3 a	1.7±0.5 a	12.0±2.3 a	3.7±0.5 b
ANOVA P-values									
FR	0.005	0.001	< 0.001	< 0.001	< 0.001	0.005	NS	NS	NS
ST	0.019	0.013	< 0.001	NS	< 0.001	NS	0.026	< 0.001	<0.001
FR × ST	NS	NS	NS	NS	NS	NS	NS	NS	NS

Values are means ± standard deviation (n = 6 or n = 18).
NS: no significant (P>0.05).
Means followed by the same letter for a given factor are not significantly different (P<0.05; Turkey's HSD test where there are more than two treatment levels).
[£] Actinobaceria, Gemmatimonadetes, Nitrospira were not included in this analysis as they had only one class.
[§] Fertilizer regimes as described in Table 1.

Statistical analysis

An OTU-based analysis was performed to calculate the richness, diversity and coverage in our samples with a cutoff of 3% dissimilarity per 4,000 sequences. Richness indices, the Chao1 estimator [42] and the abundance-based coverage estimator (ACE) [43] were calculated to estimate the number of observed OTU that were present in the sampling assemblage. The diversity within each individual sample was estimated using the non-parametric Shannon diversity index [44] and the Simpson diversity index [45]. Good's non-parametric coverage estimator [46] was used to estimate the percentage of the total species that were sequenced in each sample. Rarefaction curves generated using Mothur [36] were also used to compare relative levels of bacterial OTU diversity across all fertilizer treatments in each season. Significant differences were considered when there was no overlap between the 95% confidence intervals.

In all tests, a P-value <0.05 was considered statistically significant. All data were tested for normality and homogeneity and the data were $\log_2$ (x+1)-transformed when necessary to meet the criteria for a normal distribution. A multiple analysis of variance (MANOVA) using PASW Statistics 18 (SPSS Inc.) was used to determine the effects of fertilizer treatment and sampling time on the dependent variables, soil characteristics, bacterial abundance, microbial biomass, relative abundance of abundant taxa and the α-diversity indices. If the multivariate F was significant, we then proceeded with the individual univariate analysis.

To describe the changes in soil bacterial assemblages among treatments and seasons without biases resulting from differences in sequencing depth, all samples were first rarefied to 3,000 counts

using the rrarefy function in the vegan [47] package of R (version 2.15.0) [48] based on the retained OTUs tables. A permutational multivariate analysis of variance [49] was performed to assess the effect of fertilizer regime, sampling time and its interaction on bacterial community structure (abundance of OTUs and phyla) using the adonis function of the R vegan [47] package with 999 permutations. To compare bacterial community structures across all samples based on the OTU composition, hierarchical cluster analysis was carried out using the hclust function with the average linkage algorithm in the stats package of R [48]. Then, the Bray-Curtis and Euclidean distances were used to construct a community dissimilarity matrix and an environmental dissimilarity matrix, respectively. The effects of rare species were downweighted by applying the Hellinger transformation for community data [50]. A Mantel test, using the mantel function in the vegan [47] package of R [48], was used to calculate the correlation between the Bray-Curtis distances of bacterial community and soil characteristics. To visualize the relationship between bacterial communities and environmental factors, a redundancy analysis (RDA) was carried out using the rda function, and the environmental factors were fitted to the ordination plots using the envfit function of the vegan [47] package in R with 999 permutations. Pearson correlations between soil characteristics and abundant phyla (classes) were calculated using PASW Statistics 18 (SPSS Inc.). The BioEnv procedure was used to select the subset of environmental properties best correlated (highest Pearson correlation) with bacterial assemblage dissimilarity, and they were used to construct the soil property matrix for variation partitioning analysis (VPA) in the vegan [47] package of R [48].

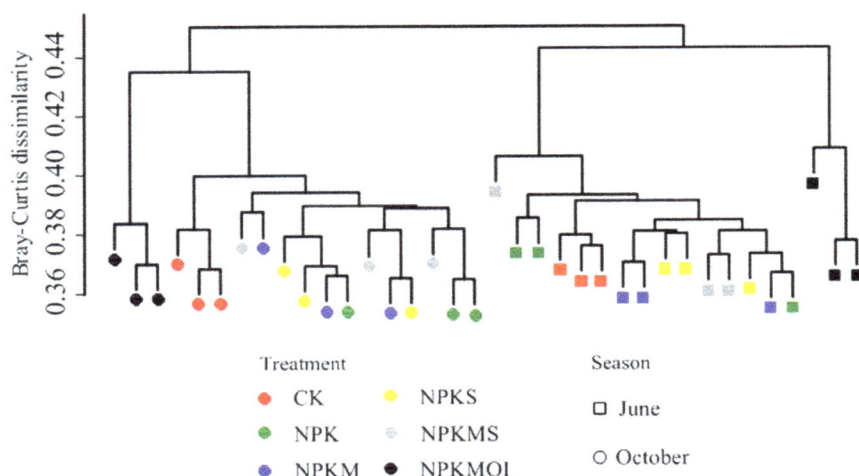

Figure 1. Hierarchical cluster dendrogram of bacterial communities. Pairwise Bray-Curtis dissimilarity of samples collected from six fertilizer treatments in June (square) and October (circle). OTU counts were rarefied to 3,000 counts per sample, and only OTUs ≥20 counts summed across all samples were included in the analysis with the Hellinger-transformation. Symbols of fertilizer regimes were as described in Table 1.

Sequence accession numbers

Sequence data have been deposited in the NCBI Sequence Read Archive (SRA) database with accession number SRA073640.

Results

Soil physicochemical characteristics

The soil pH, available N (AN) and available P (AP) were significantly ($P < 0.05$) affected by different fertilizer treatments, while organic matter (OM) and total P (TP) were significantly ($P < 0.001$) affected by sample time (Table 1). The soil OM content was higher in October (29.3 g/kg) than in June (24.0 g/kg), while the soil TP content was higher in June (1.09 g/kg) than in October (0.68 g/kg). Both treatment and season showed a significant ($P < 0.01$) effect on soil total N (TN). In addition, the treatment, sample time and treatment × sample time interaction terms were all significant affected soil available K (AK) content and separate analyses for each season showed that soil AK content was highest in treatment NPKMOI (118.4 mg/kg, 54.6 mg/kg) and lowest in treatment CK (71.4 mg/kg, 42.7 mg/kg) both in June and October (Table S1). No significant differences of soil total K

Table 4. Permutational multivariate analyses of variance of the Bray dissimilarity between bacterial communities.

Source	df	Abundance of OTUs		Abundance of phylum	
		Sums of sqs	Pseudo-F	Sums of sqs	Pseudo-F
Fertilizer regime (FR)	5	0.6079	1.7276***	0.014495	2.6929**
Sample time (ST)	1	0.40962	5.8206***	0.033321	30.9527***
FR × ST	5	0.38064	1.0818	0.006292	1.1690
Residuals	24	1.68897		0.025836	

*indicate significant correlations ($P < 0.05$); ** indicate significant correlations ($P < 0.01$); *** indicate significant correlations ($P < 0.001$).

(TK) were observed between different treatments, seasons and the interaction terms.

Microbial biomass

No significant differences ($P < 0.05$) of soil MBC and MBN were observed between the fertilizer treatment × sample time interactions (Table 2). The soil MBC changed significantly ($P < 0.001$) between different treatments, being highest in treatment NPKMOI (541 mg/kg, 402 mg/kg) and lowest in treatment CK (359 mg/kg, 234 mg/kg) both in June and October (Table S2). Soil MBC also varied significantly ($P < 0.001$) over time, being higher in June (479 mg/kg dry soil) compared with October (325 mg/kg dry soil). The greatest differences ($P < 0.001$) in MBN were between different treatments, with season having a smaller but significant ($P = 0.043$) effect. The soil MBN was higher in June (32.8 mg/kg dry soil) than in October (29.7 mg/kg dry soil). The highest and lowest MBN were also observed in treatment NPKMOI (42 mg/kg, 36 mg/kg) and CK (21 mg/kg, 19 mg/kg) both in June and October, respectively (Table S2).

Bacterial community composition

Across all soil samples, we obtained 262,299 quality sequences in total, with 4,366–12,538 sequences per sample (mean = 7,284) and were able to classify 86.5% of those sequences at the phylum level. The dominant phyla across all samples were *Proteobacteria*, *Acidobacteria*, *Chloroflexi*, *Bacteroidetes*, *Actinobacteria*, *Gemmatimonadetes*, *Verrucomicrobia* and *Nitrospira* (>1%), accounting for more than 74% of the bacterial sequences from each of the soils (Table S3; Table S4). In addition, *WS3*, *Firmicutes*, *Armatimonadetes*, *TM7*, *Planctomycetes* and *Chlorobi* were in low abundance, and 10 other rare phyla were identified (Table S3).

No significant ($P < 0.05$) interaction between treatment and time was observed for abundant phyla (>1%) (Table S4), while the phylum distribution fluctuated under different fertilizer treatments and seasons. *Chloroflexi*, *Verrucomicrobia* and *Nitrospira* were significantly ($P < 0.001$) affected by sample time, while only phylum *Gemmatimonadetes* was significantly ($P < 0.05$) affected by fertilizer treatments. In addition, phyla *Acidobacteria*, *Bacteroidetes* and were affected by both treatment and time. No significant differences of

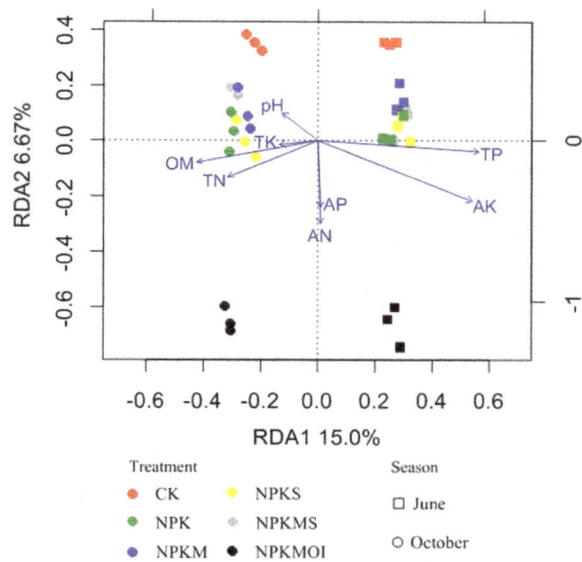

Figure 2. Redundancy analysis of soil bacterial communities and soil characteristics for individual samples. Samples from different fertilizer regimes both in June (square) and October (circle) were marked by different colors. Symbols of fertilizer regimes were as described in Table 1.

Proteobacteria and *Actinobacteria* were observed between different treatments, seasons and the interaction terms.

The relative abundance of abundant bacterial classes (subgroups) demonstrating a significant difference under different fertilizer treatments and times is outlined in Table 3. Likewise, no significant ($P<0.05$) interactions between treatment and time were observed for abundant classes (subgroups). Specifically, we observed that both treatment and time had a significant ($P<0.05$) effect on the *α-Proteobacteria*, *β-Proteobacteria*, and *γ-Proteobacteria*, while the phylum *Proteobacteria* was not affected by treatment, time or the interaction term. Moreover, the relative abundances of different proteobacterial classes were different in response to different fertilizer regimes or seasons. Classes *β-Proteobacteria* and *γ-Proteobacteria* were higher in June, while the *α-Proteobacteria* was higher in October (5.7%) than in June (4.5%). The treatment NPKMOI cultures higher relative abundances of *α-Proteobacteria* and *γ-Proteobacteria*, while the *β-Proteobacteria* was higher in treatment NPK, NPKM, NPKS. Within the *Acidobacteria*, subgroup 3 was only affected by time, while subgroups 4 and 6 were affected by treatment. This is not consistent with the phylum *Acidobacteria*, which is affected by both treatment and time. We also observed that the relative abundance of *Acidobacteria* subgroup 3 was higher in October than in June, which was opposite in the phylum level. In addition, *Anaerolineae* and *Sphingobacteria* were the most abundant subgroups in phyla *Chloroflexi* and *Bacteroidetes*, occupying the majority relative abundance in each phylum, respectively. So the effects of the treatment, time and interaction term were consistent at the phylum and class level. However, we also found that the phylum *Bacteroidetes* was significant ($P<0.05$) affected by both treatment and time, while the subgroups *Sphingobacteria* was only affected by time.

Bacterial α-diversity

Bacterial diversity and richness in individual samples under different fertilizer regimes of both seasons were calculated based on 4,000 sequences (Table S5). Statistically significant differences

in richness and diversity ($P≤0.01$) were observed only at the season level for observed OTUs, coverage, ACE, Chao1 and Simpson but not for Shannon. The number of OTUs, ACE, Chao1 and Simpson were higher in October in comparison to June, while the percentage showed the opposition pattern, being higher in June (80%) than October (76%). Although the fertilizer regimes had no significant effect on the bacterial diversity and richness, a higher number of observed OTUs, ACE, Chao1 and Shannon were observed with treatment NPKMS and NPKMOI in both seasons, indicating that rational fertilization can help increase the biodiversity in agricultural soils. Further analysis of rarefaction curves demonstrated that the number of observed OTUs for treatment NPKMS and NPKMOI were significantly higher than for other fertilizer treatments in June and treatment NPKMOI had a highest observed OTU numbers in October (Fig. S1), calculated based on 10,000 sequences randomly selected from the pooled sequences of each treatment in each season.

Bacterial community structure

A permutational multivariate analysis of variance confirmed that treatment and time were significant factors of variation for the composition of bacterial community in terms of both the relative abundance of OTUs and relative abundance of phylum (Table 4). No significant interaction between treatment and time were observed. Hierarchical cluster analysis of the similarity of bacterial community further confirmed that sampling time is the major determinant of bacterial community structure (Fig. 1). We also found that the treatment NPKMOI was separated from other treatments in both seasons, indicating this fertilizer treatment may have the greatest effect on the bacterial community.

The Mantel test showed significant ($r = 0.44$, $P<0.001$) correlation between bacterial community and soil properties. The RDA (redundancy analysis) biplot clearly showed the relationship between soil characteristics and bacterial community structure (Fig. 2). The first two axes of RDA can explain 15.0% and 6.67% of the total variation in the data. The June bacterial communities were grouped separately from the October bacterial communities along the first axis, with the June bacterial communities more relating to higher contents of soil AK and TP, while the October bacterial communities were associated with higher contents of soil pH, OM, TK and TN. Along the second RDA axis, the bacterial communities of treatment NPKMOI and CK were separated from the other fertilizer treatments in both seasons. The treatments NPK, NPKM, NPKS and NPKMS were grouped together in both seasons, indicating that they may have approximately the same effect on the bacterial assembles. The effect of soil properties on bacterial community is shown by the direction and the length of the vectors. The results showed that the soil OM, TP, AK and TN had a significant ($P<0.01$) correlation between each variable and the ordination scores.

In addition, we used Pearson's correlation coefficient to evaluate relationships between abundant phyla and classes (subgroups) (Table 5). Most of the abundant phyla or classes were positively or negatively correlated with soil OM, AK, TN and TP, while only phylum *Verrucomicrobia* showed a significant positive correlation with soil pH and TN. We also found that soil AP content had no significant correlation with any of the abundant taxa. The relative abundances of phylum *Proteobacteria* and class *γ-Proteobacteria* were positively correlated with soil AN and AK, while the *γ-Proteobacteria* was positively correlated with soil TP. The abundance of *Acidobacteria* was negatively correlated with soil OM, AN, TN and TP while positively correlated with soil AK content.

Table 5. Pearson's correlation coefficient between soil characteristics and abundant phyla (classes).

Phylum (class)	Correlation							
	pH	Organic matter	Available N	Available P	Available K	Total N	Total P	Total K
Proteobacteria	0.203	0.091	**0.393**[*]	0.221	**0.479**[**]	0.178	0.327	0.100
Alphaproteobacteria	0.159	0.220	0.042	0.214	−0.156	0.116	−0.282	−0.003
Betaproteobacteria	0.096	−0.103	0.066	−0.083	0.197	−0.041	0.282	0.058
Deltaproteobacteria	−0.046	−0.023	0.056	−0.160	0.101	−0.032	0.187	−0.018
Gammaproteobacteria	−0.079	−0.152	**0.407**[*]	0.269	**0.680**[**]	0.064	**0.472**[**]	0.030
Acidobacteria	−0.304	**−0.451**[**]	**−0.455**[**]	−0.209	**0.420**[*]	**−0.418**[*]	**−0.541**[**]	−0.237
Acidobacteria_subgroups6	0.060	−0.146	−0.127	−0.182	0.025	−0.172	0.151	−0.060
Acidobacteria_subgroups4	−0.212	**−0.673**[**]	−0.107	−0.105	**0.644**[**]	**−0.569**[**]	**0.787**[**]	−0.262
Acidobacteria_subgroups7	−0.155	−0.130	−0.277	−0.193	−0.070	−0.159	0.079	0.066
Acidobacteria_subgroups3	−0.042	0.261	−0.093	0.036	−0.292	0.288	**−0.343**[*]	−0.063
Chloroflexi	0.035	**0.452**[**]	−0.145	−0.153	**−0.753**[**]	**0.342**[*]	**−0.720**[**]	0.019
Bacteroidetes	−0.138	**−0.639**[**]	0.160	0.175	**0.836**[**]	**−0.461**[**]	**0.898**[**]	−0.220
Actinobacteria	0.093	−0.179	0.118	0.197	0.170	−0.201	0.046	−0.037
Gemmatimonadetes	−0.272	0.120	0.102	0.108	−0.240	0.041	−0.248	0.021
Verrucomicrobia	**0.354**[*]	**0.368**[*]	−0.095	0.120	**−0.580**[**]	0.261	**−0.591**[**]	**0.550**[**]
Nitrospira	−0.152	**−0.568**[**]	−0.011	0.073	**0.782**[**]	**−0.447**[**]	**0.838**[**]	−0.202

Values in boldface type indicate significant correlations with P values indicated in superscript.
*indicate significant correlations (P<0.05); ** indicate significant correlations (P<0.01).

Linking the bacterial communities to soil characteristics, sample time and fertilizer regime

Variance partitioning analysis (VPA) was used to determine the relative contributions of sample time, fertilizer regime and soil characteristics to the bacterial communities. A subset of environmental parameters (OM, TP, AK and TK) was selected by the BioEnv procedure, which provides the highest Pearson correlation with bacterial communities. The variation in bacterial community structure was partitioned among soil characteristics, sample time and fertilizer regime, as well as interactions between them. These variables explained 23.58% of the observed variation, leaving 76.42% of the variation unexplained (Fig. 3). Soil characteristics and sample time explained small portions of the observed variation, which accounted for 0.67% ($P = 0.18$) and 0.33% ($P = 0.255$), respectively, while the fertilizer regime could explain 7.43% ($P = 0.005$) of the total variation. The variation was mostly explained by interactions between soil characteristics and sample time, which accounted for 14.79%. The interactions between soil characteristics and fertilizer regime and sample time and fertilizer regime accounted for 2.81% and 0.40% of the variation, respectively. Thus, the fertilizer regimes are important factors in explaining the shifts in bacterial community structures.

Discussion

The present study attempts to assess the effect of fertilizer regime and sampling time on bacterial communities in a rice-wheat cropping system. The results demonstrate that both fertilizer treatment and season were found to impact soil chemical parameters, microbial biomass, bacterial abundance and bacterial community structure. Soil pH was lower in treatment NPK when compared to other fertilizer treatments and CK (Table 1). This might be due to the application of chemical fertilizer alone, which could acidify the soil [51], while the application of crop straw or

pig manure in combination with chemical fertilizer can stabilize the soil pH [52]. We observed that soil available nutrients (AN, AP and AK) were significantly altered by fertilizer treatment in comparison to CK, with the treatment NPKMOI having the highest concentration of soil available nutrients. Application of organic-inorganic compound fertilizer can slowly release of nutrients into the soil, promote plant growth and increase crop yields [53]. However, the treatment NPKMOI did not show any significant differences in soil AN and AP contents compared to other fertilizer regimes. We also did not observe any significant difference in soil organic matter content not only between different fertilizer treatments but also between treatments and CK. The results of soil characteristics indicated that different fertilizer regimes gradually but not dramatically changed soil attributes.

Soil MBC and MBN were found to be higher in June than that in October, suggesting that the soil MBC and MBN contents may be affected by the atmospheric temperature. This result is consistent with previous studies [10,54]. Kikuchi et al. [55] and Noll et al. [56] found that bacterial community structure and diversity in a paddy soil can be affected by flooded/upland conditions and oxygen gradient. So we consider that the flooding period during rice cultivation in our study is also an important factor in soil microbial biomass shifting. We also found that the application of fertilizer can markedly affect the soil MBC and MBN both in June and October when compared with CK. This suggests the different fertilizers can also activate the microorganisms in soil.

In this study, we applied 454-pyrosequencing of the 16S rRNA gene to characterize the bacterial community under different fertilizer treatments and sampling times. The sequence analysis reveals that phyla Proteobacteria, Acidobacteria, Chloroflexi and Bacteroidetes were the most abundant phyla in all of the samples. This result roughly corresponded with previous studies demonstrating that Proteobacteria, Acidobacteria and Bacteroidetes are domi-

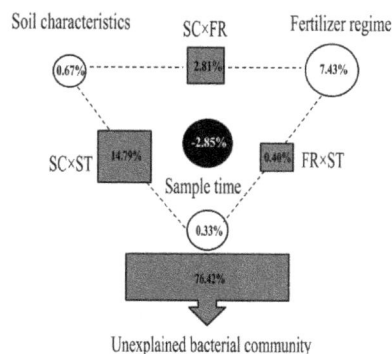

Figure 3. Variation partitioning analysis of the bacterial communities. Effects of soil characteristics (SC), fertilizer regime (FR), sample time (ST) and the interactions between them on the bacterial community structure. Circles on the edges of the triangle show the percentage of variation explained by each factor alone. The percentage of variation explained by interactions between two or three of the factors is shown as squares on the sides and as circle in the center of the triangle. The unexplained variation is depicted in square on the bottom.

nant soil bacterial taxa using sequencing of 16S rRNA gene clone libraries or pyrosequencing [57,58], while the phylum *Chloroflexi* showed an unexpectedly high relative abundance in our study. Specifically, a higher average abundance of *Chloroflexi* was observed in October (12.37%) than in June (8.66%) (Table S3), which suggests that the practice of periodic flooding (resulting in an anaerobic environment) in the rice season can significantly impact this facultative anaerobic phylum. Significant differences in soil bacterial community composition were observed both among fertilizer treatments and between sampling times. In particular, the highest relative abundance of α-*Proteobacteria* and γ-*Proteobacteria*, but the lowest of β-*Proteobacteria,* was observed in treatment NPKMOI in both seasons. β-*Proteobacteria* are considered as copiotrophic bacteria and always flourish in soils with large amounts of available nutrients [59], while the available nutrients in the treatment NPKMOI are higher than other fertilizer treatments and CK. Therefore, lower relative abundance of β-*Proteobacteria* in the treatment NPKMOI could be due to other factors. Interestingly, the phylum and its classes or subgroups had different patterns in response to the different treatments and seasons. For instance, classes α-*Proteobacteria*, β-*Proteobacteria* and γ-*Proteobacteria* were all affected by both treatment and season, while the phylum *Proteobacteria* was not affected by treatment, season or the interaction between fertilizer and season. In contrast, the phylum *Acidobacteria* was affected by both treatment and season, subgroups 6, 7 and 3 were only affected by treatment or season. This indicates that finer taxa may have more sensitive responses to different treatments and seasons due to its own characteristics.

Many studies have shown that environmental factors determine the soil microbial community [5,60,61], especially the soil pH, which has been demonstrated in several studies to be the strongest factor shaping microbial community structures [60,62–64]. Chu et al. [65] and Rousk et al. [60] have reported that the relative abundances of α-*Proteobacteria* and γ-*Proteobacteria* increase with higher soil pH, but we did not find the same trend, which may be due to the mild fluctuations (range from 6.86 to 7.37 in all samples) in the soil pH in our study. Jones et al. [66] reported that relative abundances of Acidobacterial subgroups 4, 6 and 7 were positively correlated with soil pH, which appears inconsistent with our results, being higher in treatment CK than that in treatment

NPKMOI. The soil pH values varied significantly from 3 to 9 in those studies allowing insight into the relationship between pH and soil bacterial communities besides that the soil factors assessed in this study were not measured in the studies of Chu et al., [65], Roush et al., [60] and Jones et al., [66]. Thus, we hypothesize that there are important factors other than pH in shaping soil bacterial communities, which was similar with Navarrete et al., [67], demonstrating that abiotic soil factors (not only related to soil acidity) such as Al, Ca, Mg, Mn and B can also drive the acidobacterial populations. Further Pearson correlation analysis showed that only phylum *Verrucomicrobia* had a small correlation with soil pH. In contrast, we found that most of the proportions of abundant phyla and classes (subgroups) were highly correlated with soil OM, AK, TN and TP contents (Table 5). Lauber et al. [62] considered that the soil pH may not directly alter bacterial community structure but may instead function as a variable that provides an integrated index of soil conditions. In fact, there are a number of soil characteristics (e.g., nutrient availability, organic C characteristics, soil moisture regimen and salinity) that are often directly or indirectly related to soil pH [68], and these factors may drive the observed changes in community composition as the hydrogen ion concentration varies by many orders of magnitude across the soils [60,62–64]. However, the soil pH may not represent an integrating variable in our study due to the mild variation of soil pH, which suggests that soil characteristics such as soil nutrients, organic matter, soil moisture, etc. rather than soil pH can shape the bacterial communities by themselves. In addition, Nacke et al. [5] found that the phylum *Actinobacteria* was positively correlated with TN and class δ-*Proteobacteria* was negatively correlated with organic C. However, these two bacterial taxa had no significant correlation with any soil properties in our study (Table 5). Hence, we hypothesize that soil nutrients can also drive the bacterial community composition and that the correlation patterns are varied in different soils.

Soil microbial diversity is considered to be critical to the integrity, function and long-term sustainability of soil ecosystems [69]. Moreover, greater biodiversity in soil can lead to a more stable system and enhance the combination of vital microbial functions and processes [9]. In the present study, higher biodiversity was observed in treatment NPKMOI at both seasons, indicating that combined application of chemical fertilizer and organic-inorganic compound fertilizer can maintain a higher biodiversity. It may result in a more stable agroecosystem and may contribute to sustainable crop production.

Considerable temporal variation in soil bacterial communities has been previously been reported for agricultural soils [24,26,27]. In the present study, the greatest community variation was also found to be temporal (Fig. 1). All the soil samples showed temporal clustering and higher similarity was observed within each season. Bremer et al. [70] considered that plant presence may result in seasonal dynamics of bacterial communities. In our study, we collected soil samples after crop harvest to avoid this bias. However, we also found that within each season, the bacterial community structure of treatment NPKMOI differed from other treatments or CK (Fig. 1 and Fig. 2), indicating that fertilization is also an important factor in shaping bacterial community. This was verified by the permutational multivariate analysis of variance of the bacterial community (Table 4). In addition, soil properties were also significantly correlated with bacterial communities in the present study. Further analysis revealed that sampling time could only explain a smaller variation (0.33%) in the total bacterial community. In contrast, fertilizer regime can explain a greater variation (7.43%) than sampling time and soil properties.

Moreover, only fertilizer regime can significantly ($P = 0.005$) explain the variation in bacterial community structure.

In conclusion, fertilizer practice and seasonal changes can affect soil properties, microbial biomass, and bacterial community structure. We observed a significant and distinct seasonal shift for bacterial community while the fertilizer type also lead to a significant but much smaller variation in bacterial community structure. The treatment NPKMOI significantly increased the microbial biomass and biodiversity and may be more suitable for sustainable crop production.

Further study should pay more attention to the impact of the changes of bacterial community induced by fertilization and seasonal fluctuations on soil functionality.

Supporting Information

Figure S1 Rarefaction curves of bacterial communities for June (A) and October (B) based on the number of observed OTUs at 3% distance calculated from the randomly selected 10,000 pooled sequences of each treatment. Error bars indicate 95% confidence intervals.

Table S1 Soil available K of all samples from different fertilizer regimes both in June and October.

Table S2 Soil microbial biomass C (MBC) and microbial biomass N (MBN) of all samples from different fertilizer regimes both in June and October.

Table S3 Relative average abundances of phyla across all soils and soils grouped into fertilizer regime and sampling time categories, respectively (values represent % of total non-redundant sequences). Asterisks mean values less than 0.01.

Table S4 Effects of fertilizer regime, sample time and the interaction between them on the abundant phyla (relative abundance >1%).

Table S5 Effects of fertilizer regime, sample time and the interaction between them on the OTUs, coverage, richness and diversity calculated using a random 4,000 sequences per sample.

Acknowledgments

We would like to thank the Soil and Fertilizer Technical Guidance Station of Jintan City for managing the field experiment. We also gratefully acknowledge the anonymous reviewers for their helpful and constructive comments.

Author Contributions

Conceived and designed the experiments: WR BS RZ QS. Performed the experiments: JZ TN WX YL. Analyzed the data: JZ. Contributed reagents/materials/analysis tools: RZ QS. Wrote the paper: JZ RZ QS.

References

1. Brodie EL, DeSantis TZ, Parker JPM, Zubietta IX, Piceno YM, et al. (2007) Urban aerosols harbor diverse and dynamic bacterial populations. PNAS 104: 299–304.
2. Nannipieri P, Ascher J, Ceccherini M, Landi L, Pietramellara G, et al. (2003) Microbial diversity and soil functions. Eur J Soil Sci 54: 655–670.
3. Gans J, Wolinsky M, Dunbar J (2005) Computational Improvements Reveal Great Bacterial Diversity and High Metal Toxicity in Soil. Science 309: 1387–1390.
4. Glick BR (1995) The enhancement of plant growth by free-living bacteria. Can J Microbiol 41: 109–117.
5. Nacke H, Thürmer A, Wollherr A, Will C, Hodac L, et al. (2011) Pyrosequencing-based assessment of bacterial community structure along different management types in German forest and grassland soils. PLoS One 6: e17000.
6. Bell T, Newman JA, Silverman BW, Turner SL, Lilley AK (2005) The contribution of species richness and composition to bacterial services. Nature 436: 1157–1160.
7. Naeem S, Li S (1997) Biodiversity enhances ecosystem reliability. Nature 390: 507–509.
8. Cardinale BJ, Srivastava DS, Duffy JE, Wright JP, Downing AL, et al. (2006) Effects of biodiversity on the functioning of trophic groups and ecosystems. Nature 443: 989–992.
9. Chaer G, Fernandes M, Myrold D, Bottomley P (2009) Comparative resistance and resilience of soil microbial communities and enzyme activities in adjacent native forest and agricultural soils. Microb Ecol 58: 414–424.
10. Gu Y, Zhang X, Tu S, Lindström K (2009) Soil microbial biomass, crop yields, and bacterial community structure as affected by long-term fertilizer treatments under wheat-rice cropping. Eur J Soil Bio 45: 239–246.
11. Saha S, Prakash V, Kundu S, Kumar N, Mina BL (2008) Soil enzymatic activity as affected by long term application of farm yard manure and mineral fertilizer under a rainfed soybean–wheat system in NW Himalaya. Eur J Soil Bio 44: 309–315.
12. Islam MR, Chauhan PS, Kim Y, Kim M, Sa T (2011) Community level functional diversity and enzyme activities in paddy soils under different long-term fertilizer management practices. Biol Fert Soils 47: 599–604.
13. Chu H, Lin X, Fujii T, Morimoto S, Yagi K, et al. (2007) Soil microbial biomass, dehydrogenase activity, bacterial community structure in response to long-term fertilizer management. Soil Biol Biochem 39: 2971–2976.
14. Yu WT, Bi ML, Xu YG, Zhou H, Ma Q, et al. (2013) Microbial biomass and community composition in a Luvisol soil as influenced by long-term land use and fertilization. CATENA.
15. Esperschütz J, Gattinger A, Mäder P, Schloter M, Fließbach A (2007) Response of soil microbial biomass and community structures to conventional and organic farming systems under identical crop rotations. FEMS Microbiol Ecol 61: 26–37.
16. Orr CH, Leifert C, Cummings SP, Cooper JM (2012) Impacts of Organic and Conventional Crop Management on Diversity and Activity of Free-Living Nitrogen Fixing Bacteria and Total Bacteria Are Subsidiary to Temporal Effects. PLoS One 7: e52891.
17. Ebhin Masto R, Chhonkar P, Singh D, Patra A (2006) Changes in soil biological and biochemical characteristics in a long-term field trial on a sub-tropical inceptisol. Soil Biol Biochem 38: 1577–1582.
18. Lv M, Li Z, Che Y, Han F, Liu M (2011) Soil organic C, nutrients, microbial biomass, and grain yield of rice (Oryza sativa L.) after 18 years of fertilizer application to an infertile paddy soil. Biol Fert Soils 47: 777–783.
19. Ge G, Li Z, Fan F, Chu G, Hou Z, et al. (2010) Soil biological activity and their seasonal variations in response to long-term application of organic and inorganic fertilizers. Plant Soil 326: 31–44.
20. Plaza C, Hernandez D, Garcia-Gil J, Polo A (2004) Microbial activity in pig slurry-amended soils under semiarid conditions. Soil Biol Biochem 36: 1577–1585.
21. Ramirez KS, Craine JM, Fierer N (2012) Consistent effects of nitrogen amendments on soil microbial communities and processes across biomes. Global Change Biol 18: 1918–1927.
22. Zhong W, Cai Z (2007) Long-term effects of inorganic fertilizers on microbial biomass and community functional diversity in a paddy soil derived from quaternary red clay. Applied Soil Ecol 36: 84–91.
23. Griffiths RI, Whiteley AS, O'Donnell AG, Bailey MJ (2003) Influence of depth and sampling time on bacterial community structure in an upland grassland soil. FEMS Microbiol Ecol 43: 35–43.
24. Schutter M, Sandeno J, Dick R (2001) Seasonal, soil type, and alternative management influences on microbial communities of vegetable cropping systems. Biol Fert Soils 34: 397–410.
25. Kennedy NM, Gleeson DE, Connolly J, Clipson NJ (2005) Seasonal and management influences on bacterial community structure in an upland grassland soil. FEMS Microbiol Ecol 53: 329–337.
26. Hannula SE, de Boer W, van Veen J (2012) A 3-year study reveals that plant growth stage, season and field site affect soil fungal communities while cultivar and GM-trait have minor effects. PLoS One 7: e33819.
27. Kennedy N, Brodie E, Connolly J, Clipson N (2006) Seasonal influences on fungal community structure in unimproved and improved upland grassland soils. Canadian J Microbiol 52: 689–694.

28. Bending GD, Turner MK, Rayns F, Marx MC, Wood M (2004) Microbial and biochemical soil quality indicators and their potential for differentiating areas under contrasting agricultural management regimes. Soil Biol Biochem 36: 1785–1792.

29. Sparling G, Pankhurst C, Doube B, Gupta V (1997) Soil microbial biomass, activity and nutrient cycling as indicators of soil health. Biological indicators of soil health: 97–119.

30. Acosta-Martinez V, Dowd S, Sun Y, Allen V (2008) Tag-encoded pyrosequencing analysis of bacterial diversity in a single soil type as affected by management and land use. Soil Biol Biochem 40: 2762–2770.

31. Singh BK, Campbell CD, Sorenson SJ, Zhou J (2009) Soil genomics. Nat Rev Microbiol 7: 756–756.

32. Brookes P, Landman A, Pruden G, Jenkinson D (1985) Chloroform fumigation and the release of soil nitrogen: a rapid direct extraction method to measure microbial biomass nitrogen in soil. Soil Biol Biochem 17: 837–842.

33. Vance E, Brookes P, Jenkinson D (1987) An extraction method for measuring soil microbial biomass C. Soil Biol Biochem 19: 703–707.

34. Joergensen RG, Mueller T (1996) The fumigation-extraction method to estimate soil microbial biomass: Calibration of the k_{EN} value. Soil Biol Biochem 28: 33–37.

35. Joergensen RG (1996) The fumigation-extraction method to estimate soil microbial biomass: Calibration of the k_{EC} value. Soil Biol Biochem 28: 25–31.

36. Schloss PD, Westcott SL, Ryabin T, Hall JR, Hartmann M, et al. (2009) Introducing mothur: Open-Source, Platform-Independent, Community-Supported Software for Describing and Comparing Microbial Communities. Appl Environ Microbiol 75: 7537–7541.

37. Schloss PD, Gevers D, Westcott SL (2011) Reducing the Effects of PCR Amplification and Sequencing Artifacts on 16S rRNA-Based Studies. PloS one 6: e27310.

38. Quince C, Lanzen A, Davenport RJ, Turnbaugh PJ (2011) Removing noise from pyrosequenced amplicons. BMC Bioinformatics 12: 38.

39. Pruesse E, Quast C, Knittel K, Fuchs BM, Ludwig W, et al. (2007) SILVA: a comprehensive online resource for quality checked and aligned ribosomal RNA sequence data compatible with ARB. Nucleic Acids Res 35: 7188–7196.

40. Edgar RC, Haas BJ, Clemente JC, Quince C, Knight R (2011) UCHIME improves sensitivity and speed of chimera detection. Bioinformatics 27: 2194–2200.

41. Wang Q, Garrity GM, Tiedje JM, Cole JR (2007) Naïve Bayesian Classifier for Rapid Assignment of rRNA Sequences into the New Bacterial Taxonomy. Appl Environ Microbiol 73: 5261–5267.

42. Chao A (1984) Nonparametric estimation of the number of classes in a population. Scand J Stat: 265–270.

43. Eckburg PB, Bik EM, Bernstein CN, Purdom E, Dethlefsen L, et al. (2005) Diversity of the human intestinal microbial flora. Science 308: 1635–1638.

44. Washington H (1984) Diversity, biotic and similarity indices: a review with special relevance to aquatic ecosystems. Water Res 18: 653–694.

45. Simpson EH (1949) Measurement of diversity. Nature 163: 688.

46. Bunge J, Fitzpatrick M (1993) Estimating the number of species: a review. J Am Stat Assoc: 364–373.

47. Oksanen J, Kindt R, Legendre P, O'Hara B, Simpson GL, et al. (2009) Vegan: Community ecology package. R package version 1.15–2.

48. R Development Core Team (2012) R: A Language and Environment for Statistical Computing. Vienna, Austria: R Foundation for Statistical Computing.

49. Anderson MJ (2001) A new method for non-parametric multivariate analysis of variance. Austral Ecol 26: 32–46.

50. Legendre P, Gallagher ED (2001) Ecologically meaningful transformations for ordination of species data. Oecologia 129: 271–280.

51. van Diepeningen AD, de Vos OJ, Korthals GW, van Bruggen AH (2006) Effects of organic versus conventional management on chemical and biological parameters in agricultural soils. Appl Soil Ecol 31: 120–135.

52. Li R, Khafipour E, Krause DO, Entz MH, de Kievit TR, et al. (2012) Pyrosequencing reveals the influence of organic and conventional farming systems on bacterial communities. PloS one 7: e51897.

53. Chunyu S, Fudao Z, Shuqing Z, Hui L, Chenggao F (2004) Effects of organic-inorganic slow release fertilizers on yield and nitrogen recovery in tomato. Plant Nutrition and Fertilizer Science 10: 584.

54. Diaz-Ravina M, Acea M, Carballas T (1995) Seasonal changes in microbial biomass and nutrient flush in forest soils. Biol Fert Soils 19: 220–226.

55. Kikuchi H, Watanabe T, Jia Z, Kimura M, Asakawa S (2007) Molecular analyses reveal stability of bacterial communities in bulk soil of a Japanese paddy field: estimation by denaturing gradient gel electrophoresis of 16S rRNA genes amplified from DNA accompanied with RNA. Soil Sci Plant Nutr 53: 448–458.

56. Noll M, Matthies D, Frenzel P, Derakshani M, Liesack W (2005) Succession of bacterial community structure and diversity in a paddy soil oxygen gradient. Environ Microbiol 7: 382–395.

57. Roesch LF, Fulthorpe RR, Riva A, Casella G, Hadwin AK, et al. (2007) Pyrosequencing enumerates and contrasts soil microbial diversity. The ISME J 1: 283–290.

58. Janssen PH (2006) Identifying the dominant soil bacterial taxa in libraries of 16S rRNA and 16S rRNA genes. Appl Environ Microbiol 72: 1719–1728.

59. Fierer N, Bradford MA, Jackson RB (2007) Toward an ecological classification of soil bacteria. Ecology 88: 1354–1364.

60. Rousk J, Bååth E, Brookes PC, Lauber CL, Lozupone C, et al. (2010) Soil bacterial and fungal communities across a pH gradient in an arable soil. The ISME J 4: 1340–1351.

61. Lauber CL, Strickland MS, Bradford MA, Fierer N (2008) The influence of soil properties on the structure of bacterial and fungal communities across land-use types. Soil Biol Biochem 40: 2407–2415.

62. Lauber CL, Hamady M, Knight R, Fierer N (2009) Pyrosequencing-based assessment of soil pH as a predictor of soil bacterial community structure at the continental scale. Appl Environ Microbiol 75: 5111–5120.

63. Xiong J, Liu Y, Lin X, Zhang H, Zeng J, et al. (2012) Geographic distance and pH drive bacterial distribution in alkaline lake sediments across Tibetan Plateau. Environ Microbiol 14: 2457–2466.

64. Shen C, Xiong J, Zhang H, Feng Y, Lin X, et al. (2012) Soil pH drives the spatial distribution of bacterial communities along elevation on Changbai Mountain. Soil Biol Biochem 57:204–211.

65. Chu H, Fierer N, Lauber CL, Caporaso J, Knight R, et al. (2010) Soil bacterial diversity in the Arctic is not fundamentally different from that found in other biomes. Environ Microbiol 12: 2998–3006.

66. Jones RT, Robeson MS, Lauber CL, Hamady M, Knight R, et al. (2009) A comprehensive survey of soil acidobacterial diversity using pyrosequencing and clone library analyses. The ISME J 3: 442–453.

67. Navarrete AA, Kuramae EE, Hollander M, Pijl AS, Veen JA, et al. (2012) Acidobacterial community responses to agricultural management of soybean in Amazon forest soils. FEMS Microbiol Ecol 83: 607–621.

68. Brady NC, Weil RR (1996) The nature and properties of soils: Prentice-Hall Inc. 740p.

69. Kennedy A, Smith K (1995) Soil microbial diversity and the sustainability of agricultural soils. Plant Soil 170: 75–86.

70. Bremer C, Braker G, Matthies D, Beierkuhnlein C, Conrad R (2009) Plant presence and species combination, but not diversity, influence denitrifier activity and the composition of nirK-type denitrifier communities in grassland soil. FEMS Microbiol Ecol 70: 377–387.

21

Is the Inherent Potential of Maize Roots Efficient for Soil Phosphorus Acquisition?

Yan Deng[1], Keru Chen[1], Wan Teng[2], Ai Zhan[3], Yiping Tong[2], Gu Feng[1], Zhenling Cui[1], Fusuo Zhang[1], Xinping Chen[1]*

1 Center for Resources, Environment and Food Security, China Agricultural University, Beijing, China, 2 State Key Laboratory for Plant Cell and Chromosome Engineering, Institute of Genetics and Developmental Biology, Chinese Academy of Sciences, Beijing, China, 3 State Key Laboratory of Soil Erosion and Dryland Farming on the Loess Plateau, Institute of Soil and Water Conservation, Chinese Academy of Sciences and Ministry of Water Resource, Yangling, China

Abstract

Sustainable agriculture requires improved phosphorus (P) management to reduce the overreliance on P fertilization. Despite intensive research of root adaptive mechanisms for improving P acquisition, the inherent potential of roots for efficient P acquisition remains unfulfilled, especially in intensive agriculture, while current P management generally focuses on agronomic and environmental concerns. Here, we investigated how levels of soil P affect the inherent potential of maize (Zea mays L.) roots to obtain P from soil. Responses of root morphology, arbuscular mycorrhizal colonization, and phosphate transporters were characterized and related to agronomic traits in pot and field experiments with soil P supply from deficiency to excess. Critical soil Olsen-P level for maize growth approximated 3.2 mg kg^{-1}, and the threshold indicating a significant environmental risk was about 15 mg kg^{-1}, which represented the lower and upper levels of soil P recommended in current P management. However, most root adaptations involved with P acquisition were triggered when soil Olsen-P was below 10 mg kg^{-1}, indicating a threshold for maximum root inherent potential. Therefore, to maintain efficient inherent potential of roots for P acquisition, we suggest that the target upper level of soil P in intensive agriculture should be reduced from the environmental risk threshold to the point maximizing the inherent potential of roots.

Editor: Malcolm Bennett, University of Nottingham, United Kingdom

Funding: This work was supported by the National Natural Science Foundation of China (30890133, 30971872), National Maize Production System in China (CARS-02-24), the Innovative Group Grant of NSFC (30971892), and the Innovation Fund for Graduate Student of China Agricultural University (KYCX2011032). The funders had no role in study design, data collection and analysis, decision to publish or preparation of the manuscript.

Competing Interests: The authors have declared that no competing interests exist.

* E-mail: chenxp@cau.edu.cn

Introduction

Modern high-intensity agriculture strongly relies on phosphorus (P) fertilization [1], [2], but sustainable agriculture requires improved P management. On one hand, there is increasing concern about P scarcity because the world's main source of P fertilizers, phosphate rock, is a limited and non-renewable resource [3], [4], [5], [6]. On the other hand, P fertilization can harm the environment by contributing to the eutrophication of water bodies [7], [8]. Inappropriate P fertilization accelerates the soil P imbalance in croplands worldwide [9].

P is readily fixed in most soils and has low availability to plants [10]. To enhance P acquisition, plants and their root-associated microbes have evolved a series of strategies that include modified root growth and functioning. Common strategies about root growth are increased root/shoot ratio [11], [12], modified root architecture [13], [14], decreased root diameter [15], enhanced specific root length (root length per unit root mass) [16], higher root hair length and/or density [17], [18], and production of aerenchyma [13], [19]. These morphological adaptations can greatly enhance the volume of soil root will exploit, and/or benefit exploitation of P-rich patches [20]. Also associations with arbuscular mycorrhizal (AM) fungi greatly extend the soil exploration space beyond the roots for many higher plant species [21]. Besides increasing soil volume exploited, roots and associated microbes can increase P availability from touched inorganic and organic sources by enhancing synthesis and exudation of organic acids and phosphatases [15], [22]. Increased P-uptake capacity by enhancing expression of high-affinity phosphate (Pi) transporters is another typical response of root functioning to facilitate P acquisition [23].

Current P management in intensive agriculture focuses on agronomic and environmental concerns, aiming to maintain soil P level between critical values that maximize crop yield but minimize P loss [24]. Most research, which explores the inherent potential of roots for efficient P acquisition, has focused on the adaptive mechanisms of P-efficient plants under P deficiency in natural ecosystems or low-yielding agricultural systems [13], [25], [26], [27]. In intensive agriculture, where soil P supply is increased by fertilization, the inherent potential of roots for efficient P acquisition is unfulfilled. In addition, interactions of P between soil and plants have often been studied in controlled and short term experiments not representative of field cropping systems.

Here, we hypothesized that the inherent potential of roots can be manipulated by managing soil P level to achieve a soil P supply that maximizes root uptake of P, optimizes crop yield and minimizes P loss. In pot and field experiments, we investigated root morphology, AM colonization, and expression of Pi transporter genes in maize (Zea mays L.) with different levels of P supply. The responses of these root traits were related to agronomic traits, and the optimum soil P supply was estimated.

Materials and Methods

Pot experiment

A pot experiment was carried out in the glasshouse at China Agricultural University from April to June, 2011. Maize (NE15, a test-cross variety) plants were planted in a calcareous silt loam soil which was collected from the same experimental site that was used for the field study (see next section). The initial soil properties were: pH 8.35 (1 : 5 soil : water ratio), organic matter 7.09 g kg^{-1}, total N 0.51 g kg^{-1}, Olsen-P 1.19 mg kg^{-1}, and exchangeable K 90 mg kg^{-1}. Seven P application rates (0, 12.5, 15, 50, 75, 100 and 300 mg P kg^{-1} soil) were used with P added as calcium superphosphate. In addition, N and K (as urea and potassium sulfate, respectively) were each applied at 150 mg kg^{-1} soil. Soil was air-dried and ground to pass through a 2-mm sieve. The added nutrients were mixed well with 8 kg of soil and filled into 4.5-L pots, with six replicate pots per treatment. Maize seeds were surface sterilized (30 min in a 10% H_2O_2 solution), rinsed, imbibed (8 h in a saturated $CaSO_4$ solution), and germinated in a dark and humid environment for 24 h. Two germinated seeds were planted per pot, and the seedlings were thinned to one per pot at the 3-leaf stage.

Plants were harvested 56 days after planting (DAP) at the 8-leaf stage. In each treatment, three pots were randomly chosen for shoot, root and soil sampling. Shoots were removed, oven-dried at 60°C for three days, weighed, and ground for nutrient analysis. All visible roots in each pot were carefully picked out by hand and stored in an ice box before transferring to the laboratory, after which soil samples were taken. In the laboratory root samples were carefully cleaned with tap water and frozen at −20°C before measurement of root morphology and AM colonization. Soil samples were air-dried and ground to pass through a 1-mm sieve for analysis of soil Olsen-P and CaCl$_2$-P. The remaining three pots in each treatment (except pots treated with 75 mg P kg^{-1} soil) were sampled for analysis of expression of Pi transporters in roots. Plant roots were gently taken out and immediately washed and then stored in liquid nitrogen for later RNA extraction.

Field experiment

The field experiment was conducted at the Shangzhuang Experimental Station of China Agricultural University (40°8′27″N, 116°10′39″E) in Beijing. This site is located in the northern North China Plain and has a typical semi-humid monsoon climate of the warm temperate zone. The annual average temperature ranges from 11 to 13°C, and annual rainfall ranges from 480 to 580 mm with precipitation mainly occurring from June to August. Annual mean sunshine is 2750 h, and 180–200 days are frost-free. Similar to the pot experiment, the soil in the field experiment was silt loam and with these properties at 0–30 cm depth: pH 8.00 (1 : 5 soil : water ratio), organic matter 8.02 g kg^{-1}, total N 0.37 g kg^{-1}, Olsen-P 1.82 mg kg^{-1}, and exchangeable K 82 mg kg^{-1}.

The treatments consisted of eight P application rates (0, 12.5, 25, 50, 75, 100, 150 and 300 kg P ha^{-1}) with four replicate plots per rate in a randomized complete block design. Each plot had an area of 17.5 m^2 (5 m×3.5 m). Before sowing, the entire quantity of P (as calcium superphosphate), 75 kg N ha^{-1} (as urea), and 62 kg K ha^{-1} (as potassium sulfate) were broadcast and mixed into top 20 cm of soil by disking. At 6-leaf and 13-leaf stages, an additional 150 kg N ha^{-1} as urea was top-dressed to each plot, with 75 kg N ha^{-1} each time. The same maize variety as that in

pot experiment was used and seeds were planted at 67,500 plants ha^{-1} on 6 June, 2011.

At the flowering stage (57 DAP), three plants per plot were harvested from all plots and separated into shoots and roots. The shoot samples were used for determination of shoot dry weight and P content. Root samples were washed for assessment of AM colonization. Root samples for analysis of morphological traits were collected using the monolith method (see below) in all replicates of the following six P treatments: 0, 12.5, 25, 50, 100 and 300 kg P ha^{-1}. With the shoot at the center, each monolith measured 40 cm perpendicular to the rows (row spacing was 60 cm)×20 cm parallel with the rows (plant spacing was 25 cm)×30 cm deep. Each monolith was subdivided into three depths (0–10, 10–20, and 20–30 cm). All visible roots in each soil layer were carefully collected by hand and stored in an ice box before transport to the laboratory for washing. Additional roots for analysis of Pi transporter genes were taken from the same plots for root morphology assessment. Roots of three plants per plot were carefully excavated and washed with water, separated from stems, and combined into one sample. The samples were preserved in liquid nitrogen for transport to laboratory. At the grain maturity stage (111 DAP), three plants per plot were harvested for determination of total shoot dry weight and P content. Grain yield was determined by manually harvesting and drying (at 60°C) ears from two rows per plot. At flowering and grain maturity stages, topsoil (0–20 cm) was sampled from each plot for analysis of soil Olsen-P and CaCl$_2$-P.

Sample analysis

We used Olsen-P as the indicator of plant available P in the calcareous soils used here [28] and CaCl$_2$-P as the indicator of P loss risk [29]. Soil Olsen-P level was determined by the molybdo-vanadophosphate method based on extraction of air-dried soil with 0.5 M NaHCO$_3$ (pH 8.5) at 25°C [28]. Soil CaCl$_2$-P was measured by extracting air-dried soil with 0.01 M CaCl$_2$ according to Hesketh and Brookes [29]. Plant P concentration was measured by the molybdo-vanadophosphate method after samples were digested with concentrated H_2SO_4 and H_2O_2 [30]. Plant P uptake was then calculated from plant dry weight and P concentration.

To measure root morphological traits, cleaned root samples were dispersed in water in a transparent array (30 cm×20 cm×2 cm) and imaged with a scanner (Epson Expression 1600, Seiko Epson, Nagano, Japan) at a resolution of 800 dpi. The images were analyzed by WinRhizo software (Regent Instrument Inc., Quebec, QC, Canada) to determine root length. Root dry weight was determined by weighing the oven-dried samples after scanning. Specific root length was calculated from root length and root dry weight, and root/shoot ratio was assessed from root dry weight and shoot dry weight, respectively.

AM colonization was measured in 1-cm fine root segments that had been thoroughly mixed. Roots were cleared with 10% KOH at 90°C for 1 h and stained with Trypan blue [31]. The percent root colonization by AM fungi was assessed by examining 30 randomly selected stained root segments at 100–400 × magnification with a light microscope according to Trouvelot *et al.* [32].

Six Pi transporters of the Pht1 family have been reported for maize: from *ZEAma;Pht1;1* to *ZEAma;Pht1;6* [33]. For simplification, these genes are presented as *ZmPht1;1* to *ZmPht1;6* here. Expression of these Pi transporter genes was analyzed in maize root samples by real-time quantitative RT-PCR (qRT-PCR) [34]. Total RNA was isolated from frozen root tissue using the Trizol reagent (cat. no. 15596018, Invitrogen, USA). The isolated RNA was treated with the RNase-Free DNase Set (cat. no. 79254,

Table 1. Agronomic traits of pot maize plants in response to different P application rates.

P rate (mg P kg^{-1} soil)	Olsen-P (mg kg^{-1})	CaCl$_2$-P (mg kg^{-1})	Shoot dry weight (g plant^{-1})	Shoot P concentration (g kg^{-1})	P uptake (mg plant^{-1})
0	1.21 (0.11)^d	0.15 (0.00)^b	1.9 (0.1)^d	1.06 (0.12)^e	2.1 (0.2)^d
12.5	1.71 (0.12)^d	0.15 (0.00)^b	8.3 (0.5)^c	1.14 (0.12)^{de}	9.4 (0.9)^{cd}
25	2.39 (0.24)^d	0.14 (0.01)^b	12.4 (0.7)^b	1.42 (0.04)^d	17.6 (1.4)^c
50	4.80 (0.88)^c	0.13 (0.01)^b	17.8 (1.3)^a	1.94 (0.11)^c	34.6 (2.5)^b
75	6.86 (0.15)^b	0.13 (0.01)^b	18.1 (1.0)^a	2.24 (0.05)^{bc}	40.7 (3.0)^{ab}
100	7.65 (0.87)^b	0.15 (0.01)^b	18.4 (2.2)^a	2.43 (0.18)^b	45.2 (8.3)^{ab}
300	35.77 (0.88)^a	0.83 (0.07)^b	16.2 (0.7)^a	3.18 (0.13)^a	51.7 (4.3)^a

Plants were sampled at the 8-leaf stage (56 days after planting).
Each value is the mean ($\pm$ SE) of three replicates.
Values in a column followed by different letters are significantly different at $p < 0.05$.

Qigen, Germany) to eliminate genomic DNA contamination before it was cleaned further with the RNeasy Plant Mini Kit (cat. no. 74904, Qigen, Germany). The first-strand cDNA was synthesized using the PrimeScript® RT reagent Kit Perfect Real Time (cat. no. DRR037A, Takara, Dalian) according to the manufacturer's protocol. Then the qRT-PCR was performed on a Mastercycler Realplex4 Real Time PCR System (Eppendorf, Germany) based on the protocol of the SYBR® Premix EX TaqTM (cat. no. DRR041A, Takara) in 20 µl reaction volume, which contained 10 µl of SYBR Green PCR mix, 0.4 µM of each forward and reverse primer, 0.4 µg of diluted cDNA template, and the appropriate amounts of sterile double distilled water. The applied program was set as initial polymerase activation at 95°C,

30 s, then 40 cycles at 95°C, 5 s; 60°C, 35 s. The specificity of the PCR amplification was evaluated with a melt curve analysis from 60°C to 95°C following the final cycle of the PCR. All reactions were set up using four biological replicates. We used UBQ2 as the internal control gene as reported by Calderon-Vazquez et al. [34], and the transcription levels of each gene were normalized to that of UBQ2 by the $2^{-\triangle\triangle Ct}$ method. The sequences of the gene-specific primers used for the six transporters and UBQ2 were provided by L. Z. Long, pers. comm.:

ZmPht1;1 primers: - 5′-GACCCAGATGGTGTAGAATCGAA-CAT-3′, and

- 5′-TCACTTACTTTCCCGCCTATAACACACA-3′.

Table 2. Agronomic traits of field maize plants in response to different phosphorus application.

DAP	P rate (kg P ha^{-1})	Olsen-P (mg kg^{-1})	CaCl$_2$-P (mg kg^{-1})	Shoot dry weight (Mg ha^{-1})	Grain yield (Mg ha^{-1})	Shoot P concentration (g kg^{-1})	Grain P concentration (g kg^{-1})	P uptake (kg ha^{-1})
57	0	1.02 (0.11)^e	0.03 (0.00)^c	4.01 (0.02)^c		1.57 (0.14)^c		6.29 (0.60)^d
	12.5	1.39 (0.03)^e	0.03 (0.00)^c	4.43 (0.13)^c		1.83 (0.21)^{bc}		8.16 (1.05)^{cd}
	25	1.67 (0.11)^e	0.04 (0.00)^c	6.02 (0.16)^b		1.89 (0.11)^{bc}		11.39 (0.69)^{bc}
	50	3.04 (0.45)^{de}	0.05 (0.00)^c	6.15 (0.11)^b		2.04 (0.15)^{bc}		12.57 (1.09)^{ab}
	75	5.97 (0.14)^{cd}	0.06 (0.01)^c	6.69 (0.29)^a		2.12 (0.13)^b		14.21 (1.20)^{ab}
	100	7.03 (0.49)^c	0.04 (0.00)^c	6.62 (0.17)^a		2.15 (0.30)^{ab}		14.26 (2.14)^{ab}
	150	11.57 (2.61)^b	0.13 (0.04)^b	6.70 (0.11)^a		2.23 (0.23)^{ab}		15.01 (1.77)^{ab}
	300	18.00 (2.50)^a	0.19 (0.03)^a	5.87 (0.14)^b		2.68 (0.13)^a		15.75 (0.83)^a
111	0	1.21 (0.01)^d	0.02 (0.00)^c	13.44 (0.68)^e	2.97 (0.04)^d	0.47 (0.02)^{cd}	1.89 (0.05)^c	10.57 (0.56)^e
	12.5	1.53 (0.25)^d	0.03 (0.00)^{bc}	13.95 (1.19)^{de}	5.06 (0.01)^c	0.42 (0.02)^d	1.95 (0.18)^c	12.20 (0.52)^e
	25	1.54 (0.25)^d	0.02 (0.00)^c	14.26 (0.50)^{cde}	4.66 (0.04)^c	0.41 (0.00)^d	1.85 (0.01)^c	12.60 (0.18)^e
	50	2.12 (0.23)^d	0.03 (0.01)^{bc}	16.20 (0.20)^{bcd}	6.08 (0.30)^b	0.40 (0.02)^d	2.25 (0.06)^b	17.69 (0.61)^d
	75	4.43 (0.17)^c	0.03 (0.01)^{bc}	16.36 (1.01)^{bc}	7.12 (0.25)^a	0.56 (0.03)^{bc}	2.45 (0.11)^a	22.73 (1.83)^c
	100	5.41 (0.46)^c	0.06 (0.01)^{bc}	16.59 (0.95)^b	7.41 (0.22)^a	0.62 (0.06)^b	2.41 (0.03)^{ab}	23.54 (0.66)^{bc}
	150	9.82 (0.67)^b	0.09 (0.02)^{ab}	18.93 (0.42)^a	7.35 (0.58)^a	0.62 (0.02)^b	2.50 (0.07)^a	25.48 (1.05)^b
	300	14.12 (0.80)^a	0.13 (0.06)^a	19.29 (0.88)^a	7.74 (0.20)^a	0.75 (0.04)^a	2.66 (0.07)^a	29.07 (0.80)^a

Each value is the mean ($\pm$ SE) of four replicates.
Within each column and for each sampling time, values followed by different letters are significantly different at $p < 0.05$.
Abbreviation: DAP: days after planting.

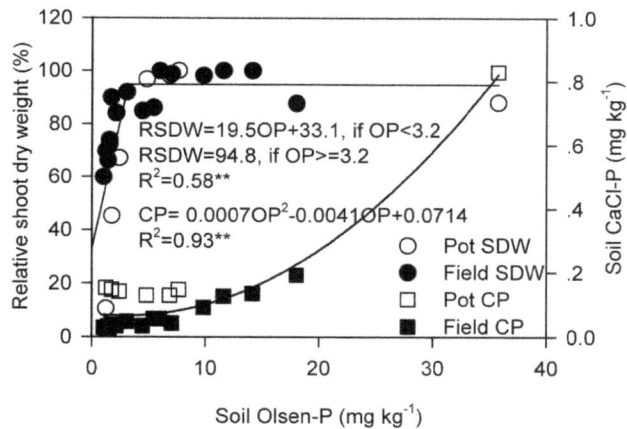

Figure 1. Maize growth and P loss risk in response to increasing soil P supply. Maize growth was presented as relative shoot dry weight, which was expressed relative to the highest mean value at each sampling time in each experiment. P loss risk was presented as soil CaCl$_2$-P level. Abbreviations: RSDW: relative shoot dry weight; OP: Olsen-P; CP: CaCl$_2$-P.

ZmPht1;2 primers: - 5'-GTCTGGTGAGGCTGAAGACTCA-GAGG-3', and

- 5'-ACATGATAGCCCACCATGTGCAGTGC-3'.

ZmPht1;3 primers: - 5'-TGTTTCCGTTCTGTCTGGTGCTT-GTG-3', and

- 5'-TCCCGACGGTGACCTCCGATTATTTA-3'.

ZmPht1;4 primers:- 5'-GAGACCCAGATGGTGTAGAGA-ATCG-3', and

- 5'-CATCAAAACACAGCCAGGGTTGACT-3'.

ZmPht1;5 primers: - 5'-CCAAAGGTAAGTCGCTGGAAGA-GAT-3', and

- 5'-CCATTGCGTGCAACAAACAGTGAC-3'.

ZmPht1;6 primers: - 5'-CGGACGTGAGCAAGGATGACAA-3', and

- 5'-GGATTCCACACCCCCTGTGTAGT-3'.

ZmUBQ2 primers: -5'- CTTTGCTGCTGCACGGGAG-GAATG- 3', and

-5'- ATGGACGCACGCTGGCTGACTA-3'.

Statistical analysis

Data are presented as means and standard errors (SE). One-way ANOVA (SPSS 13.0, USA) was conducted, and significant differences among means were determined by LSD at the $p <$ 0.05 probability level. To explore the relationship between shoot dry weight and soil Olsen-P, we used relative shoot dry weight by normalizing shoot dry weight data according to the maximum value obtained at each sampling time for each experiment; the data were then fitted to a linear-plateau model using SAS statistical software (SAS 8.1, USA) [35]. The relationships between soil CaCl$_2$-P, root morphological traits, AM colonization, expression of the six Pi transporter genes and soil Olsen-P were plotted using SigmaPlot statistical software (SigmaPlot 10.0, USA).

Results

Plant growth, P uptake, and risk of soil P loss

P application significantly increased soil P supply in terms of Olsen-P in both experiments, with the highest Olsen-P level in pot experiment being almost twice of that in the field (Table 1 and 2). Accordingly, shoot dry weight of 8-leaf plants in pot experiment

increased fast with increasing P application up to 50 mg kg^{-1} soil, then increased only slightly and were reduced with further P supply (Table 1). Similar results were obtained for field plants at the flowering stage, with the critical P application rate for best shoot growth being 75 kg P ha^{-1} (Table 2). At maturity stage in the field, total shoot dry weight increased with increasing P supply, while grain yield plateaued when P application rate exceeded 75 kg P ha^{-1} (Table 2). Taken together, the result show that shoot dry weight was positively correlated with soil Olsen-P up to a threshold of 3.2 mg kg^{-1}, at which value almost 95% of the relative shoot dry weight had been achieved (Figure 1).

In the pot experiment, shoot P concentration were similar until P application rate reached 50 mg kg^{-1} soil, above which it increased significantly with further P supply (Table 1). Contrarily, shoot P uptake in aboveground parts increased rapidly below P application rate of 50 mg kg^{-1} soil and then more slowly with higher P applications (Table 1). Similar responses were observed for field plants at both sampling stages, with a critical P application rate of 50 kg P ha^{-1} at flowering stage and 75 kg P ha^{-1} at maturity stage, respectively (Table 2).

The risk of soil P loss was indicated by the level of soil CaCl$_2$-P. In the pot experiment, soil CaCl$_2$-P level remained unchanged with P application rates between 0 and 100 mg P kg^{-1} soil and significantly increased only with the highest application rate (Table 1). Similarly, soil CaCl$_2$-P in the field did not increase until P was applied at 150 kg P ha^{-1} at both sampling stages (Table 2). Overall, soil CaCl$_2$-P level remained unchanged with soil Olsen-P level < 15 mg kg^{-1} (Figure 1).

Responses of root morphology, AM colonization and expression of Pi transporter genes to soil P supply

In the pot experiment, root dry weight increased as soil P supply increased, but plateaued once soil Olsen-P level reached 5 mg kg^{-1} (Figure 2a). In the field experiment, root dry weight also initially increased with increasing soil P supply, peaked when soil Olsen-P was about 2.5 mg kg^{-1}, and then gradually declined to a plateau at an Olsen-P level around 10 mg kg^{-1} (Figure 2a). The responses of root length to P supply in both experiments were very similar to that of root dry weight of field plants, with the critical Olsen-P level indicating a plateau at 8 mg kg^{-1} (Figure 2b). As soil Olsen-P increased from very low levels, specific root length and root/shoot ratio of pot plants declined substantially at first, and then gradually reached a plateau when Olsen-P exceeded 8 mg kg^{-1}; field plants showed similar responses, but decreases were not so pronounced compared with those of pot plants, and both traits reached the plateau at a lower critical Olsen-P level about 5 mg kg^{-1} (Figure 2c, d). Generally, specific root length and root/shoot ratio were much higher for pot-grown plants than field-grown plants.

In the pot experiment, root AM colonization of plants at the 8-leaf stage initially increased with increasing soil P supply, peaked at an Olsen-P level about 2.5 mg kg^{-1}, and then tended to gradually decrease until Olsen-P level reached 10 mg kg^{-1}, above which AM colonization remained stable at an average value of 70% with further P supply (Figure 3). In the field at the flowering stage, with an increase of soil Olsen-P root AM colonization declined rapidly at first, and gradually plateaued at 40% when Olsen-P level exceeded 5 mg kg^{-1} (Figure 3). AM colonization rates were generally higher for pot plants than field plants.

We detected expressions of all the six Pht1 Pi transporter genes in both experiments. For pot plants, the six genes responded similarly to soil P supply, i.e. their transcript levels initially decreased rapidly with increasing P supply until soil Olsen-P level reached about 10 mg kg^{-1}, above which their expression kept very

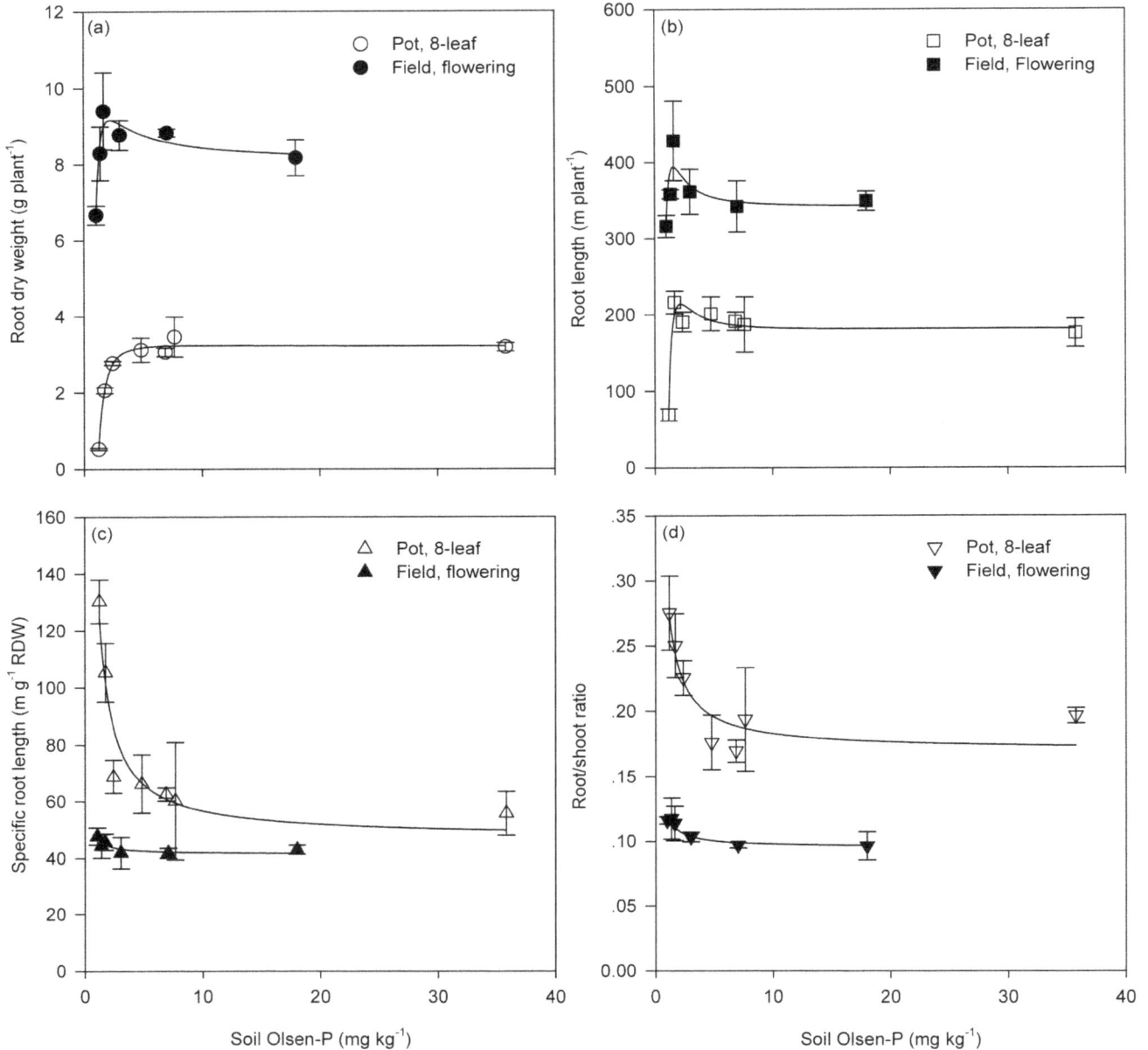

Figure 2. Root morphological traits in response to increasing soil P supply. In the pot experiment (open symbols), data were collected when plants were at the 8-leaf stage, and all visible roots in each pot were collected. In the field experiment (closed symbols), data were collected at the flowering stage, and roots were collected in a soil volume of 40 cm (row spacing) × 20 cm (plant spacing) × 30 cm (depth). Each symbol represents the mean (± SE) of three replicates for the pot experiment and four replicates for the field experiment, respectively. Abbreviation: RDW: root dry weight.

low and stable (Figure 4). Results were more complex in the field experiment. Expressions of *ZmPht1;1* to *ZmPht1;4* had similar responses to increase of soil Olsen-P, which were in accord with their expressions in pot plants, with the critical Olsen-P level indicating up-regulation approximating 5 mg kg^{-1} (Figure 4a–c). The transcript level of *ZmPht1;5* increased with increasing soil Olsen-P, peaked around 2.5 mg kg^{-1}, and then gradually declined to a stable status at Olsen-P level about 8 mg kg^{-1} (Figure 4e). Expression response of *ZmPht1;6* was similar to that of *ZmPht1;5* when soil Olsen-P was below 8 mg kg^{-1}, but declined further above this level. Generally, the expression responses of the six

genes to soil P supply were more significant in the pot experiment than those in the field experiment.

Discussion

Many studies have reported that as soil P supply increased from initially low levels, crop yield increases quickly at first and then more slowly to an asymptote [36], [37], [38]. Based on this typical response, it has been proposed that there is a critical level of soil P for optimum crop yield [38], [39]. Under the conditions of our study, the critical level of soil P for maize production was 3.2 mg kg^{-1} (Olsen-P) (Figure 1). This value is relatively low as the critical

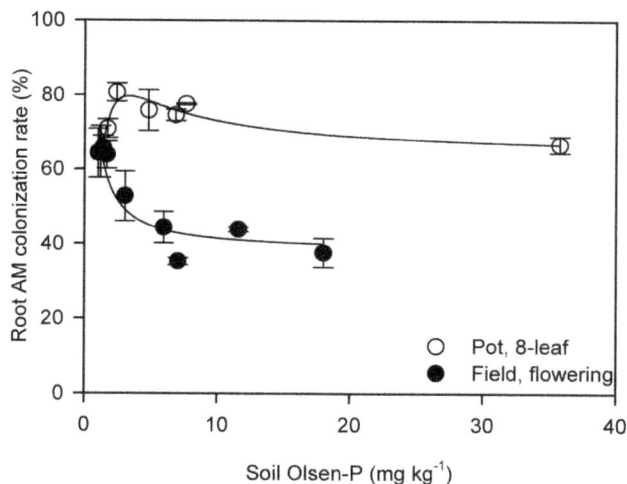

Figure 3. Root AM colonization in response to increasing soil P supply. In the pot experiment (open symbols), data were collected when plants were at the 8-leaf stage. In the field experiment (closed symbols), data were collected at the flowering stage. Each symbol represents the mean (± SE) of three replicates for the pot experiment and four replicates for the field experiment, respectively. Abbreviation: AM: arbuscular mycorrhizal.

P values reported in other studies have ranged from 3.9 to 17.3 mg kg^{-1} for maize production [35], [39], [40], [41].

To maximize crop yield, farmers tend to increase soil P and maintain it at a level that is greater than the critical value required for optimum yield [24]. Soil, however, cannot retain unlimited quantities of P. When soil P rise to a certain point, the environmental risk threshold, the risk of P loss increases significantly; and P loss may cause water pollution [7], [42]. As a result, P management has developed to embraces both agronomic and environmental goals, with the environmental risk threshold used as the building-up threshold for soil P [43], [44], [45]. Using CaCl$_2$-P as an indicator of P loss risk, we found that the environmental risk threshold in terms of Olsen-P, which resulted in a significant increase in CaCl$_2$-P, was about 15 mg kg^{-1} in this study (Figure 1). This value is in the range of 10–119 mg kg^{-1} previously reported for a range of soils [29]. Therefore, based on the current P management goals of maximizing crop yield and minimizing adverse environmental effects, our results indicate that soil Olsen-P level should be maintained between 3.2 and 15 mg kg^{-1} for maize production under our experimental conditions.

When soil Olsen-P decreased from 10 to 3.2 mg kg^{-1}, shoot growth was maintained but root length (and probably root dry weight) gradually increased (Figure 2a, b), so did root surface area (data not shown), whilst specific root length and root/shoot ratio remained relatively low and stable. This suggests that maize root growth was more sensitive than shoot growth to the reduction in soil P supply, and that maize root morphology acclimated to a reduced P supply firstly by increasing root length (and probably root dry weight and root surface area). When soil P supply was further reduced below 3.2 mg kg^{-1}, increases in specific root length and in the root/shoot ratio appeared to be the main morphological adaptation (Figure 2c, d). These results are consistent with commonly observed responses of plants to P limitation: plants allocate more biomass to root and produce more root length per unit of metabolic investment in root tissue in order to improve the capacity of roots to explore soil [22], [46]. Soil

Olsen-P level higher than 10 mg kg^{-1} showed an inhibitory effect on all the morphological traits we measured.

For AM plants like maize, there are two P uptake pathways. One is the direct P uptake via root epidermis and root hairs. Among the six Pi transporters in Pht1 family reported for maize, *ZmPht1;1–ZmPht1;4* are considered to facilitate the direct uptake pathway [33], although the affinity for Pi of these four transporters remains uncertain. Phylogenetic analysis shows that *ZmPht1;1*, *ZmPht1;2*, and *ZmPht1;4* are closely related with rice *OsPT8* while *ZmPht1;3* closely clusters with rice *OsPT6* [33]. Both *OsPT6* and *OsPT8* are identified high-affinity Pi transporters [47]. Thus we speculate that *ZmPht1;1* to *ZmPht1;4* may also play the role of high-affinity Pi transporters. In the current study we found that these four genes responded similarly to soil P supply, and the up-regulation of their expressions was induced at an Olsen-P level (10 and 5 mg kg^{-1} in pot and field experiments, respectively) higher than the critical level for shoot growth (Figure 1, 4a–d), suggesting that root response on the molecular level to reduced P supply occurred at a higher P level than shoot growth. The up-regulation of the expressions for the four genes coincided with the increase of root length until soil Olsen-P was reduced to 2.5 mg kg^{-1}, from which root length began to decrease with further P supply reduction (Figure 2b, 4a–d). Such responses probably indicate the cooperation between root morphology and physiology for efficient P acquisition in the direct pathway when soil P supply reduced from sufficiency to deficiency. We noticed the overall up-regulation levels of these four genes were significantly higher in pot-grown plants than field-grown plants. One reason may be that the pot plants suffered relatively more P deficiency at lower P supply as indicated by the lower shoot P concentrations of pot plants than those of field plants (Table 1 and 2).

The other P uptake pathway is through AM symbioses. The rate of root colonization by AM fungi is considered an important factor for the extent to which AM symbioses contribute to P uptake and growth response of AM plants [48]. The soil P level plays an important role in influencing AM colonization [49], [50], [51]. In the current study, we found the influence of soil P on AM colonization was strongest when soil Olsen-P level was below 10 mg kg^{-1} (Figure 3), indicating a threshold of soil P supply for AM colonization as reported by others [51], [52]. As Olsen-P reduced from 10 to 3.2 mg kg^{-1}, the increasing root AM colonization suggests that the mycorrhizal pathway was enhanced to increase P acquisition without loss of shoot growth.

There are two types of Pht1 transporter genes whose expression can be induced in mycorrhizal roots: AM-specific Pi transporter genes which are expressed strictly in response to AM symbioses, and AM-inducible Pi transporter genes which can be strongly induced by AM symbioses but have a basal expression in non-mycorrhizal roots [53]. *ZmPht1;6* has been identified as an AM-inducible Pi transporter gene for maize [33]. Currently there is no report regarding an association between *ZmPht1;5* and AM symbioses. *ZmPht1;5*, however, clusters with *OsPT13* on the phylogenetic tree [33], [53], and *OsPT13* is an AM-inducible gene in rice [54], possibly indicating a role of *ZmPht1;5* in the mycorrhizal pathway. Coinciding with the response of root AM colonization, the expression levels of the two genes were gradually up-regulated when soil Olsen-P was reduced from 10 to 3.2 mg kg^{-1} in both experiments (Figure 4e, f). When soil P supply further decreased below the critical level for shoot growth, we noticed that both AM colonization and expressions of the two genes in the pot experiment were different from those in the field study. This may be partly in relation to the dependency on mycorrhizal pathway for P uptake under P deficiency when maize plants were at different growth stages. In the early growth season, root growth

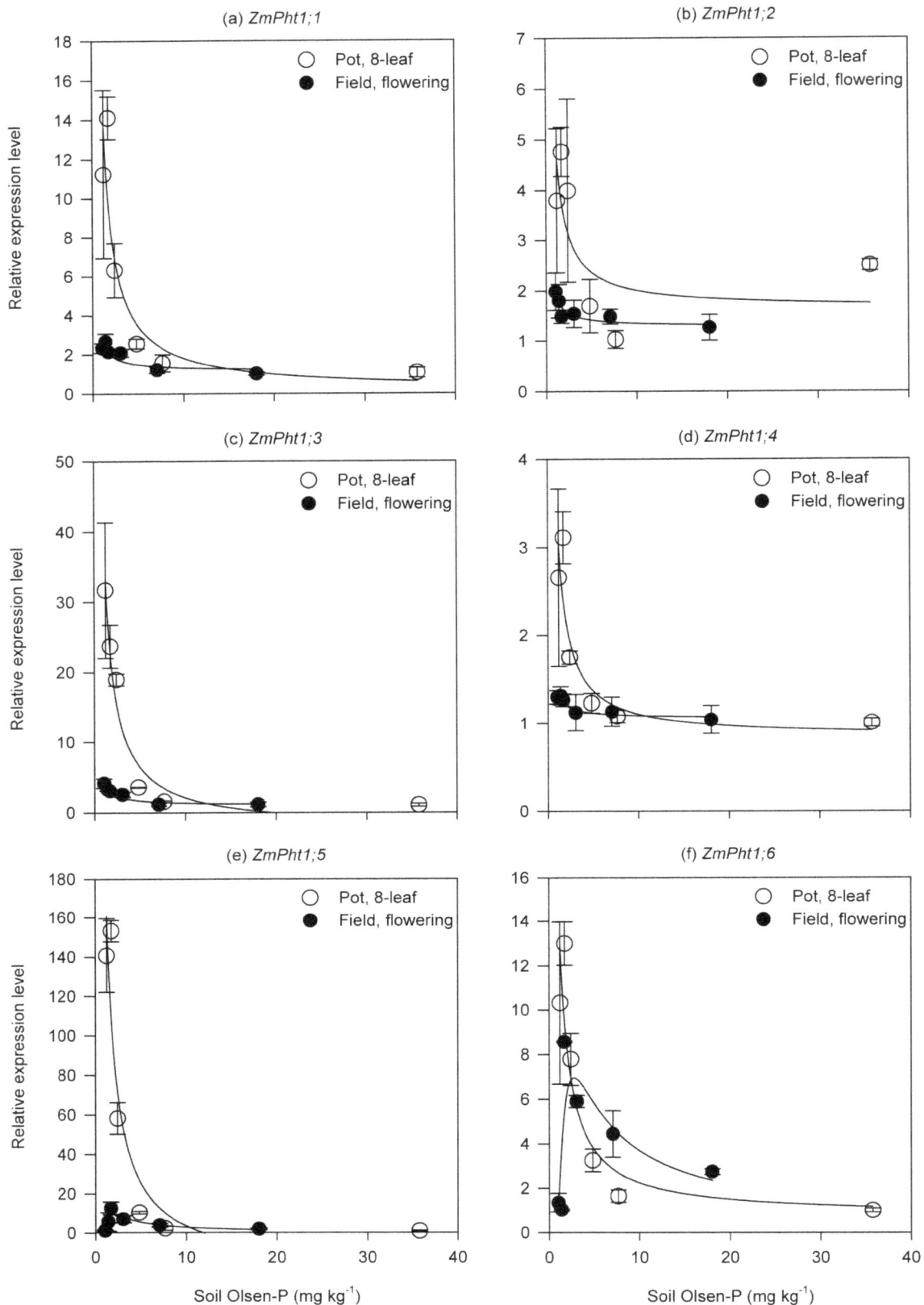

Figure 4. Expression of root Pht1 transporter genes in response to increasing soil P supply. Roots were sampled at the 8-leaf stage for pot plants (open symbols) and at the flowering stage for field plants (closed symbols). Gene relative expression level was measured by real-time quantitative RT-PCR. UBQ2 was used as the internal control. For each gene, the lowest expression level was set equal to 1.0. Each symbol represents the mean (± SE) of three replicates for the pot experiment and four replicates for the field experiment, respectively.

and development is relatively poor, but plant requirement of P is high to support fast growth [55], [56]. Thus an effective AM association may help to improve P nutrition for young plants. Our data showed that pot plants at the 8-leaf stage had much less root length (Figure 2b) and root surface area (data not shown) for P acquisition compared with field plants at the flowering stage. To compensate the direct pathway, the younger plants in the pot experiment might depend more on mycorrhizal pathway through enhancing expression of AM-inducible Pi transporter genes but limiting fungal growth as represented by reduced AM colonization, possibly a strategy to save carbon for efficient P acquisition. Besides, the much smaller soil volume for root growth and exploration in the pot experiment compared with field conditions may trigger pot plants to depend more on the mycorrhizal pathway under P deficiency, as indicated by the overall higher AM colonization rates and up-regulation levels of AM-inducible Pi transporter genes.

In summary, under the conditions of our study, most root adaptations (root morphology, AM colonization and expression of Pi transporter genes) involved with P acquisition were triggered at Olsen-P level < 10 mg kg^{-1}. Therefore, to maximize the inherent potential of maize roots to obtain P, we suggest that the upper level of optimum soil P supply in present study should be reduced to 10 mg kg^{-1}, rather than the environmental risk threshold of 15 mg kg^{-1}.

Soil P status varies across the world's croplands, and P deficits occur on 30–40% of the world's arable land [57], especially in Africa and South Asia. Sustainable crop production on these soils requires some level of P fertilization, a level that can optimize agronomic goals while maximizing inherent potential of roots to obtain P efficiently. In contrast to areas with P-depleted soils, most areas with intensive agriculture (Western Europe, North America, and East Asia) have the opposite problem in that levels of soil P have reached or exceeded the levels required by crops. In these soils, management should prevent further increases in P level or reduce P level not only to reduce environmental risk but also to better use the inherent potential of roots to acquire P. Although the importance of developing cultivars with an increased ability to obtain and utilize P is now recognized, our study illustrates that improvement of agronomic soil P management will also be important to explore the inherent potential of roots for efficient P acquisition.

Acknowledgments

We are very grateful to Prof. Fanjun Chen for supplying the plant material. We also thank Prof. Chunqin Zou and Prof. Susanne Schmidt for comments on an earlier version of the manuscript.

Author Contributions

Conceived and designed the experiments: YD YT FZ XC. Performed the experiments: YD KC WT AZ YT GF ZC XC. Analyzed the data: YD. Wrote the paper: YD XC.

References

1. Tilman D (1999) Global environmental impacts of agricultural expansion: The need for sustainable and efficient practices. Proceedings of the National Academy of Sciences of the United States of America 96: 5995–6000.
2. Tilman D, Cassman KG, Matson PA, Naylor R, Polasky S (2002) Agricultural sustainability and intensive production practices. Nature 418: 671–677.
3. Cordell D, Drangert J-O, White S (2009) The story of phosphorus: Global food security and food for thought. Global Environmental Change 19: 292–305.
4. Gilbert N (2009) Environment: The disappearing nutrient. Nature 461: 716.
5. Vaccari D (2009) Phosphorus: A looming crisis. Scientific American Magazine 300: 54–59.
6. Cordell D, Rosemarin A, Schröder J, Smit A (2011) Towards global phosphorus security: A systems framework for phosphorus recovery and reuse options. Chemosphere 84: 747–758.
7. Sharpley AN, Chapra S, Wedepohl R, Sims J, Daniel TC, et al. (1994) Managing agricultural phosphorus for protection of surface waters: Issues and options. Journal of Environmental Quality 23: 437–451.
8. Conley DJ, Paerl HW, Howarth RW, Boesch DF, Seitzinger SP, et al. (2009) Controlling eutrophication: Nitrogen and phosphorus. Science 323: 1014–1015.
9. MacDonald GK, Bennett EM, Potter PA, Ramankutty N (2011) Agronomic phosphorus imbalances across the world's croplands. Proceedings of the National Academy of Sciences of the United States of America 108: 3086–3091.
10. Hinsinger P (2001) Bioavailability of soil inorganic P in the rhizosphere as affected by root-induced chemical changes: A review. Plant and Soil 237: 173–195.
11. Hermans C, Hammond JP, White PJ, Verbruggen N (2006) How do plants respond to nutrient shortage by biomass allocation? Trends in Plant Science 11: 610–617.
12. Hill J, Simpson R, Moore A, Chapman D (2006) Morphology and response of roots of pasture species to phosphorus and nitrogen nutrition. Plant and Soil 286: 7–19.
13. Lynch JP (2007) Roots of the second green revolution. Australian Journal of Botany 55: 493–512.
14. Vance CP, Uhde-Stone C, Allan DL (2003) Phosphorus acquisition and use: Critical adaptations by plants for securing a nonrenewable resource. New Phytologist 157: 423–447.
15. Raghothama K (1999) Phosphate acquisition. Annual Review of Plant Biology 50: 665–693.
16. Richardson AE, Hocking PJ, Simpson RJ, George TS (2009) Plant mechanisms to optimise access to soil phosphorus. Crop and Pasture Science 60: 124–143.
17. Bates T, Lynch J (1996) Stimulation of root hair elongation in *Arabidopsis thaliana* by low phosphorus availability. Plant, Cell & Environment 19: 529–538.
18. Ma Z, Bielenberg D, Brown K, Lynch J (2001) Regulation of root hair density by phosphorus availability in *Arabidopsis thaliana*. Plant, Cell & Environment 24: 459–467.
19. Fan M, Zhu J, Richards C, Brown KM, Lynch JP (2003) Physiological roles for aerenchyma in phosphorus-stressed roots. Functional Plant Biology 30: 493–506.
20. Brown L, George T, Dupuy L, White P (2013) A conceptual model of root hair ideotypes for future agricultural environments: What combination of traits should be targeted to cope with limited P availability? Annals of Botany 112 (2): 317–330.
21. Bolan N (1991) A critical review on the role of mycorrhizal fungi in the uptake of phosphorus by plants. Plant and soil 134: 189–207.
22. Lambers H, Shane MW, Cramer MD, Pearse SJ, Veneklaas EJ (2006) Root structure and functioning for efficient acquisition of phosphorus: Matching morphological and physiological traits. Annals of Botany 98: 693–713.
23. Ramaekers L, Remans R, Rao IM, Blair MW, Vanderleyden J (2010) Strategies for improving phosphorus acquisition efficiency of crop plants. Field Crops Research 117: 169–176.
24. Li H, Huang G, Meng Q, Ma L, Yuan L, et al. (2011) Integrated soil and plant phosphorus management for crop and environment in China: A review. Plant and Soil 349: 157–167.
25. Coelho GT, Carneiro NP, Karthikeyan AS, Raghothama KG, Schaffert RE, et al. (2010) A phosphate transporter promoter from *Arabidopsis thaliana* AtPHT1; 4 gene drives preferential gene expression in transgenic maize roots under phosphorus starvation. Plant Molecular Biology Reporter 28: 717–723.
26. Chin JH, Gamuyao R, Dalid C, Bustamam M, Prasetiyono J, et al. (2011) Developing rice with high yield under phosphorus deficiency: Pup1 sequence to application. Plant Physiology 156: 1202–1216.
27. Gamuyao R, Chin JH, Pariasca-Tanaka J, Pesaresi P, Catausan S, et al. (2012) The protein kinase Pstol1 from traditional rice confers tolerance of phosphorus deficiency. Nature 488: 535–539.
28. Sims JT (2000) Soil test phosphorus: Olsen P. In: Pierzynski GM, editor. Methods of phosphorus analysis for soils, sediments, residuals, and waters. Raleigh, nc: North Carolina State University. pp. 20–21.
29. Hesketh N, Brookes P (2000) Development of an indicator for risk of phosphorus leaching. Journal of Environmental Quality 29: 105–110.
30. Shi R (1986) Soil and agricultural chemistry analysis. Beijing: China Agricultural Press.
31. Phillips J, Hayman D (1970) Improved procedures for clearing roots and staining parasitic and vesicular-arbuscular mycorrhizal fungi for rapid assessment of infection. Transactions of British Mycological Society 55: 158–161.
32. Trouvelot A, Kough JL, Gianinazzi-Pearson V (1986) Mesure du taux de mycorhization VA d'un système radiculaire. Recherche de méthodes d'estimation ayant une signification fonctionnelle. In: Gianinazzi-Pearson V, Gianinazzi S, editors. Physiological and genetical aspects of mycorrhizae. Paris: Institut geNational de la Recherche Agronomique. pp. 217–221.
33. Nagy R, Vasconcelos M, Zhao S, McElver J, Bruce W, et al. (2006) Differential regulation of five Pht1 phosphate transporters from maize (*Zea mays* L.). Plant Biology 8: 186–197.

34. Calderon-Vazquez C, Ibarra-Laclette E, Caballero-Perez J, Herrera-Estrella L (2008) Transcript profiling of *Zea mays* roots reveals gene responses to phosphate deficiency at the plant-and species-specific levels. Journal of Experimental Botany 59: 2479–2497.

35. Tang X, Ma Y, Hao X, Li X, Li J, et al. (2009) Determining critical values of soil Olsen-P for maize and winter wheat from long-term experiments in China. Plant and Soil 323: 143–151.

36. Johnston A, Dawson C (2005) Phosphorus in agriculture and in relation to water quality. Peterborough: Agricultural Industries Confederation.

37. Howard AE (2006) Agronomic thresholds for soil phosphorus in Alberta: A review. In: Alberta soil phosphorus limits project volume 5: Background information and reviews. Alberta, Canada: Alberta Agriculture, Food and Rural Development. 42 p.

38. Kirkby EA, Johnston AEJ (2008) Soil and fertilizer phosphorus in relation to crop nutrition. In: White PJ, Hammond JP, editors. The ecophysiology of plant-phosphorus interactions. Dordrecht, the Netherlands: Springer. pp. 177–223.

39. Mallarino A, Blackmer A (1992) Comparison of methods for determining critical concentrations of soil test phosphorus for corn. Agronomy Journal 84: 850–856.

40. Mallarino AP, Atia AM (2005) Correlation of a resin membrane soil phosphorus test with corn yield and routine soil tests. Soil Science Society of America Journal 69: 266–272.

41. Colomb B, Debaeke P, Jouany C, Nolot J (2007) Phosphorus management in low input stockless cropping systems: Crop and soil responses to contrasting P regimes in a 36-year experiment in southern France. European Journal of Agronomy 26: 154–165.

42. Heckrath G, Brookes P, Poulton P, Goulding K (1995) Phosphorus leaching from soils containing different phosphorus concentrations in the Broadbalk experiment. Journal of Environmental Quality 24: 904–910.

43. Higgs B, Johnston A, Salter J, Dawson C (2000) Some aspects of achieving sustainable phosphorus use in agriculture. Journal of Environmental Quality 29: 80–87.

44. Sharpley AN, Naniel T, Sims T, Lemunyon J, Stevens R, et al. (2003) Agricultural phosphorus and eutrophication, 2nd ed. Washington, DC: U.S. Department of Agriculture, Agricultural Research Service.

45. Elliott H, O'Connor G (2007) Phosphorus management for sustainable biosolids recycling in the United States. Soil Biology and Biochemistry 39: 1318–1327.

46. Lynch JP, Brown KM (2008) Root strategies for phosphorus acquisition. In: White PJ, Hammond JP, editors. The ecophysiology of plant-phosphorus interactions. Dordrecht, the Netherlands: Springer. pp. 83–116.

47. Nussaume L, Kanno S, Javot H, Marin E, Pochon N, et al. (2011) Phosphate import in plants: Focus on the PHT1 transporters. Frontiers in Plant Science 2: 1–12.

48. Smith SE, Smith FA (2011) Roles of arbuscular mycorrhizas in plant nutrition and growth: New paradigms from cellular to ecosystem scales. Annual Review of Plant Biology 62: 227–250.

49. Abbott L, Robson A (1991) Factors influencing the occurrence of vesicular-arbuscular mycorrhizas. Agriculture, Ecosystems & Environment 35: 121–150.

50. Grant C, Bittman S, Montreal M, Plenchette C, Morel C (2005) Soil and fertilizer phosphorus: Effects on plant P supply and mycorrhizal development. Canadian Journal of Plant Science 85: 3–14.

51. Covacevich F, Echeverría HE, Aguirrezabal LA (2007) Soil available phosphorus status determines indigenous mycorrhizal colonization of field and glasshouse-grown spring wheat from Argentina. Applied Soil Ecology 35: 1–9.

52. Fernandez M, Gutierrez B, Flavio H, Rubio F (2009) Arbuscular mycorrhizal colonization and mycorrhizal dependency: A comparison among soybean, sunflower and maize. Proceedings of the International Plant Nutrition Colloquium XVI. Department of Plant Science, Univ. Calf., Davis, USA.

53. Javot H, Pumplin N, Harrison MJ (2007) Phosphate in the arbuscular mycorrhizal symbiosis: Transport properties and regulatory roles. Plant, Cell & Environment 30: 310–322.

54. Glassop D, Godwin RM, Smith SE, Smith FW (2007) Rice phosphate transporters associated with phosphate uptake in rice roots colonized with arbuscular mycorrhizal fungi. Botany 85: 644–651.

55. Römer W, Schilling G (1986) Phosphorus requirements of the wheat plant in various stages of its life cycle. Plant and soil 91: 221–229.

56. Grant C, Flaten D, Tomasiewicz D, Sheppard S (2001) The importance of early season phosphorus nutrition. Canadian Journal of Plant Science 81: 211–224.

57. Von Uexküll, HR, Mutert E (1995) Global extent, development and economic impact of acid soils. Plant and Soil 171: 1–15.

22

Pollination and Plant Resources Change the Nutritional Quality of Almonds for Human Health

Claire Brittain[1]*, Claire Kremen[2], Andrea Garber[3], Alexandra-Maria Klein[4]

1 Institute of Ecology, Ecosystem Functions, Leuphana University of Lüneburg, Germany and Department of Entomology, University of California Davis, Davis, California, United States of America, 2 Environmental Sciences Policy and Management, University of California, Berkeley, Berkeley, California, United States of America, 3 Division of Adolescent Medicine, University of California San Francisco, San Francisco, California, United States of America, 4 Institute of Ecology, Ecosystem Functions, Leuphana University of Lüneburg, Germany and Chair of Nature Conservation and Landscape Ecology, University of Freiburg, Germany

Abstract

Insect-pollinated crops provide important nutrients for human health. Pollination, water and nutrients available to crops can influence yield, but it is not known if the nutritional value of the crop is also influenced. Almonds are an important source of critical nutrients for human health such as unsaturated fat and vitamin E. We manipulated the pollination of almond trees and the resources available to the trees, to investigate the impact on the nutritional composition of the crop. The pollination treatments were: (a) exclusion of pollinators to initiate self-pollination and (b) hand cross-pollination; the plant resource treatments were: (c) reduced water and (d) no fertilizer. In an orchard in northern California, trees were exposed to a single treatment or a combination of two (one pollination and one resource). Both the fat and vitamin E composition of the nuts were highly influenced by pollination. Lower proportions of oleic to linoleic acid, which are less desirable from both a health and commercial perspective, were produced by the self-pollinated trees. However, higher levels of vitamin E were found in the self-pollinated nuts. In some cases, combined changes in pollination and plant resources sharpened the pollination effects, even when plant resources were not influencing the nutrients as an individual treatment. This study highlights the importance of insects as providers of cross-pollination for fruit quality that can affect human health, and, for the first time, shows that other environmental factors can sharpen the effect of pollination. This contributes to an emerging field of research investigating the complexity of interactions of ecosystem services affecting the nutritional value and commercial quality of crops.

Editor: Tianzhen Zhang, Nanjing Agricultural University, China

Funding: We acknowledge the Alexander von Humboldt, Hellmann and the McDonnell 21st Century Foundations, the Chancellor's Partnership Fund of the University of California, Berkeley, and the Germany Science Foundation (KL 1849/4-1) for financial support. The funders had no role in study design, data collection and analysis, decision to publish, or preparation of the manuscript.

Competing Interests: The authors have declared that no competing interests exist.

* E-mail: cabrittain@ucdavis.edu

Introduction

As the global population grows, so does the demand for food [1]. A balanced diet containing a broad spectrum of nutrients is important for human health [2,3]. Animal-pollinated crop species provide key nutrients valuable for human health, including an estimated 74% of all globally produced lipids and 35–65% of vitamin E [4]. The benefits of animal pollination for crop production have been documented for many crop species [5–7], but the effects of animal pollination on the nutritional composition of crops and other measures of quality such as fruit shelf life and therefore its commercial value have just begun to be investigated [8,9].

Animal pollination may become a limiting resource if the growth of crops reliant on pollination outpaces the growth in the number of honey bee hives [10] and agricultural intensification and habitat loss continue to negatively affect wild bees [11–13]. Human population growth and climate change are predicted to increase the strain on the resources required for food production (particularly water) [14]. While the effects of plant resource shortages have been documented on the yield and development of crops [15–17], the effects of the availability of water and other plant resources on nutritional composition are largely unknown. How pollination and resource availability both singly and jointly impact a crop's nutritional composition is thus a pressing question, particularly in light of the growing demand for certain crops based on their purported nutritional value. Almond (*Prunus dulcis* [Mill.] DA Webb) is an example of a commodity where advertising has focused on the potential health benefits of its consumption [18]. Clinical trials have shown almonds to be cardio-protective [19], an effect that is attributed to their high monounsaturated fat content. The primary monounsaturated fatty acid present in California-grown almonds is oleic acid, an omega-9 fatty acid accounting for 58–74% of the total fat content [20]. Almonds also contain linoleic acid, an omega-6 fatty acid, and very small amounts of linolenic, an omega-3 fatty acid [20]. The health effects of omega-6 fatty acids are controversial, with evidence to suggest that the high omega-6 to omega-3 ratio of the Western diet is a contributor to cardiovascular disease, cancer and diabetes [21]. Almonds are also valued for their vitamin E (α-Tocopherol) content as it is an antioxidant which helps protect cell membranes from peroxidative damage [22].

Since almond is a highly pollinator-dependent crop and a large consumer of water and fertilizer [23], its production may be sensitive to fluctuations in these resources. Furthermore, approximately eighty percent of the world's almonds are produced in California [24], a state where climate change is expected to reduce water availability [25,26]. The aim of this study was to investigate if pollination and plant resource availability (water and fertilizer) influence the nutritional composition of almonds. We exposed whole almond trees to different pollination treatments (the exclusion of pollinators to allow for self-pollination only and hand cross-pollination with compatible pollen to achieve high levels of cross-pollination) and different resource treatments (reduced water and no fertilizer). We investigated whether self-pollination and reduced plant resources (fertilizer and water), when experienced by whole trees, influenced the nutritional composition of almonds, and if there were interacting effects. Since almond is highly pollinator dependent [23], we hypothesized that the impact of self-pollination on the nuts' nutritional content would be of similar strength to that of the availability of water and fertilizer, and that pollen and resource availability would have interacting effects.

Materials and Methods

The study was conducted in 2008, in a 3.2 ha almond orchard in the Sacramento Valley, Northern California (122°2′1.925″W, 38°55′19.372″N, WGS 1984: the owner of the land gave permission to conduct the study at this site). Our study orchard contained trees of the most popular variety for production, Nonpareil, grafted onto peach rootstock (*Prunus persica* (L.) Batsch) in 2005 and planted in 2006 (third leaf planting). Other tree varieties compatible with Nonpareil were located 100–300 meters away, including Mission (100% compatible), Carrion and Wood Colony (50% compatible). Honey bee hives were placed in the orchard during bloom with the eight nearest hives being 300–350 meters away. As part of the grower's pollination strategy, all hives in the orchard had Padre pollen (100% compatible) placed at the hive entrance to maximize the transport of compatible pollen to the Nonpareil trees.

Treatments

Whole trees were exposed to different pollination and resource treatments (Table 1). The treatments were assigned randomly to individual trees in the orchard and were replicated five times in adjacent rows (n = 40 experimental trees). Hand-pollination was conducted from 20–28th February. When flowers opened, Padre pollen was applied to the stigmas using small brushes. The pollen had been harvested before bud opening and stored at −20°C and was thawed before immediate use. All open flowers were hand-pollinated every two to three days until approximately 90% of all buds had opened (the last 10% of buds that opened often had deformed or missing parts). The trees exposed to self-pollination were covered with cages (1.5×2.0 m²) with a mesh size of 0.8–1.0 mm from the end of January until mid-March when flowering had ceased.

For the three months before bloom, trees were irrigated when necessary and no fertilizer was applied. Once flowering began, three out of the four water emitters for each reduced water tree were closed. Reduced water trees received 27 L water every third day, while the other trees received 108 L. No nutrients were applied to the no fertilizer trees during bloom. Once flowering began, the other trees each received the following: 521.6 g nitrate, 344.7 g of potassium, 244.9 g of sulphur, 158.8 g of calcium, 158.8 g of phosphorus, 54.4 g of magnesium, 27.22 g of boron, 27.22 g of iron, 27.22 g of manganese, and zinc, cobalt, molybdenum, and various other micronutrients (amounts given

Table 1. The different treatments whole almond trees received (5 trees each, for details see methods).

Treatment	Pollination	Fertilizer	Water
NF no fertilizer	open-pollination	none	normal
RW reduced water	open-pollination	normal	reduced
SP self-pollination	exclusion of pollinators	normal	normal
CP cross-pollination	hand cross-pollination	normal	normal
SP+NF	exclusion of pollinators	none	normal
SP+RW	exclusion of pollinators	normal	reduced
CP+NF	hand cross-pollination	none	normal
CP+RW	hand cross-pollination	normal	reduced

are per tree, for that growing year). All the experimental trees were similar in height and number of main branches.

On 2nd July, 48 fruits were harvested from each experimental tree. Fruits were randomly selected from the main branches (12, 16 or 24 fruits per branch, corresponding to trees with 4, 3 or 2 main branches). The harvested fruits were dried on the ground for seven days (mimicking harvesting practices), with metal cages protecting them from bird and mammal predation. After the seven days, the hulls were removed from the fruits and the shells were cracked. The almonds were placed in a fridge at 4°C prior to nutritional analyses.

Nutritional Analyses

All nutritional analyses were performed by NP Analytical Laboratories, St. Louis, MO, U.S.A. and followed the Official Methods of Analysis of the Association of Analytical Communities (AOAC). A single analytical sample was a handful of almonds collected from several branches of an individual tree, and each tree within a treatment group was a replicate. Nuts from trees which had received the four single treatments were analyzed for their full nutritional composition (Table 2). Details of the analytical methods used in the nutritional analysis are given in the supporting information S1. Although each treatment group had five tree replicates in the orchard, some of the trees produced insufficient nuts for nutritional analysis so two treatments (no fertilizer and self-pollination) only had four replicate trees.

The trees which received two treatments (one pollination plus one plant resource) were analyzed for fats and vitamin E composition only, since the levels of these nutrients were found to vary most between samples (see supporting information S2) and almonds are valued for these nutrients. The trees which received two treatments (SP+RW, SP+NF, CP+RW, CP+NF) produced sufficient nuts for 4, 3, 5, and 3 replicates respectively.

Statistical Analyses: Single Treatments

From the full nutritional analysis, 26 nutrients in almonds were quantified (Table 2) for the four main treatment groups (SP, CP, RW, NF). The samples did not contain any lactose, maltose or trans fatty acids so these were removed from all analyses. We performed non-metric multi-dimensional scaling (NMDS with two dimensions) of the nutritional data, to visualize differences in the nutritional composition of the nuts. To test if differences in the nutritional composition were related to the treatment group, permutational multivariate analysis of variance using Bray-Curtis distances was performed (adonis, R package vegan). P values were calculated using F-tests based on sequential sums of squares from

Table 2. The range of nutrients that were quantified in almonds from trees exposed to a single resource or pollination treatment.

Nutrient	Method of quantification	Units
Vitamin E		
Alpha-Tocopherol acetate	HPLC with fluorescent detector	IU/100 g
Fat composition¶		
Oleic, linoleic and linolenic fatty acids, monounsaturated fatty acids, polyunsaturated fatty acids, saturated fatty acids, trans fatty acids, total fat	gas chromatography	g/100 g
Vitamin B		
B1 thiamine	manual fluorescence	ppm
B2 riboflavin	semi-automated fluorometric	ppm
B3 niacin	turbidimetric microbiological	ppm
Minerals		
Calcium, copper, iron, magnesium, manganese, phosphorus, potassium, sodium, zinc	inductively coupled plasma mass spectrometry	ppm
Sugar profile		
Fructose, glucose, lactose, maltose, sucrose	high pressure liquid chromatography	percent

¶Oleic acid is included in the total monounsaturated fatty acids; linoleic and linolenic acid are included in the total polyunsaturated fatty acids.

999 permutations of the data. Differences in variation of the nutritional composition between treatment groups were tested using Bray-Curtis distances to the treatment group's centroid. Analysis of variance of the distances was performed, with treatment group as the explanatory variable (betadisper, R package vegan). P values were calculated from 999 permutations of model residuals which were used to generate a permutation distribution of F under the null hypothesis of there being no difference in dispersion between the groups.

Statistical Analyses: Combined Treatments

The fat and vitamin E composition was analyzed for variation between the eight different treatments: SP, CP, RW, NF, SP+RW, SP+NF, CP+RW, CP+NF. An analysis of variance was carried out for each nutrient separately, with the treatment group as the explanatory variable. The proportion oleic to linoleic fatty acids was also tested as it is an indicator of almond kernel quality [27]. Where treatment was significant ($P \leq 0.05$), Tukey's HSD (honestly significant difference) was used to compare between treatments.

A linear mixed effect model was run for each nutrient with the treatment as the explanatory variable. We used a restricted likelihood ratio test to determine if the variance of the random effect of orchard row differed from zero (R package RLRsim). Orchard row was not included as a random effect as the test found that for all nutrients, the variance of the orchard rows did not differ significantly from zero ($P > 0.05$). All nutrients were tested for correlation.

Results

Linoleic acid (an omega-6) is the predominant polyunsaturated fatty acid in almonds, so the values for linoleic acid and total polyunsaturated fatty acids were highly correlated (supporting information S3). Similarly, as the predominant monounsaturated fat, levels of oleic acid were correlated with the total amount of monounsaturated fat and total fat. We will therefore present the results for linoleic and oleic acid but not for monounsaturated fatty acids, polyunsaturated fatty acids or total fat. Levels of linolenic acid were below the threshold for accurate quantification (< 0.04 g/100 g).

The overall nutritional composition of the almonds differed between treatment groups (Fig. 1, stress 10%). This was confirmed by the permutational multivariate analysis of variance, which found that 36% of the difference in nutritional composition was related to the treatment group ($R^2 = 0.36$, $F_3 = 2.7$, $P = 0.026$). The variation in nutritional composition did not differ between treatment groups ($F_3 = 1.3$, $P = 0.339$).

We found differences between the eight treatment combinations in the levels of vitamin E ($F_7 = 20.4$, $P < 0.001$), oleic acid ($F_7 = 5.0$, $P = 0.001$) and linoleic acid ($F_7 = 5.0$, $P = 0.001$). Vitamin E was highest in the self-pollinated nuts and was lowest in the nuts from treatments combined with hand cross-pollination, especially cross-pollinated with no fertilizer (Fig. 2a). However, oleic acid was lower in the self-pollinated nuts of the single treatments (Fig. 2b), and lower still in self-pollinated nuts combined with reduced water and no fertilizer. Linoleic acid was highest in the self-pollinated nuts combined with no fertilizer and lowest in nuts from cross-pollinated trees (Fig. 2c). The proportion oleic to linoleic acid ($F_7 = 7.1$, $P < 0.001$) varied between treatments (Fig. 2d). It was lower in self-pollinated nuts and decreased further when self-pollination was combined with no fertilizer.

Discussion

While the adverse effects of pollination limitation and resource scarcity on fruit set and crop development have been documented [15,17,28], this is the first study to our knowledge, to test if pollination and plant resources act alone or in combination on the nutritional composition of a crop. The effect of self-pollination on the nutritional composition of Nonpareil almonds was greater than the effect of reduced water and fertilizer. The health benefits of almonds are attributed to their fat composition and vitamin E, and these were both affected by the pollination treatments.

We found the highest oleic to linoleic ratio in almonds in cross-pollinated trees, and the lowest ratio in pollinator-excluded trees. Almonds with a high oleic to linoleic ratio would be most favorable to consumers seeking health benefits. The high oleic acid content is credited for the cardio-protective effect of almonds [19,29]. Although linoleic acid is an essential fatty acid that the body cannot synthesize, rising dietary levels of omega-6 fatty acids are

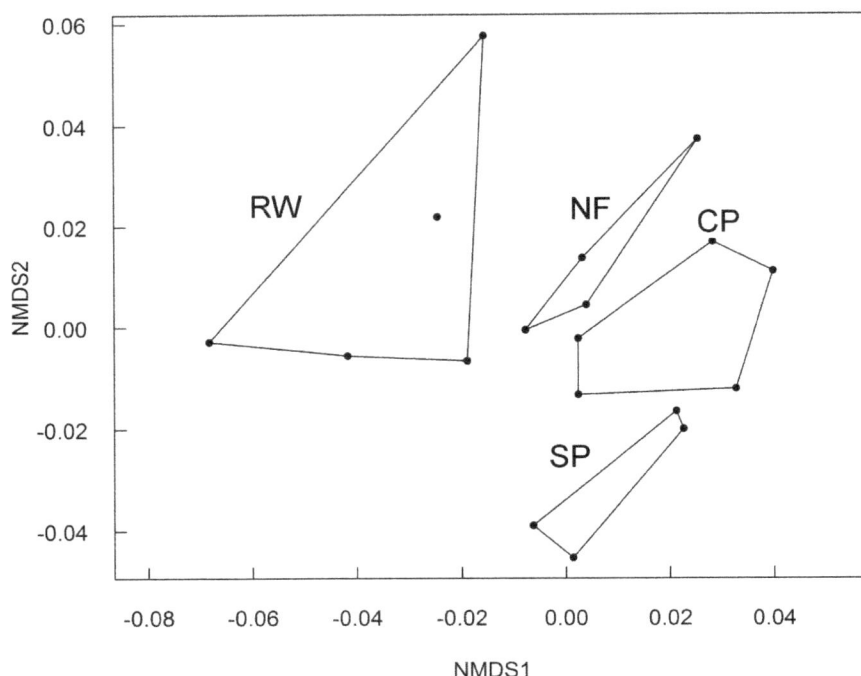

Figure 1. Non-metric multi-dimensional scaling (NMDS) of the nutritional composition of almonds. Almonds were collected from trees that had received one of four treatments: reduced water (RW), no fertilizer (NF), cross-pollination (CP) and self-pollination (SP). The content of a wide range of nutrients (listed in Table 2) was quantified in the laboratory (methods detailed in the supporting information S1). Permutational multivariate analysis of variance using Bray-Curtis distances was performed on the nutritional content of the almonds, with the treatment as the grouping factor, replicated at the tree level. Each point on the graph represents an individual tree that was exposed to one of the treatments and the hulls link trees exposed to the same treatment. The further apart the points are on the graph, the more different the nuts' nutritional composition.

believed to contribute to chronic diseases in Western cultures [21]. In terms of crop characteristics, a high oleic (mono-unsaturated fatty acid) to linoleic (an omega-6 polyunsaturated fat) acid ratio is desirable as it improves the stability of the fats against rancidity and therefore increases the almonds' shelf life [27]. A positive effect of bee-pollinated flowers on shelf life was recently demonstrated for strawberry production, since bee pollination leads to firmer fruits that last longer [8]. Further, Klatt *et al.* [8] show that bee pollination increased fruit weight and redness and also reduced sugar-acid ratios and therefore produced higher commercial grades. Together these two effects increased the value attributed to bee-pollinated fruits. Other studies showed that oil content in oilseed rape (*Brassica napus* var. SW Stratos™) [30] and sugar content in mandarin orange (*Citrus reticulata* Blanco) [31] were improved by pollination. Negative effects of limited pollination have included lowered calcium concentrations (associated with storage disorders) in Braeburn apples (*Malus domestica* Borkh) [32], although this effect seems to be variety specific [9] and lowered fruit weight and length in self-pollinated *Cucurbita moschata*, cv. Piramoita [33]. Consistent with previous studies, our overall findings were that cross-pollinated trees bore fruit with the most favorable nutritional content. A notable exception in our study was vitamin E (α Tocopherol), which was higher in self-pollinated almonds.

The mechanism(s) for the impact of cross- *versus* self-pollination on nutrients is as yet unknown. Self-pollination in almond may alter the nutritional composition of the nuts through a reduced development rate. Higher levels of linoleic acid are found in the earlier stages of oil accumulation in almond, with levels declining in the later developmental stages as oleic acid increases [34]. The

self-pollinated almonds may develop more slowly than the cross-pollinated almonds. Compatible pollen tubes were found to grow more quickly in almond than self-pollen tubes, and reached the ovary earlier [35]. In addition, self-pollination was found to delay megasporocyte differentiation [36]. If self-pollinated almonds develop more slowly, at harvest time they may not have accumulated as much oleic acid or lost as much linoleic acid as cross-pollinated almonds. Given that self-pollen tubes grow more slowly [35], in self-pollinated flowers there may also be an effect of embryo sac degeneration on the fruit's development and nutritional composition. While the mesh cages used to exclude pollinators from the trees could also have played a role in slowing fruit development by reducing the amount of sunlight the trees received, we would expect this to be minor due to the relatively short length of time the cages were covering the trees. Klatt *et al.* [8] explained their findings of increased strawberry shelf life with bee pollination as deriving from the increased amount of fertilized achenes per fruit (strawberry is an aggregate accessory fruit, consisting of many achenes each containing a single seed) when pollinated by bees. These achenes both create firmness structurally and produce the plant hormones that prevent fruit softening. It remains to be tested for any kind of crop if plant hormone production differs in self- versus cross-pollinated fruits or seeds.

In contrast to the pollination treatment, the fertilizer and water treatments had the potential to alter the resources available to the trees after the flowers had finished blooming. However, the limited impact of the fertilizer regime on almond's nutritional composition suggests that sufficient resources may have been stored in the soil from prior fertilizations. An effect of reduced resources such as fertilizer on almond composition may only be detected after

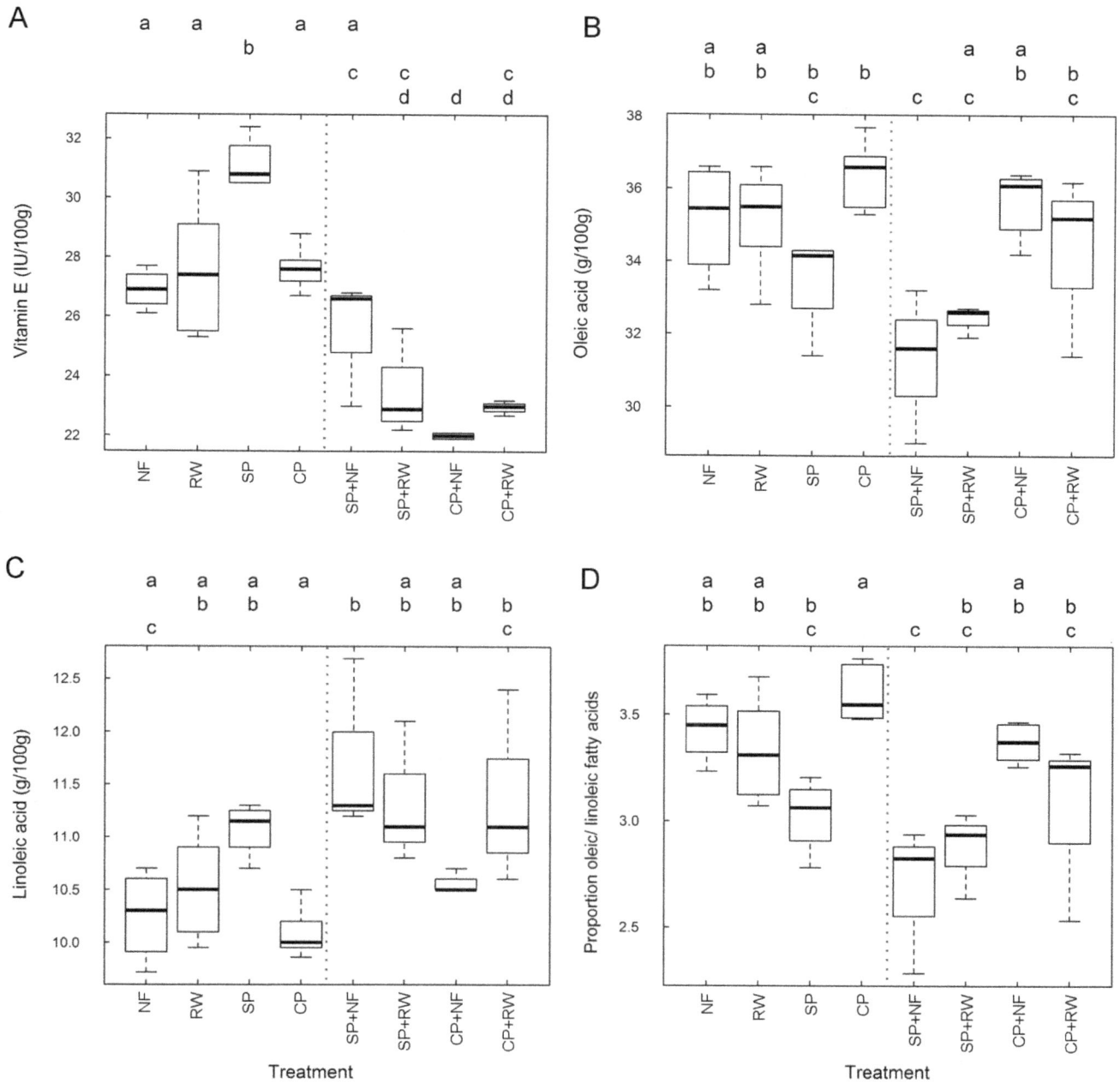

Figure 2. The nutritional composition of almonds from trees that received different resource treatments alone and in combination.
Treatments: no fertilizer (NF), reduced water (RW), self-pollination (SP) and cross-pollination (CP). Single and combined treatments are separated by a dashed line. The boxes show the median, 25th and 75th percentiles and the letters a,b,c,d, indicate significant differences from Tukey's HSD test at $P<$ 0.05.

multiple years with no fertilizer; especially as perennial almond trees may also store resources [17]. Reductions in irrigation have been found to reduce almond production and yield during the drought year [23] and in subsequent years [37]. The lack of impact of reduced water on the nutritional content of the almonds is supported by a review of almond composition studies, which found that irrigation had little effect on the oil content [38].

We did not analyze nutrients of almonds from open-(bee) pollinated trees not treated with reduced water or no fertilizer in the same study year. In a previous study [23], Klein *et al.*

investigated the weight and number of nuts produced by these same almond trees as well as open-pollinated trees and found that self-pollinated trees produced the fewest, heaviest nuts. Open pollinated nuts were intermediate with respect to weight and number, meaning that hand cross-pollinated trees produced larger numbers of smaller, lighter nuts. Nut weight and number was strongly related to the pollination treatments and the plant resource treatments had little influence [23]. This led us to hypothesize that the effect of self- versus cross-pollinated nuts could be indicated by nut size. We therefore analyzed the same nutrients for small and large open-pollinated nuts collected randomly in the experimental orchard under normal water and fertilizer regimes a year after the experiment took place. However, contrary to our expectations we found that the nutrient content of large and small nuts were not significantly different (data not shown). We conclude that although pollination can influence fruit weight, the weight of a fruit seems not to influence the amount of nutrients per unit weight in the nuts. Since these two processes are decoupled, this suggests that pollination directly influences the ratio of these nutrients that are important for human health.

The nutritional composition of crops may vary under a variety of interacting factors including agricultural practices, pollination and climate. The combined resource manipulations in this study showed little interactive effect on the fat composition. For vitamin E however, the combined resource and pollination treatments showed different levels of vitamin E than would have been expected from the effects of the single treatments. This suggests that changes in pollination could have unexpected effects on crop nutrition due to interactions with other inputs. In tomato (*Lycopersicon esculentum* L.), the benefits of arbuscular mycorrhizal fungal colonization for fruit quality (lower acidity, greater ascobrbic acid content) were found to be more pronounced when the plants were water stressed [39].

The present study demonstrates that pollination can positively change the nutritional value and commercial quality of almonds and that the effects are even stronger in combination with reduced availability of water and especially of nutrients applied to the soil. The potential nutritional benefits from cross-pollination deserve further study and may increase the current estimates of the economic and health value of pollination service to crops [4,8,40,41].

Supporting Information

Supporting information S1   Details of the methods used for the nutritional analysis of almonds.

Supporting information S2   The results of analyses of variance of the nutrients in almonds (listed in Table 2) from trees exposed to a single resource or pollination treatment.

Supporting information S3   A correlation matrix of the different nutrients quantified in almonds.

Acknowledgments

We thank Drew Scofield for assistance and the use of his orchard, Elisabeth Eilers and Miriam Voss for assistance in the field, Neal Williams for valuable discussions and two anonymous reviewers for valuable comments on the manuscript.

Author Contributions

Conceived and designed the experiments: AMK CK. Performed the experiments: AMK CK CB. Analyzed the data: CB. Wrote the paper: CB. interpreted the results, discussed and revised the manuscript: CB AMK CK AG.

References

1. Tilman D, Balzer C, Hill J, Befort BL (2011) Global food demand and the sustainable intensification of agriculture. Proc Natl Acad Sci U S A 108: 20260–20264.

2. Greenwald P, Clifford CK, Milner JA (2001) Diet and cancer prevention. Eur J Cancer 37: 948–965.

3. Hu FB (2003) Plant-based foods and prevention of cardiovascular disease: an overview. Am J Clin Nutr 78: 544S–551S.

4. Eilers EJ, Kremen C, Smith Greenleaf S, Garber AK, Klein A-M (2011) Contribution of pollinator-mediated crops to nutrients in the human food supply. Plos One 6: e21363.

5. Klein A-M, Vaissière BE, Cane JH, Steffan-Dewenter I, Cunningham SA, et al. (2007) Importance of pollinators in changing landscapes for world crops. Proc R Soc Lon B Biol Sci 274: 303–313.

6. Aizen MA, Garibaldi LA, Cunningham SA, Klein AM (2009) How much does agriculture depend on pollinators? Lessons from long-term trends in crop production. Ann Bot 103: 1579–1588.

7. Lautenbach S, Seppelt R, Liebscher J, Dormann CF (2012) Spatial and temporal trends of global pollination benefit. PLoS ONE 7: e35954.

8. Klatt BK, Holzschuh A, Westphal C, Clough Y, Smit I, Pawelzik E, Tscharntke T (2014) Bee pollination improves crop quality, shelf life and commercial value. Proc R Soc B 281: 20132440.

9. Garratt MPD, Breeze TD, Jenner N, Polce C, Biesmeijer SG (2013) Avoiding a bad apple: Insect pollination enhances fruit quality and economic value. Agr Ecosyst Environ 184: 34–40.

10. Aizen MA, Harder LD (2009) The global stock of domesticated honey bees is growing slower than agricultural demand for pollination. Curr Biol 19: 915–918.

11. Ricketts TH, Regetz J, Steffan-Dewenter I, Cunningham SA, Kremen C, et al. (2008) Landscape effects on crop pollination services: are there general patterns? Ecol Lett 11: 499–515.

12. Winfree R, Aguilar R, Vazquez DP, LeBuhn G, Aizen MA (2009) A meta-analysis of bees' responses to anthropogenic disturbance. Ecology 90: 2068–2076.

13. Kennedy CM, Lonsdorf E, Neel MC, Williams NM, Ricketts TH, et al. (2013) A global quantitative synthesis of local and landscape effects on wild bee pollinators in agroecosystems. Ecol Lett 16: 584–599.

14. Vörösmarty CJ, Green P, Salisbury J, Lammers RB (2000) Global Water Resources: Vulnerability from climate change and population growth. Science 289: 284–288.

15. Greven M, Neal S, Green S, Dichio B, Clothier B (2009) The effects of drought on the water use, fruit development and oil yield from young olive trees. Agric Water Manag 96: 1525–1531.

16. Castillo EG, Buresh RJ, Ingram KT (1992) Lowland rice yield as affected by timing of water deficit and nitrogen fertilization. Agron J 84: 152–159.

17. Groeneveld JH, Tscharntke T, Moser G, Clough Y (2010) Experimental evidence for stronger cacao yield limitation by pollination than by plant resources. Perspect Plant Ecol Evol Syst 12: 183–191.

18. Carman HF (2007) California farmers adapt mandated marketing programs to the 21st century. Calif Agric 61: 177–183.

19. Jalali-Khanabadi BA, Mozaffari-Khosravi H, Parsaeyan N (2010) Effects of almond dietary supplementation on coronary heart disease lipid risk factors and serum lipid oxidation parameters in men with mild hyperlipidemia. J Altern Complement Med 16: 1279–1283.

20. Sathe SK, Seeram NP, Kshirsagar HH, Heber D, Lapsley KA (2008) Fatty acid composition of California grown almonds. J Food Sci 73: C607–614.

21. Simopoulos AP (2008) The importance of the omega-6/omega-3 fatty acid ratio in cardiovascular disease and other chronic diseases. Exp Biol Med (Maywood) 233: 674–688.

22. Packer L (1991) Protective role of vitamin E in biological systems. Am J Clin Nutr 53: 1050S–1055S.

23. Klein AM, Hendrix SD, Clough Y, Scofield A, Kremen C (2014) Interacting effects of pollination, water and nutrients on fruit tree performance. Plant Biol, accepted.

24. Almond Board of California (2011) 2011 almond almanac. Available: http://www.almondboard.com/AbouttheAlmondBoard/Documents/ALM110600_Almanac2011_LR.pdf. Accessed 7 June 2013.

25. Vicuna S, Maurer EP, Joyce B, Dracup JA, Purkey D (2007) The sensitivity of California water resources to climate change scenarios. J Am Water Resour As 43: 482–498.

26. Hayhoe K, Cayan D, Field CB, Frumhoff PC, Maurer EP, et al. (2004) Emissions pathways, climate change, and impacts on California. Proc Natl Acad Sci U S A 101: 12422–12427.

27. Kodad O, Company RSI (2008) Variability of oil content and of major fatty acid composition in almond (*Prunus amygdalus* Batsch) and its relationship with kernel quality. J Agric Food Chem 56: 4096–4101.

28. Bénard C, Gautier H, Bourgaud F, Grasselly D, Navez B, et al. (2009) Effects of low nitrogen supply on tomato (*Solanum lycopersicum*) fruit yield and quality with special emphasis on sugars, acids, ascorbate, carotenoids, and phenolic compounds. J Agric Food Chem 57: 4112–4123.

29. Kris-Etherton PM, Pearson TA, Wan Y, Hargrove RL, Moriarty K, et al. (1999) High–monounsaturated fatty acid diets lower both plasma cholesterol and triacylglycerol concentrations. Am J Clin Nutr 70: 1009–1015.

30. Bommarco R, Marini L, Vaissière B (2012) Insect pollination enhances seed yield, quality, and market value in oilseed rape. Oecologia 1: 1–8.

31. Wallace HM, Lee LS (1999) Pollen source, fruit set and xenia in mandarins J Hortic Sci Biotech 74: 82–86.

32. Volz RK, Tustin DS, Ferguson IB (1996) Pollination effects on fruit mineral composition, seeds and cropping characteristics of 'Braeburn' apple trees. Sci Hortic-Amsterdam 66: 169–180.

33. Cardoso AII (2004) Depression by inbreeding after four sucessive self-pollination squash generations. Sci Agric (Piracicaba, Braz.) 61: 224–227.

34. Soler L, Canellas J, Saura-Calixto F (1988) Oil content and fatty acid composition of developing almond seeds. J Agric Food Chem 36: 695–697.

35. Certal A, Almeida R, Bošković R, Oliveira M, Feijó J (2002) Structural and molecular analysis of self-incompatibility in almond (*Prunus dulcis*). Sex Plant Reprod 15: 13–20.

36. Pimienta E, Polito VS (1983) Embryo sac development in almond [*Prunus dulcis* (Mill.) D. A. Webb] as affected by cross-, self- and non-pollination. Ann Bot-London 51: 469–479.

37. Goldhamer DA, Smith TE (1995) Single-season drought irrigation strategies influence almond production. Calif Agric 49: 19–22.

38. Yada S, Lapsley K, Huang GW (2011) A review of composition studies of cultivated almonds: macronutrients and micronutrients. J Food Compos Anal 24: 469–480.

39. Subramanian KS, Santhanakrishnan P, Balasubramanian P (2006) Responses of field grown tomato plants to arbuscular mycorrhizal fungal colonization under varying intensities of drought stress. Sci Hort 107: 245–253.

40. Gallai N, Salles JM, Settele J, Vaissière B (2009) Economic valuation of the vulnerability of world agriculture confronted to pollinator decline. Ecol Econ 68: 810–821.

41. Breeze TD, Bailey AP, Balcombe KG, Potts SG (2011) Pollination services in the UK: how important are honeybees? Agr Ecosyst Environ 142: 137–143.

Permissions

All chapters in this book were first published in PLOS ONE, by The Public Library of Science; hereby published with permission under the Creative Commons Attribution License or equivalent. Every chapter published in this book has been scrutinized by our experts. Their significance has been extensively debated. The topics covered herein carry significant findings which will fuel the growth of the discipline. They may even be implemented as practical applications or may be referred to as a beginning point for another development.

The contributors of this book come from diverse backgrounds, making this book a truly international effort. This book will bring forth new frontiers with its revolutionizing research information and detailed analysis of the nascent developments around the world.

We would like to thank all the contributing authors for lending their expertise to make the book truly unique. They have played a crucial role in the development of this book. Without their invaluable contributions this book wouldn't have been possible. They have made vital efforts to compile up to date information on the varied aspects of this subject to make this book a valuable addition to the collection of many professionals and students.

This book was conceptualized with the vision of imparting up-to-date information and advanced data in this field. To ensure the same, a matchless editorial board was set up. Every individual on the board went through rigorous rounds of assessment to prove their worth. After which they invested a large part of their time researching and compiling the most relevant data for our readers.

The editorial board has been involved in producing this book since its inception. They have spent rigorous hours researching and exploring the diverse topics which have resulted in the successful publishing of this book. They have passed on their knowledge of decades through this book. To expedite this challenging task, the publisher supported the team at every step. A small team of assistant editors was also appointed to further simplify the editing procedure and attain best results for the readers.

Apart from the editorial board, the designing team has also invested a significant amount of their time in understanding the subject and creating the most relevant covers. They scrutinized every image to scout for the most suitable representation of the subject and create an appropriate cover for the book.

The publishing team has been an ardent support to the editorial, designing and production team. Their endless efforts to recruit the best for this project, has resulted in the accomplishment of this book. They are a veteran in the field of academics and their pool of knowledge is as vast as their experience in printing. Their expertise and guidance has proved useful at every step. Their uncompromising quality standards have made this book an exceptional effort. Their encouragement from time to time has been an inspiration for everyone.

The publisher and the editorial board hope that this book will prove to be a valuable piece of knowledge for researchers, students, practitioners and scholars across the globe.

List of Contributors

Paul Szpak, Christine D. White and Jean-François Millaire
Department of Anthropology, The University of Western Ontario, London, Ontario, Canada

Fred J. Longstaffe
Department of Earth Sciences, The University of Western Ontario, London, Ontario, Canada

Víctor F. Vásquez Sánchez
Centro de Investigaciones Arqueobioló gicas y Paleoecoló gicas Andinas (ARQUEOBIOS), Trujillo, Peru

David P. Janos and Catalina Aristizábal
Department of Biology, University of Miami, Coral Gables, Florida, United States of America

John Scott
Research Institute for the Environment and Livelihoods, Charles Darwin University, Darwin, Northern Territory, Australia

David M. J. S. Bowman
School of Plant Science, The University of Tasmania, Hobart, Tasmania, Australia

Xinshu Liang, Yinan Gao, Xiaoying Zhang, Yongqiang Tian, Zhenxian Zhang and Lihong Gao
Beijing Key Laboratory of Growth and Developmental Regulation for Protected Vegetable Crops, Department of Vegetable Science, China Agricultural University, Beijing, P.R. China

Enke Liu, Changrong Yan, Xurong Mei, Yanqing Zhang
Institute of Environment and Sustainable Development in Agriculture, Chinese Academy of Agricultural Sciences, Beijing, China
Key Laboratory of Dryland Farming g Agriculture, Ministry of Agriculture of the People's Republic of China (MOA), Beijing, China

Tinglu Fan
Dryland Agricultural Institute, Gansu Academy of Agricultural Sciences, Lanzhou, Gansu, China

Liqun Zhu, Naijuan Hu and Zhengwen Zhang
College of Agriculture, Nanjing Agricultural University, Nanjing, China

Minfang Yang and Xinhua Zhan
College of Resources and Environmental Science, Nanjing Agricultural University, Nanjing, China

Maria del Mar Alguacil, Emma Torrecillas and Antonio Roldán
CSIC-Centro de Edafología y Biología Aplicada del Segura, Department of Soil and Water Conservation, Campus de Espinardo, Murcia, Spain

Zenaida Lozano
Universidad Central de Venezuela (UCV), Facultad de Agronomía, Instituto de Edafología, El Limón, Campus Universitario, Maracay, Venezuela

Maria Pilar Torres
Departamento de Biología Aplicada, Area de Botánica, Universidad Miguel Hernández, Elche, Alicante, Spain

Fannie I. Parra-Cota, Juan J. Peñ a-Cabriales, Norma A. Martínez-Gallardo and John P. Délano-Frier
Centro de Investigación y de Estudios Avanzados-Unidad Irapuato, Irapuato, Guanajuato, México

Sergio de los Santos-Villalobos
Departamento de Ciencias del Agua y del Medio Ambiente, Instituto Tecnoló gico de Sonora, Ciudad Obregó n, Sonora, México

Zhihong Du
Department of Ecology, College of Urban and Environmental Sciences, and Key Laboratory for Earth Surface Processes of the Ministry of Education, Peking University, Beijing, China
Key Laboratory for Urban Habitat Environmental Science and Technology, Peking University Shenzhen Graduate School, Shenzhen, China

Wei Wang and Wenjing Zeng
Department of Ecology, College of Urban and Environmental Sciences, and Key Laboratory for Earth Surface Processes of the Ministry of Education, Peking University, Beijing, China

Hui Zeng
Key Laboratory for Urban Habitat Environmental Science and Technology, Peking University Shenzhen Graduate School, Shenzhen, China

Marty R. Schmer, Gary E. Varvel and Virginia L. Jin
Agroecosystem Management Research Unit, United States Department of Agriculture-Agricultural Research Service (USDA-ARS), Lincoln, Nebraska, United States of America

Kenneth P. Vogel and Robert B. Mitchell
Grain, Forage and Bioenergy Research Unit, United States Department of Agriculture-Agricultural Research Service (USDA-ARS), Lincoln, Nebraska, United States of America

Ronald F. Follett
Soil-Plant Nutrient Research Unit, United States Department of Agriculture-Agricultural Research Service (USDA-ARS), Ft. Collins, Colorado, United States of America

Liming Zhang
College of Resource and Environment, Fujian Agriculture and Forestry University, Fuzhou, China
State Key Laboratory of Soil and Sustainable Agriculture, Institute of Soil Science, Chinese Academy of Sciences, Nanjing, China

Dongsheng Yu, Xuezheng Shi, Shengxiang Xu and Yongcong Zhao
State Key Laboratory of Soil and Sustainable Agriculture, Institute of Soil Science, Chinese Academy of Sciences, Nanjing, China

Shihe Xing
College of Resource and Environment, Fujian Agriculture and Forestry University, Fuzhou, China

Guanghua Yin, Jian Gu, Fasheng Zhang and Zuoxin Liu
Institute of Applied Ecology, Chinese Academy of Sciences, Shenyang, China

Liang Hao and Peifei Cong
Institute of Applied Ecology, Chinese Academy of Sciences, Shenyang, China
University of Chinese Academy of Sciences, Beijing, China

Jun Wang
College of Urban and Environmental Sciences, Northwest University, Xian, Shaanxi Province, China,

Upendra M. Sainju
U.S. Department of Agriculture, Agricultural Research Service, Northern Plains Agricultural Research Laboratory, Sidney, Montana, United States of America

Anja Madelen Ö dman and Pål Axel Olsson
Biodiversity, Department of Biology, Lund University, Lund, Sweden

Jifu Li, Jianwei Lu, Xiaokun Li, Tao Ren, Rihuan Cong and Li Zhou
College of Resources and Environment, Huazhong Agricultural University, Wuhan, China
Key Laboratory of Arable Land Conservation (Middle and Lower Reaches of Yangtse River), Ministry of Agriculture, Wuhan, China

John Hamblin and Katia Stefanova
Institute of Agriculture, University of Western Australia, Crawley, Western Australia, Australia

Tefera Tolera Angessa
School of Plant Biology, University of Western Australia, Crawley, Western Australia, Australia

Riikka Rinnan
Department of Biology, Lund University, Lund, Sweden
Department of Biology, University of Copenhagen, Copenhagen, Denmark
Center for Permafrost (CENPERM), University of Copenhagen, Copenhagen, Denmark

Anders Michelsen
Department of Biology, University of Copenhagen, Copenhagen, Denmark
Center for Permafrost (CENPERM), University of Copenhagen, Copenhagen, Denmark

Erland Bååth
Department of Biology, Lund University, Lund, Sweden

Leilei Tang
Key Laboratory of Aquatic Botany and Watershed Ecology, Wuhan Botanical Garden, Chinese Academy of Sciences, Wuhan, China
Hainan Modern Agriculture Inspection and Testing Precaution & Control Center, Agricultural Department of Hainan Province, Haikou, China

Chuanpeng Cheng1, Yong Tao, Junfeng Pan and Juan Xie
Key Laboratory of Aquatic Botany and Watershed Ecology, Wuhan Botanical Garden, Chinese Academy of Sciences, Wuhan, China

Kaiyuan Wan
Key Laboratory of Aquatic Botany and Watershed Ecology, Wuhan Botanical Garden, Chinese Academy of Sciences, Wuhan, China
Ecological Restoration (ECORES) Lab, Chengdu Institute of Biology, Chinese Academy of Sciences, Chengdu, China

Ruhai Li
Institute of Plant Protection and Soil Science, Hubei Academy of Agricultural Sciences, Wuhan, China

Daozhong Wang
Institute of Soil and Fertilizer Science, Anhui Academy of Agricultural Sciences, Hefei, China

Fang Chen
Key Laboratory of Aquatic Botany and Watershed Ecology, Wuhan Botanical Garden, Chinese Academy of Sciences, Wuhan, China
China Program of International Plant Nutrition Institute (IPNI), Wuhan, China

Matthew H. Meisner
Department of Evolution and Ecology, University of California Davis, Davis, California, United States of America
Department of Statistics, University of California Davis, Davis, California, United States of America

Jay A. Rosenheim
Department of Entomology and Nematology, University of California Davis, Davis, California, United States of America

Chang Wang, Chichao Huang, Jian Qian, Jian Xiao, Huan Li, Yongli Wen, Qirong Shen, Guanghui Yu and Wei Ran
National Engineering Research Center for Organic-based Fertilizers, Jiangsu Collaborative Innovation Center for Solid Organic Waste Resource Utilization, Nanjing Agricultural University, Nanjing, PR China,

Xinhua He
School of Plant Biology, University of Western Australia, Crawley, Australia
Jun Zhao, Tian Ni, Wu Xiong, Wei Ran and Biao Shen

Key Laboratory of Plant Nutrition and Fertilization in Low-Middle Reaches of the Yangtze River, Ministry of Agriculture, Nanjing, Jiangsu, China
Jiangsu Key Lab and Engineering Center for Solid Organic Waste Utilization, Nanjing Agricultural University, Nanjing, Jiangsu, China

Yong Li
Soil and Fertilizer Technical Guidance Station of Jintan City, Jintan Agricultural and Forestry Bureau, Jintan, Jiangsu, China

Qirong Shen and Ruifu Zhang
Key Laboratory of Plant Nutrition and Fertilization in Low-Middle Reaches of the Yangtze River, Ministry of Agriculture, Nanjing, Jiangsu, China
Jiangsu Key Lab and Engineering Center for Solid Organic Waste Utilization, Nanjing Agricultural University, Nanjing, Jiangsu, China
Jiangsu Collaborative Innovation Center for Solid Organic Waste Resource Utilization, Nanjing Agricultural University, Nanjing, Jiangsu, China

Yan Deng, Keru Chen, Gu Feng, Zhenling Cui, Fusuo Zhang and Xinping Chen
Center for Resources, Environment and Food Security, China Agricultural University, Beijing, China

Wan Teng and Yiping Tong
State Key Laboratory for Plant Cell and Chromosome Engineering, Institute of Genetics and Developmental Biology, Chinese Academy of Sciences, Beijing, China,

Ai Zhan
State Key Laboratory of Soil Erosion and Dryland Farming on the Loess Plateau, Institute of Soil and Water Conservation, Chinese Academy of Sciences and Ministry of Water Resource, Yangling, China

Claire Brittain
Institute of Ecology, Ecosystem Functions, Leuphana University of Lüneburg, Germany and Department of Entomology, University of California Davis, Davis, California, United States of America

Claire Kremen
Environmental Sciences Policy and Management, University of California, Berkeley, Berkeley, California, United States of America

Andrea Garber
Division of Adolescent Medicine, University of
California San Francisco, San Francisco, California,
United States of America

Alexandra-Maria Klein
Institute of Ecology, Ecosystem Functions, Leuphana
University of Lüneburg, Germany and Chair of
Nature Conservation and Landscape Ecology,
University of Freiburg, Germany

Index

www.ingramcontent.com/pod-product-compliance
Lightning Source LLC
Chambersburg PA
CBHW061248190326
41458CB00011B/3617